T0302091

Surface Structure Determination by LEED and X-rays

This timely text covers the theory and practice of surface and nanostructure determination by low-energy electron diffraction (LEED) and surface X-ray diffraction (SXRD): it is the first book on such quantitative structure analysis in over 30 years. It provides a detailed description of the theory, including cutting-edge developments and tested experimental methods. The focus is on quantitative techniques, while the qualitative interpretation of the LEED pattern without quantitative I(V) analysis is also included. Topics covered include the future study of nanoparticles, quasicrystals, thermal parameters, disorder and modulations of surfaces with LEED, with introductory sections enabling the non-specialist to follow all the concepts and applications discussed. With numerous figures throughout, this text is ideal for undergraduate and graduate students and researchers, whether experimentalists or theorists, in the fields of surface science, nanoscience and related technologies. It can serve as a textbook for graduate-level courses of one or two semesters.

Wolfgang Moritz is Professor Emeritus in the Department of Earth and Environmental Sciences at the University of Munich. He has contributed extensively to the theory and practice of both LEED and SXRD.

Michel A. Van Hove is Professor Emeritus in the Department of Physics at Hong Kong Baptist University. He has contributed extensively to the theory and practice of LEED, photoelectron diffraction and other techniques of surface science.

Surface Structure Determination by LEED and X-rays

WOLFGANG MORITZ
University of Munich

MICHEL A. VAN HOVE
Hong Kong Baptist University

CAMBRIDGE
UNIVERSITY PRESS

CAMBRIDGE
UNIVERSITY PRESS

University Printing House, Cambridge CB2 8BS, United Kingdom

One Liberty Plaza, 20th Floor, New York, NY 10006, USA

477 Williamstown Road, Port Melbourne, VIC 3207, Australia

314–321, 3rd Floor, Plot 3, Splendor Forum, Jasola District Centre, New Delhi – 110025, India

103 Penang Road, #05–06/07, Visioncrest Commercial, Singapore 238467

Cambridge University Press is part of the University of Cambridge.

It furthers the University's mission by disseminating knowledge in the pursuit of
education, learning, and research at the highest international levels of excellence.

www.cambridge.org
Information on this title: www.cambridge.org/9781108418096
DOI: 10.1017/9781108284578

© Wolfgang Moritz and Michel A. Van Hove 2022

First published 2022

A catalogue record for this publication is available from the British Library.

Library of Congress Cataloging-in-Publication Data
Names: Moritz, Wolfgang, 1943- author. | Van Hove, M. A. (Michael A.), 1947– author.
Title: Surface structure determination by LEED and X-rays / Wolfgang Moritz, University of Munich,
     Michel A. Van Hove, Hong Kong Baptist University.
Description: First edition. | United Kingdom ; New York, NY, USA : Cambridge University Press,
     [2022] | Includes bibliographical references and index.
Identifiers: LCCN 2021058150 (print) | LCCN 2021058151 (ebook) | ISBN 9781108418096
     (hardback) | ISBN 9781108284578 (epub)
Subjects: LCSH· Surface chemistry. | Surfaces (Technology)–Analysis. | Low energy electron
     diffraction. | X-ray diffraction imaging. | Chemical structure.
Classification: LCC QD506 .M66 2022 (print) | LCC QD506 (ebook) | DDC 541/.33–dc23/eng/20220126
LC record available at https://lccn.loc.gov/2021058150
LC ebook record available at https://lccn.loc.gov/2021058151

ISBN 978-1-108-41809-6 Hardback

Contents

Preface

In recent decades, surface science has vastly widened its relevance and capabilities, ranging from the original clean metal and semiconductor surfaces to highly complex interfaces between solid and soft matter or between solid and fluid matter (gaseous or liquid). Surface science has also expanded from 2-dimensional surfaces to 1-dimensional structures (surface steps, nanowires, nanotubes, molecular contacts) and to 0-dimensional particles (quantum dots, nanoparticles, nanomachines), leading to 'nanoscience'.

Arguably, knowledge of atomic-scale structure is central to understanding and designing such structures. In concert with the expansion of surface science, the capabilities of many of its fundamental techniques for structure analysis have been amplified and widened to cover interfaces, 1-dimensional defects and nanostructures, among others.

Among these techniques, low-energy electron diffraction (LEED) and surface X-ray diffraction (SXRD) have undergone vast developments over the last decades. It is therefore timely to bring together the state of the art of these two related but complementary techniques. Indeed, no other books comprehensively cover these advances. In particular, the last monographs devoted to LEED were published in the mid-1980s and have long been out of print.

There is no doubt that other techniques for surface structure determination have also evolved considerably in these decades: they deserve books of their own and can only be alluded to here (cf. Chapter 1).

The wide variety of 'competing' techniques has generated a very dynamic field of surface and nanostructure investigations. This activity was also amplified by the growing capabilities of ab initio calculations, which in turn found in surfaces and nanostructures new challenging test cases to validate their considerable theoretical and computational progress.

The last decades have seen major advances in LEED and SRXD that justify this new book:

- LEED instrumentation has been developed to measure LEED patterns with high angular resolution (as in SPA-LEED) or with high spatial resolution on the nanometre scale in the surface image (as in LEEM).
- LEED instruments were tailored to specific new applications, such as electron-beam-sensitive surfaces and nanometre-scale structures.

- Methods have been introduced to determine angles of incidence in LEED with greater accuracy.
- Improvements have been made to remove background intensities in the LEED pattern.
- LEED structure determination has been partially automated to speed up the process and allow the solution of much more complex structures, including both local and global search methods.
- Holographic LEED and NanoLEED have been developed. They have potential for further extensions and applications to surface structure determination.
- Diffuse LEED has been developed and applied to a few disordered surface structures and provides an additional method that provides complementary information to photoelectron diffraction.
- LEED methods have been developed to study thermal parameters, including anisotropic vibrations and static displacements, interatomic correlations and librations.
- It is now possible to examine modulated structures and determine modulation amplitudes with LEED.
- Quasicrystal surfaces are now amenable to structural analysis by LEED.
- SXRD has made substantial progress in the last two decades by the development of 2-D area detectors, which enormously speed up the measuring time, increase the data accuracy and enable measurements that were impossible with a point detector. Measurement with a point detector is very time consuming and requires stable and inert surfaces.
- Diffraction chambers have been developed which allow measurement at different environments up to high pressures in the context of combination with UHV diffractometers. (By high pressures we mean the range of some mbar to several bar.)
- The development of XMCD (X ray magnetic circular dichroism) and RSXS (resonant soft X-ray scattering) has opened up new fields of research. These methods are not discussed in this book but represent an important field in surface properties in connection with surface structures.
- The development of direct methods should greatly facilitate the surface structure solution.
- The study of surface morphology, adsorbed molecules and nanoparticles has made substantial progress in experimental methods as well as in data analysis by the development of GISAXS (grazing incidence small angle X-ray scattering).

The objectives of this book are manifold: (1) to provide an up-to-date and comprehensive account of current capabilities in LEED and SXRD (repeating only the relevant essentials from earlier LEED books, since they are no longer in print); (2) to describe the advances of the last decades with relatively more detail; (3) to include practical information that can save the practitioner much valuable time; (4) to educate new practitioners and students in the art and science of crystallography of surfaces and nanostructures; and (5) to offer a comprehensive reference to useful literature, software and databases.

The focus of this book is thus on practical experimental and theoretical aspects that enable using LEED and/or SXRD in surface structure determination. The basic physics are only briefly covered, while more emphasis is placed on experimental requirements and methods, on theoretical developments and computational approaches, and on structural analysis methods and capabilities. Included are 'imperfect' surface structures, such as stepped surfaces, faceted surfaces, antiphase boundaries, modulated layers and disordered surfaces, as well as thermal effects, quasicrystals and nanostructures.

This book took over a dozen years to crystallise through the vagaries of life. We are heavily indebted to our many colleagues for their expert input at many stages of preparation. We wish to particularly mention John Rundgren and Klaus Hermann for numerous and very constructive discussions. We are also grateful to colleagues who have contributed substantially to our computer codes, including Angelo Barbieri, Rüdiger Döll, Yves Gauthier, George Mihai Gavaza, Martin Gierer, Jürgen Landskron, Herbert Over, Philip J. Rous, John Rundgren, Adrian Wander and Zhengji Zhao. Our sincere gratitude goes to the professional and patient staff at Cambridge University Press. The authors, however, carry full responsibility for all remaining imperfections.

WM dedicates this book to the lasting memory of Anna Pietsch. MAVH dedicates this book to the continuous support of Sheng-Wei Wang.

W. Moritz and M. A. Van Hove
München and Hong Kong 2022

List of Abbreviations

AD	atom diffraction
ADP	atomic displacement parameter
AED	Auger electron diffraction
AES	Auger electron spectroscopy
AFM	atomic force microscopy
ALICISS	alkali-ion impact collision ion scattering spectroscopy
ALISS	alkali-ion scattering spectroscopy
ANKA	Angströmquelle Karlsruhe
APS	Advanced Photon Source
APS	appearance potential spectroscopy
ARAES	angle-resolved Auger electron spectroscopy
ARPED	angle-resolved photoelectron diffraction
ARPEFS	angle-resolved photoelectron extended fine structure
ARPES	angle-resolved photoelectron spectroscopy
ARUPS	angle-resolved ultraviolet photoelectron spectroscopy
ARXPD	angle-resolved X-ray photoelectron diffraction
ARXPS	angle-resolved X-ray photoelectron spectroscopy
BOBYQA	bounded optimisation by quadratic approximation
CBLEED	convergent beam LEED
CCD	charge-coupled device
CG	conjugate gradient
CMA-ES	covariance matrix adaptation evolutionary strategy
CMTA	constant momentum transfer averaging
COBRA	coherent Bragg rod analysis
CPD	contact-potential difference (work function change)
CTDS	correlated thermal diffuse scattering
CTR	crystal truncation rod (in SXRD)
DAPS	disappearance potential spectroscopy
DCAF	difference map using the constraints of atomicity and film shift
DE	differential evolution
DESY	Deutsches Elektronen Synchrotron
DFT	density functional theory
DLEED	diffuse LEED
EAL	effective attenuation length

EAPFS	extended appearance potential fine structure
EAs	evolutionary algorithms
ED	electron diffraction
EDAX	energy dispersive X-ray analysis
EELFS	electron energy loss fine structure
EELS	electron energy loss spectroscopy
EH	electron holography
ELEED	elastic low-energy electron diffraction
ELNES	electron energy loss near-edge structure
ELS	energy loss spectroscopy
EM	electron microscopy
EMTO	exact muffin-tin orbital theory
ESCA	electron spectroscopy for chemical analysis
ESD	electron stimulated desorption
ESDIAD	electron stimulated desorption ion angular dependence
ESR	electron spin resonance
ESRF	European Synchrotron Radiation Facility
EXAFS	extended X-ray absorption fine structure
EXELFS	extended electron energy loss fine structure
EXFAS	extended fine Auger structure
FEED	field emission energy distribution
FEM	field emission microscopy
FFT	fast Fourier transform
FT	Fourier transform
FIM	field-ion microscopy
FOM	figure of merit in SXRD
FSA	fast simulated annealing
FYNES	fluorescence-yield near-edge structure
GAs	genetic algorithms
GISAXS	grazing-incidence small angle X-ray scattering
GIXD	grazing-incidence X-ray diffraction
GIXRD	grazing-incidence X-ray diffraction
GIXS	grazing-incidence X-ray scattering
GPS	generalised pattern search
GSA	generalised simulated annealing
GSS	generating set search
HEED	high-energy electron diffraction
HEIS	high-energy ion scattering
HESXRD	high-energy surface X-ray diffraction
HREELS	high-resolution electron energy loss spectroscopy
HR-LEED	high-resolution LEED
ICISS	impact collision ion scattering spectroscopy
IETS	inelastic electron tunnelling spectroscopy

IFT	inverse Fourier transform
IID	ion impact desorption
ILEED	inelastic LEED
IMFP	inelastic mean free path
INS	ion neutralisation spectroscopy
IPE	inverse photoemission
IRAS or **IRS**	infrared absorption spectroscopy
IS	ion scattering
ISS	ion scattering spectroscopy
KEH	Kikuchi electron holography
KKR	Korringa–Kohn–Rostoker
LEED	low-energy electron diffraction
LEEM	low-energy electron microscopy
LEIS	low-energy ion scattering
LEPD	low-energy positron diffraction
LID	laser induced desorption
LIF	laser induced fluorescence
MBE	molecular beam epitaxy
MCP	microchannel plate
MEED	medium-energy electron diffraction
MEIS	medium-energy ion scattering
MSF	modulus sum function
MT	muffin tin
ND	neutron diffraction
NEXAFS	near-edge X-ray absorption fine structure
NGA-LEED	novel genetic algorithm for LEED
NIS	neutron inelastic scattering
NMR	nuclear magnetic resonance
NN	nearest neighbour
NPD	normal photoelectron diffraction
NSLS	National Synchrotron Light Source
OPD	off-normal photoelectron diffraction
PARADIGM	phase and amplitude recovery and diffraction image generation method
PD or **PhD**	photoelectron diffraction
PDF	probability density function
PED	photoelectron diffraction
PEEM	photoemission electron microscopy
PEH	photoelectron holography
PES	photoelectron spectroscopy
PEXAFS	photoemission extended X-ray absorption fine structure
PIES	Penning ionisation electron spectroscopy
PLEED	polarised low-energy electron diffraction

PSDIAD	photon stimulated desorption ion angular distributions
RAIRS	reflection–absorption infrared spectroscopy
RBS	Rutherford backscattering
RFS	renormalised forward scattering
RHEED	reflection high-energy electron diffraction
RoI	region of interest (in SXRD)
RSXS	resonant soft X-ray scattering
SA	simulated annealing
SAES	scanning Auger electron spectroscopy
SAM	scanning Auger microscopy
SARIS	scattering and recoiling imaging spectroscopy
SEE	secondary electron emission
SEELFS	surface extended energy loss fine structure
SEM	scanning electron microscopy
SERS	surface-enhanced Raman scattering
SEXAFS	surface extended X-ray absorption fine structure
SIMS	secondary ion mass spectroscopy
SLS	Swiss Light Source
SMCG	sparse matrix canonical grid
SOLEIL	Source Optimisée de Lumière d'Energie Intermédiaire du LURE
SPA-LEED	spot profile analysis LEED
SPI	surface Penning ionisation
SPIES	surface Penning ionisation electron spectroscopy
SPLEED	spin-polarised LEED
SPLEEM	spin-polarised LEEM
SPM	scanning probe microscopies
SPRING-8	Super Photon ring-8 GeV
SSIMS	static secondary-ion mass spectroscopy
SSR	superstructure rod (in SXRD)
STEM	scanning transmission electron microscopy
STM	scanning tunnelling microscopy
SVD	singular value decomposition
SXAPS	soft X-ray appearance potential spectroscopy
SXPS	soft X-ray photoelectron spectroscopy
SXRD	surface X-ray diffraction
TDMS	thermal desorption mass spectroscopy
TDS	thermal desorption spectroscopy
TDS	thermal diffuse scattering
TEAS	thermal energy atomic scattering
TED	transmission electron diffraction
TEM	transmission electron microscopy
THEED	transmission high-energy electron diffraction
TOF-SARS	time-of-flight scattering and recoiling spectrometry

TPD	temperature-programmed desorption
TPR	temperature-programmed reaction
UHV	ultra-high vacuum
UPS	ultraviolet photoelectron spectroscopy
VLEED	very-low-energy electron diffraction
WFC	work function change
XAES	X-ray-stimulated Auger electron spectroscopy
XAFS	X-ray absorption fine structure
XANES	X-ray absorption near-edge spectroscopy
XAS	X-ray absorption spectroscopy
XMCD	X-ray magnetic circular dichroism
XPD	X-ray photoelectron diffraction
XPS	X-ray photoelectron spectroscopy
XRD	X-ray diffraction
XSW	X-ray standing waves

List of Major Symbols

0-D, 1-D, 2-D, 3-D	zero-, one-, two-, three-dimensional space or translational symmetry
$a, b, c, \alpha, \beta, \gamma$	lattice constants in 3-D direct space
a, b, c	lattice translation vectors (basis vectors) in 3-D direct space
$\mathbf{a}^*, \mathbf{b}^*, \mathbf{c}^*$	reciprocal lattice translation vectors in 3-D reciprocal space
$a(L_1, L_2, L_3)$	Clebsch–Gordan or Gaunt coefficients
a, b	lattice translation vectors (basis vectors) in 2-D direct space
$\mathbf{a}^*, \mathbf{b}^*$	reciprocal lattice translation vectors in 2-D reciprocal space
$\mathbf{a}', \mathbf{b}'$	superlattice (superstructure) basis vectors in 2-D direct space
$\mathbf{a}'^*, \mathbf{b}'^*$	superlattice (superstructure) basis vectors in 2-D reciprocal space
c	speed of light
$C^l(l'm', l''m'')$	Clebsch–Gordan or Gaunt coefficients
d_{hkl}	d-spacing (layer spacing) between layers of orientation $[hkl]$
e	electron charge (>0)
E	energy of electron (in vacuum)
$f(\vartheta)$	atomic scattering amplitude or factor as function of scattering angle
$F(\mathbf{k}', \mathbf{k}_0)$	structure factor of one unit cell in LEED
$\mathbf{F}(\mathbf{k}_{g'}, \mathbf{k}_g)$	tensor in Tensor LEED
$F(hkl)$	structure factor in SXRD
F	Fourier transform
G	metric matrix for vector calculations in oblique systems
$\mathbf{G}, G_{ij}$	Green's function
$\mathbf{g}, \mathbf{g}_{hk}, \mathbf{g}(hk)$	two-dimensional reciprocal lattice vector
$G_{hkl} = g(h)g(k)g(l)$	lattice factor in SXRD
$h_l^{(1)}, h_l^{(2)}$	Hankel functions of the first and second kinds
$h = 2\pi\hbar$	Planck's constant
h	Miller index, as in (hkl)
$(hk) = (h,k) = (h, k)$	Miller indices, as in (hk) beam
$(hkl) = (h,k,l) = (h, k, l)$	Miller indices, as in (hkl) plane
$[hkl] = [h,k,l] = [h, k, l]$	Miller indices, as in $[hkl]$ direction in direct space

$(hkil) = (h,k,i,l) =$ (h, k, i, l)	Bravais–Miller indices, as in $(hkil)$ plane
$I, I(V)$	intensity (as a function of voltage V or energy eV)
I_0	incident (or primary) intensity
i	imaginary unit, $\sqrt{-1}$
i	Miller–Bravais index, as in $(hkil)$
$\mathbf{I}$	unit matrix
j_l	spherical Bessel function
J_l	Bessel function
k_B	Boltzmann constant
k	Miller index, as in (hkl)
k	wave number $k = |\mathbf{k}|$
$\mathbf{k}$	electron wave vector
$\mathbf{k}^i = \mathbf{k}_i = \mathbf{k}_{in} =$ $\mathbf{k}_0 = \mathbf{k}_0{}^+$	wave vector incident from vacuum
$\mathbf{k}' = \mathbf{k}_f = \mathbf{k}_s =$ $\mathbf{k}_{out} = \mathbf{k}_{\mathbf{g}}{}^-$	wave vector scattered/diffracted out into vacuum
$\mathbf{k}_{\mathbf{g}}{}^+, \mathbf{k}_{\mathbf{g}}{}^-$	wave vector of beam $\mathbf{g}^+, \mathbf{g}^-$
l	Miller index, as in (hkl)
l	angular momentum quantum number
l_{max}	cutoff angular momentum in partial-wave expansion
L	'angular momentum' quantum numbers $L = (l, m)$
L	Lorentz factor for X-rays
m	magnetic quantum number
m_a	mass of atom or molecule
m_e	mass of electron
M	exponent in Debye–Waller factor e^{-2M}
$\mathbf{M} = \begin{pmatrix} m_{11} & m_{12} \\ m_{21} & m_{22} \end{pmatrix}$	matrix notation for superlattices
$\mathbf{M}^* = \begin{pmatrix} m_{11}^* & m_{12}^* \\ m_{21}^* & m_{22}^* \end{pmatrix}$	matrix notation for reciprocal superlattices
$\mathbf{M}^{\pm\pm}, M_{\mathbf{g}'\mathbf{g}}^{\pm\pm}$	layer diffraction matrix (in reciprocal space), matrix element
n	refractive index for X-rays
$\mathbf{p}$	linear momentum
P	Patterson function
P	polarisation factor for X-rays
P_l	Legendre polynomials
$p(\mathbf{u})$	probability density function (PDF)
$P_{\mathbf{g}}^{\pm}, p_{\mathbf{g}}^{\pm}$	plane-wave propagators
$\mathbf{q}$	momentum transfer (scattering vector)
$\mathbf{r} = (x, y, z)$	general position vector
$\mathbf{r}, r_{\mathbf{g}'\mathbf{g}}^{\pm\pm}$	layer reflection matrix (in reciprocal space), matrix element
$R(\alpha_i)$	reflection coefficient in SXRD

$R_l(r)$	radial atomic wave function
RR	double-reliability factor, used in calculating error estimates in structural determination
s	momentum transfer (scattering vector)
S	significance factor, used in calculating error estimates in structural determination
T	instrumental broadening function
T	temperature
$T(\mathbf{q})$	Debye–Waller factor
T	T-matrix (in angular momentum space) for a layer consisting of several Bravais-lattice planes of atoms
$T(\alpha_i)$	transmission coefficient in SXRD
$\mathbf{t}, t_{\mathbf{g'g}}^{\pm\pm}$	layer transmission matrix (in reciprocal space), matrix element
$\mathbf{t}, t_l$	t-matrix (in angular momentum space) for single atom, radial matrix element
V_0	real part of inner potential or muffin-tin zero level
V_{0i}	imaginary part of inner potential
$(\mathbf{W}, \mathbf{w})$	point operation $\mathbf{W}$ and translation vector $\mathbf{w}$ of general symmetry operation in 3-D
x, y	coordinates parallel to the surface
Y	Pendry Y-function (used in Pendry R-factor and DLEED)
$Y_{lm}(\vartheta, \varphi)$	spherical harmonic function
z	coordinate perpendicular to the surface (may point into or out of the surface)
Z	atomic number (number of electrons in an atom)
α, β, γ	angular lattice constants in 3-D (see $a, b, c, \alpha, \beta, \gamma$)
α_c	critical angle for total external reflection of X-rays
α, α_i	angle of incidence relative to the surface plane in SXRD
α_f	exit angle relative to the surface plane in SXRD
α	layer-to-layer electron or X-ray attenuation coefficient
β	imaginary part of refractive index of X-rays
β_{kl}	thermal parameters of thermal ellipsoid
δ	real part of refractive index of X-rays
δ_l	phase shift
$\delta_l(T)$	temperature-dependent phase shift
$\delta_{ll'}$	Kronecker symbol, with value 1 when $l = l'$ or 0 otherwise
ϑ	scattering angle in LEED
ϑ	polar angle of incidence or emergence relative to the surface normal in LEED
ϑ	Bragg angle
2ϑ	scattering angle in XRD
Θ_D	Debye temperature
λ	wavelength of photon or electron
Λ	penetration depth of X-rays

Λ_{eff}	effective scattering depth of X-rays
μ_l	absorption coefficient of X-rays
ρ	atomic electron charge density
$\frac{d\sigma}{d\Omega}(\Omega_{\text{sc}})$	differential cross-section into solid angle Ω_{sc}
σ_e	photon-electron scattering cross-section
$\boldsymbol{\tau}$	t-matrix (in angular momentum space) for a Bravais-lattice plane of atoms
φ	azimuthal angle of incidence or emergence in LEED
$\parallel$	2-D component parallel to surface
$\perp$	1-D component perpendicular to surface

Glossary

This glossary defines frequently used concepts and interrelates common terminology. It refers to book sections where further discussions of these concepts can be found. The subject index of this book provides access to additional details in the text.

Absorption: *See* inelastic mean free path.

Adatom: *See* adsorption.

Admolecule: *See* adsorption.

Adsorbate: *See* adsorption, domains.

Adsorption (2.1.11): Adsorbates, including adatoms and admolecules (of different elemental composition than the substrate), can adsorb on a surface substrate as overlayers or substitutionally or interstitially within the substrate by chemisorption or physisorption (also called physical adsorption); coadsorption of different elements or molecules may occur. The adsorbates may form disordered structures or ordered (2-D periodic) structures; ordered structures may have superlattices, which may form disconnected 2-D islands, or out-of-phase or antiphase domains (4.1.2, 4.6), or incommensurate or modulated (4.7) structures. Adsorbates may cause relaxations and/or reconstructions in the substrate.

Anharmonic vibration: *See* thermal effects.

Anisotropic vibration: *See* thermal effects.

Antiphase domains: *See* domain.

Atomic displacement parameter: *See* thermal effects.

Atomic potential: *See* potential.

Attenuation: *See* inelastic mean free path.

Automation of structural determination: *See* R-factors.

Beam: Represents a stream of electrons or X-rays, which may be incident on or diffracted from a sample (with X-rays, the incident beam is often called primary beam). Beams may be specular (reflected as if from a mirror), collimated (traveling in a single direction), convergent (focused on one point or region; 5.1, 5.2) or divergent (expanding from one point or region; 5.6). A beam may have fractional-order (3.2, 4.7) vs. integer-order indices when diffracted from a surface with a superlattice (2.1.4). A beam has finite width (3.4), but is usually described theoretically as a plane wave of infinite width (2.2, 5.1). A beam has an angular profile that depends on inelastic effects (electron–phonon, electron–electron or electron–plasmon scattering; 5.1, 5.5) and on static defects (vacancies, impurities, domain boundaries, islands; 3.3.1, 3.6.1, 4.4, 4.6), in addition to instrumental effects (through the instrumental response function; 3.4). A beam also has a finite energy width, but is usually considered elastic and mono-energetic in the theory (having the nominal energy of the incident beam).

Bragg points and peaks: Bragg points define, for 3-D lattices, the directions where the scattering vector fulfill the conditions for interference maxima (Laue conditions). For surfaces with 2-D translation symmetry, reflections occur when two components of the scattering vector are fulfilled: the reciprocal lattice then consists of Bragg rods; this situation applies to both LEED and SXRD (2.2.1). Bragg peaks, for LEED, are maxima in the I(V) curves near Bragg points of the 3-D substrate lattice. For SXRD, Bragg peaks are reflections arising from the 3-D periodic bulk lattice in the crystal truncation rods (CTR) (7.1).

Bravais-lattice 2-D layer: Contains one atom per 2-D unit cell.

Chemisorption: *See* adsorption.

Cluster of atoms (5.2): Group of atoms that has no internal periodicity, such as a molecule, fragment of molecule or nanoparticle.

Coadsorption: *See* adsorption.

Collimated beam: *See* beam.

Commensurate layers (2.1.11, 4.7): Two periodic layers may have different 2-D lattices that coincide periodically with a larger unit cell than either of the two layers. Such commensurability often leads to rational ratios of lattice constants (i.e., integer ratios or ratios of two finite integers). Superlattices due to reconstruction or adsorption often are commensurate. The Wood and matrix notations of surface periodicity assume commensurability, which contrasts with incommensurability.

Composite 2-D layer: Contains more than one atom per 2-D unit cell.

Convergent beam: *See* beam.

Correlated vibrations: *See* thermal effects.

Damping: *See* inelastic mean free path.

de Broglie: *See* wave.

Debye temperature: *See* thermal effects.

Debye–Waller factor: *See* thermal effects.

Defects: Deviations from ideal crystallinity include static defects (e.g., disordered atoms or molecules, substitutional or interstitial atoms, and vacancies; 3.3.1, 3.6.1, 4.4, 4.6) and dynamic thermal defects (e.g., thermal vibration and diffusion; 5.1, 5.5). Static and thermal defects can be quantified with atomic displacement parameters (5.5).

Detectors (3.1, 7.2): LEED electrons and diffraction patterns are detected and/or displayed with Faraday cups, fluorescent or phosphorescent screens, microchannel plates, position sensitive detectors (5.4.3) or video cameras (5.1). Software is used to digitise the beam intensities into I(V) curves. X-rays in SXRD can be detected with scintillation crystals in the case of point detectors, but these are currently rarely used. Most common are area detectors of different types, primarily CCD (charge coupled device) detectors (7.2).

Diffraction (2.2, 5.2): Waves (of electrons or X-rays) are scattered by individual atoms, clusters of atoms, 2-D layers of atoms and crystal surfaces. Coherent scattering (which requires elastic scattering to maintain relative phases) causes diffraction that is sensitive to relative atomic positions and permits determining those relative positions (crystallography) in LEED and SXRD. In LEED, multiple scattering is strong, while single scattering dominates in SXRD.

Diffraction pattern (4): Diffracted intensities vary greatly with direction of diffraction, with two angular variables (polar and azimuthal). Ordered structures, such as periodic and quasicrystalline surfaces, produce sharp scattered beams (and sharp spots on LEED detectors) that form a diffraction pattern. This pattern reflects directly the surface periodicity and symmetry (4.1); in particular, it can exhibit the presence of arrays of more-or-less ordered domains (4.1), steps (4.3), facets (4.4) and modulated surface layers (4.6, 5.7). Multiple scattering can create new fractional or split-off diffracted beams or enhance weak ones. Disordered surfaces and inelastic effects (electron–photon and electron–phonon interactions) cause diffuse background intensities between beams and spots (3.3.3), which can be analysed in diffuse LEED (DLEED) (5.4.3).

Diffuse LEED (DLEED) (5.4.3): With diffuse LEED, diffuse scattering between sharp beams can be used to analyse the local structure of disordered adsorbates, impurities or other defects.

Diffuse scattering and diffuse intensity (5.1, 5.2, 5.4.3, 5.5): Scattering in directions between sharp beams (also called background intensity), most of which is incoherently scattered by static defects, thermal vibrations and other inelastic effects. The local structure of disordered adsorbates, impurities or other defects can be studied with diffuse LEED.

Diffusion of atoms: *See* thermal effects.

Direct methods: In LEED, no generally applicable direct methods have been developed so far (6.1). Methods with limited applicability include: averaging methods such as the constant momentum transfer averaging (CMTA) and the Patterson function (6.1.1.1), electron holography (6.1.1.2), and correlated thermal diffuse scattering (CTDS) (6.1.1.3). In SXRD, several methods exist (7.3). The phases of the crystal truncation rod intensities can be determined using the known phases near the Bragg points (7.3.3). For large unit cells, phase estimates for superstructure reflections can be calculated (7.3.4). The electron density in the surface slab can be directly determined using the interference between bulk and surface contribution to the structure factor (7.3.5).

Direct space: *See* reciprocal space, reciprocal lattice.

Disorder: Any departure from 2-D crystallinity or quasicrystallinity constitutes disorder. This includes static disorder (primarily lattice gas disorder of adsorbates or substitutional atoms, treatable by diffuse LEED, other point defects, overlayer islands and line or plane defects, the latter two causing domains or microcrystallites, respectively; 3.3.1, 3.6.1, 4.4, 4.6, 5.4.3) and dynamic disorder (mainly thermal motion, but also 2-D atomic diffusion; 5.1, 5.5). Disorder causes diffuse background diffraction into directions between sharp beams and spots due to the ordered structures.

Divergent beam: *See* beam.

Domain (also called mosaic) (4.1, 4.5): 2-D periodicity at surfaces is normally spatially limited to exist only within microcrystallites of the substrate, finite sized islands of adsorbates or individual terraces of stepped surfaces. These regions form separate domains that do not match up according to the substrate periodicity and, therefore, produce out-of-phase or in particular antiphase structures. Domains are particularly frequent at reconstructed surfaces and in adsorbate layers which have superlattices, which can break the substrate periodicity and symmetry and broaden beam and spot profiles. Domain boundaries can have different local atomic relaxations and can locally change multiple scattering.

Dynamical theory: *See also* multiple scattering. In SXRD (as in XRD), dynamical (or Darwin or Ewald) theory includes refraction (change of direction due to a change of index of refraction between outside and inside the sample) and absorption (attenuation) (7.1). In LEED, dynamical theory additionally includes multiple scattering (5.2), that is, successive scatterings of electrons from more than one atom (refraction is included through a change of potential energy at the surface). Dynamical theory contrasts with kinematic or single scattering theory.

Effective attenuation length (6.1.6.5): Similar to the inelastic mean free path, but the effective attenuation length also includes elastic attenuation due to diffraction (elastic scattering) in addition to the inelastic effects.

Energy: In LEED, normally refers to the total energy of a LEED electron in vacuum. Experimentally, there is an energy spread, given by the electron gun (3.1), the detector's energy selection grids (3.1) and inelastic effects (5.1), while theoretically a single energy value is assumed. This energy is equal to the kinetic energy in vacuum near to the surface, but is increased by the inner potential within the surface.

Error estimates: *See* R-factor.

Expansion of crystal lattice: *See* thermal effects.

Faraday cup: *See* detectors.

Fluorescent screen: *See* detectors.

Fractional-order beam: *See* beam.

Full potential: *See* potential.

Glancing angle: Refers to a grazing incidence angle, nearly parallel to the surface, in SXRD.

Glide symmetry: *See* symmetry.

Harmonic vibration: *See* thermal effects.

High-Miller-index surface: *See* stepped surface.

Image potential: *See* potential.

Imaginary part of potential: *See* inelastic mean free path.

Incommensurate layers (2.1.11): Two periodic layers with 2-D lattices that do not coincide periodically within a finite region are incommensurate. The Wood and matrix notations of surface periodicity cannot be used for incommensurate layers. Such incommensurability translates to irrational ratios of lattice constants, in contrast to commensurate layers. Incommensurability, strictly understood, requires an infinite 2-D lattice extent; in practice, due to finite terraces and microcrystallite sizes as well as non-zero lateral interactions between layers, strict incommensurability is unlikely to occur.

Inelastic effects: *See* inelastic mean free path.

Inelastic mean free path (IMFP): In LEED, electrons lose energy due to inelastic effects (mainly electron–electron scattering and excitations) (2.2.3). This causes damping, also called absorption or attenuation, of the elastic electron beams and causes a finite inelastic mean free path (6.1.6.5) and penetration depth. This effect also dominates peak widths in I(V) curves (5.1) and is represented in the theory by an imaginary part of the potential (5.1), also called optical potential. When elastic diffraction (i.e., elastic scattering) is included, one speaks of effective attenuation length. The inelastic mean free path of electrons in surfaces depends strongly on energy, but only weakly on surface composition, giving rise to the Universal Curve (2.2.3, 6.1.3, 6.1.6.5) for penetration. In SXRD, grazing incidence angles provide shallow penetration, which minimises the background intensity arising from the substrate.

Instrumental response function (3.4): *See also* resolution. Describes the resolution of the LEED instrument and depends on the incident beam profile. It is determined by the aperture of the electron gun and the energy spread of the incident beam, as well as by the resolution of the detector or display screen, among other factors. It affects the structural resolution attainable in structure determination. In SXRD, the resolution is not defined as instrumental limit but related to the measurement as the minimum distance d_{min} which has been measured at the highest diffraction angle at a given wavelength according to Bragg's law, $2d\sin\vartheta = n\lambda$.

Integer-order beam: *See* beams.

Intensity (3.3, 5.6, 7.2.2): *See also* I(V) curve. In LEED, the electron beam current is normally expressed as intensity, without particular unit, leading to the term I(V) curve (or equivalently I-V or IV curve). For diffracted electron beams, the term reflectivity is frequently used, which is obtained by dividing their intensity by the incident beam intensity. In the theory, intensity is derived from the wave amplitude. Besides beam intensity, there is also diffuse intensity (also called background intensity), most of which is incoherently scattered. Intensity is obtained from beam current or spot brightness. In SXRD, the measured intensity is usually the integrated intensity of a reflection given on an arbitrary scale defined by the counting rate in the detector. The measured intensity has to be corrected by several factors due to the diffraction geometry and detector aperture. In the analysis the intensity is defined as the square of the modulus of the structure factor. The absolute intensity of the incident beam (also called primary beam) is usually not measured or not used in the analysis.

Interstitial adsorption: *See* adsorption.

Interstitial potential: *See* potential.

Island: *See* adsorption, domains.

Isotropic vibration: *See* thermal effects.

I(V) curve (3.3): *See also* intensity. In LEED, intensity is normally measured and displayed as a function of accelerating voltage, hence the term I(V) curve, or equivalently I-V or IV curve. Often, voltage and energy are used interchangeably, hence I(E) curve, etc., and the expression 'electron energy' is often used as abscissa when plotting I(V) curves.

Kinematic or single scattering theory (2.2, 7.1): In LEED and SXRD, kinematic theory implies the neglect of multiple scattering. The kinematic theory contrasts with the dynamical or multiple scattering theory.

Kinks and kinked surface: *See* stepped surface.

Lattice (2.1): Planar, perfectly truncated 2-D surfaces of 3-D crystals generally exhibit 2-D translational periodicity in the surface plane (exceptions are stepped surfaces with aperiodic terrace widths and quasicrystal surfaces). Such 2-D lattices have 2-D unit cells with two basis vectors and two lattice constants; the third real-space dimension, perpendicular to the surface, has an infinitely large lattice constant. Within the resulting 3-D surface-adapted unit cell, atomic positions are defined or to be determined. The surface structure may deviate from the ideally truncated substrate structure (due to reconstruction or adsorption of other atoms): if the surface is ordered (periodic), the 2-D surface lattice may coincide with that of the substrate or have a different periodicity, often called a superlattice; the latter may be commensurate or incommensurate with the substrate.

Laue conditions: *See* Bragg points.

Layer (2.1): A surface plane defines 2-D infinite atomic layers parallel to that plane. Of particular interest are interlayer spacings (often called d-spacings or simply spacings) and interlayer registries (i.e., lateral shifts parallel to the surface). Such atomic layers are especially useful in theoretical methods that construct a surface by stacking simpler layers on top of each other (5.2.6). The simplest layers have a Bravais lattice (with one atom per 2-D unit cell); layers with multiple atoms per 2-D cell are sometimes called composite layers. Overlayers are composed of adsorbed adatoms and admolecules, different from the substrate material.

Libration: *See* thermal effects.

Matrix notation: *See* superlattice.

Mean free path: *See* inelastic mean free path.

Microchannel plate: *See* detectors.

Microfacet: *See* stepped surface.

Miller and Miller–Bravais indices (2.1.3): The orientation of a lattice plane is described by its Miller indices (h, k, l), which are defined in reciprocal space; they are therefore also very useful to describe momentum change during diffraction in relation to the crystal lattice (2.2.1). Miller–Bravais indices (h, k, i, l) are convenient for describing lattice planes in hexagonal lattices, for example, hexagonal close packed. Miller indices are used to define the surface plane: low-Miller-index surfaces tend to be atomically flat and have small 2-D unit cells, while high-Miller-index surfaces are stepped (exposing alternating terraces and atomic-height steps, each of which have lower Miller indices; 2.1.12).

Mirror symmetry: *See* symmetry.

Modulated surfaces (4.6, 6.3): Superstructures with 2-D superlattices (different from the substrate 2-D lattice) are common with overlayers and surface reconstructions.

Such layers may modulate one another, that is, impress a different lattice on another layer. Modulated layers produce characteristic diffraction patterns with additional satellite beams.

Muffin-tin model: *See* potential.

Multiple scattering (5.2): Describes the occurrence of successive scatterings of a wave (representing electrons or X-rays) by different atoms. Multiple scattering is an ingredient of dynamical theory and is generally very strong in LEED but very weak in SXRD. Multiple scattering complicates the determination of surface structure by LEED.

Optical potential: *See* inelastic mean free path.

Out-of-phase domains: *See* domain.

Overdetermination (6.1.4): Inclusion of more experimental data than strictly needed to determine the unknown structural (and possibly non-structural) parameters. In practice many times more data points than parameters are needed for a reliable and accurate structure determination.

Overlayer: *See* adsorption.

Partial-wave expansion: *See* wave.

Particle-wave duality: *See* wave.

Peak width in I(V) curves: *See* inelastic mean free path.

Penetration depth: *See* inelastic mean free path.

Phase shifts: *See* scattering.

Phosphorescent screen: *See* detectors.

Physisorption or physical adsorption: *See* adsorption.

Plane waves: *See* waves.

Point group symmetry: *See* symmetry.

Position sensitive detector: *See* detectors.

Potential: In LEED theory, electron scattering is based on the crystal potential (2.2, 5.2, 6.1.6), which includes: the image potential linking the vacuum potential (often taken as zero potential energy) to the first layer of atoms; possible potential steps linking layers of different atoms (e.g., between an overlayer and a substrate or between inequivalent layers in compounds); an interstitial potential between atoms (usually constant within each layer in the muffin-tin model), often called muffin-tin constant or muffin-tin zero; and more or less spherically symmetric atomic potentials centred on nuclei (made spherically-symmetric in the muffin-tin model, possibly mutually overlapping or allowed to be non-spherical in the 'full' potential model). The potential typically has a real part (including Coulomb and exchange-correlation terms) representing elastic scattering, and an imaginary part (5.1, 6.1.6.5) representing inelastic scattering (to describe the electron penetration depth, that is, absorption, attenuation, damping, inelastic mean free path). In SXRD as in 3-D XRD, photons are primarily scattered by electrons, as expressed by scattering factors rather than potentials. The photons are scattered by the electron density based on the observation that the electromagnetic field of the photon causes the electron to oscillate and in turn to emit radiation (7.1).

Potential step: *See* potential.

Quantum theory: *See* scattering.

Quasicrystal (6.2): 3-D atomic structure having no long-range translational period-icity, but exhibiting orientational order, including unusual 5-fold, 8-fold, 10-fold or 12-fold rotational axes which are not compatible with translation symmetry. Quasicrystals produce sharp diffraction reflections in non-periodic diffraction patterns. Their 2-D surfaces often have similar properties.

R-factor (6.1.2): Used to objectively (i.e., computationally) quantify the level of agreement between theoretical and experimental diffraction intensities, so as to establish the reliability of structural determination. R-factors (also called residuals or residues) also permit the automation of structural determination. A double-reliability factor RR and a significance factor S can be used to determine error estimates (6.1.3) of optimised structural parameters.

Real space: *See* reciprocal space and reciprocal lattice.

Reciprocal space and reciprocal lattice: As with 3-D lattices, a 3-D reciprocal lattice can be defined for 2-D surface lattices (2.1.5); one speaks of reciprocal space vs. direct or real space. This 3-D reciprocal lattice is very useful for defining diffraction conditions, beams and directions (2.2.1, 2.2.4, 3.2, 4.1, 4.2, 5.1, 5.3.1, 6.2.3, 6.3, 7.1), often with Miller indices (2.1.3). The 3-D reciprocal unit cell has finite basis vectors parallel to the surface, and an infinitely short basis vector perpendicular to the surface. In SXRD, this gives rise to crystal truncation rods perpendicular to the surface, whose intensities are frequently measured as a function of momentum transfer perpendicular to the surface (7.1).

Reconstruction (2.1.4, 6.3.2, 7.1, 7.3): Describes the property that many clean (i.e., adsorbate-free) surfaces adopt a different structure from the crystalline substrate, including in particular periodic (or near-periodic) superstructures with superlattices that differ from the 2-D surface lattice of the bulk.

Reflectivity: *See* intensities.

Relaxation (6.1.4, 7.1.4): Describes the property that atoms at surfaces can change bond lengths and interlayer spacings relative to the corresponding bulk structure of the substrate. This includes multilayer relaxations within a surface, as well as adsorbate-induced or reconstruction-induced relaxations of the positions of sub-strate atoms.

Reliability of structural determination: *See* R-factors.

Residual: *See* R-factors.

Residue: *See* R-factors.

Resolution: *See also* instrumental response function. Defines the smallest structural detail resolvable in the structure analysis and depends on the range in reciprocal space which has been measured. Resolution normal and parallel to the surface can be distinguished. The resolution defines minimum error bars, while the actual error bars depend on the errors in the intensity measurement. For LEED a clear definition of the resolution is not given as it is influenced by multiple scattering. For SXRD the resolution is defined as the minimum value d_{min} resulting from the maximum scattering angle in the data set.

Rotation, thermal: *See* thermal effects.

Rotational symmetry: *See* symmetry.

Scattering: Describes the collision of electrons or X-rays with atoms and assemblies of atoms (clusters and surfaces), as given by Schrödinger's quantum theory (LEED; 2.2, 3.2, 5.2, 5.5) or Thomson scattering (SXRD) (2.2.2, 7.1). Multiple scattering occurs when a wave scatters successively from more than one atom; it is sensitive to multiple atom correlations. Kinematic theory assumes single scattering and pair correlation functions (as is normal in SXRD). The structure factor (2.2.1) describes the strength and angular distribution of scattering by a unit cell, while the atomic scattering amplitude (LEED) or form factor (XRD) describes scattering by a single atom. In LEED, the scattering is derived from the electron–atom and electron–crystal interaction potential and usually parametrised by phase shifts (which depend on chemical element and energy; 2.2.3); inelastic scattering results from electron–phonon and electron–electron interactions. In SXRD, the scattering factor is described by parameters derived from the atomic electron density and near absorption edges by an additional imaginary part derived from the absorption coefficient. Inelastic processes are Compton scattering and electronic excitations like photo-emission, interband transitions and phonon scattering (7.1). In both LEED and SXRD, thermal effects on elastic scattering are represented by Debye–Waller factors or related factors.

Single scattering: *See* scattering.

Space group symmetry: *See* symmetry.

Spacing: *See* layer.

Specular reflection: Mirror-like reflection, with equal incident and reflected angles.

Spherical harmonics: *See* wave, and **t**-matrix.

Spherical wave: *See* wave.

Splitting of spots: *See* stepped surface.

Spot (3.1): Visual mark of an electron beam on a display, forming a diffraction pattern. The spot profile reflects the degree of ordering of the surface, as well as inelastic effects.

Step: *See* stepped surface.

Step bunching: *See* stepped surface.

Stepped surface (2.1.12, 5.2.6.5): Surfaces cut along planes with high Miller indices typically exhibit an alternation of terraces and steps, each of which have lower Miller indices: they are often called high-Miller-index or vicinal surfaces. The steps, which are frequently of mono-atomic height, can themselves be stepped, that is, kinked, exhibiting kinks, depending on the Miller indices. The terraces, steps, and kinks can be viewed as crystalline microfacets. Multi-atom-height steps result from step bunching. Stepped surfaces produce diffraction patterns with characteristic splitting of spots, relative to the surface having the low-Miller indices of the terrace plane.

Structure determination (6.1, 7.3): Derivation from experimental data of atomic positions and, optionally, atomic properties such as vibration amplitudes, chemical identities, occupation fractions, etc. In LEED, mostly a trial and error approach is used, coupled with structure refinement. In SXRD, the same approach is applied and in addition direct methods can be used for model finding.

Structure factor: *See* scattering.

Substitutional adsorption: *See* adsorption.

Substrate: *See* adsorption.

Superlattice: Surface structures often exhibit a 2-D surface periodicity different from that of the perfectly truncated 3-D substrate (2.1.11, 3.2, 4.1, 5.1, 6.3). The resulting 2-D superlattices are particularly common with overlayers and surface reconstructions. The 2-D periodicity of the superlattice may be commensurate or incommensurate with the substrate (2.1.11); in either case, modulated structures can result (4.6, 6.3). Superlattices are often described quantitatively using the matrix notation or the Wood notation. Superlattices give rise to fractional superlattice beams and spots in the diffraction pattern (2.1.10).

Symmetry: The structural symmetry of surfaces is normally derived from that of the bulk structure of the substrate (2.1). Surface structures can also exhibit, besides periodic translational symmetry (i.e., 2-D periodicity; 2.1.4), rotation, mirror and glide symmetries (in real/direct space) (4.1): these are described by space group symmetry that combines a periodic lattice with point group symmetry. Quasicrystals are characterised by unusual symmetries (5-fold, 8-fold, 10-fold or 12-fold rotation axes) without translational symmetry (6.2). The incident beam may have a direction that matches symmetry elements of the surface (e.g., a rotation axis and/or mirror planes and/or glide planes). The combined symmetry of incident beam and surface gives rise to symmetry in the diffraction pattern, specifically symmetry-equivalent beam intensities (3.3.2). Experimentally, equivalent beam intensities can be used to verify the beam incidence direction (3.3.2) and are frequently averaged together (3.2.2). In LEED, surface and beam symmetries are often exploited to vastly accelerate the theoretical computations of diffracted beam intensities (5.3).

t-matrix, τ-matrix, T-matrix: In LEED theory of scattering, electron waves are often represented by spherical harmonics: **t**-matrices are then used to calculate their scattering by individual atoms (5.2.1), by clusters of atoms (5.2.3) or by layers of atoms; in this book the notation **τ**-matrix is used for scattering by Bravais-lattice layers (5.2.5.1) and **T**-matrix for composite layers (5.2.5.4).

Terrace: *See* stepped surface, domain.

Thermal effects: Dynamic motion of atoms or molecules (5.5) includes thermal vibrations (5.5.1), rotation and libration (5.5.4) of relatively stiff molecules; vibrations can be isotropic or anisotropic (described by a thermal ellipsoid; 5.5.5, 5.5.6), harmonic (5.5.3) or anharmonic (5.5.4) and/or correlated among neighbouring atoms (5.5.5). These motions can be quantified with atomic displacement parameters (5.5). Simple thermal effects on diffraction are normally modeled by a Debye temperature (5.5.2) and a Debye–Waller factor (5.2.1, 5.5.1), also in the multiple scattering formalism of LEED. Dynamic defects such as diffusing atoms are generally ignored, as is thermal expansion of the crystal lattice.

Thomson scattering: *See* scattering.

Transfer width (3.4): *See also* instrumental response function. Defined for LEED instruments, the transfer width gives the maximum distance which can be measured by a beam profile analysis.

Translational periodicity: *See* symmetry.

Truncation rod: In SXRD, intensities are often measured continuously along truncation rods perpendicular to the surface, due to the lack of periodicity perpendicular to the surface.

Ultra-high vacuum: A high level of vacuum is needed to keep the surface clean of foreign atoms or molecules for the duration of the experiment, which may last for many hours (3.1). In LEED, the electron beams (incident and diffracted) also require a good vacuum to avoid attenuation and chemical effects on the background gas.

Universal Curve: *See* inelastic mean free path.

Vacuum: *See* ultra-high vacuum.

Vacuum potential: *See* potential.

Vibration: *See* thermal effects.

Vicinal surface: *See* stepped surface.

Video camera: *See* detectors.

Voltage: *See* I(V) curve.

Wave: Collimated beams are conveniently represented by plane waves in theory (2.2, 5.1), while convergent or diffracted beams can also be expressed in terms of spherical waves (more precisely spherical harmonics in the partial-wave expansion; 5.2.3, 5.2.4). With de Broglie's particle–wave duality, electrons can be assigned wavelengths and wave vectors.

Wavelength: *See* wave.

Wave vector: *See* wave.

Wood notation: *See* superlattice.

1 Introduction

1.1 Brief Historical Background

X-ray crystallography of 3-D bulk materials opened entire new fields of discovery in the twentieth century, from elemental crystals to DNA molecules. Similarly, the determination of atomic-scale structure of 2-D surfaces of condensed matter has achieved fundamental new understanding and generated powerful techniques that helped spawn new areas of research, from catalysis to nanotechnology, for the twenty-first century. Specific examples include, among many others: new catalysts; various new carbon structures (such as buckminsterfullerenes, nanotubes and graphene); quantum dots used in optoelectronic displays (including television displays); molecules allowing electron transport and switching; nanoparticles enabling targeted drug delivery; and nanomachines for future manufacturing and medical applications.

Over 100 years ago, in 1912, von Laue discovered X-ray diffraction (XRD) [1.1], providing experimental confirmation of Einstein's wave–particle duality for photons. This led one year later to the development of the highly successful technique of X-ray crystallography by W. L. Bragg [1.2]. Due to the weak scattering of X-rays, however, it was difficult to make the technique surface-sensitive and suitable for the ultra-high vacuum (UHV) needed to control surface composition at the monolayer level. One approach, developed in the early 1980s, uses fluorescent scattering from X-ray standing waves generated near a Bragg total-reflection condition [1.3]. Another uses glancing incidence to limit the penetration depth of the X-rays into the sample [1.4]: the latter method is now commonly used, primarily in connection with synchrotron radiation that enhances the photon flux and allows continuously varying the wavelength. As a result, surface X-ray diffraction (SXRD) became a routine tool for surface crystallography in the 1990s and 2000s.

In 1927, C. Davisson and L. H. Germer [1.5; 1.6] discovered low-energy electron diffraction (LEED), followed rapidly by the work of G. P. Thomson and A. Reid [1.7], thereby experimentally establishing the wave–particle duality for electrons. After the introduction of ultra-high-vacuum technology in the 1960s, LEED was used primarily for surface characterisation. Despite the inherent surface sensitivity of LEED, due mainly to strong absorption of low-energy electrons, it proved difficult to use this technique for surface structure determination. That goal required theoretical methods which were introduced in the 1960s and 1970s to take into account the 'multiple scattering' of electrons that is so characteristic of LEED. This effort gave rise to effective surface structure determination methods by the late 1980s, after the addition of automated structure optimisation and other important methodologies.

Starting around 1980, a variety of other techniques sensitive to surface structure emerged. They include, in no particular order: photoelectron diffraction (PED, which also goes by a variety of other names), ion scattering (IS), transmission electron microscopy (TEM) and low-energy electron microscopy (LEEM), atom diffraction (AD), angle-resolved Auger electron spectroscopy (ARAES), extended appearance potential fine structure (EAPFS), surface extended energy loss fine structure (SEELFS), X-ray absorption fine structure (XAFS), including surface extended XAFS (SEXAFS) and near-edge XAFS (NEXAFS), as well as scanning probe microscopies (SPM), which include scanning tunnelling microscopy (STM) and atomic force microscopy (AFM), among others. These and further techniques of surface science are mentioned in the list of abbreviations and acronyms, provided at the beginning of this book.

These new techniques often contributed competing or complementary structural information about surfaces, sparking healthy and constructive debates that led to further improvements in the respective methodologies. In parallel, predictive computational ab initio methods were developed that provided a further opportunity for cross-checking surface structures. In particular, experimental structural results have helped validate a long list of computer codes based on density functional theory. As a result, we can today increasingly rely on ab initio predictions of many types of structures (2-D, 1-D and 0-D). It must be added that no theoretical method is currently able to reliably identify the 'globally best' structure; on the experimental side, multiple techniques must be used in parallel to try to identify that 'globally best' structure.

Appendices A and B provide lists of selected books and websites that may be helpful to find further information on many aspects of surface science and surface structure, as well as related techniques and LEED codes. In addition, we refer to the only other monographs written on LEED [1.8–1.10]. SXRD has been described only in a few review articles [1.11–1.13].

1.2   Physical Basis of LEED and SXRD

In this book we will not repeat the underlying physics of LEED and XRD, but only briefly sketch their foundations here and the specific conditions for surface X-ray diffraction. Both LEED and XRD, like other diffraction techniques, are based on the interference of waves scattered from different atoms: the interference depends on the relative positions of the atoms, which positions are then extracted from measured beam intensities by suitable analysis methods. In LEED, electrons scatter strongly from atoms of a surface and can therefore suffer 'multiple scattering' from two or more atoms in succession: this complicates the analysis considerably. Many attempts to 'average' out multiple scattering effects (including holographic schemes) have been made, but with only limited success in selected weakly-scattering materials. In XRD the photon scattering is much weaker and 'single scattering' usually dominates overwhelmingly; this allows simple Fourier-like transformations to obtain atomic positions, except for the famous phase problem, due to the measurement of wave intensities rather than wave amplitudes.

Incidentally, calling multiple scattering a 'succession' of scatterings is a convenience of speech stemming from a picture of particles bouncing around. However, in the wave picture relevant to diffraction, all scattering in the time independent Schrödinger equation occurs simultaneously rather than in a time sequence. Nonetheless, many calculation schemes of LEED in effect expand the multiple scattering process into a sequence of successive 'single' scatterings from different atoms, as if they were collisions ordered in time.

While the structure analysis is thus relatively difficult in LEED, the experiment is more difficult in XRD. This is due to the more exacting requirement on the incident beam alignment with the sample; another factor is the need to use synchrotron radiation at large synchrotron facilities, due to the relative weakness of the surface signal which requires high intensities and especially the possibility to vary the wavelength. The LEED experiment can be performed in a small university laboratory; the strong source current and scattering enable the accumulation of relatively large databases of measured intensities for structure analysis.

In LEED, electrons follow the laws of quantum mechanics. For most cases, the Schrödinger equation is used. The Dirac equations can be invoked for heavy atoms and magnetic materials, in which cases electron spin and relativistic effects are also taken into account. In XRD, photons are scattered according to the laws of electromagnetism. Nevertheless, it is possible to treat the physics of XRD with a simple scalar wavefunction, because the index of refraction n is very close to 1 (see, for example [1.14]). The scattered X-ray intensity arises from oscillation of electrons excited by the incident electromagnetic wave, while protons are too heavy to contribute significantly. X-ray diffraction therefore determines the electron density of a material, in contrast to LEED where the main contribution of the scattering potential comes from the protons and the core electrons.

Progress in recent decades, especially on the theoretical side in the case of LEED, has made possible the structure determination of more complex structures with LEED and SXRD. These include disordered surfaces, stepped surfaces, faceted surfaces, antiphase boundaries, modulated layers, quasicrystals and nanostructures.

1.3   Organisation of This Book

We start in Chapter 2 with basic elements of surface geometry and diffraction from surfaces, including translational and other symmetries; in particular, we point out similarities and differences between LEED and SXRD.

Chapter 3 addresses the LEED experiment, including various recent developments and useful practical information.

Chapter 4 covers the interpretation of the diffraction pattern as seen in LEED, paying particular attention to symmetries and 'imperfect' surface structures: stepped surfaces, faceted surfaces, antiphase boundaries and modulated layers.

This is followed by Chapter 5 on the basic theory of LEED, which covers a wide range of issues and methods, old and new, partly to address the consequences and treatment of multiple scattering; included in particular are the use of symmetry in calculations, newer approximations, and developments in studying nanoparticles, disordered surfaces and thermal effects.

Chapter 6 covers LEED applications, starting with a wide range of structure analysis methods. In particular, the influence of the phase shift calculation is discussed in detail and it is shown that some improvements can be achieved by using potential models in which steps between muffin tin spheres are avoided. These methods are followed by studies of quasicrystals and modulated surfaces.

Chapter 7 is devoted to X-ray diffraction at surfaces: it points out advantages and disadvantages relative to LEED and describes the two-step process of 'direct methods' to determine structural models, for subsequent 'structural refinement'. After summarising the relatively simpler theory of X-ray diffraction with emphasis on surfaces, the specific properties of grazing incidence X-ray diffraction are discussed. Next, the main experimental methods and experimental set-ups appropriate for surfaces are described, both at synchrotron facilities and in the laboratory, with specific examples. Methods of data analysis are presented next, from the relatively simple Patterson function approach to a series of more recent advanced direct methods that produce a unique model structure for further refinement.

1.4   Comparison of Some Surface Structure Techniques

It may be useful here to compare LEED and SXRD to the other techniques available for surface structure determination. (For the meaning of the technique abbreviations, please see the list provided at the beginning of this book; some of these are also given in Section 1.1.

We classify the techniques according to the type of structure information they are able to provide, starting from the most detailed information.

A number of techniques are sensitive to the three coordinates of surface atoms. These techniques provide essentially the complete surface structural information averaged over many 2-D unit cells. This information goes down to a depth that is determined by the penetration of the electrons, photons or other particles used, but depends on whether the analysis actually optimised the positions and other parameters (thermal vibrations, chemical identification, occupancy, etc.) of atoms at various depths below the surface. This category includes: SXRD, LEED, MEED, RHEED, MEIS, HEIS, ARPES, ARAES, NEXAFS/XANES, HREELS (if done with a normal-mode analysis or I-V curve interpretation) and SEXAFS (when contrasting results are obtained with polarised radiation).

Of these techniques, several are mainly sensitive to layer spacings (because they involve angle-resolved measurements), namely MEED, RHEED and HREELS (with I(V) curve interpretation).

Several techniques are mainly sensitive to bond lengths and, in some cases, bond orientation: SEXAFS (without polarisation dependence), NEXAFS/XANES, PED and EAPFS.

Some techniques are sensitive to bond orientation only: ESDIAD, HREELS (when applying only dipole selection rules) and IRAS (with selection rules).

Other techniques are only sensitive to overall conformation of the surface, without giving detailed positions, bond lengths or bond orientations: FIM, ESDIAD, SPI, atom diffraction, EM, HREELS (with selection rules only), IETS, STM, UPS and IRAS.

There are techniques that are primarily sensitive to the low-electron-density contour of the surface: LEIS and atom diffraction.

STM is mainly sensitive to the surface density of states (of both the surface and the STM tip), and less directly to atomic positions; in certain circumstances, STM can detect that some atoms have a chemical identity that is different from the substrate atoms.

A few techniques are sensitive to foreign atoms only (i.e., they are not sensitive to the geometry within the substrate surface), such as: SEXAFS and EAPFS.

Several spectroscopic techniques are sensitive to molecular conformations at surfaces and interfaces, through molecular vibration frequencies: HREELS, IRAS/IRS and IETS.

Very useful as well are analytical techniques that identify chemical elements (including sometimes ionisation states and bonding configurations): AES, XPS/ESCA, ISS and SIMS.

2   Basic Elements

The description of crystal surfaces requires some basic knowledge of crystallography. Therefore, this chapter presents a short overview of crystal lattices and their classification due to symmetry. This knowledge is required to understand the substrate structure and the orientation of the surface. However, the 3-D point groups, space groups and the mathematical description of symmetry operations in three dimensions are not described here: for a more detailed explanation the reader is referred to the International Tables of Crystallography [2.1], which is the standard reference book, or a number of textbooks on crystallography published by the International Union of Crystallography [2.2–2.5]. The 2-D space groups and symmetry operations are explained with somewhat more detail here because these are frequently used in surface structure determination. A very detailed description of the geometry of crystal surfaces is given in a recent book by K. Hermann [2.6]. A short introduction into the kinematic theory of diffraction and into diffraction at 2-D periodic lattices is also included here.

2.1   Surface Geometry

2.1.1   3-D Lattices

The characteristic property of a periodic crystal is the translational symmetry, that is, periodicity (quasicrystals do not have periodicity, as described in Section 6.2). The ideal periodic crystal is assumed to consist of an infinite repetition of a unit cell in three directions. The crystal lattice is defined by three translation vectors, often called basis vectors, and the angles between them. The lattice constants are:

$$\mathbf{a}, \mathbf{b}, \mathbf{c} \quad \text{and} \quad \alpha, \beta, \gamma.$$

The angles are defined such that:

$$\mathbf{a} \cdot \mathbf{b} = |\mathbf{a}||\mathbf{b}| \cos \gamma, \quad \mathbf{b} \cdot \mathbf{c} = |\mathbf{b}||\mathbf{c}| \cos \alpha, \quad \mathbf{a} \cdot \mathbf{c} = |\mathbf{a}||\mathbf{c}| \cos \beta. \tag{2.1}$$

The position vector $\mathbf{R}$ of each unit cell in the crystal is then given by:

$$\mathbf{R} = n_1 \mathbf{a} + n_2 \mathbf{b} + n_3 \mathbf{c}, \tag{2.2}$$

where one of the unit cells is taken as origin. The atomic positions inside the unit cell are usually described in relative coordinates, that is, in length units of the translation vectors. Thus,

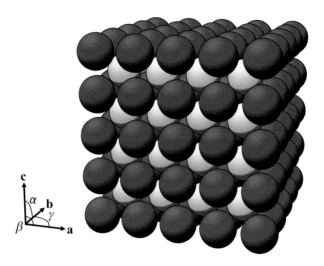

Figure 2.1 An example of a crystal lattice: the CsCl lattice (Cs in red, Cl in green), drawn with ball radii proportional to ionic radii. Adapted from K. Hermann, private communication.

$$\mathbf{r}_i = x_i\mathbf{a} + y_i\mathbf{b} + z_i\mathbf{c}, \tag{2.3}$$

where $0 \leq x_i, y_i, z_i < 1$ for positions inside the unit cell. For example, the CsCl lattice is shown in Figure 2.1, with $|\mathbf{a}| = |\mathbf{b}| = |\mathbf{c}| = 0.4123$ nm $= 4.123$ Å, $\alpha = \beta = \gamma = 90°$, $x_{Cs} = y_{Cs} = z_{Cs} = (0, 0, 0)$ and $x_{Cl} = y_{Cl} = z_{Cl} = (0.5, 0.5, 0.5)$.

The atom coordinates are in general described in an oblique coordinate system, that is, as relative coordinates based on *non-rectangular* translation vectors, or, for lattices with suitable symmetry, as relative coordinates based on *rectangular* translation vectors in rectangular coordinates with different unit lengths – except for the cubic system, which has identical unit lengths. The use of oblique coordinate systems is mostly unfamiliar to the physicist who prefers vector calculations in Cartesian coordinates. An oblique system has, however, substantial advantages over the Cartesian coordinates in the diffraction theory so that oblique coordinate systems are used in all diffraction methods. Vector calculations in oblique coordinates are more complex than in rectangular coordinates and require the definition of a reciprocal coordinate system or covariant and contravariant systems: some rules for vector calculations in oblique coordinate systems are given in Appendix C.

The periodicity of the crystal allows the description of the positions of all atoms once the positions are known for only the atoms within a single unit cell. This is used later in the calculation of diffraction intensities where the sum over all atoms is divided into two parts, the sum over atoms inside the unit cell and the sum over all unit cells.

It should be noted that the definition of the unit cell is not unique: there are many possibilities to define a unit cell in a lattice. In Figure 2.2, four different unit cells are shown, the volumes of which are identical; these unit cells belong to a primitive

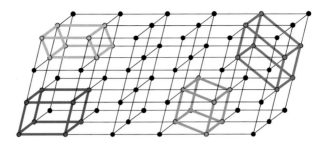

Figure 2.2 Different choices for the unit cell, also called 'settings'

lattice. Larger cells are also possible and are used in some cases; frequently used are centred lattices or non-primitive lattices, which contain more than one lattice point.

To make the choice of the unit cell more unique there are conventional rules to choose the translation vectors and angles. The first rule is to choose the unit cell according to the symmetry of the crystal if there is one. If there is no symmetry other than an inversion centre, then the general rule is to order $|\mathbf{a}| \le |\mathbf{b}| \le |\mathbf{c}|$ and to arrange for α, β, γ to be as close to 90° as possible. Deviations are allowed and frequently used, for example to compare a structure with related structures having different symmetry or to compare two phases below and above a transition temperature. These are then called unconventional choices of the unit cell.

There are seven distinguishable crystal systems, illustrated in Figure 2.3, which differ in their symmetry properties. In the tetragonal, trigonal and hexagonal lattices the vector $\mathbf{c}$ is chosen along the highest-order rotation axis; in the monoclinic case it is either the vector $\mathbf{b}$ or $\mathbf{c}$; and in the orthorhombic case all three translation vectors should be in the direction of the 2-fold axes. The cubic system has either 2- or 4-fold axes in the directions of the translation vectors; the characteristic of this system is a 3-fold axis in the direction of $\mathbf{a} + \mathbf{b} + \mathbf{c}$. These rules are summarised in Table 2.1.

2.1.1.1   Symmetry Operations

The symmetry operations occurring in 3-D lattices are: the inversion $\bar{1}$, mirror planes m, glide planes g, rotational axes 2, 3, 4 and 6, the rotoinversion axes $\bar{3}, \bar{4}$ and $\bar{6}$, and 11 n-fold screw axes $2_1, 3_1, 3_2, 4_1, 4_2, 4_3, 6_1, 6_2, 6_3, 6_4$ and 6_5. The subscript of the screw axes indicates the fraction of the full rotation which represents a single screw operation, that is, 4_3 means ¾ of a full rotation: repeating this operation four times corresponds to three full rotations; 4_3 can be also understood as 4_1 in the opposite direction. The rotoinversion axis $\bar{2}$ is identical to a mirror plane. There is an alternative description of the rotoinversion by rotoreflection; both descriptions are completely equivalent. In this book the Hermann–Mauguin symbol is used, which is based on the rotoinversion; the Schoenflies symbol based on the rotoreflection is still used by chemists and is also listed here.

The symmetry operations in 3-D space, their mathematical description, the combination with translation lattices and the space groups are not explained here. Further

Table 2.1 Crystal systems

Crystal system	Lattice parameters	Orientation
Triclinic	$a \neq b \neq c, \alpha \neq \beta \neq \gamma$	
Monoclinic I or II	$a \neq b \neq c, \alpha = \beta = 90° \neq \gamma$	$\mathbf{c} \parallel 2$
	$a \neq b \neq c, \alpha = \gamma = 90° \neq \beta$	$\mathbf{b} \parallel 2$
Orthorhombic	$a \neq b \neq c, \alpha = \beta = \gamma = 90°$	$\mathbf{a}, \mathbf{b}, \mathbf{c} \parallel 2$
Tetragonal	$a = b \neq c, \alpha = \beta = \gamma = 90°$	$\mathbf{c} \parallel 4$
Trigonal (rhombohedral)	$a = b = c, \alpha = \beta = \gamma \neq 90°$	$(\mathbf{a} + \mathbf{b} + \mathbf{c}) \parallel 3$
Hexagonal	$a = b \neq c, \alpha = \beta = 90°, \gamma = 120°$	$\mathbf{c} \parallel 6$
Cubic	$a = b = c, \alpha = \beta = \gamma = 90°$	$(\mathbf{a} + \mathbf{b} + \mathbf{c}) \parallel 3$

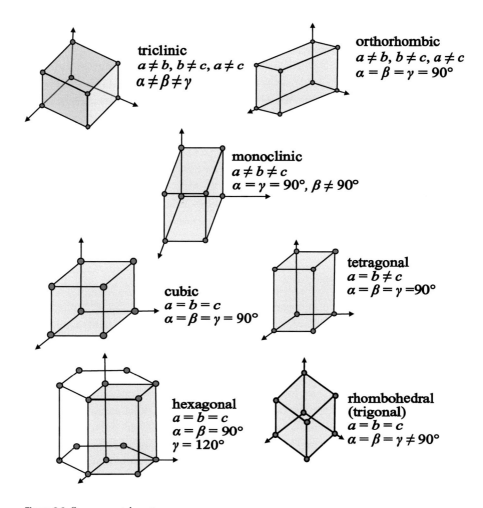

triclinic
$a \neq b, b \neq c, a \neq c$
$\alpha \neq \beta \neq \gamma$

orthorhombic
$a \neq b, b \neq c, a \neq c$
$\alpha = \beta = \gamma = 90°$

monoclinic
$a \neq b \neq c$
$\alpha = \gamma = 90°, \beta \neq 90°$

cubic
$a = b = c$
$\alpha = \beta = \gamma = 90°$

tetragonal
$a = b \neq c$
$\alpha = \beta = \gamma = 90°$

hexagonal
$a = b = c$
$\alpha = \beta = 90°$
$\gamma = 120°$

rhombohedral
(trigonal)
$a = b = c$
$\alpha = \beta = \gamma \neq 90°$

Figure 2.3 Seven crystal systems

information can be found in the International Tables [2.1] and in textbooks on crystallography, for example in some publications of the International Union of Crystallography [2.2–2.5]. Here only the symmetrically different lattice types and the orientation of the lattice planes by Miller indices are described to provide the necessary information for the understanding of surface structures. The symmetry operations in 2-D lattices which are necessary to set up surface structure models are explained in somewhat more detail.

2.1.2   Bravais Lattices

In 3-D, in addition to the seven crystal systems, seven centred lattice types can be distinguished which have a smaller primitive unit cell with an oblique lattice but with additional symmetry elements. There are overall 14 distinguishable symmetrically different lattice types: the so-called Bravais lattices shown in Figure 2.4.

For practical reasons, it is not always the smallest unit cell that is chosen. The calculation of the coordinates of two atoms related by a symmetry operation are simpler if one coordinate axis is parallel to a rotation axis or normal to a mirror plane. Therefore, all coordinate systems are chosen with at least two angles of $90°$, except in the triclinic and trigonal systems. The combination of these 14 Bravais lattices with the different symmetry operations leads to 230 space groups in 3-D. A complete description of the space groups, the symmetry symbols and diffraction conditions can be found in the International Tables [2.1]. The diffraction conditions for the different space groups are of less importance for LEED structure analysis because of the limited penetration depth, but they are necessary for the interpretation of surface X-ray diffraction data where the contributions from the substrate and surface interfere in the so-called truncation rod. The diffraction conditions for 3-D lattices will therefore be discussed in Chapter 7.

2.1.3   Miller Indices

The orientation of a lattice plane is described by its Miller indices (hkl), which are often also written as $(h\ k\ l)$ or (h,k,l) or (h, k, l); negative indices are often written with a bar above the index, as in $(\bar{h}, \bar{k}, \bar{l}) = (-h, -k, -l)$. Figure 2.5 gives examples. The Miller indices (hkl) are also used to label the reflection from the plane (hkl). As is the convention in the crystallographic literature, a direction in the crystal is written with square brackets, for example $[hkl]$ (where h, k and l need not be integers and thus can indicate arbitrary directions), to distinguish it from indices or lattice planes which are vectors in reciprocal space (see Section 2.1.5) and written with parentheses, for example (hkl). Curly brackets $\{hkl\}$ denote the family of planes that are symmetrically equivalent to the plane (hkl), while $\langle hkl \rangle$ denotes the family of directions that are symmetrically equivalent to the direction $[hkl]$. Atom positions are also written with parentheses: $\mathbf{r} = (x, y, z)$.

The Miller indices define the direction of the normal to the lattice plane. In an infinite lattice there exists an infinite number of parallel lattice planes all having the

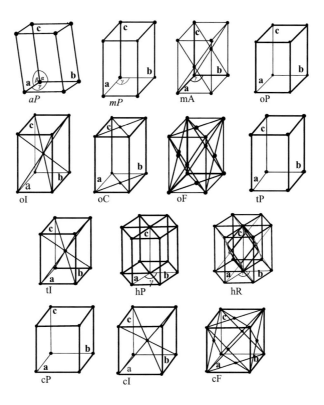

Figure 2.4 The 14 Bravais lattices in 3-D. The label under the unit cells indicates the Bravais type. The first character (lower case) signifies the crystal system, see Figure 2.3; the second character (upper case) gives the Bravais type, P: primitive, A: A-face-centred, C: C-face-centred, F: face-centred, I: body-centred, R: rhombohedral. For hR, the primitive unit cell is the rhombohedron inside the hexagonal cell.

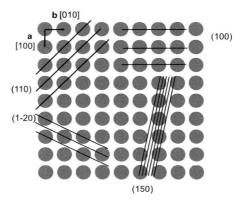

Figure 2.5 Examples of Miller indices based on a square lattice (here the third index is always 0, corresponding to lattice planes that are perpendicular to the plane of the figure)

same Miller indices. The Miller indices are calculated as follows: a lattice plane intersects the coordinate axes at an integer number of translation vectors. The reciprocal values of these numbers are taken and multiplied by a common factor such that the smallest set of three integers is obtained: these are the Miller indices. Let A, B, C be distances from the origin to the intersections of the lattice plane with the coordinate axes in fractions of translation vectors. The Miller indices are then:

$$(h, k, l) = \left(\frac{n}{A}, \frac{n}{B}, \frac{n}{C} \right). \tag{2.4}$$

A note should be added: the indices (hkl) and (nh, nk, nl), where n is an integer, indicate the same plane because a common factor can be eliminated. The indices (hkl) are also used as reflection indices, that is, as distinctive labels for reflections in different directions (often called reflected 'beams'), because the diffraction maxima from 3-D lattices can be interpreted as reflections from a lattice plane. The reflection indices, in contrast to the indices of a lattice plane, can have a common factor and refer to different reflections, for example, the reflections (120) and (240) are different reflections: the first and second order reflections from the same plane.

A further remark may be useful regarding the Miller indices in hexagonal lattices. The trigonal or rhombohedral lattice has a primitive unit cell with three equal translation vectors, three equal angles $\neq 90°$ and the Bravais lattice type hR. The space groups belonging to this crystal system have an R in the first place of the Hermann–Mauguin symbol, that is, it is one of the seven groups R3, $R\bar{3}$, R3m, R3c, R32, $R\bar{3}$ $2/m$ and $R\bar{3}$ $2/c$; the corresponding Schoenflies symbols are C_3^4, C_{3i}^2, C_{3v}^5, C_{3v}^6, D_{3v}^7, D_{3d}^5 and D_{3d}^6. The structures having these space groups can be described with rhombohedral lattice vectors and a primitive cell, as well as with hexagonal lattice vectors and a non-primitive cell, which is three times as large (see Figure 2.6). Both choices of the unit cell

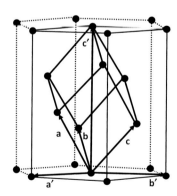

Figure 2.6 Hexagonal (**a′**, **b′**, **c′**) and rhombohedral (**a**, **b**, **c**) 'settings' (choices) of a rhombohedral lattice. The so-called obverse setting (**a**, **b**, **c**) is shown, which is standard. A second symmetrically equivalent setting is possible, called the reverse setting, using the vectors –**a**, –**b**, **c** as a basis. The centring points in the obverse setting are (2/3, 1/3, 1/3) and (1/3, 2/3, 2/3), while in the reverse setting the centring points are (1/3, 2/3, 1/3) and (2/3, 1/3, 2/3).

are commonly used for the same structure and both can be found in the crystallographic databases, where they are called 'settings'. Which setting is used depends on the choice of the author. The number of symmetrically independent atoms is of course the same, but the atomic coordinates as well as the reflection indices are different as they refer to different translation vectors. In order to distinguish the indices of lattice planes referring to hexagonal axes from those referring to rhombohedral axes without explicitly stating each time to which axis system they refer, a four-index symbol is used for all hexagonal lattices and a three-index symbol is used if the rhombohedral setting is used.

The Miller indices for lattice planes in hexagonal lattices are $(hkil)$, with the condition that $h + k + i = 0$, as they are not linearly independent; this four-index notation is often called Miller–Bravais indices. When the indices refer to a rhombohedral lattice, only the three indices (hkl) are used. The base plane of a hexagonal lattice is therefore (0001); the same plane in the rhombohedral setting is the (111) plane. When four indices occur with 3-D lattices, we know that a hexagonal lattice is used. It should also be kept in mind that not all hexagonal lattices can be described with a rhombohedral unit cell; the frequently occurring hexagonal close packed lattice is a primitive lattice with two atoms per unit cell, space group $P6_3/mmc$ or D_{6h}^4, and it is not rhombohedral. A further advantage of four indices is the easy calculation of the indices of symmetrically equivalent reflections by permutation of the first three indices h, k and i, which corresponds to a 3-fold rotation.

The Miller indices of a crystal surface do not uniquely define the structure of the truncated crystal. In many cases the crystal surface structures on both sides of a sectioning plane are different and the (hkl) and $(-h, -k, -l)$ surfaces are therefore inequivalent. When one index is 0, the plane is parallel to the corresponding lattice vector, cf. for example Figure 2.5. In nearly all cases, except in mono-atomic crystals, it is possible to intersect the crystal at different points in the unit cell and thereby create inequivalent surface structures, often called surface terminations. This case is illustrated in Figure 2.7 for the corundum structure α-Al_2O_3, where the (0001) plane could be located just above an oxygen layer (giving an O termination) or just above a pair of Al layers or in between two Al layers (giving two distinct Al terminations).

Most metal crystals have mono-atomic and close-packed structures where the surfaces are uniquely defined. The low-Miller-index surfaces of metal crystals, as well as adsorption structures on these surfaces, have been intensively studied in the past and some low-index planes of face-centred cubic (fcc), body-centred cubic (bcc) and hexagonal close-packed (hcp) lattices are shown in Figure 2.8.

2.1.4   2-Dimensional Lattices

The 3-D crystal is in general described by three translation vectors and the symmetry group is one of the 230 space groups (if magnetic properties such as orientation of magnetic moment are neglected). The truncation of a crystal reduces the translation symmetry to two vectors parallel to the surface; the third vector may be normal or oblique to the surface plane. The two translation vectors parallel to the surface are in general not the translation vectors of the 3-D crystal except for some low-index

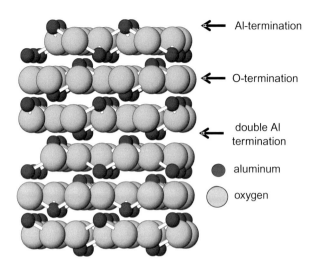

Figure 2.7 Choice of lattice truncation for the (0001) surface of α-Al$_2$O$_3$ (space group R$\bar{3}$c, full symbol is R$\bar{3}$2/c)

planes. The calculation of the translation vectors for a general orientation of the surface plane has been discussed by K. Hermann [2.6] and is treated here for stepped surfaces in Section 2.1.12. We assume that the surface is a lattice plane and the appropriate translation vectors have been chosen. Five symmetrically different lattices can be distinguished, cf. Figure 2.9. As for the case of 3-D crystals, the choice of the unit cell is not unique.

It is the convention to choose the shortest translation vectors and the angle between the vectors to be as close as possible to 90°, and for rectangular or oblique lattices $|\mathbf{a}| < |\mathbf{b}|$. Combining translational, rotational, reflection and glide symmetries limits the possible geometries of a periodic surface to the five 2-D Bravais lattices shown in Figure 2.9.

At surfaces, different layers can have different 2-D translation lattices: a surface reconstruction or an overlayer of adatoms or admolecules may adopt a different lattice from the underlying substrate surface. Often, the new 2-D lattice is simply related to the substrate 2-D lattice, forming a superlattice that has a larger unit cell area than the 2-D substrate lattice. Examples are given in Sections 2.1.9 and 2.1.10, and are further discussed in Section 2.1.11.

A remark should be made here regarding the use of centred cells. The centred rectangular unit cell in two dimensions is the only one that contains two lattice points, that is, the unit cell is twice as large as necessary, and the primitive cell could be chosen with $a' = b'$, $\gamma' \neq 60°$, 90° or 120°, as shown in Figure 2.10. The primitive cell is the smallest possible cell.

The reason why the centred cell is often chosen instead of a primitive cell is a symmetry property of this lattice. It possesses at least one glide plane and possibly

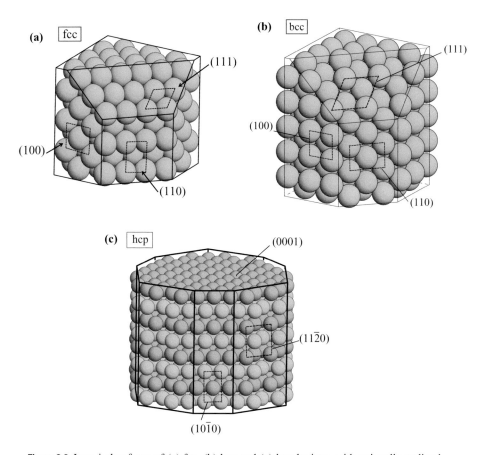

Figure 2.8 Low-index faces of (a) fcc, (b) bcc and (c) hcp lattices, with unit cells outlined using dashed lines. To illustrate the ABAB... stacking sequence in the hcp lattice the two sites are differently shaded, although both sites are occupied by the same atoms and both terminations (0001) and (000$\bar{1}$) are equivalent.

also mirror planes and the mathematical description of the glide operation is much easier in rectangular lattices than in oblique lattices. LEED multiple scattering calculations require the inclusion of all possible scattering paths within a radius around each atom of the unit cell. This is done more easily using the primitive lattice. Therefore, for the calculation of LEED intensities the primitive cell is normally used in the computer programs, independent of the choice of the unit cell by the experimentalist.

For primitive square lattices, a centred unit cell is not used (however, centred unit cells are used for square and rectangular superlattices, as in Ni(100)+c(2×2)–O, for example). The (100) surface of the fcc lattice has a centred lattice if the bulk lattice constants are used. The convention for the surface structures is here to use the primitive cell; the centred square cell does not offer any advantages, since the primitive cell already has rectangular coordinates.

In general, the unit cell must be adapted to the symmetry, that is, when a rotational symmetry exists the origin must be chosen in the rotation axis, while in space groups

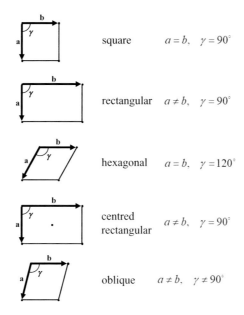

Figure 2.9 The five symmetrically different 2-D translation lattices, the so-called 2-D Bravais lattices

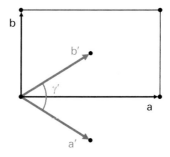

Figure 2.10 Rectangular centred cell; $\mathbf{a}'$, $\mathbf{b}'$ are the translation vectors of the primitive cell

where only a mirror plane exists the origin must lie on the mirror plane; without symmetry the choice of the origin is free.

2.1.5   Reciprocal Lattice

The crystal lattice is defined by the translation vectors $\mathbf{a}$, $\mathbf{b}$ and $\mathbf{c}$. The coordinate system to describe the atomic positions is chosen accordingly with $\mathbf{a}$, $\mathbf{b}$ and $\mathbf{c}$ as coordinate axes, and is therefore in general not rectangular. Vector calculations and symmetry operations in oblique coordinate systems are greatly simplified by using a

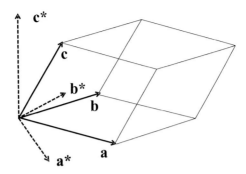

Figure 2.11 Reciprocal lattice $\mathbf{a}^*$, $\mathbf{b}^*$, $\mathbf{c}^*$ of a crystal lattice $\mathbf{a}$, $\mathbf{b}$, $\mathbf{c}$

second coordinate system: the reciprocal lattice with the basis vectors $\mathbf{a}^*$, $\mathbf{b}^*$ and $\mathbf{c}^*$, shown in Figure 2.11. Another, more practical reason for introducing the reciprocal lattice is to conveniently identify the diffracted beams and determine their propagation directions; the reciprocal lattice vector also connects directly to the linear momentum components k_i of electrons in a particular diffracted beam, as described in Section 2.2. The basis vectors of the reciprocal coordinate system fulfil the following conditions:

$$\mathbf{a}^* \cdot \mathbf{b} = \mathbf{a}^* \cdot \mathbf{c} = 0, \quad \mathbf{b}^* \cdot \mathbf{a} = \mathbf{b}^* \cdot \mathbf{c} = 0, \quad \mathbf{c}^* \cdot \mathbf{a} = \mathbf{c}^* \cdot \mathbf{b} = 0,$$
$$\mathbf{a}^* \cdot \mathbf{a} = 1, \quad \mathbf{b}^* \cdot \mathbf{b} = 1, \quad \mathbf{c}^* \cdot \mathbf{c} = 1, \tag{2.5}$$
$$\mathbf{a}^* = \frac{1}{V}(\mathbf{b} \times \mathbf{c}), \quad \mathbf{b}^* = \frac{1}{V}(\mathbf{c} \times \mathbf{a}), \quad \mathbf{c}^* = \frac{1}{V}(\mathbf{a} \times \mathbf{b}).$$

Here V is the volume of the original unit cell, with $V = \mathbf{a} \cdot (\mathbf{b} \times \mathbf{c})$, while $V^* = \frac{1}{V}$ is the volume of the unit cell in the reciprocal lattice. Thus, reciprocal vector $\mathbf{a}^*$ is perpendicular to both vectors $\mathbf{b}$ as $\mathbf{c}$ as well as to the plane spanned by vectors $\mathbf{b}$ and $\mathbf{c}$; corresponding relationships hold for $\mathbf{b}^*$ and $\mathbf{c}^*$ by simple permutation.

To differentiate the reciprocal lattice from the crystal lattice, the crystal lattice is often called the 'direct' lattice or sometimes the 'real' lattice. Furthermore, one uses the corresponding terms reciprocal space versus 'direct' space or 'real' space. As mentioned in Section 2.1.3, vectors in reciprocal space are written with parentheses, for example, (hkl), while vectors in direct space are written with square brackets, for example, [hkl].

Vector components in the reciprocal lattice have the unit of reciprocal length, for example, nm^{-1} or Å^{-1}. We note that in the physics literature the reciprocal lattice is usually multiplied by a factor 2π relative to the definition given in Eq. (2.5). In the crystallographic literature the reciprocal lattice is in general defined without the factor 2π, as in Eq. (2.6). Either way has its advantages: the calculation of inter-atomic vectors and distances in oblique lattices is easier without the factor 2π and in the physics literature the factor 2π is left out in the phase factors, thus simplifying the notation, as it is contained in the diffraction vector. We use in the following the crystallographic notation.

The basis vectors of the reciprocal coordinate system can be taken as translation vectors forming a lattice: this is the reciprocal lattice. The importance of this lattice results from the fact that each lattice point marks a diffraction condition allowing an easy calculation and geometric construction of the condition (notably direction and energy or wavelength) at which a reflection can be observed, as will be discussed in this subsection.

In two dimensions the reciprocal lattice is defined by means of a vector $\mathbf{n}$ of unit length normal to the surface. Using two lattice vectors $\mathbf{a}$ and $\mathbf{b}$ in the plane of the surface, we can then define in reciprocal space:

$$\mathbf{a}^* = \frac{\mathbf{b} \times \mathbf{n}}{V}, \quad \mathbf{b}^* = \frac{\mathbf{n} \times \mathbf{a}}{V}, \quad \gamma^* = \pi - \gamma, \quad \text{with } V = (\mathbf{a} \times \mathbf{b}) \cdot \mathbf{n}. \tag{2.6}$$

As in Eq. (2.5), in the physics literature a factor 2π would appear in Eq. (2.6) in the formulas for $\mathbf{a}^*$ and $\mathbf{b}^*$. Here γ^* is the angle between $\mathbf{a}^*$ and $\mathbf{b}^*$. Due to the lack of periodicity normal to the surface a third reciprocal vector $\mathbf{c}^*$ cannot be defined, and the reciprocal lattice consists of continuous lattice 'rods' normal to both $\mathbf{a}$ and $\mathbf{b}$ (these lattice rods can be understood as infinitely dense lines of reciprocal lattice points due to a unit cell that is infinitely long perpendicular to the surface).

The lattice points in the reciprocal lattice are usually denoted by (hkl) and are related to the Miller indices of lattice planes. The perpendicular distance between neighbouring lattice planes, often called layer spacing or d-spacing, is given by:

$$|d_{hkl}| = \frac{1}{|h\mathbf{a}^* + k\mathbf{b}^* + l\mathbf{c}^*|}. \tag{2.7}$$

The calculation of such d-spacings from the lattice vectors for the seven crystal systems is described in Appendix D. The X-ray reflections in 3-D lattices can be interpreted as reflections from lattice planes, as is well known from Bragg's equation, see Section 2.2. A lattice point of the reciprocal lattice fulfilling the diffraction conditions is therefore also known as Bragg point. The diffraction conditions for centred lattices are given in Table 2.2.

Each vector in the crystal lattice can be defined in either coordinate system; the corresponding coordinates of course will change accordingly:

$$\mathbf{r} = x\mathbf{a} + y\mathbf{b} + z\mathbf{c} = x'\mathbf{a}^* + y'\mathbf{b}^* + z'\mathbf{c}^*. \tag{2.8}$$

The scalar product of two vectors is easily calculated in an orthonormal coordinate system, but vector calculations in oblique systems must use the metric matrix as described in Appendix C. This is not necessary for a scalar product of a vector in real space with one in reciprocal space. We can define a vector $\mathbf{q}$ in the reciprocal space using the coordinates h, k, l for the vector $\mathbf{q}$:

$$\mathbf{q} = 2\pi(h\mathbf{a}^* + k\mathbf{b}^* + l\mathbf{c}^*). \tag{2.9}$$

Table 2.2 General reflection conditions for the centred 3-D Bravais lattices

Bravais lattice	General reflection conditions
Monoclinic A-centred (mA)	$k + l =$ even
Orthorhombic body-centred (oI)	$h + k + l =$ even
Orthorhombic C-centred (oC)	$h + k =$ even
Orthorhombic face-centred (oF)	$h, k, l =$ all even or all odd
Tetragonal body-centred (tI)	$h + k + l =$ even
Rhombohedral, hexagonal setting (hR)	obverse representation: $-h + k + l = 3n$, n integer reverse representation: $h - k + l = 3n$, n integer; or $h + k - l = 3n$, n integer
Cubic body-centred, bcc (cI)	$h + k + l =$ even
Cubic face-centred, fcc (cF)	$h, k, l =$ all even or all odd

See Figure 2.6 for the definition of 'reverse' and 'obverse'. The two possible settings of the hexagonal lattice hR refer to the same lattice and space group but differ in a 60° rotation. For example, the translation vectors **a** and **b** can be chosen as well as **a** + **b** and −**a**.

Then the scalar product

$$\mathbf{q} \cdot \mathbf{r} = 2\pi(hx + ky + lz) \tag{2.10}$$

becomes similar to a scalar product in a Cartesian system in which both the direct and the reciprocal lattices coincide.

The reciprocal lattice is connected to the choice of the lattice vectors in real space. If, for example in centred lattices, a larger unit cell has been chosen than necessary, the reciprocal lattice has more lattice points than necessary. The excessive points must be omitted in all cases where only the properties of the primitive lattice count, that is, for diffraction or in calculations of inter-atomic distances. The points in reciprocal lattice can be identified with possible reflections in diffraction from 3-D lattices – these are the Bragg points – but the occurrence of reflections cannot depend on the choice of the unit cell. Therefore, general reflection conditions can be formulated for centred lattices, as listed in Table 2.2. There exist additional special reflection conditions for each space group due to the existence of glide planes and screw axes, which will not be discussed here and can be found in the International Tables [2.1].

2.1.6   2-D Space Groups

When a crystal surface is generated by truncation of the 3-D crystal lattice, the only symmetry operations that remain are rotation axes normal to the surface and mirror planes having their normal in the plane of the surface. Of the glide planes, only those which have both their normal and the glide vector in the plane of the surface remain. All other symmetry operations are destroyed by the surface. The number of space groups is therefore limited to the 17 plane groups which are shown in Figure 2.12.

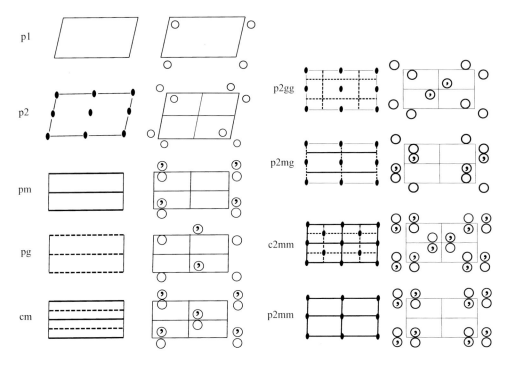

Figure 2.12 The 17 2-D space groups. The symmetry elements are shown in the left panel for each space group, while the general equivalent sites are shown on the right. The circles with a comma represent sites generated by a mirror operation. (*continued on opposite page*)

The symmetry operations occurring in 2-D lattices are: 2-, 3-, 4- and 6-fold rotational axes, as well as mirror planes and glide planes. Screw axes and inversion axes are not allowed; the inversion points correspond in 2-D space to a 2-fold axis. The graphical symbols of the symmetry operations are shown in Table 2.3.

2.1.7   Atomic Positions

The atomic positions in the unit cell are usually described in fractions of cell vectors, which means that in general an oblique coordinate system is used for the crystallographic calculations. The position of the n-th atom is then described as

$$\mathbf{r}_n = x_n\mathbf{a} + y_n\mathbf{b} + z_n\mathbf{c}, \tag{2.11}$$

with coordinates $0 \leq x_n, y_n, z_n < 1$. As no translation vector exists normal to the surface due to the missing half crystal, the vector $\mathbf{c}$ must be explicitly defined. Frequent choices are vectors between substrate layers for simple cases, the substrate lattice vector $\mathbf{c}$ or a vector of unit length (1 nm or 1 Å) normal to the surface. In the latter case the z-coordinates are given in nm or Å.

The use of fractions of translation vectors has the advantage that the coordinates are independent of the orientation of the unit cell; it also corresponds to the convention in the crystallographic databases. It is nevertheless often necessary to have the atomic

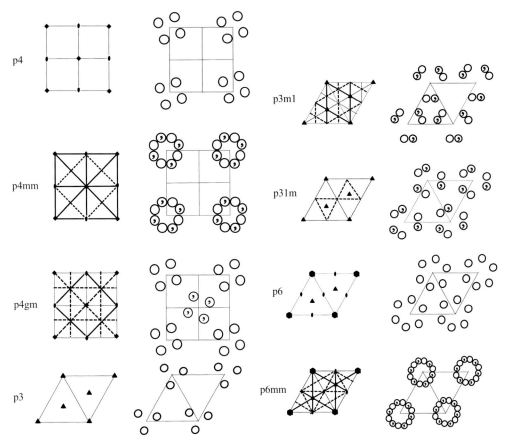

Figure 2.12 (*cont.*)

coordinates in a Cartesian coordinate system with units such as nm, Å or atomic units. The multiple scattering equations described in Chapter 5 use the atomic coordinates and the wave vectors of the incoming and reflected beams are expressed in a fixed Cartesian coordinate system, which means that the orientation of the unit cell with respect to that system is required. We give here the transformation equations from one system to the other. We assume the vector **c** to be normal to the surface. The orientation of the unit cell is given by the Cartesian coordinates of the cell vectors, that is, (a_x, a_y), (b_x, b_y) and c_z in nm or Å.

$$\mathbf{a} = (a_x, a_y, 0), \quad \mathbf{b} = (b_x, b_y, 0) \quad \text{and} \quad \mathbf{c} = (0, 0, c_z). \tag{2.12}$$

The coordinates of an atom i in fractions of translation vectors are $\mathbf{r}_i = (x_i, y_i, z_i)$ and the Cartesian coordinates of the vector $\mathbf{r}_i$ are (x_{ci}, y_{ci}, z_{ci}), as illustrated in Figure 2.13. The calculation of the coordinates follows then from:

$$\begin{pmatrix} x_{ci} \\ y_{ci} \\ z_{ci} \end{pmatrix} = \begin{pmatrix} a_x & b_x & 0 \\ a_y & b_y & 0 \\ 0 & 0 & c_z \end{pmatrix} \begin{pmatrix} x_i \\ y_i \\ z_i \end{pmatrix}, \tag{2.13}$$

Table 2.3 Symmetry symbols

Symmetry operation	Graphical symbol	Printed symbol
Identity	None	None
2-fold rotation axis	⬥	2
3-fold rotation axis	▲	3
4-fold rotation axis	◆ ■	4
6-fold rotation axis	⬡	6
Mirror plane	———	m
Glide plane	– – – – ·	g

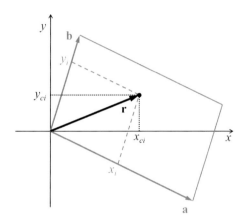

Figure 2.13 Coordinates in relative units (x_i, y_i) and absolute (Cartesian) units (x_{ci}, y_{ci})

that is,

$$\mathbf{r}_{ci} = \mathbf{M}\mathbf{r}_i \quad \text{and} \quad \mathbf{r}_i = \mathbf{M}^{-1}\mathbf{r}_{ci}. \tag{2.14}$$

The columns of the transformation matrix contain the components of the basis vectors.

The calculation of inter-atomic distances and bond angles does not require the definition of a Cartesian system and needs only lattice constants, that is, the cell vectors and the angles between them. The vector calculation in oblique coordinate systems or coordinate systems with different unit lengths is described in Appendix C.

2.1.8   Symmetry Operators in 2-D Space Groups

A detailed description of symmetry operations and lattice transformations in 3-D space is given in the International Tables [2.1]. Here we give a short overview of the symmetry operations in 2-D lattices which are a subset of the 3-D symmetry operations. A thorough description of the symmetry of lattice planes and superlattices using Cartesian coordinates has been given by K. Hermann [2.6].

A symmetry operation in direct space may consist of a point operation $\mathbf{W}$ and a translation vector $\mathbf{w}$. The translation vector occurs only for glide operations as screw axes are not allowed in 2-D space groups. The transformation of an atomic position is then written as:

$$\mathbf{r}' = (\mathbf{W}, \mathbf{w}) \cdot \mathbf{r} = \mathbf{W}\mathbf{r} + \mathbf{w}. \tag{2.15}$$

With the point operation

$$\mathbf{W} = \begin{pmatrix} w_{11} & w_{12} & w_{13} \\ w_{21} & w_{22} & w_{23} \\ w_{31} & w_{32} & w_{33} \end{pmatrix} \tag{2.16}$$

and translation vector $\mathbf{w} = \left(w_x, w_y, w_z \right)$, we obtain the new coordinates

$$\begin{aligned} x' &= w_{11}x + w_{12}y + w_{13}z + w_x \\ y' &= w_{21}x + w_{22}y + w_{23}z + w_y \\ z' &= w_{31}x + w_{32}y + w_{33}z + w_z. \end{aligned} \tag{2.17}$$

The determinant of $\mathbf{W}$ is $+1$ for rotations and -1 for mirror operations and glide operations. While this formulation is quite general, the symmetry matrices become very simple when using units of cell vectors: only the values 1, -1 and 0 occur then. In 2-D space groups, screw axes do not occur, while a shift vector only occurs for a glide operation, in which case the shift vector is one half of the translation vector. The possible symmetry matrices and the shift vector for glide planes are listed in Table 2.4.

If the rotation point is not at the origin or the mirror plane does not go through the origin, the symmetrically equivalent points must be calculated by first shifting the rotation point or the mirror plane, then applying the symmetry operation and finally shifting the point backwards with the transformed shift vector.

If the atomic positions are given in Cartesian coordinates, the symmetrically equivalent points are most conveniently calculated by first transforming the initial point to relative units by Eq. (2.14), then applying the symmetry operation and in a final step transforming back using the inverse of Eq. (2.14), instead of defining the symmetry operations in Cartesian coordinates.

2.1.9 Lattice Transformations

A frequently occurring transformation is the choice of a new unit cell while the structure remains unchanged, for example, transforming atomic coordinates related to a centred cell to coordinates related to a primitive unit cell (this may occur when coordinates come from a database). Another example is the transformation of the coordinates of the bulk structure to the lattice vectors of the surface cell of a plane

Table 2.4 Selected symmetry operations occurring in plane groups, with symmetrically equivalent positions and transformation matrices for coordinates

Symmetry operation	Mirror/glide plane orientation	Symmetrically equivalent positions	Plane groups	Transformation matrix for point operation W	Translation vector for glide operation w
m	[100] $\perp$ **a**	(x, y, z) $(-x, y, z)$	pm p2mm p4mm	$\begin{pmatrix} -1 & 0 & 0 \\ 0 & 1 & 0 \\ 0 & 0 & 1 \end{pmatrix}$	–
m	[010] $\perp$ **b**	(x, y, z) $(x, -y, z)$	pm p2mm p4mm	$\begin{pmatrix} 1 & 0 & 0 \\ 0 & -1 & 0 \\ 0 & 0 & 1 \end{pmatrix}$	–
m	[110] $\perp$ (**a** + **b**)	(x, y, z) (y, x, z)	p4mm p3m1 p6mm	$\begin{pmatrix} 0 & 1 & 0 \\ 1 & 0 & 0 \\ 0 & 0 & 1 \end{pmatrix}$	–
m	[1–10] $\perp$ (**a** − **b**)	(x, y, z) $(-y, -x, z)$	p4mm p6mm p31m	$\begin{pmatrix} 0 & -1 & 0 \\ -1 & 0 & 0 \\ 0 & 0 & 1 \end{pmatrix}$	–
m	[100] $\perp$ **a**	(x, y, z) $(-x + y, y, z)$	p3m1 p6mm	$\begin{pmatrix} -1 & 1 & 0 \\ 0 & 1 & 0 \\ 0 & 0 & 1 \end{pmatrix}$	–
m	[010] $\perp$ **b**	(x, y, z) $(x, x - y, z)$	p3m1 p6mm	$\begin{pmatrix} 1 & 0 & 0 \\ 1 & -1 & 0 \\ 0 & 0 & 1 \end{pmatrix}$	–
m	[120] $\perp$ (**a** + 2**b**)	(x, y, z) $(-x + y, y, z)$	p6mm p31m	$\begin{pmatrix} -1 & 1 & 0 \\ 0 & 1 & 0 \\ 0 & 0 & 1 \end{pmatrix}$	–
m	[210] $\perp$ (2**a** + **b**)	(x, y, z) $(x, x - y, z)$	p6mm p31m	$\begin{pmatrix} 1 & 0 & 0 \\ 1 & -1 & 0 \\ 0 & 0 & 1 \end{pmatrix}$	–
2_z	[001]	(x, y, z) $(-x, -y, z)$	p2 p4 p6	$\begin{pmatrix} -1 & 0 & 0 \\ 0 & -1 & 0 \\ 0 & 0 & 1 \end{pmatrix}$	–
3_z	[001]	(x, y, z) $(-y, x - y, z)$ $(-x + y, -x, z)$	p3 p6 p6mm p3m1 p31m	$3_z^{+}: \begin{pmatrix} 0 & -1 & 0 \\ 1 & -1 & 0 \\ 0 & 0 & 1 \end{pmatrix}$ $3_z^{-}: \begin{pmatrix} -1 & 1 & 0 \\ -1 & 0 & 0 \\ 0 & 0 & 1 \end{pmatrix}$	–

Table 2.4 (*cont.*)

Symmetry operation	Mirror/glide plane orientation	Symmetrically equivalent positions	Plane groups	Transformation matrix for point operation W	Translation vector for glide operation w
4_z	[001]	(x, y, z) $(-y, x, z)$ $(-x, -y, z)$ $(y, -x, z)$	p4 p4mm	$4_z^+: \begin{pmatrix} 0 & -1 & 0 \\ 1 & 0 & 0 \\ 0 & 0 & 1 \end{pmatrix}$ $4_z^-: \begin{pmatrix} 0 & 1 & 0 \\ -1 & 0 & 0 \\ 0 & 0 & 1 \end{pmatrix}$	–
6_z	[001]	(x, y, z) $(x - y, x, z)$ $(-y, x - y, z)$ $(-x, -y, z)$ $(y - x, -x, z)$ $(y, y - x, z)$	p6 p6mm	$6_z^+: \begin{pmatrix} 1 & -1 & 0 \\ 1 & 0 & 0 \\ 0 & 0 & 1 \end{pmatrix}$ $6_z^-: \begin{pmatrix} 0 & 1 & 0 \\ -1 & 1 & 0 \\ 0 & 0 & 1 \end{pmatrix}$	–
g	[100] $\perp \mathbf{a}$	(x, y, z) $(-x, y + 1/2, z)$	pg p2gg cm c2mm	$\begin{pmatrix} -1 & 0 & 0 \\ 0 & 1 & 0 \\ 0 & 0 & 1 \end{pmatrix}$	$(0, 1/2, 0)$
g	[010] $\perp \mathbf{b}$	(x, y, z) $(x + 1/2, -y, z)$	pg p2gg cm c2mm	$\begin{pmatrix} 1 & 0 & 0 \\ 0 & -1 & 0 \\ 0 & 0 & 1 \end{pmatrix}$	$(1/2, 0, 0)$
g	[110] $\perp (\mathbf{a} + \mathbf{b})$	(x, y, z) $(-x + 1/2, -y + 1/2, z)$	p4gg p4gm p4mm	$\begin{pmatrix} 0 & -1 & 0 \\ -1 & 0 & 0 \\ 0 & 0 & 1 \end{pmatrix}$	$(1/2, -1/2, 0)$
g	[1–10] $\perp (\mathbf{a} - \mathbf{b})$	(x, y, z) $(y + 1/2, x + 1/2, z)$	p4gg p4gm p4mm	$\begin{pmatrix} 0 & 1 & 0 \\ 1 & 0 & 0 \\ 0 & 0 & 1 \end{pmatrix}$	$(1/2, 1/2, 0)$

W is the transformation matrix for the point operation; **w** is the translation vector for the glide operation. For the mirror planes and glide planes, the orientations normal to $\mathbf{a}$, $\mathbf{b}$, $\mathbf{a} + \mathbf{b}$ or $\mathbf{a} - \mathbf{b}$ are chosen. It is assumed that the origin lies on the mirror or glide planes and on the highest-order rotation axis if there is one.

(*hkl*). Two possible vectors $\mathbf{a}'$ and $\mathbf{b}'$ of the plane can be chosen from the definition of the Miller indices (Eq. 2.4), by the difference vectors between the intersection points on the coordinate axes. This choice is not unique and usually leads to an inappropriate combination of the smallest translation vectors. These are found by a combination $m\mathbf{a}' + n\mathbf{b}'$ and elimination of common factors in the components, using, for example, the Minkowski reduction [2.6] which is a 2-D version of the

Buerger/Niggli reduction [2.7]. The 2-D translation vectors should be symmetry adapted if there is any symmetry. The vectors determined in that way refer to the 3-D lattice; in the case of centred lattices a further reduction is possible. A formal solution for the determination of the 2-D lattice vectors of any plane (hkl) is discussed in detail in Section 2.1.12.

The need for a lattice transformation frequently occurs when LEED I(V) curves should be calculated for structure models obtained from, for example, a DFT calculation in which another setting of the unit cell or slightly different cell vectors have been used. In such cases it is convenient to first calculate the matrix describing the lattice tranformation and then calculate the new coordinates. New lattice vectors are obtained from:

$$\mathbf{a}' = m_{11}\mathbf{a} + m_{12}\mathbf{b} + m_{13}\mathbf{c}$$
$$\mathbf{b}' = m_{21}\mathbf{a} + m_{22}\mathbf{b} + m_{23}\mathbf{c} \tag{2.18}$$
$$\mathbf{c}' = m_{31}\mathbf{a} + m_{32}\mathbf{b} + m_{33}\mathbf{c}.$$

A simple example is the transformation of a primitive cell to a centred cell while the c-coordinate remains unchanged. $\mathbf{M}$ is not a unitary matrix: its determinant $|\mathbf{M}|$ indicates the size of the superlattice unit cell.

$$\begin{aligned}\mathbf{a}' &= \mathbf{a} + \mathbf{b} \\ \mathbf{b}' &= -\mathbf{a} + \mathbf{b} \qquad \mathbf{M} = \begin{pmatrix} 1 & 1 & 0 \\ -1 & 1 & 0 \\ 0 & 0 & 1 \end{pmatrix}. \\ \mathbf{c}' &= \mathbf{c}, \end{aligned} \tag{2.19}$$

The atomic coordinates related to the new unit cell $\mathbf{a}'$, $\mathbf{b}'$ are given by:

$$\mathbf{r}' = \left(\mathbf{M}^{\mathrm{T}}\right)^{-1}\mathbf{r} \tag{2.20}$$

(the superscript T denotes the transpose of the matrix). If a shift $\mathbf{s}$ of the origin is required, the shift must be applied first to $\mathbf{r}$ and the transformation must be applied to the shifted $\mathbf{r}$:

$$\mathbf{r}' = \left(\mathbf{M}^{\mathrm{T}}\right)^{-1}(\mathbf{r} - \mathbf{s}). \tag{2.21}$$

A remark may be added here regarding the notation used in the International Tables [2.1] and in some textbooks [2.2]. There, for lattice transformations the matrix $\mathbf{P} = \mathbf{M}^{\mathrm{T}}$ is defined which must be applied to a (1×3) row matrix of the basis vectors, and the matrix $\mathbf{Q} = \mathbf{P}^{-1}$ transforms the coordinates. This usage has the advantage of having complete reciprocal symmetry between lattice transformations and coordinate transformations and a simplified description of general transformations including linear shifts of the origin. To keep the more familiar form of the matrix calculation, we have used Eq. (2.18) here to define the matrix $\mathbf{M}$.

While the transformation of coordinates is in many cases simple, it becomes tedious in structures with many atoms and requires a computer program. An example is given in Figure 2.14 showing the Fe_3O_4(100) surface with the truncated bulk structure. There are in total 52 atoms in the unit cell of the 3-D structure, of which only three are

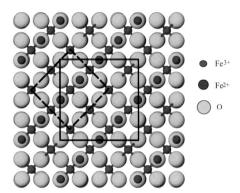

Figure 2.14 Top-down view of the (100) surface of the truncated bulk structure of magnetite, Fe_3O_4, showing the projection of one half of the bulk unit cell, with origin at the centre. The centred bulk unit cell is marked by the solid line. The dashed line marks the primitive surface unit cell required for LEED calculations, with the mirror lines along the translation vectors and the origin in the 2-fold axis. The shift of the origin between the two unit cells is (1/8, 1/8, 0) and 45° rotation.

symmetrically independent; the symmetry group is $Fd\bar{3}m$. The calculation of all coordinates in a surface unit cell from the coordinates of the three atoms listed in the database first requires the application of all symmetry operations in the cubic unit cell and then the reduction to a primitive setting with a tetragonal unit cell. This is illustrated in Figure 2.14.

The solid black line in Figure 2.14 shows the centred unit cell of the fcc lattice; the origin is chosen at the inversion point of this structure. This choice of the origin is commonly used in crystal structure databases but is not well suited for LEED calculations which require a primitive cell, the mirror planes parallel to the **a**- and **b**-axes and the origin in the 2-fold axis at (1/8, 5/8, 0) of the 3-D unit cell in the plane of the Fe^{2+} atoms. Programs are available to calculate the coordinate transformations. References and examples for lattice transformations are given in Appendix E.

2.1.10   Transformation of Reflection Indices

The matrix **M** of Eqs. (2.18) and (2.19) can also be used to transform the reflection indices. In LEED structure analysis of structures with superlattices, reflection indices related to the substrate lattice are frequently used. This leads to fractional indices for superlattice reflections. For small unit cells this is acceptable, but for large unit cells it becomes very impractical: the indices are then more easily located and identified relative to the superlattice cell. The indices are then integer. For X-ray diffraction (XRD) this is the common usage. When different superlattices occur at the same substrate surface, it may be useful to compare them to the substrate reflections and to relate the indices to the substrate lattice. For X-ray diffraction it is also often necessary to transform the indices from the surface setting to the substrate structure in order to identify the Bragg reflections. The transformation matrix for indices can be shown to be the same as that for lattice vectors:

$$\begin{pmatrix} h' \\ k' \\ l' \end{pmatrix} = \mathbf{M} \begin{pmatrix} h \\ k \\ l \end{pmatrix}. \tag{2.22}$$

When the vector $\mathbf{c}$ is normal to the surface, the matrix $\mathbf{M}$ is then the (2×2) matrix used to describe the superlattice in matrix notation, extended for (001) in the third column and row. An example is the transformation from a primitive cell to a centred cell. The primitive cell may be the substrate cell while the superlattice has the centred cell (see Eq. (2.19)). Then, the reflection $(1/2, 1/2, l)$ becomes $(1, 0, l)$, for example.

2.1.11   Description of Superlattices

In studies of clean, reconstructed or adsorbate covered surfaces, 2-D unit cells larger than that of the truncated bulk structure are frequently observed. To describe such superlattices, it is common to refer the observed unit cell to the unit cell of the ideally truncated bulk structure. Conventionally, the so-called Wood notation [2.8] is used, which takes the form $(n \times m)$, or $(n \times m)R\alpha°$, whenever the observed unit cell allows this description with integers n, m or square roots of integers; here, $R\alpha°$ denotes a rotation of the superlattice unit cell by angle $\alpha°$ with respect to the substrate unit cell [2.8]; $R\alpha°$ is omitted when $\alpha = 0$. This notation also allows the forms $p(n \times m)$ and $c(n \times m)$, where p and c stand for primitive and centred, respectively: p is usually omitted, being redundant, while c can be very useful to visualise an oblique unit cell as a centred rectangular unit cell. Some examples are given in Figure 2.15.

In more general cases a matrix notation is used. The matrix $\mathbf{M}$ describes the relation between the lattice vectors of the superlattice cell $\mathbf{a'}$, $\mathbf{b'}$ and the lattice vectors of the substrate cell $\mathbf{a}$, $\mathbf{b}$.

$$\begin{pmatrix} \mathbf{a'} \\ \mathbf{b'} \end{pmatrix} = \mathbf{M} \begin{pmatrix} \mathbf{a} \\ \mathbf{b} \end{pmatrix}, \text{ with } \mathbf{M} = \begin{pmatrix} m_{11} & m_{21} \\ m_{12} & m_{22} \end{pmatrix}. \tag{2.23}$$

With integer numbers m_{ij} the superlattice is commensurate with the substrate, meaning that the superlattice and the substrate lattice coincide periodically on a common coincidence lattice with finite lattice constants, as in Figure 2.15. The area of the superlattice cell is then an integer multiple of the substrate cell:

$$A = m_{11}m_{22} - m_{21}m_{12}. \tag{2.24}$$

As an example, the structure of thiouracil on Ag(111) is shown in Figure 2.16 [2.9]. The superlattice cell is here $\begin{pmatrix} 8 & 6 \\ -1 & 2 \end{pmatrix}$. This structure has symmetry p1 and the origin of the unit cell can be chosen arbitrarily: it is chosen here at the bridge site of the fcc (111) surface. It should be noted that this description is not unique and depends on the choice of the substrate cell vectors. The choice of the unit cell with the solid lines in

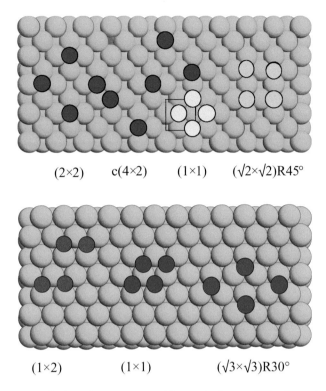

| (2×2) | c(4×2) | (1×1) | (√2×√2)R45° |

| (1×2) | (1×1) | (√3×√3)R30° |

Figure 2.15 Examples of adsorption structures with superlattices and Wood notation on a square substrate lattice (upper panel) and on a triangular substrate lattice (lower panel)

Figure 2.16 is such that the angle γ between $\mathbf{a}'$ and $\mathbf{b}'$ is close to 90°. The same structure can be described as $\begin{pmatrix} 4 & -5 \\ 2 & 3 \end{pmatrix}$ with an oblique angle γ close to 120° and referring to a substrate unit cell rotated by 60° compared to the first notation [2.10], as drawn with dashed lines in Figure 2.16.

Note that the Wood notation can always be replaced by the more general matrix notation, as exemplified in Table 2.5, but the reverse is not true: not all superlattices can be represented by the Wood notation.

Common superlattices on low-index planes of cubic crystals are listed in Table 2.5. The maximum symmetry of a single domain is given in the table; the actual symmetry of a specific adsorbate structure may be lower due to the atomic arrangement within the superlattice cell.

The symmetry group of the superlattice is a subgroup of the symmetry group of the substrate. Consequently, there occur symmetrically equivalent domains, which may be called twin domains if the point group of the superlattice has lower symmetry than the point group of the substrate surface. Due to the subset of translation vectors there necessarily exist antiphase domains in addition. Twin domains occur always unless by specific experimental conditions a single domain is generated.

Incommensurate superlattices are also possible. For these, at least one number m_{ij} in the matrix $\mathbf{M}$ of Eq. (2.23) is irrational, that is, it is neither an integer nor a ratio of

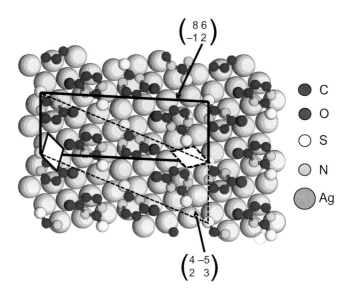

Figure 2.16 Example of a superlattice described by matrix notation, for thiouracil on Ag(111), illustrating that the notation is not unique. Two possible settings of the superlattice cell are marked by solid and dashed lines. The corresponding matrices are given and refer to two different settings of the substrate unit cell marked white. The two substrate unit cells are rotated by 60°. In the substrate unit cells conventionally a 120° angle between the substrate vectors **a** and **b** is used. (Color code: Ag light grey, C dark blue, N light blue, O red, S yellow, H is not shown.)

two integers. Then the superlattice and substrate lattice do not coincide periodically in one or more surface directions. An example is a graphene overlayer on a metal substrate surface: the two materials have unrelated lattice constants for any relative orientation, so they cannot coincide periodically. Strictly speaking, any weak lateral interaction between the overlayer and the substrate can slightly change their lattice constants to make them commensurate and lower their total energy; furthermore, no actual surface extends to infinity in two dimensions, so that the concept of strict incommensurability cannot be realised. No Wood notation is available for incommensurate superlattices.

2.1.12   Stepped Surfaces

High-Miller-index surfaces are often stepped surfaces with terraces that consist of relatively 'flat' low-Miller-index planes, separated by atomic height steps; the steps themselves can be 'kinked', that is, can exhibit a regular zigzag structure. When the high-index surface is close in orientation to a low-index surface, one speaks of a 'vicinal' surface, in which relatively large terraces exist with low-index orientations, separated by steps with or without kinks.

The Miller indices (hkl) of a crystal plane usually refer to the conventional 3-D lattice. We use this convention here, not the primitive setting of the unit cell used in surface structure analyses. The indices (hkl) define the orientation of the surface on a macroscopic scale. The morphology on the atomic scale is not uniquely defined by the

Table 2.5 Some common superlattices, symmetries and numbers of symmetrically equivalent domains for the (100) and (111) surfaces of the fcc and bcc lattices and the (0001) surface of the hcp lattice. The number of possible antiphase (i.e., laterally shifted) domains is not listed. Other superlattice notations are possible, depending on the choice of cell vectors.

Substrate	Substrate symmetry	Superlattice	Superlattice symmetry, number of domains	Unit cells of some different domains, primitive setting of centred cells
fcc (100), bcc (100)	p4mm	$(\sqrt{2} \times \sqrt{2})R45°$ or c(2 × 2) or $\begin{pmatrix} 1 & -1 \\ 1 & 1 \end{pmatrix}$	p4mm 1 domain	
		(2 × 1) or (1 × 2), $\begin{pmatrix} 2 & 0 \\ 0 & 1 \end{pmatrix}, \begin{pmatrix} 1 & 0 \\ 0 & 2 \end{pmatrix}$	p2mm 2 domains	
		(2 × 2) or $\begin{pmatrix} 2 & 0 \\ 0 & 2 \end{pmatrix}$	p4mm 1 domain	
		$(\sqrt{5} \times \sqrt{5})R26.7°$ or $\begin{pmatrix} 2 & 1 \\ -1 & 2 \end{pmatrix}$	p4 2 domains	
		c(2 × 8) or $\begin{pmatrix} 4 & 1 \\ 0 & 2 \end{pmatrix}$ or $\begin{pmatrix} 4 & -1 \\ 4 & 1 \end{pmatrix}$	c2mm 2 domains	
fcc (111), bcc (111), hcp(0001)	p3m1	(2 × 2) or $\begin{pmatrix} 2 & 0 \\ 0 & 2 \end{pmatrix}$	p3m1 1 domain	
		(2 × 1) or $\begin{pmatrix} 2 & 0 \\ 0 & 1 \end{pmatrix}$	p1 6 domains	
		$(\sqrt{3} \times \sqrt{3})R30°$ or $\begin{pmatrix} 2 & 1 \\ -1 & 1 \end{pmatrix}$	p31m 1 domain	
		$(\sqrt{7} \times \sqrt{7})R19.1°$ or $\begin{pmatrix} 1 & 2 \\ -3 & 1 \end{pmatrix}$	p3 2 domains	

Miller indices: it also depends on the material, temperature or further conditions, due to different truncations for structures with multiple atoms in the 3-D unit cell, reconstructions, atomic position relaxations, adsorbates, etc.

The morphology of the surface can be described in a more intuitive manner than with high-index Miller indices, namely in terms of easily visualised terraces and steps; this can be done quantitatively, so as to gauge the terrace width and step density, as well as the step structure (e.g., kinked or not). The general case, for an arbitrary underlying 3-D crystal structure, is analysed thoroughly in the book by K. Hermann [2.6]. We consider here some examples of stepped surfaces of mono-atomic lattices, since steps have been investigated mostly on metal surfaces, while only relatively fewer studies exist on stepped surfaces of semiconductors or compounds.

For steps on the simpler mono-atomic lattices, common for metal surfaces, two different notations have been proposed. The first is the 'step notation', which counts the atomic rows on a terrace between two steps [2.11], described by:

$$(hkl) = n_t(h_t, k_t, l_t) \times (h_s, k_s, l_s). \tag{2.25}$$

Here, h_t, k_t, l_t are the Miller indices of the low-index terrace plane, while h_s, k_s, l_s give the direction of the low-index plane formed by the atoms in the step edge which may be seen as an inclined microfacet, and n_t is the number of atomic rows in the terrace. This notation is applicable to all lattices; it is, however, not very practical for surfaces with kinked steps.

In order to define the morphology more precisely, a second notation has been proposed: the 'microfacet notation' [2.12]. This is based on the fact that Miller indices are vectors in reciprocal space and can be decomposed into two or three vectors of low-index planes. This decomposition is not unique, and the specific choice characterises the morphology. The surface notation reads then, again in the case of two low-index planes:

$$(hkl) = n_t(h_t, k_t, l_t) + n_s(h_s, k_s, l_s), \tag{2.26}$$

where

$$h = n_t h_t + n_s h_s, \qquad k = n_t k_t + n_s k_s, \qquad l = n_t l_t + n_s l_s. \tag{2.27}$$

This notation can be directly generalised from two to three microfacet orientations in order to describe so-called kinked steps, as illustrated later in this section (see Figure 2.20). It has been discussed in detail in [2.6] and [2.12] for all cubic lattices: fcc, bcc, simple cubic and compound structures like zincblende, NaCl, CsCl, as well as other lattices. The factors n_t and n_s suggest the relative size of terrace and step face: typically, n_t will be large, indicating a wide terrace, while n_s will be 1, indicating a step of mono-atomic height.

However, to obtain a precise size comparison of terrace versus step height requires a proper indexing of the microfacets (h_t, k_t, l_t) and (h_s, k_s, l_s), to be described in the next paragraph. This indexing depends on the lattice types, differing between simple cubic (sc), face-centred cubic (fcc) and body-centred cubic (bcc). Figure 2.17(a) illustrates this necessity with the example fcc(311): using the usual simple cubic notation for fcc lattices, we can directly decompose $(311) = 2(100) + 1(111)$, correctly suggesting a (100) terrace and a (111) step face. However, when we examine the atomic structure of fcc(311), we find that the (100) 'terrace' has a width of a *single* atomic row, while the 'step face' has a height of a *single* atomic row as well: thus both terrace and step are one atom wide, as opposed to two versus one as suggested by the decomposition $(311) = 2(100) + 1(111)$. The cause of this discrepancy is the presence of multiple (four) atoms in the cubic cell of the fcc lattice: the additional atoms change the terrace/step structure in ways that depend on the surface orientation. The same issue arises for stepped surfaces of the bcc lattice, which has two atoms in the cubic cell. Another issue exists for non-mono-atomic bulk structures, like hcp, graphite, diamond, or compounds like zincblende, wurtzite, etc.: these cases are further complicated by the existence of different surface terminations due to cutting the bulk at different depths (cf. Figure 2.7); we will not discuss these further, also because little work has been performed on stepped surfaces of these lattices.

The decompositions into terraces and steps given and illustrated in the following paragraphs are for an fcc lattice. To correctly quantify the terrace versus step sizes, we apply the usual simple cubic notation for the Miller indices [2.6]: this means in particular that, for a correct decomposition, h, k and l must be all even or all odd; if h, k and l are not all even or all odd, they must be multiplied by 2 to yield all even or all odd indices. The corresponding rule for the decomposition into terraces and steps for bcc lattices is that $h + k + l$ must be even; otherwise they should be multiplied by 2. (For comparison, the sc lattice has no such constraints, but only the element polonium has the sc bulk structure.)

Simple examples of the microfacet notation are shown in Figures 2.17–2.19 for the fcc lattice, where the choice of (100), (110) and (111) terraces and step facets is obvious. In the examples for the (701), (771) and (779) surfaces the decompositions are:

(701) oriented surface : $(701) \times 2 = (14, 0, 2) = 6(200) + (202) = 7(200) + (002);$
(771) oriented surface : $(771) = 3(220) + (111);$
(779) oriented surface : $(779) = 7(111) + (002).$

The two decompositions given for the (701) surface reflect an arbitrariness in defining the step orientation: either as a (101) facet or as a (001) facet; the (101) step facet is inclined (by 45°) relative to the terrace, while the (001) step facet stands perpendicular to the terrace, which is mathematically equally valid. This decomposition implies, for the (701) surface, six (100) unit cells in the (100) terraces and one (101) unit cell for the step or, alternatively and equivalently, seven (100) unit cells in the terraces and one (001) unit cell for the step. For the (771) surface, we have three unit cells in the (110) terraces and one (111) unit cell for the step, while for (779) there are seven unit cells in the (111) terraces and one (001) unit cell for the step (all these steps thus have a height of one atomic layer): the unit cells can be counted in Figures 2.17–2.19. In the final step we deduce the surface translation vector with a simple rule described further in Sections 2.1.12.1–2.1.12.3.

The step edge direction in the (701) surface is along [010] with next nearest neighbour distances in the edge direction. Depending on the material, the step edge might become rough and form zigzag lines along close packed rows of atoms in the [110] direction. Straight step edges might occur on surfaces with (110) or (111) terraces: their indices are then of the form (hhl) and ($h0l$).

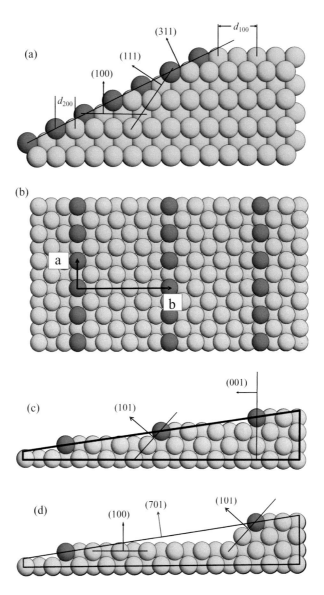

Figure 2.17 (a) Side view of the (311) surface of the fcc lattice illustrating the decomposition of the Miller indices for correctly counting the number of atomic rows in the terraces and the step in the microfacet notation, see text for explanation. (b) and (c) Top view and side view of the (701) surface of the fcc lattice with single height steps. The decomposition 7(200) + (002) extends the terrace, by one unit cell, by using a step of orientation (001) that is perpendicular to the terrace versus the more intuitive inclined (101) step that extends to double height in (d). (d) (701) surface with double height steps, based on the decomposition (14, 0, 2) = 6(200) + (202), using terrace orientation (100) and step orientation (101). The surface has a primitive 2-D unit cell in the (701) plane. The translation vectors for the surface with single steps are shown in (b).

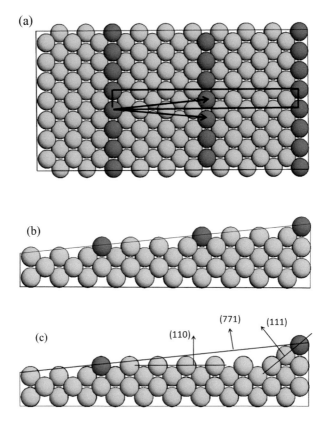

Figure 2.18 (a) Top view of the fcc (771) surface with (110) terraces and (111)-oriented steps, based on the decomposition (771) = 3(220) + (111). The factor 3 counts the number of long rectangular (110) unit cells between steps. (b) and (d) Side views of the fcc (771) surface for single- and double-height steps, respectively. The terrace plane and the step plane are indicated in (c). The 2-D unit cell of the (771) surface with single steps is centred as shown in (a).

The surfaces (hhl) exhibit a centred cell if all indices are odd, cf. Figure 2.18; otherwise, with mixed even and odd indices one obtains a primitive cell. With l close to h, the terrace plane is (111); an example is shown in Figure 2.19 for the (779) surface.

For general vicinal surfaces (hkl), the steps have kinks. An example for the fcc (11, 9, 8) plane with a decomposition (11, 9, 8) × 2 = (22, 18, 16) = 17(111) + 2(200) + (1, 1, −1) is shown in Figure 2.20: note the generalisation of the microfacet notation to three microfacets, namely one terrace, a dominant step face and a smaller kink face. The resulting large unit cell contains 17 (111) unit cells in the terrace, while the kinked steps have, respectively, two facet unit cells of orientation (100) and one facet unit cell of orientation (1, 1, −1), resulting in straight steps two atoms in length alternating with kinks of one atom in length.

(a)

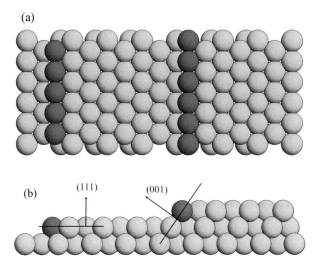

(b)

Figure 2.19 Top view (a) and side view (b) of the (779) surface of the fcc lattice with (111) terraces and (001) step faces, based on the decomposition (779) = 7(111) + (002). The (779) surface unit cell is centred, as in the case of Figure 2.18.

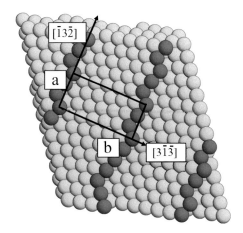

Figure 2.20 Top view of the (11, 9, 8) kinked surface of the fcc lattice with (111) terraces: dark blue atoms form the kinked step edges. The general direction of the steps is [−1, 3, −2] (see text); however, the kink sites are not necessarily regularly spaced on an actual imperfect surface, as illustrated in the bottom left. The directions between nearest neighbours in the steps are [−1, 1, 0] and [0, 1, −1] with step faces (100) and (1, 1, −1). The 2-D unit cell and the direction of the translation vectors are shown. The vector **a** is actually $a/2[1, -3, 2]$. A right-handed coordinate system is chosen with the z-component pointing outward.

2.1.12.1 Direction of Step Edges

There remains the question how to decompose the vector (*hkl*) in reciprocal space to determine the terrace/step/kink structure and how to calculate the direction of the step edges in real space. There is an easy way to calculate the direction of the step edges and the edge microfacets. The calculation outlined in the following holds for all lattice planes (*hkl*) and all lattices, including oblique lattices. The example shown here refers to the fcc lattice; the calculation for an oblique lattice is shown in Appendix F [2.13].

At first the terrace plane must be defined. This is the plane with the dominant factor if the plane (*hkl*) is decomposed into a low-index plane and one other plane with the difference indices (which may be a simple straight step or a kinked step). We assume that low-index planes are the energetically more stable ones; this is not necessarily the case for compounds, but for the metals with close packed structures the surface energy is generally lower for the low-index planes. For example, the (11, 9, 8) plane can be decomposed into 9(111) + (2, 0, −1). The terrace plane is therefore the (111) plane since it has the dominant factor. The Miller indices (*hkl*) of the surface plane and (h_t, k_t, l_t) of the terrace plane are vectors in reciprocal space; the vector product of two vectors in reciprocal space yields a vector in real space parallel to the line common to both planes, which has the step edge direction, cf. Appendix F. The direction in direct space is written with square brackets to distinguish it from indices or lattice planes which are vectors in reciprocal space, see Section 2.1.3:

$$[s_1, s_2, s_3] = (h_t, k_t, l_t) \times (h, k, l). \tag{2.28}$$

Here $[s_1, s_2, s_3]$ is the direction of the step edge. For the (11, 9, 8) surface with (111) terraces the direction of the step edge is thus:

$$(1, 1, 1) \times (11, 9, 8) = [-1, 3, -2].$$

If a common integer factor exists for the three indices, we take the smallest index triple as the direction of the step vector. The direction of the step edge can be decomposed into a sum of vectors between nearest neighbours in the (111) plane. In the fcc lattice these are ±[0, 1/2, −1/2], ±[1/2, 0, −1/2] and ±[1/2, −1/2, 0]. The vector [−1, 3, −2] can therefore be written as the sum of two vectors in the direction of nearest neighbours:

$$[-1, 3, -2] = [-1, 1, 0] + 2[0, 1, -1] = s_1 + 2s_2.$$

The ideal step edges [obtained by a mathematical cut along the macroscopic (11, 9, 8) surface plane] may be realised by an edge with high kink density, with one nearest neighbour bond in the [−1, 1, 0] direction and with two such bonds in the [0, 1, −1] direction, as shown in Figure 2.20. The actual edge line may also form larger zigzag sections, analogous to multiple-atom-height steps. It is in any case useful to visualise the surface geometry by a structure plot program, for example Balsac [2.14] or

CRYSCON [2.15]. The programs require the data of the 3-D unit cell and for Balsac the Miller indices of the surface, for CRYSCON the transformation matrix from the 3-D unit cell to the surface unit cell (see also Appendix E).

2.1.12.2 Determination of Step Microfacets

So far, the indices of the terrace and the step direction have been determined. In the case of kinked steps, we wish to also determine the two different planes of the two step microfacets. For this purpose, two vectors in real space are needed. The vector product gives the Miller indices of the step plane:

$$(h_{s1}, k_{s1}, l_{s1}) = [t_1, t_2, t_3] \times [d_1, d_2, d_3]. \tag{2.29}$$

The vector **t** points to the nearest neighbour along the edge, while **d** denotes the vector to the nearest neighbour in the layer below.

The Miller indices of the step facets can be formally calculated using the vectors displayed in Figure 2.21(a) and (b). For simplicity we take the points in the (111) plane at $\mathbf{r_1} = (1/2, 1/2, 0)$, $\mathbf{r_2} = (1/2, 0, 1/2)$ and $\mathbf{r_3} = (0, 1/2, 1/2)$ as three points in a (111) terrace. The atom at the origin (000) is in the plane of the (111) terrace below. The two vectors defining the facet of the (100)-like step can be easily identified from Figure 2.21(a): $\mathbf{t_1} = \mathbf{r_1} - \mathbf{r_2}$ along the terrace edge and $\mathbf{d_1} = \mathbf{r_3}$ up the step to the next

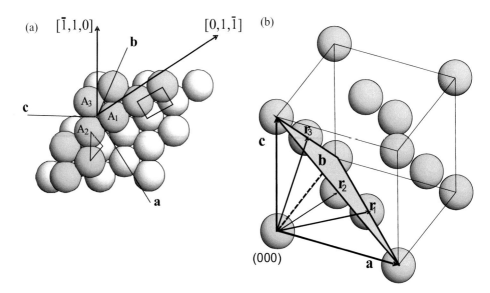

Figure 2.21 (a) Top view of fcc (111) terraces separated by a kinked step: the blue atoms are in the top terrace and the grey atoms in the terrace below it, extending toward the bottom right. Two different inclined step facets are shown between them: one step facet is (111)-like (with a triangular arrangement of atoms) and the other is (100)-like (with a square arrangement of atoms). The lattice directions are shown in the figure and refer to the calculation described in the text as well as to the translation vectors shown in (b). (b) Unit cell of the fcc lattice illustrating the three vectors $\mathbf{r_1}$, $\mathbf{r_2}$ and $\mathbf{r_3}$ which point at the atoms A_1, A_2 and A_3 in the (111) plane in the upper terrace; these three atoms are labelled in (a).

terrace. The facet normal is given by the vector product $\mathbf{t}_1 \times \mathbf{d}_1 = (1/2, 0, 0)$, implying that the Miller indices of the step facet are (100). Similarly, the two vectors defining the (111)-like step facet are: $\mathbf{t}_2 = \mathbf{r}_2 - \mathbf{r}_3$ and $\mathbf{d}_2 = \mathbf{r}_3$. The vector product is $\mathbf{t}_2 \times \mathbf{d}_2 = (-1/4, 1/4, 1/4)$ and the Miller indices of the step facet are therefore $(-1, 1, 1)$. The Miller indices (hkl) and $(-h, -k, -l)$ define the same plane and the appropriate signs should be chosen by inspection of the figure to take care of the direction of the vectors and the right-hand rule of the vector product. In this case the (111)-like step facet is properly $(1, 1, -1)$.

2.1.12.3   Determination of Step-to-Step Translation Vectors

The 2-D lattice constants of the plane (hkl) can be calculated in different ways. A general analytic solution has been described by K. Hermann [2.6], showing that the possible translation vectors can be calculated from the transformation matrix which relates the 3-D translation vectors to the 2-D translation vectors of the net plane (hkl). This transformation matrix must have integer components, leading to three Diophantine equations with an infinite set of solutions. The symmetry-adapted shortest translation vectors must then be determined by a Minkowski reduction, which leads to a unique solution.

We describe here a simpler approach, using the shortest vector along the step edge as one translation vector. It applies to cases where it is appropriate to define a terrace and a step direction. The direction of the step edge also defines the vector between two lattice points of the surface plane; in Figure 2.20 the unit cell of the surface plane is marked. It should be noted that the vector calculated by Eq. (2.28) may be a multiple of the translation vector in the case of centred lattices, depending on the direction. The shortest translation vectors in centred lattices will be discussed later in this subsection; for the moment we refer to the cubic unit cell with integer indices. One translation vector is known: it is the vector of the step edge. The second vector can be calculated from the condition that the vector product of the two translation vectors define the Miller indices of the plane. Let $\mathbf{a}_s$, $\mathbf{b}_s$ be the translation vectors in the plane (hkl) with $\mathbf{a}_s$ the known vector in the direction of the step edge. The vector product $\mathbf{a}_s \times \mathbf{b}_s$ defines the plane (hkl):

$$[a_{s1}, a_{s2}, a_{s3}] \times [b_{s1}, b_{s2}, b_{s3}] = (hkl). \tag{2.30}$$

Here $\mathbf{a}_s$ and $\mathbf{b}_s$ are vectors in real space and (hkl) is a vector in reciprocal space, see Appendix C. For cubic lattices we can take (hkl) as a vector in real space, since the basis vectors of the reciprocal lattice and the real lattice coincide. The vector (hkl) is normal to the lattice plane and for cubic lattices no problem occurs.

In the case of oblique lattices, we can make use of the fact that the vector product of two vectors in reciprocal space define a direction in real space. Therefore, the vector $\mathbf{a}_s = [a_{s1}, a_{s2}, a_{s3}]$ in real space must be transformed into a vector in reciprocal space $\mathbf{a}_s^*$ using the metric matrix, see Appendix C. From the cross product we obtain at first $\mathbf{w}'$, a vector normal to the step edges which may be a multiple of the shortest distance $\mathbf{w}$ between steps.

$$a_s^* \times (h, k, l) = \left[w_1', w_2', w_3'\right],$$ (2.31)

$$\left[w_1', w_2', w_3'\right] = n[w_1, w_2, w_3].$$ (2.32)

In real space $\mathbf{w}'$ is a lattice point and has integer components, while the components of $\mathbf{w}$ are not necessarily integer and may take real values. In order to determine the factor n, we can make use of the fact that the volume of the unit cell is given by

$$
\begin{aligned}
V &= |\mathbf{a_s}||\mathbf{b_s}| \sin \gamma_s \cdot d_{hkl} \\
&= |\mathbf{a_s}||\mathbf{w}|d_{hkl} = |\mathbf{a_s}| \frac{|\mathbf{w}'|}{n} \\
&= a_0{}^3.
\end{aligned}
$$ (2.33)

The last line applies for cubic lattices in which the distance between lattice planes is given by

$$d_{hkl} = \frac{a_0}{\sqrt{h^2 + k^2 + l^2}},$$ (2.34)

where a_0 is the cubic lattice constant. A list of d-spacings for the other lattices is given in Appendix D. We can set $a_0 = 1$; $\mathbf{a_s}$, $\mathbf{w}'$ and d_{hkl} are known and the factor n is given by:

$$n = \frac{\sqrt{a_{s1}^2 + a_{s2}^2 + a_{s3}^2} \cdot \sqrt{w_1'^2 + w_2'^2 + w_3'^2}}{\sqrt{h^2 + k^2 + l^2}}.$$ (2.35)

Here n must be an integer, which is always the case if h, k and l have integer values. The translation vector $\mathbf{b_s}$ is then obtained from the condition that all three vector components b_{si} are integer:

$$\mathbf{b_s} = \frac{1}{n}(\mathbf{w}' + m\mathbf{a_s}),$$ (2.36)

as illustrated in Figure 2.20. Vector $\mathbf{b_s}$ is a lattice point with integer components, n is known, a_{si}, w_i' are integer and $\mathbf{b_s}$ is chosen with the smallest value of m that matches the condition in Eq. (2.36). When two values for m are possible, the vector with the angle γ' closer to 90° should be chosen.

The translation vectors $\mathbf{a_s}$, $\mathbf{b_s}$ determined in the way described earlier in this subsection refer to the 3-D lattice constants. For centred lattices the true translation vectors may be smaller. For fcc lattices, $\mathbf{a_s}$ and $\mathbf{b_s}$ can be divided by 2 if two indices are odd and one index is even; no reduction is possible if all three indices are odd, while if all indices are even a division by 2 is possible anyway.

For the example of the (11, 9, 8) plane shown in Figure 2.20, the edge direction obtained from Eq. (2.28) is [1, –3, 2]. The smallest translation vector actually is $\mathbf{a_s} = (1/2, -3/2, 1)$, because $(1/2, -1/2, 0)$ corresponds to a nearest neighbour distance and $(1/2, -1/2, 0) + (0, -1, 1)$ is also a lattice point. As we have a cubic lattice, we can take (hkl) as a vector in real space and use $\mathbf{a_s} \times (h, k, l) = \left[w_1', w_2', w_3'\right]$ instead of the reciprocal vector $\mathbf{a_s^*}$ in Eq. (2.31). We obtain $\mathbf{w}' = [-3, 1, 3]$. Since $\mathbf{w}'$ is normal to $\mathbf{a_s}$

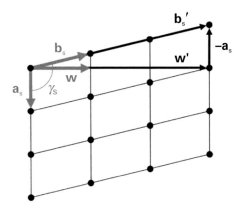

Figure 2.22 Illustration of the determination of 2-D translation vectors of a stepped surface. Vector $\mathbf{a_s}$ is the direction of the step edge calculated via Eq. (2.28), $\mathbf{w}'$ is determined from Eq. (2.30), n is determined from Eq. (2.35) ($n = 3$ in the figure) and $\mathbf{b_s}$ is a step-to-step vector, chosen as the smallest vector fulfilling Eq. (2.36) with integer m, in this example $m = -1$.

and $\mathbf{w}' \times \mathbf{a_s} = (11, 9, 8)$, it follows that $m = 0$ and $n = 1$ in Eq. (2.36). Therefore, $\mathbf{a_s}$ and $\mathbf{w}'$ define a rectangular lattice, and their vector product gives directly the desired (hkl), the indices of the stepped surface. With $\mathbf{b_s} = \mathbf{w} = [-3, 1, 3]$, the terrace width is $|\mathbf{w}| = \sqrt{19}a_0 = 4.36a_0$ where a_0 is the cubic lattice constant. In Figure 2.22 an example of an oblique lattice is shown where $\mathbf{w}' \neq \mathbf{w}$ and $m = -1$.

The angle γ_s between the translation vectors $\mathbf{a_s}$ and $\mathbf{b_s}$ of the surface lattice is generally calculated from

$$\cos(\gamma_s) = \frac{\mathbf{a_s} \cdot \mathbf{b_s}}{|\mathbf{a_s}||\mathbf{b_s}|}. \tag{2.37}$$

The integer vector components t_1, t_2, t_3 of the step edge, determined from Eq. (2.28), where a common integer factor has been removed and the step distance vector, from Eq. (2.36), may be multiples of the shortest translation vectors. The shortest translation vectors in centred lattices are listed in Table 2.6.

For bcc lattices the shortest translation vector (in units of the lattice constants) is half as long if the sum of the indices is odd, otherwise its length is not changed. Similar rules exist for centred tetragonal lattices and rhombohedral lattices in the hexagonal setting. The rules to reduce the translation vectors are summarised in Table 2.6. Examples for stepped surfaces in other lattices are given in Appendix F.

It should be noted that the translation vectors determined in the way described here are not necessarily the smallest vectors, as the vector $\mathbf{a_s}$ is chosen in the step direction and that depends on the choice of the decomposition of the plane (hkl) into two terraces, which is not unique. The analytic determination of the shortest vectors has been described by K. Hermann [2.6].

The calculation outlined in this subsection applies to cases where a dominant terrace exists and a general step direction can be defined. A formal extension to facetted surfaces with a more complicated morphology is possible but is not discussed

Table 2.6 Determination of the shortest translation vectors in centred 3-D lattices

Crystal system	Bravais type	Translation vector (t_1, t_2, t_3)	Shortest translation vector
Cubic, orthorhombic	F face-centred	all even	$t_1/2, t_2/2, t_3/2$
		all odd	no reduction
		$t_1 + t_2 = 2n, t_3 = 2n$	$t_1/2, t_2/2, t_3$
		$t_1 + t_3 = 2n, t_2 = 2n$	$t_1/2, t_2, t_3/2$
		$t_2 + t_3 = 2n, t_1 = 2n$	$t_1, t_2/2, t_3/2$
Cubic, tetragonal, orthorhombic	I body-centred	all even	$t_1/2, t_2/2, t_3/2$
		all odd	$t_1/2, t_2/2, t_3/2$
		$t_1 + t_2 + t_3 = 2n + 1$	
		$t_1 + t_2 + t_3 = 2n$	no reduction
Orthorhombic	C C-face-centred	all even	$t_1/2, t_2/2, t_3/2$
		$t_1 + t_2 = 2n + 1$	$t_1/2, t_2/2, t_3$
Monoclinic	A A-face-centred	all even	$t_1/2, t_2/2, t_3/2$
		$t_1 + t_3 = 2n + 1$	$t_1/2, t_2, t_3/2$
Rhombohedral with hexagonal obverse setting of the unit cell	R	$t_1 + t_3 = 3n$, and $-t_1 + t_2 + t_3 = 3n$, and $-t_2 + t_3 = 3n$	$t_1/3, t_2/3, t_3/3$

Columns 1 and 2 contain the crystal system(s) and the Bravais types possible in these systems. The Bravais type occurs as an upper-case letter in the first position of the Hermann–Mauguin symbol. The A and B face centring that is possible in orthorhombic lattices is not considered to represent different Bravais types, but different choices of the unit cell. Column 3 gives the conditions under which a translation vector can be shortened. A translation vector connects two unit cells of the centred lattice and has integer components. The shortest translation vectors connect two lattice points; the possible reduction is shown in column 4.

here as the surface morphology of facetted surfaces is usually not studied by diffraction; a direct image obtained with scanning tunnelling microscopy (STM) is more frequently used instead. The determination of the orientation of facets from the LEED diffraction pattern, however, is possible and is discussed in Chapter 4. Once the translation vectors of the stepped surface have been determined, the calculation of all atomic coordinates within a slab can proceed via Eq. (2.19) or (2.20) using $\mathbf{a_s}$, $\mathbf{b_s}$ and $\mathbf{c_s} = \mathbf{a_s} \times \mathbf{b_s}$ as new translation vectors or with another appropriate choice of the vector $\mathbf{c}$. The transformation to Cartesian coordinates is easily possible using the metric tensor, cf. Appendix C.

2.2   Diffraction from Surfaces

The diffraction geometry is different between LEED and surface X-ray diffraction due to large contrasts in scattering cross-sections and therefore also in penetration lengths into the samples. Nevertheless, the diffraction from a 2-D periodic lattice is common to both techniques. Therefore, the diffraction geometry and general features of diffraction from 2-D lattices will be very briefly outlined here with a common perspective, in order to provide the basic knowledge necessary to understand the experiments and the interpretations of the diffraction patterns described in Chapters 3 and 4.

In LEED, the calculation of diffracted intensities requires a multiple scattering theory (which is treated in Chapter 5), while for X-ray diffraction in many cases a kinematic theory is sufficient (cf. Chapter 7). The so-called kinematic theory assumes that an incident photon or electron is scattered once by a single atom and then leaves the crystal. The term kinematic is synonymous with 'single scattering', in contrast to the dynamical or 'multiple scattering' theory, in which the attenuation of the incident wave due to creation of diffracted beams as well as interference among diffracted beams is included, thereby allowing successive scatterings by two or more atoms. For X-rays and small crystals in 3-D structure analysis, the kinematic theory is usually sufficient. For surface structure analysis by X-ray diffraction, the kinematic theory is used, with slight modifications due to refraction of X-rays which becomes important at small incident or exit angles at surfaces. In some applications of surface X-ray diffraction a dynamical theory is also required. Here only some basic relations are explained which are necessary for the understanding of the following chapters.

The central differences between LEED and X-ray diffraction stem from differing scattering processes, even though both X-rays and electrons can be described as propagating waves. In both cases, we can represent the incoming beam of X-rays or electrons as a plane wave of the form $\psi(\mathbf{k}_0, \mathbf{r}) = \exp(i\mathbf{k}_0\mathbf{r})$, where $\mathbf{k}_0$ is the wave vector. Elastic (energy conserving) scattering of that plane wave by a single atom causes a scattered *spherical* wave of the form $\psi_s(\mathbf{k}', \mathbf{r}) = f(\vartheta_{\text{scatt}}) \exp(ik'r)/r$, where $k' = |\mathbf{k}'| = |\mathbf{k}_0|$ for energy conservation, $r = |\mathbf{r}|$ is the distance from the atom and ϑ_{scatt} is the scattering angle, cf. Figure 2.23. In X-ray diffraction theory, ϑ_{scatt} is $2\vartheta_{\text{inc}}$, where ϑ_{inc} is the incidence (and also reflection) angle relative to the macroscopic reflection plane. In LEED, ϑ_{scatt} is the local scattering angle at each scattering by an individual atom, so that ϑ_{scatt} can vary from one scattering event to the next in a chain of multiple scattering. Scattered waves from an array of periodically arranged atoms will recreate *plane* waves $\psi_s(\mathbf{k}_0, \mathbf{r}) = f(\vartheta_{\text{scatt}}) \exp(i\mathbf{k}'\mathbf{r})$ far from those atoms (ignoring some prefactors here, see Section 5.2.5). The factor $f(\vartheta_{\text{scatt}})$ is called atomic scattering amplitude, while $|f(\vartheta_{\text{scatt}})|^2 = d\sigma/d\Omega$ is called differential cross-section; when integrating the differential cross-section over all angles ϑ_{scatt}, we obtain the total cross-section σ_{tot}, which measures the total scattering strength of the atom. The total cross-section is on the order of 10^{-7} to 10^{-8} nm^2 (10^{-5} to 10^{-6} Å^2) for X-rays and about 10^{-2} nm^2 (1 Å^2) for electrons in the energy range used for LEED: this large contrast is crucial to the different behaviour and the different required theoretical modelling of X-ray diffraction versus LEED.

X-rays are scattered by the electrons attached to the sample's atoms. The distribution of the density of these electrons can be determined by analysis of the diffracted intensities, yielding the crystal structure of the surface. The atomic scattering amplitude for X-rays is calculated as the Fourier transform of the electron density of the atom:

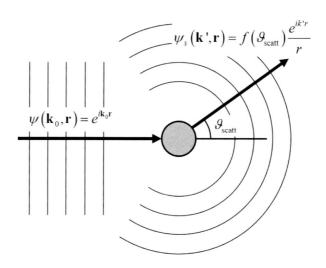

$$\psi_s\left(\mathbf{k'},\mathbf{r}\right) = f\left(\vartheta_{scatt}\right)\frac{e^{ik'r}}{r}$$

$$\psi\left(\mathbf{k}_0,\mathbf{r}\right) = e^{ik_0r}$$

Figure 2.23 Scattering geometry of a plane wave by one atom, representing either X-rays or electrons. In X-ray diffraction theory, $\vartheta_{scatt} = 2\,\vartheta_{inc}$, where $\vartheta_{inc} = \vartheta_{refl}$ is both the incidence and reflection angle relative to the reflecting plane (not shown). In LEED theory, ϑ_{scatt} is the deviation angle at the individual scattering atom.

$$f(|\mathbf{q}|) = \int_{V_{at}} \rho(\mathbf{r})e^{i\mathbf{qr}}d\mathbf{r}, \qquad (2.38)$$

where $\mathbf{q} = \mathbf{k'} - \mathbf{k}_0$ is the momentum transfer. For a spherically symmetric electron density around an atom and for elastic scattering, the scattering factor is real and depends only on the scattering angle ϑ.

Electrons are scattered by the atomic potential which is dominated by the electrostatic potential of the atomic core, even though the core potential is partially screened by the atomic electrons and there is also an interaction of the incoming electrons with the bound electrons. The atomic scattering amplitude is more complicated than for X-rays: it is calculated by solution of the Schrödinger equation, as described in Sections 2.2.3 and 5.2.1. In contrast to X-ray scattering, the atomic scattering amplitude for electrons is a complex number. For the energies used in LEED, typically in the range 30–500 eV, the way in which the atomic scattering potential is modelled plays an important role due to the complicated effect of the screening electrons, but the dominant part remains the core potential, especially at the higher energies. The analysis of the LEED intensities yields the positions of the atomic cores and hence the surface structure.

2.2.1   Kinematic Theory of Diffraction

In both LEED and X-ray diffraction, the incident beam is supposed to be well collimated and nearly mono-energetic. The incident beam is described as a plane wave with wave vector $\mathbf{k}_0$ (with the same unit as a reciprocal lattice vector, 1/length) and wavelength λ:

$$\psi(\mathbf{k}_0, \mathbf{r}) = Ae^{i\mathbf{k}_0\mathbf{r}}, \ |\mathbf{k}_0| = |\mathbf{k}'| = k = \frac{2\pi}{\lambda}. \tag{2.39}$$

The wave function is written as a complex function, but it must be kept in mind that this is chosen for practical reasons to allow easy calculation of the two real quantities characterising a plane wave, namely its amplitude and its phase. The amplitude A is usually set to 1 for the incident beam.

In Eq. (2.39), $\mathbf{k}'$ is the wave vector of the diffracted beam: its magnitude is the same as that of $\mathbf{k}_0$, because we include only elastic scattering. To calculate the scattering amplitude from a crystal we must sum up the scattering amplitudes from all atoms; we assume here that the diffracted waves can be described as plane waves far from the scattering centre. The following description is quite general and applies to LEED as well as to X-ray diffraction.

Let us first consider the scattering from two atoms. Coherent summation of the elastically scattered waves from the two different scattering centres leads to the combined scattered amplitude

$$A(\mathbf{k}', \mathbf{k}_0) = \sum_{j=1}^{2} f_j(\mathbf{k}', \mathbf{k}_0)e^{i\rho_j}. \tag{2.40}$$

Here, ρ_j are the phase differences between the incident and scattered waves with respect to an arbitrary origin, due to a path length difference, to be calculated in the next paragraph. One of the scattering centres can be taken as origin and with $\mathbf{k}_0$ taken as principal axis an azimuth φ can be defined around that axis. The scattering amplitude of the j-th atom can be written as a function of the scattering angles $f_j(\mathbf{k}', \mathbf{k}_0) = f_j(\vartheta_{\text{scatt}}, \varphi)$; for spherically symmetric scattering centres it depends only on the angle ϑ_{scatt}. The scattering amplitude relates the amplitude of the diffracted wave (generally complex in LEED, but usually real with X-rays) to the amplitude of the incident wave; far away from the atom we can describe the incident and the scattered waves as plane waves. Scattering from two centres is illustrated in Figure 2.24.

The phase difference $\rho_2 - \rho_1$ between the two scattered waves in Eq. (2.40) comes from the sum over the two distances s and t in Figure 2.24. With $|\mathbf{k}_0| = |\mathbf{k}| = \frac{2\pi}{\lambda}$, $s = \frac{\lambda}{2\pi}\mathbf{k}_0\mathbf{r}$ and $t = -\frac{\lambda}{2\pi}\mathbf{k}'\mathbf{r}$, we obtain

$$\rho = s + t = \frac{\lambda}{2\pi}\mathbf{r}(\mathbf{k}' - \mathbf{k}_0). \tag{2.41}$$

Here, $\mathbf{q} = \mathbf{k}' - \mathbf{k}_0$ is called the scattering vector.

We now consider a complete periodic crystal; $\mathbf{q}$ can then be described with coordinates h, k, l in reciprocal space: $\mathbf{q} = 2\pi(h\mathbf{a}^* + k\mathbf{b}^* + l\mathbf{c}^*)$. The scattering amplitude from the whole crystal is given by:

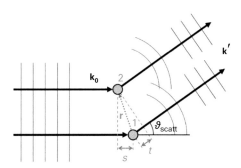

Figure 2.24 Coherent superposition of scattered waves from two atoms. The diffraction angle ϑ_{scatt} is defined in Figure 2.23 for both X-ray and electron diffraction.

$$A(\mathbf{k}', \mathbf{k_0}) = \sum_{n=1}^{N_c} f_n(\mathbf{k}', \mathbf{k_0}) \cdot e^{i\mathbf{q}\mathbf{r}_n}$$

$$= \sum_{n=1}^{N_c} f_n(\mathbf{k}', \mathbf{k_0}) \cdot e^{2\pi i(hx_n + ky_n + lz_n)}. \qquad (2.42)$$

The sum over n means here the sum over all atoms N_c of the crystal. This sum can be divided into a sum over all N atoms within a unit cell and a sum over all M unit cells, so that $N_c = NM$. The total scattering amplitude then can be written as

$$A(\mathbf{k}', \mathbf{k_0}) = \sum_{j=1}^{N} f_j(\mathbf{k}', \mathbf{k_0}) e^{2\pi i(hx_j + ky_j + lz_j)} \sum_{m=1}^{M} e^{i\mathbf{q}\mathbf{r}_m}. \qquad (2.43)$$

The sum over j runs over all atoms in the unit cell and the sum over m runs over all unit cells; $\mathbf{r}_m$ is a vector between the origin and all lattice points in the direct lattice: $\mathbf{r}_m = m_1\mathbf{a} + m_2\mathbf{b} + m_3\mathbf{c}$ with integer m_1, m_2, m_3. The sum over lattice vectors in the last term of Eq. (2.43) gives the so-called Laue function and, with $\mathbf{q} = 2\pi(h\mathbf{a}^* + k\mathbf{b}^* + l\mathbf{c}^*)$ and a finite lattice with $M_1M_2M_3$ lattice points, we obtain:

$$\sum_{m=1}^{M} e^{i\mathbf{q}\mathbf{r}_m} = \sum_{m_1=-(M_1-1)/2}^{(M_1-1)/2} e^{2\pi ihm_1} \sum_{m_2=-(M_2-1)/2}^{(M_2-1)/2} e^{2\pi ihm_2} \sum_{m_3=-(M_3-1)/2}^{(M_3-1)/2} e^{2\pi ihm_3}$$

$$= \frac{\sin(\pi M_1 h)}{\sin(\pi h)} \frac{\sin(\pi M_2 k)}{\sin(\pi k)} \frac{\sin(\pi M_3 l)}{\sin(\pi l)}$$

$$= G(h)G(k)G(l). \qquad (2.44)$$

The Laue function may be approximated by a sum of δ-functions in the limit of an infinite lattice:

$$\sum_{m} e^{i\mathbf{q}\mathbf{r}_m} = \sum_{g=-\infty}^{\infty} \delta(\mathbf{k}' - \mathbf{k_0} - \mathbf{g}) = \sum_{g=-\infty}^{\infty} \delta(\mathbf{q} - \mathbf{g})$$

$$= \sum_{h,k} \delta(\mathbf{q} - h\mathbf{a}^* - k\mathbf{b}^* - l\mathbf{c}^*). \qquad (2.45)$$

Eq. (2.45) defines the three Laue conditions for the diffraction vector $\mathbf{q}$ at which a reflection from a 3-D crystal occurs. For 2-D infinite lattices, only two of the three Laue conditions remain:

$$\mathbf{a} \cdot (\mathbf{k}_0 - \mathbf{k}') = 2\pi h,$$
$$\mathbf{b} \cdot (\mathbf{k}_0 - \mathbf{k}') = 2\pi k. \tag{2.46}$$

For the 2-D case, with integer reflection indices h and k, $\mathbf{g} = h\mathbf{a}^* + k\mathbf{b}^*$ is a vector of the 2-D reciprocal lattice, and the third index l is a continuous function of the two indices h and k; now l is given by the condition of elastic scattering

$$|\mathbf{k}_0| = |\mathbf{k}'| = |\mathbf{k}_0 + \mathbf{g}|, \tag{2.47}$$

which defines the z-component of the diffracted beam, where the Cartesian components of $\mathbf{k}_0$ and $\mathbf{g}$ are required. The z-component must be perpendicular to the surface:

$$k'_z = \sqrt{|\mathbf{k}_0|^2 - (\mathbf{k}_{0x} + g_x)^2 - (\mathbf{k}_{0y} + g_y)^2}. \tag{2.48}$$

The intensity diffracted from the direction of the incident wave $\mathbf{k}_0$ into the scattered wave $\mathbf{k}'$ is then given by

$$I(\mathbf{k}', \mathbf{k}_0) = |F(\mathbf{k}', \mathbf{k}_0)|^2 |G(h)|^2 |G(k)|^2$$
$$= \left| \sum_{j=1}^{n} f_j(\mathbf{k}', \mathbf{k}_0) e^{2\pi i \left(hx_j + ky_j + lz_j \right)} \right|^2 \delta(\mathbf{k}' - \mathbf{k}_0 + \mathbf{g}), \tag{2.49}$$

where j runs over all atoms in one unit cell. $|F(\mathbf{k}', \mathbf{k}_0)|^2$ is called the structure factor: it represents the scattering by the atoms in one unit cell, while $|G(h)|^2 |G(k)|^2$ is called the lattice factor, which may be replaced by a δ-function for infinite lattices, as in Eq. (2.45). The separation of structure factor and lattice factor in Eq. (2.49) has the advantage that all calculations of directions of wave vectors, orientation of the crystal, etc., occurring in the atomic scattering factors $f(\mathbf{k}', \mathbf{k}_0)$ are completely decoupled from the shape and size of the unit cell.

It should be noted, for clarity, that in Eq. (2.49) all prefactors which relate the absolute measured intensity to the intensity of the incident beam are omitted. Thus, the intensity as defined in Eq. (2.49) only describes the relative intensity between diffracted beams.

For disordered lattices or small crystals, the lattice factor must be evaluated separately from Eq. (2.44) using an appropriate distribution function and cannot be approximated by a δ-function, as in Eq. (2.45). Then the reflections are broadened and the beam profile can be calculated independently from the structure factor if the profile is sufficiently narrow. When no lattice exists, for example in some applications of small angle X-ray scattering from ordered or disordered layers or from polycrystalline materials or in LEED from nanoparticles, the diffracted intensities are measured as a continuous function of diffraction angles and the use of a reciprocal lattice does not make sense. In these applications the intensities are therefore measured as a function of scattering angles (for X-rays) or also energy (in LEED).

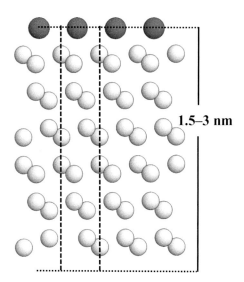

Figure 2.25 The surface unit cell for LEED and surface X-ray diffraction consists of a whole column down to a suitable depth. In the case of LEED, the thickness of the resulting surface slab depends on material, electron energy, and incidence angle: the range is frequently about 1.5–3 nm. For surface X-ray diffraction the penetration depth is very much larger, leading to Bragg reflections in lattice rods related to the substrate lattice.

Eq. (2.49) for the diffracted intensity is applicable in general, whether we have single scattering or multiple scattering. The difference lies in the definition of the scattering amplitude of the two cases. The scattering amplitude of the j-th atom in Eq. (2.42), $f_j(\mathbf{k}', \mathbf{k}_0)$, relates the amplitude of the wave leaving atom j without further scattering to the amplitude of the external incoming wave. In the case of multiple scattering, the incoming wave field consists of the primary incident wave and the diffracted waves from all surrounding atoms.

For a 2-D lattice, the unit cell consists of the whole column of atoms down to a depth where the contribution to the diffracted intensity can be neglected, cf. Figure 2.25. This is applicable for LEED with its small penetration depth as well as for surface X-ray diffraction, even with its much larger penetration depth. For practical reasons, the surface unit cell can be divided into two parts: the surface in which the structure deviates from the substrate structure; and a bulk part in which the material can be regarded as having the bulk structure.

2.2.2 X-ray Diffraction

The wavelengths of X-rays used for structure determination range roughly from 0.05 to 0.5 nm; wavelengths around 0.1 nm $= 1$ Å are common. The wavelength λ and energy E of an X-ray are $\lambda = c/\nu$ and $E = h\nu$, so that $\lambda = hc/E$, where c is the speed of light, h is Planck's constant and ν is the frequency. The (inverse) correspondence between wavelength and energy is exemplified by:

$$1 \text{ nm} \quad \leftrightarrow \quad 123.984 \text{ keV},$$
$$1 \text{ Å} \quad \leftrightarrow \quad 12.3984 \text{ keV}.$$

Surface X-ray diffraction is mostly performed with synchrotron radiation where the appropriate wavelength can be chosen. The wavelength should be sufficiently far away from an absorption edge of the elements in the sample, otherwise the background intensity may become too high. Some experiments have been performed in the laboratory using a rotating anode or micro-focus sources to obtain a high intensity in the primary beam. The experimental setup is discussed in Chapter 7.

The scattering of X-rays by electrons is called Thomson scattering. The scattering amplitude of an atom is calculated from the atomic electron density and is given in units of the scattering amplitude of a single electron, see Eq. (2.38). The atomic scattering amplitude is called 'atomic scattering factor' or 'atomic form factor' and is denoted by f. The atomic electron density is assumed to be spherically symmetric and is normally obtained by a Hartree–Fock calculation. The Fourier transform of the electron density gives the scattering factor. As the electron density distribution of an atom is approximately given by a superposition of Gaussian functions for the different orbitals, the scattering amplitude is therefore also given by a superposition of Gaussian functions. The scattering factors for all atoms and ions are tabulated in the International Tables [2.1] as function of $(\sin \vartheta)/\lambda$. Because spherical symmetry is assumed for single atoms, the only variables are the scattering angle ϑ and the wavelength λ. The maximum amplitude at forward scattering ($\vartheta = 0$) is equal to the number of electrons in the atom. The electron density in positively charged ions is slightly more compact than in the neutral atoms and the scattering factor is therefore slightly broader. Some examples are shown in Figure 2.26.

As the electron density is a real quantity, the scattering factor is also real. Above the excitation threshold near an absorption edge a phase factor occurs in the scattering

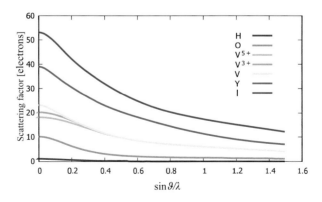

Figure 2.26 Atomic scattering factors in X-ray diffraction for some elements indicated in the figure. The scattering factors of ions become similar to those of the neutral atoms for larger diffraction angles, as illustrated for V^{3+} and V^{5+}.

amplitude, which can be represented by an imaginary component of the scattering factor, leading to resonant or anomalous scattering: these cases will be considered in Chapter 7.

Eq. (2.40) can be simplified for X-ray diffraction. In the kinematic theory of X-ray diffraction, the atoms are usually assumed to be spherically symmetric and the atomic scattering factors depend only on the magnitude of the scattering vector:

$$|\mathbf{q}| = \frac{2 \sin \vartheta}{\lambda}. \tag{2.50}$$

The diffracted intensity then reads

$$I(\mathbf{q}) = |F(\mathbf{q})|^2 |G(\mathbf{q})|^2 = \left| \sum_{j=1}^{n} f_j(\mathbf{q}) e^{2\pi i \left(hx_j + ky_j + lz_j \right)} \right|^2 |G(\mathbf{q})|^2. \tag{2.51}$$

The bulk structure analysis of 3-D crystals by X-rays therefore requires only the measured intensities, the reflection indices and the lattice constants. The direction of the incident beam is not required. The intensity depends usually only on $|\mathbf{q}|$, independent of the orientation of the incident beam relative to the crystal lattice; by 'usually' we mean when the kinematic theory can be applied. The atomic coordinates are determined as fractions of lattice vectors; the lattice constants, however, must be determined independently, mostly by powder diffraction.

The situation is the same in surface X-ray diffraction; there the lattice constants of the substrate are usually known. But the diffraction conditions differ from those for 3-D bulk crystals. Due to limited penetration depth at small incidence angles and the refraction of X-rays, the direction of the incident beam must be precisely known, as will be described in detail in Chapter 7. The direction of the incident beam must also be known in the case of LEED: here, due to multiple scattering, the intensity depends explicitly on both wave vectors of the incident and the diffracted beam, as expressed in Eq. (2.49).

For X-ray diffraction the atomic scattering factor $f(\mathbf{q})$ can usually be taken as real, which leads to Friedel's law:

$$I(h, k, l) = I(-h, -k, -l). \tag{2.52}$$

This relation is not valid for LEED. In surface X-ray diffraction the reflection with negative l cannot be directly measured, but can be derived from Friedel's law as:

$$I(h, k, -l) = I(-h, -k, l).$$

2.2.3 Electron Diffraction

The wavelength of electrons is given by the de Broglie wavelength:

$$\lambda = \frac{h}{\sqrt{2m_e E}} \approx \sqrt{\frac{150.4}{E[eV]}} [\text{Å}] \approx \sqrt{\frac{1.504}{E[eV]}} [\text{nm}]. \tag{2.53}$$

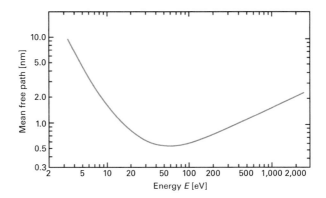

Figure 2.27 'Universal curve' of the inelastic mean free path of electrons as a function of kinetic energy, fit to experimental data for elemental surfaces. Adapted with permission from [2.16] M. P. Seah and W. A. Dench, *Surf. Interface Anal.*, vol. 1, pp. 2–11, 1979. © 1979 Wiley-VCH Verlag GmbH & Co. KGaA, Boschstr. 12, 69469 Weinheim, Germany.

The energies used for LEED are about 20–500 eV, which implies that the wavelengths are in the range of 0.05–0.25 nm (0.5–2.5 Å) and in the same range as used for structure analysis with X-rays. Electrons in this energy range exhibit very strong interactions with matter. The large cross-section for elastic as well as for inelastic scattering leads to a very limited penetration depth, called electron attenuation length (EAL) or inelastic mean free path (IMFP), for the electrons and makes the method very surface-specific.

Average penetration depths fitted to data measured for polycrystalline elemental solids are shown in Figure 2.27, forming the 'universal curve' [2.16]. The total cross-section of each atom for the elastic scattering of electrons in this energy range is on the order of 10^{-16} cm$^2 = 1$ Å^2, which means that almost every electron is scattered when passing an atomic layer. Due to the large cross-section, the kinematic or single scattering theory cannot be applied; therefore, a dynamical or multiple scattering theory is necessary for the quantitative description of diffraction intensities. In the energy range of about 20–500 eV, the fraction of the electrons which are elastically scattered is on the order of about 1% of the primary current: this part is used for LEED; the remaining electrons are either 'absorbed' in the sample (due, for example, to energy loss to electron or phonon excitations) or scattered in other directions outside the sample (due, for example, to scattering from phonons or non-periodic surface defects).

The scattering amplitude for elastic scattering from a single atom is calculated by solving the time independent Schrödinger equation with an atomic potential $V(\mathbf{r})$, which describes the interactions between the external electron, on one hand, and the atomic cores and atomic electrons, on the other hand ($2\pi\hbar = h$ is Planck's constant and m_e the mass of the electron):

$$-\frac{\hbar^2}{2m_e}\nabla^2\psi(\mathbf{r}) + V(\mathbf{r})\psi(\mathbf{r}) = E\psi(\mathbf{r}). \tag{2.54}$$

For the heavier elements, relativistic effects can become important, so the Dirac equation is solved: this case will not be treated here.

Integration of the Schrödinger equation and transformation into the Lippman–Schwinger equation leads to

$$\psi(\mathbf{r}) = \phi(\mathbf{r}) + \int d\mathbf{r}' G(\mathbf{r} - \mathbf{r}') V(\mathbf{r}') \psi(\mathbf{r}'). \qquad (2.55)$$

With the Green function

$$G(\mathbf{r}) = -\frac{m_e}{2\pi\hbar^2} \frac{e^{i\mathbf{k}\cdot\mathbf{r}}}{r} \qquad (2.56)$$

and a t-matrix formulation as well as an expansion in spherical harmonics, Eq. (2.55) is solved in a spherical wave expansion. The solution of Eq. (2.55) is typically based on the Korringa–Kohn–Rostoker (KKR) method and the so-called muffin-tin model of the crystal potential. The theory is described in detail in Chapters 5 and 6.

The atomic scattering amplitude for electron scattering differs significantly from that for X-rays shown in Figure 2.26. In the energy range used by LEED the electron is mainly scattered from the core potential which is partially screened by the atomic electron density. The electron density of the atom is obtained from any one of a variety of first-principles calculational schemes for atoms (with adjustment for overlapping charge densities in a lattice) or for periodic bulk crystals, while the atomic potential is numerically calculated by solving the Poisson equation. An analytic solution exists for the scattering amplitude from a spherically symmetric potential in a spherical wave expansion. The scattering amplitude for a plane wave scattered from the direction of the incoming wave $\mathbf{k}_0$ into the direction $\mathbf{k}'$ is:

$$f(\mathbf{k}', \mathbf{k}_0) = -i \frac{\hbar^2}{2m} \frac{2\pi}{k} \sum_{l=0}^{\infty} (2l+1) P_l\left(\cos\left(\vartheta_{\mathbf{k}_0,\mathbf{k}'}\right)\right) \left\{ e^{2i\delta_l} - 1 \right\}, \qquad (2.57)$$

Here $\vartheta_{\mathbf{k}_0,\mathbf{k}'}$ is the angle between the incoming wave vector and the diffracted wave vector and $\delta_l(k)$ are the so-called phase shifts, as they describe the phase differences between spherical waves leaving the atom relative to the incoming waves. The phase shifts vary with the energy; the number of angular momentum values l in principle is infinite but in practice is determined by the phase shifts required in the LEED I(V) calculation; often, values from 6 to 15 are used depending on element and energy range. The diffraction amplitude depends on energy and angle and exhibits a maximum in the forward scattering direction, but also secondary maxima in other directions at certain energies; there also occurs in general a maximum in the back-scattering direction. This is illustrated for Ag in Figure 2.28. A detailed description of the calculation of phase shifts and the derivation of Eq. (2.57) is given in Section 5.2.1.

2.2.4   Ewald Construction for Diffraction from Surfaces

For both X-ray diffraction and LEED, due to the lack of periodicity in the direction normal to the surface, the reciprocal lattice consists of a 2-D array of reciprocal lattice

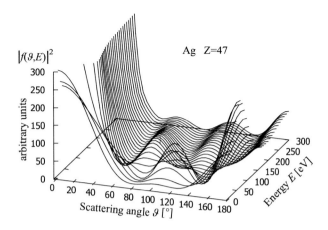

Figure 2.28 Atomic scattering amplitude for Ag and for electrons between 10 and 600 eV as a function of the scattering angle. The scattering amplitude is calculated by Eq. (2.57). For a spherical potential, $f(\mathbf{k'}, \mathbf{k})$ is displayed as $f(\vartheta, E)$.

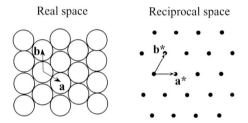

Figure 2.29 2-D lattice in real and reciprocal space

points, as illustrated in Figure 2.29. In three dimensions this becomes an array of lattice rods. Due to the small penetration depth of the electrons, no Bragg points exist along the lattice rods, that is, for all scattered wave vectors $\mathbf{k'}$ that fall on one of those rods, non-zero reflected intensity is allowed: this leads to continuous I(V) curves in LEED. In surface X-ray diffraction continuous intensity along the reciprocal lattice rod appears normal to the surface. This is a result of the truncation of the lattice and of the small thickness of an adsorbate layer if there is one. The large penetration depth of X-rays causes Bragg points from the substrate. A detailed description of X-ray diffraction from clean and adsorbate covered surfaces is given in Section 7.1.

The diffraction geometry and the Ewald construction for LEED are illustrated in Figure 2.30 and for X-rays at grazing incidence in Figure 2.31. The wave vector of the incoming wave points to the origin of the reciprocal lattice and all diffracted beams – we consider elastic diffraction only – lie on a sphere around the origin of the wave vector, forming the so-called Ewald sphere. Reflections can occur where the lattice rods intersect the Ewald sphere. The diffraction angles and the accessible range in reciprocal space for a given wavelength are then easily obtained by simple geometric considerations. The diffraction geometries of LEED and surface X-ray diffraction are

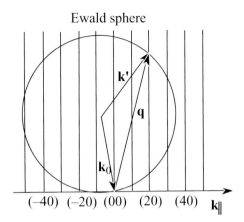

Figure 2.30 Diffraction geometry and Ewald construction for LEED

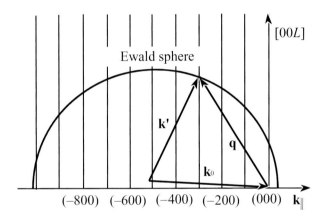

Figure 2.31 Diffraction geometry and Ewald construction for surface X-ray diffraction. The direction [00L] is normal to the surface; as in LEED, diffracted intensity is measured at each point where the Ewald sphere intersects a lattice rod: L is therefore a continuous index.

different and the scattering processes differ as well. Nevertheless, the same calculation of diffraction angles applies to both techniques.

If an incident plane wave with wave vector $\mathbf{k}_0$ is diffracted from a 2-D periodic surface into a set of diffracted waves $\mathbf{k_g}$, the two components parallel to the surface must fulfil the 2-D diffraction conditions. The incident plane wave with wave vector $\mathbf{k}_0$ is described by its Cartesian components, where the z-axis is perpendicular to the surface:

$$|\mathbf{k}_0| = \frac{2\pi}{\lambda}, \quad \mathbf{k}_0 = \left(k_{0x}, k_{0y}, k_{0z}\right),$$

$$k_{0x} = |\mathbf{k}_0| \cos \vartheta \cos \varphi, \quad k_{0y} = |\mathbf{k}_0| \cos \vartheta \sin \varphi.$$

$$(2.58)$$

Here ϑ is the polar angle between the incident beam and the surface normal, and φ is the azimuth of the incident beam around the surface normal. The 2-D Laue conditions apply:

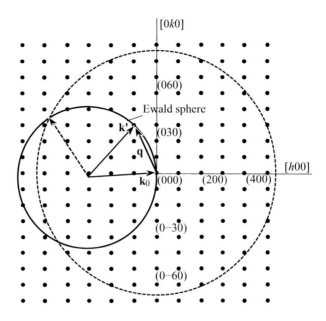

Figure 2.32 Schematic drawing of the in-plane reflections observable by surface X-ray diffraction at grazing incidence

$$(\mathbf{k}' - \mathbf{k}_0)\mathbf{a} = 2\pi h \quad \text{and} \quad (\mathbf{k}' - \mathbf{k}_0)\mathbf{b} = 2\pi k,$$
$$\mathbf{k}_g = \mathbf{k}_0 + \mathbf{g}, \quad \mathbf{g} = 2\pi(h\mathbf{a}^* + k\mathbf{b}^*), \tag{2.59}$$

where $\mathbf{g}$ is a vector of the 2-D reciprocal lattice. The intensity along the normal component of the diffracted beam is a continuous function of the wavelength or energy of the primary beam due to the 2-D periodicity of the surface. That normal component is:

$$k'_z = \sqrt{\frac{2\pi}{\lambda} - (k_{0x} + g_x)^2 - (k_{0y} + g_y)^2}. \tag{2.60}$$

The Ewald construction provides an easy method to calculate the diffraction angles. Whether non-zero diffracted intensity can be observed at each value of k'_z or not depends on the structure of surface and substrate and on the penetration depth.

The diffraction geometry for LEED is shown in Figure 2.30. In LEED usually normal or near normal incidence is chosen and backscattered reflections are observed at high momentum transfer normal to the surface.

In surface X-ray diffraction a grazing incidence angle (or a grazing exit angle) is used, allowing the measurement of reflections with high momentum transfer parallel to the surface, but smaller normal to the surface. Usually the exit angle is limited to about $60°$ and $k'_z \approx 0.85 |\mathbf{k}_0|$. The range for the in-plane reflections is illustrated in Figure 2.32.

The experimental methods of LEED are described in Chapter 3 and the specific aspects and requirements of surface X-ray diffraction at grazing incidence or exit angles are described in Section 7.1, while the experimental methods can be found in Section 7.2.

3 LEED Experiment

This chapter describes specialised equipment and techniques used to perform LEED experiments and to measure intensities of diffracted LEED beams. An overview of the most common setups for experiments will be given. The diffraction geometry is important for the comparison of experimental LEED data with theory and will, thus, be covered in some detail. For the measurement of LEED intensities, close attention will be paid in particular to the preparation of the sample, the accurate alignment of the sample and the physical properties of the detectors, such as the frequently used video cameras. The instrumental response function is one aspect of detectors that can affect the measured intensities, most notably spot profiles used to measure lateral dimensions such as island sizes and disorder. Among various LEED systems that are available on the market, two types will be addressed in relatively more detail, as they provide higher resolution (i.e., are able to detect structural correlations over larger distances along the surface): spot profile analysis LEED (SPA-LEED) and low-energy electron microscope (LEEM). Finally, instrumentation will be described that has been developed for more targeted applications, such as electron-beam sensitive surfaces, and surfaces with micro- or nanoscale structures.

3.1 Experimental Setup

LEED experiments are commonly performed in ultra-high vacuum (UHV) chambers at a base pressure in the range of 10^{-10} mbar or below. Some useful relations between frequently used pressure units are as follows: 1 mbar = 0.75 mmHg, 1 mmHg = 1 Torr, 1 hPa = 1 mbar and 1.333 mbar = 1 Torr. The vacuum conditions are required for two reasons: the electron gun requires a vacuum for its operation and the surface needs a vacuum to remain clean during the measurement. The electron diffraction experiment could be carried out in a vacuum up to about 10^{-6} mbar depending on the composition of the residual gas, but the sample is then usually contaminated or damaged by the ions created by the electron beam passing through the gas, in addition to the adsorption of the residual gas onto the surface. A close packed metal surface with about 3×10^{14} atoms in the surface monolayer would be fully covered within about 1 second at 10^{-6} mbar assuming a sticking coefficient of 1; such a high sticking coefficient applies only to some residual gases (especially oxygen and hydrogen), while most gases would cover the surface or react with it within the required experimental preparation and measurement times, which can last from many minutes to many hours. The unit 1 Langmuir = 1.33 mbar·sec = 1 torr·sec is used as a convenient unit for gas exposure; see for example [3.1].

Several experimental setups are in use for LEED studies. The most common is the display system to observe the diffraction pattern or to measure the diffraction intensities. LEED is usually applied in the backscattering geometry near normal incidence, and the backscattered diffraction spots are observed. The most frequently used experimental setup is the display system that shows the diffracted spots (corresponding to diffracted beams) either on a fluorescent screen or in a digital image using a 2-D position sensitive detector. The principle of a LEED display system is shown in Figure 3.1.

Both the sample and the first grid are grounded, so the diffracted beams propagate straight through the space between the sample and the first grid. The second and third grids act as a high pass filter to suppress the inelastically scattered electrons. The filter voltage is slightly below the electron energy, typically a few eV, and adjustable. There are three-grid and four-grid systems in use. In the case of a three-grid system there is only one grid at the retarding voltage, giving a slightly worse energy resolution than the four-grid system. The LEED system is also frequently used as a retarding field analyser for Auger electron spectroscopy (AES), and in that case the four-grid version is required. The screen is at high voltage, usually 4–7 keV, to excite the fluorescent screen. The screen is made of glass, and the diffraction spots are observed from the back without disturbance by the grids, although the screen's central portion is then obscured by the electron gun. The system is therefore also called 'reverse view LEED

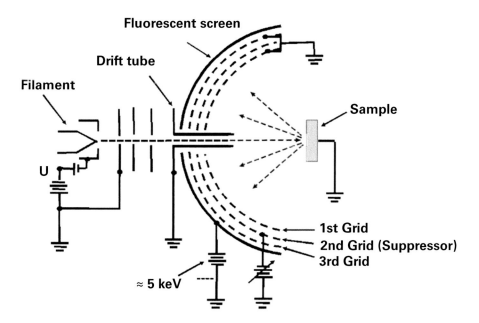

Figure 3.1 Schematic drawing of a three-grid LEED display system using a fluorescent screen. The screen is usually viewed from the left ('rear' or 'back').

optic'. Figure 3.2 shows a schematic of the reverse view arrangement – also called 'back view system' or 'rear view system' – with a spherical glass display screen. Systems with a planar phosphorescent screen are also in use. The spherical screen offers the advantage that the spot pattern has the geometry of the 2D reciprocal lattice of the surface, without distortion when viewed by a distant camera, as illustrated for a square lattice in Figure 3.2(b): the spot pattern shows the same square spot array as the crystal surface. The diffraction intensities are usually recorded by a CCD camera.

The intensity in the diffracted beams can be directly measured with a Faraday cup or a channeltron, also called a channel electron multiplier, and this approach has been used in earlier instruments. More recently, the measurement of intensities has been mostly done with a video camera, as first introduced by P. Heilmann et al. in 1976 [3.3]. Other systems use a 2-D position sensitive detector (microchannel plate) instead of the fluorescent screen, allowing the direct measurement of the intensity and electronic storage of the data.

A different experimental setup is used in high-resolution systems like SPA-LEED (spot profile analysis LEED), with which angular beam profiles are measured [3.4], and low-energy electron microscopy (LEEM), which has a high spatial resolution and allows measurement of LEED intensities from a single structural domain [3.5]. Also, spin-polarised LEED (SPLEED) systems have been developed using a spin-polarised primary beam to allow the measurement of the spin polarisation of the diffracted beams; see for example [3.6] and further references therein.

The diffracted intensities are always recorded on an arbitrary scale but must be normalised to the primary beam intensity, because this incident intensity in most instruments depends on the primary beam's energy. The primary intensity is measured by the current leaving the power supply, not by the sample current, because there are secondary electrons leaving the sample, which affects the net sample current. The current in the diffracted beam is usually on the order of 10^{-2} to 10^{-4} relative to the

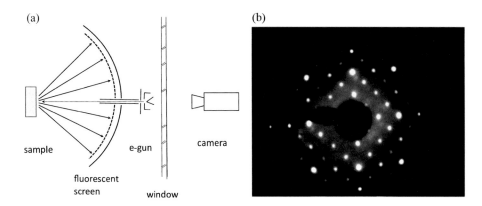

Figure 3.2 (a) Schematic drawing of the experimental 'rear view' LEED setup and the diffraction pattern; (b) Diffraction pattern of $Fe_3O_4(100)$ with a c(2×2) reconstruction. Reprinted from [3.2]: R. Pentcheva, W. Moritz, J. Rundgren, S. Frank, D. Schrupp and M. Scheffler, *Surf. Sci.*, vol. 602, pp. 1299–1305, 2008, with permission from Elsevier.

primary beam current. Modern video techniques allow the measurement of a complete set of I(V) curves in several minutes.

The distance of the surface plane from the front of the LEED system is usually between 10 and 20 mm, which is not sufficient to install further instruments. Therefore, either the sample or the LEED system must be moved for sample preparation, deposition of adsorbate layers, AES measurements or other analytical investigations. Many commercially available LEED systems therefore offer the possibility to retract the LEED system farther away from the sample. The LEED screen and the grids need to be covered during sample preparation to avoid contamination by sputtering products or material from evaporation sources.

Electrons with the energies applied in LEED are noticeably deflected by the Earth's magnetic field and stray fields from magnetic materials inside the chamber. The Earth's magnetic field therefore must be shielded either by Helmholtz coils, usually mounted outside the UHV chamber, or by a shield of μ-metal, an alloy of high magnetic permeability, placed inside the chamber. The magnetic shielding may not be necessary if only the diffraction pattern will be observed without quantitative interpretation. For I(V) measurements magnetic shielding is required in most cases.

3.2   Diffraction Geometry

The diffraction geometry in reciprocal space is displayed in Figure 3.3. The reciprocal lattice of the 2-D periodic surface is a 2-D lattice of rods oriented perpendicular to the surface plane. The Ewald construction for elastic scattering shows that a reflection occurs whenever the Ewald sphere intersects a lattice rod (see also Section 2.2).

For quantitative structure analysis, one measures the intensity along the lattice rods. This can be done in different ways as shown in Figure 3.4.

The usual approach is to measure the spot intensity with varying energy, producing I(V) curves, where eV is the electron kinetic energy. Changing the angle of incidence or rotating the sample is possible as well but rarely applied, because it induces some uncertainties when the surface is not quite homogeneous, since different patches on the surface may be hit by the primary beam when the sample is rotated. To avoid this source of error requires careful control of the sample position which is rarely reached with the commonly used manipulators. It is sometimes applied, however, to measure data sets at different angles of incidence to extend the amount of data and the reliability of the analysis. It may also be necessary to measure at different incidence angles in cases where only few beams can be measured due to experimental limitations.

At normal incidence the reciprocal lattice of the surface lattice can be directly observed on a display screen. In Figure 3.5(a), a simple hexagonal lattice of atoms is shown. The vectors **a**, **b** mark the (1×1) unit cell, while the vectors $\mathbf{a}'$, $\mathbf{b}'$ are the unit vectors of a $(\sqrt{3} \times \sqrt{3})R30°$ superlattice, or in matrix notation a $\begin{pmatrix} 2 & 1 \\ -1 & 1 \end{pmatrix}$ superlattice. In Figure 3.5(b), the diffraction pattern and the indices related to the (1×1) unit cell are given; in Figure 3.5(c), the same diffraction pattern with indices related to the

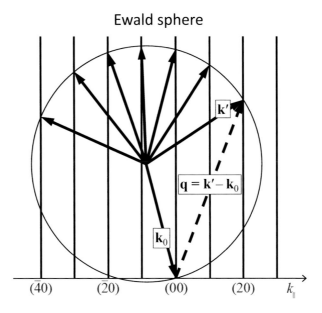

Figure 3.3 Diffraction geometry with Ewald construction of scattering directions of diffracted beams; the surface is horizontal in this representation, with $\mathbf{k}_0$ the incident beam momentum (in the plane of the figure) and $\mathbf{k}'$ one of several possible diffracted beam momenta (the Ewald sphere has radius $|\mathbf{k}_0|$); $\mathbf{q}$ gives the momentum transfer. The vertical lattice rods are given by the reciprocal lattice of the surface lattice.

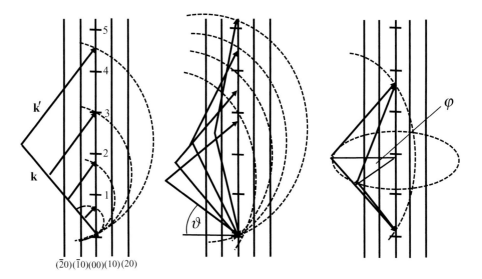

Figure 3.4 Measurement of I(V) curves (left panel), I(ϑ) curves (middle panel) and I(φ) curves (right panel). LEED intensities depend on the polar incidence angle ϑ and on the azimuthal angle φ. The same point in reciprocal space can be reached with different ϑ and change of energy and also with different azimuthal angle φ. In each case different intensities are measured, which is a multiple scattering effect. For kinematical diffraction, the usual case for X-ray diffraction, such dependence would not occur.

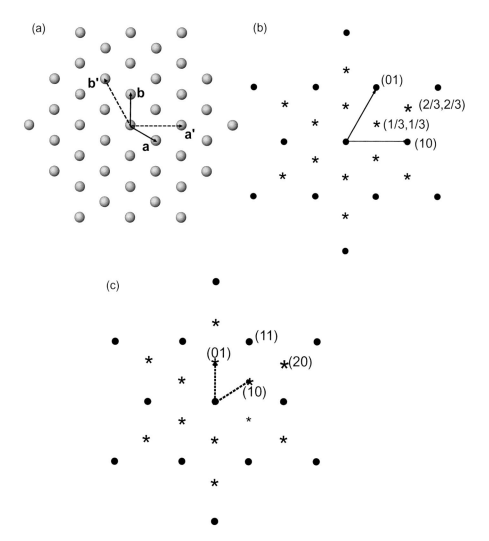

Figure 3.5 (a) Hexagonal lattice and unit cell vectors **a**, **b** for a (1×1) unit cell and **a′**, **b′** for a ($\sqrt{3}\times\sqrt{3}$)R30° superlattice; (b) reciprocal lattice with 'fractional' indices related to the (1×1) unit cell; and (c) with 'integer' indices related to the superlattice.

superlattice cell is shown. The latter is frequently more convenient for superlattices with large unit cells. The transformation matrix for the indices is the same as that for the unit cell vectors; see Section 2.1.10.

The range in reciprocal space accessible to LEED for energies between 50 and 450 eV is displayed in Figure 3.6 for a Ag(111) surface. The positions of the Bragg peaks of the 3-D substrate are indicated in the figure for comparison. It is obvious that the back-scattering geometry used in LEED only allows the measurement of reflections at high momentum transfer normal to the surface corresponding to a high index l in the

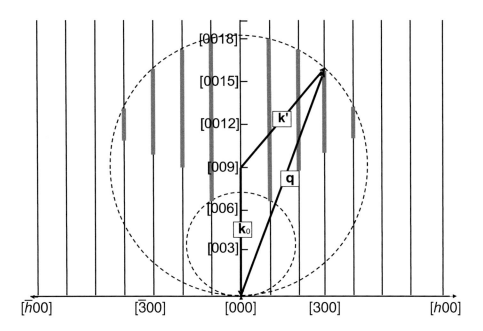

Figure 3.6 Ewald construction for the diffraction angles of the (h0) beams of the Ag(111) surface. The small and large dashed circles correspond to energies of 50 and 450 eV, respectively. The measurable range of the (h0) beams in this energy range is marked in grey on the reciprocal lattice rods. The Bragg points at the (00) beam are marked for reference; $\mathbf{q} = \mathbf{k}' - \mathbf{k}_0$ is the diffraction vector.

reciprocal lattice. LEED intensity analysis is therefore most sensitive to the z-component of the atom positions and less sensitive to the components parallel to the surface. This argument applies to the kinematic or single scattering contribution to the total diffracted intensity, but also partially to the multiple scattering contribution because the scattering amplitude has maxima in the forward and backward directions; see Figure 2.28.

3.3   Measurement of LEED Intensities

There are three basic steps in the measurement of LEED I(V) curves which we shall discuss in the following subsections.

3.3.1   Sample Preparation

While the experimental setup is relatively simple, the experiment itself is more complex to carry out. The quantitative measurement of diffracted intensities requires a planar and homogeneous surface of several square millimeters. An atomically flat and well-ordered surface is required, as reflected in a diffraction pattern with sharp spots and low background. If the surface is too rough, no diffraction pattern can be observed. It should also be possible to orient the sample to within 0.1°. The manipulator, therefore, should have two independent axes of rotation to align the sample. In many cases the preparation of the sample requires heating it to high temperatures for surface annealing and it is further desirable for intensity measurements to cool the crystal to liquid nitrogen or lower temperatures. To investigate insulating surfaces (which charge up under an incident LEED beam and thereby prevent or complicate diffraction), one may grow a thin insulating film on metal or semiconductor surfaces such that the conductivity is large enough to prevent the probe from charging.

Different methods are used to prepare clean surfaces in UHV depending on the material. The meaning of 'clean' must be defined first as it depends on the planned experiments and the experimental methods used to control the contamination. For LEED measurements on metal surfaces and adsorbate layers on metal surfaces, frequently AES is used to check that the surface is substantially free of contaminants after each preparation (except for H, which is not detectable by AES). It is also necessary to check that the surface is still not contaminated after measurement. The detection limit of AES is usually in the range of 10^{-3} monolayers. This is sufficient for LEED I(V) measurements but may not be sufficient for the preparation of well-ordered adsorbate layers. It should also be kept in mind that a sharp LEED pattern with low background does not guarantee a contamination-free surface. LEED from ordered surfaces is not very sensitive to a low concentration of defects. Therefore, the cleanliness is often controlled by scanning tunnelling microscopy (STM), with which technique single contaminant atoms may become visible. The disadvantage is that the area investigated by STM is usually much smaller than the area covered by the LEED beam. The cleanliness of a surface can also be very sensitively controlled by photoemission if this technique is available.

Different methods exist to prepare clean and well-ordered surfaces in UHV. Some materials, such as alkali halides, can be cleaved in UHV, while others can be cleaned by chemical reaction and desorption (e.g., oxidising C contaminants on the surface and subsequent desorption of CO is used to remove C from several metal surfaces or removing O on the surface by reaction with H_2 and subsequent desorption of H_2O). The main method used for preparing clean metal surfaces is ion bombardment (sputtering) and annealing. For ion bombardment, usually Ar^+ ions are used in the energy range between 500 and 1500 eV, because in this energy range the impact on the surface atoms is large enough to remove them from the surface. At higher energies the probability increases that Ar^+ ions penetrate into deeper layers and get stuck there. It depends on the material whether Ar can be removed from the surface region by annealing or not. This annealing often also brings other near-surface bulk impurities to the surface, requiring further cycles of ion bombardment and annealing and helping reduce later surface segregation of bulk impurities. With surfaces of bulk compounds, such as alloys and oxides, the aforementioned surface preparation methods can also produce non-bulk-like segregation and ordering of the bulk components near the surface, thereby altering its composition and structure.

More detailed discussions of ion bombardment are given by L. S. Dake et al. [3.7] and by H. Ibach [3.1] where the preparation of semiconductor surfaces is also summarised. A description of preparation methods of metal surfaces can be found in the overview given by C. Becker [3.8] and for transition metal compounds by C. Crotti et al. [3.9]. The preparation on thin oxide films has been described by M. Sterrer and H.-J. Freund [3.10].

3.3.2   Accurate Alignment

Normal incidence is preferred whenever the surface exhibits a rotational symmetry; when only a mirror plane or glide plane exists the incident beam should be in this plane. The proof of whether these conditions are fulfilled is made by comparing the intensities of beams that should be symmetrically equivalent. Control is usually achieved by comparing symmetrically equivalent beams at normal incidence. The general strategy is to adjust the sample orientation at first by comparing symmetrically equivalent I(V) curves at high energies (above 200 eV). The adjustment is made mechanically by means of the manipulator. It is required that the manipulator allow adjustment in two independent rotations. In a second step this position is kept and the adjustment at low energies is done by adjusting the magnetic field compensation. These steps should be repeated until finally the symmetrically equivalent I(V) curves agree in the complete energy range.

The incidence angle should be controlled with an accuracy of $<0.1°$. In Figure 3.7 the I(V) curves of five among six symmetrically equivalent beams of $Fe_2O_3(0001)$ are shown (the sixth was shadowed by the gun) [3.11]. The incident beam is aligned to the lattice plane, not to the surface normal. A slight miscut away from the lattice plane orientation does not produce significantly different intensities unless it becomes

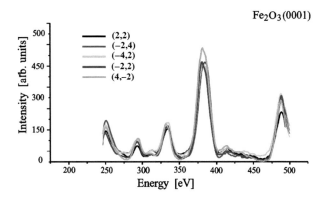

Figure 3.7 Intensities from five of six supposedly symmetrically equivalent beams near normal incidence from $Fe_2O_3(0001)$

visible by beam splitting or by an elliptical shape of the reflections. The differences in the peak maxima may be caused by remaining misalignment or inhomogeneities of the screen. These differences are tolerable: decisive are the matching peak energies, since these most directly affect atomic coordinates. The symmetrically equivalent beams are averaged together to further reduce errors and the averaged curve is used for the subsequent structural analysis.

It should be noted that the comparison of one beam between experiment and theory is not sufficient, as discussed in Section 6.1.2: the peak positions should match for all beams. In general, the I(V) curves of all beams visible on the screen should be measured. The average of symmetrically equivalent beams and the mean square deviation is then stored in a separate file to be used for the analysis. The mean square deviation can be used for the calculation of error estimates as discussed in Section 6.1.3.

There are cases where it is not possible to align the sample precisely (e.g., when the manipulator is not equipped with rotation axes or when the sample does not possess a symmetry which could be used for alignment). For example, in the case of stepped surfaces a mirror plane can often be used to align the incident beam in the mirror plane. This fixes the azimuthal angle φ of the incident beam while the polar angle ϑ (see Figure 3.8(a)) remains to be determined. It may not be possible to determine the polar angle precisely enough from the diffraction pattern: then the polar angle must be determined in the analysis by taking it as an additional parameter to be optimised.

In cases where the beams can be indexed and the lattice constants of the substrate are known, the incidence angle can be determined from the position of the beams on the screen. The method has been described previously by S. L. Cunningham and W. H. Weinberg [3.12] and by M. A. Van Hove et al. [3.13], and will only be summarised here.

The incident wave vector in the laboratory frame is $\mathbf{k}_l^i$ and in the crystal frame $\mathbf{k}_c^i$. The incident wave in the laboratory frame has only a z_l component and is given by

$$k_{lz}^i = k = \frac{\sqrt{2m_e E}}{\hbar}, \tag{3.1}$$

where E is the electron energy relative to vacuum zero, m_e is the free electron mass and $\hbar$ is Planck's constant. The transformation matrix which relates a vector $\mathbf{k}_c^i$ in the crystal frame to a vector $\mathbf{k}_l^i$ in the laboratory frame is given by

$$\mathbf{k}_c^i = \begin{pmatrix} -\sin\varphi & -\cos\vartheta\cos\varphi & \sin\vartheta\cos\varphi \\ \cos\varphi & -\cos\vartheta\sin\varphi & \sin\vartheta\sin\varphi \\ 0 & \sin\vartheta & \cos\vartheta \end{pmatrix} \mathbf{k}_l^i, \tag{3.2}$$

where $\varphi = -\psi - \pi/2$. In the crystal frame, the incident wave vector is given by

$$\begin{aligned} k_{cx}^i &= k\sin\vartheta\cos\varphi, \\ k_{cy}^i &= k\sin\vartheta\sin\varphi, \\ k_{cz}^i &= k\cos\vartheta. \end{aligned} \tag{3.3}$$

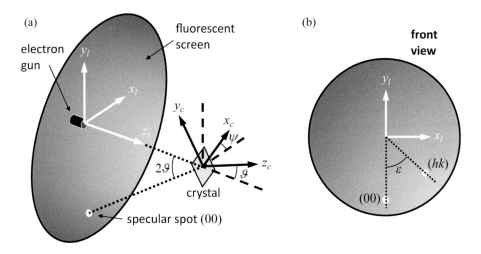

Figure 3.8 (a) Schematic of the LEED experiment showing the laboratory coordinate system (x_l, y_l, z_l), as related to the hemispherical fluorescent screen and electron gun, and the crystal coordinate system (x_c, y_c, z_c). The two systems are related by the angles ϑ and $\psi = -\varphi - \pi/2$. (b) Front view of the fluorescent screen showing the definitions of the laboratory y-axis and the angle ε for the (hk) LEED spot.

Upon scattering from the crystal, the (hk) diffracted beam has the wave vector components

$$
\begin{aligned}
k_{cx}^s &= k_{cx}^i + g_x(hk), \\
k_{cy}^s &= k_{cy}^i + g_y(hk), \\
k_{cz}^s &= -\sqrt{k^2 - \left(k_{cx}^s\right)^2 - \left(k_{cy}^s\right)^2}.
\end{aligned}
\tag{3.4}
$$

Here, $\mathbf{g}(hk)$ is the (hk) reciprocal lattice vector, and the z_c-component is determined by energy conservation. Finally, using the transpose of Eq. (3.2), the scattered wave components in the laboratory frame are given by

$$
\begin{aligned}
k_{lx}^s &= -k_{cx}^s \sin\varphi + k_{cy}^s \cos\varphi, \\
k_{ly}^s &= -k_{cx}^s \cos\vartheta \cos\varphi - k_{cy}^s \cos\vartheta \sin\varphi + k_{cz}^s \sin\vartheta.
\end{aligned}
\tag{3.5}
$$

From a LEED photograph, the angle $\varepsilon(hk)$ between the y_l axis and the (hk) spot (see Figure 3.8(b)) is related to the wave vector components by

$$
\tan\varepsilon(hk) = \frac{k_{lx}^s}{-k_{ly}^s}.
\tag{3.6}
$$

For a single spot, Eq. (3.6) is a single equation with two unknowns. Therefore, in principle, any two spots on the photograph [other than the (00) spot which defines the y_l axis] can be used to determine the angles ϑ and φ. For any two chosen spots, labelled $n = 1$ and 2, Eq. (3.6) represents two highly non-linear equations in two unknowns. These may be solved numerically by using Newton's method. The equations can be written as:

$$f_n(\vartheta, \varphi) = k_{lx}^s + k_{ly}^s \tan \varepsilon(hk) = 0, \ n = 1, 2. \tag{3.7}$$

For the i-th iteration, the (2×2) matrix equation

$$\begin{pmatrix} J_{1\vartheta} & J_{1\varphi} \\ J_{2\vartheta} & J_{2\varphi} \end{pmatrix} \begin{pmatrix} \Delta\vartheta \\ \Delta\varphi \end{pmatrix} = \begin{pmatrix} -f_1(\vartheta_i, \varphi_i) \\ -f_2(\vartheta_i, \varphi_i) \end{pmatrix} \tag{3.8}$$

is solved numerically for $\Delta\vartheta$ and $\Delta\varphi$, where J is the Jacobian matrix given by, for example,

$$J_{1\vartheta} = \left. \frac{\partial f_1(\vartheta, \varphi)}{\partial \vartheta} \right|_{\vartheta_i, \varphi_i} , \text{ etc.} \tag{3.9}$$

New values of ϑ and φ are determined by

$$\begin{aligned} \vartheta_{i+1} &= \vartheta_i + \Delta\vartheta \\ \varphi_{i+1} &= \varphi_i + \Delta\varphi, \end{aligned} \tag{3.10}$$

and the procedure is repeated until $\Delta\vartheta$ and $\Delta\varphi$ are less than, for example, 10^{-3} rad. This numerical procedure is quite rapid due partly to the fact that the Jacobian is analytic. One advantage of this technique is that the angle $\varepsilon(hk)$ does not depend on the location of the crystal: the angle $\varepsilon(hk)$ is invariant as the crystal is moved along the axis of the hemispherical screen (assuming that the electron beam is collinear with this axis). It is important, however, that the camera also be located on this axis, so care must be taken in this alignment. An example for the determination of the incidence angle at Ir(111) is given in [3.12].

This method of determining the angles of incidence in a LEED experiment is both accurate and simple. The method is accurate because the angles are determined many times from a single photograph. Absolute accuracies of less than $0.1°$ should be attainable routinely. The method is simple in that it requires only a camera. In addition, there is no need for a calibration point (such as normal incidence) to be established. However, it should be noted that the above procedure is applicable only when the electron gun is collinear with the axis of the camera. A related method was developed by G. P. Price [3.14] to determine the incident and azimuthal angles by treating the position of the electron gun and the electron energy as unknowns and using at least three LEED spots, including the specular (00) beam. This method, however, requires the crystal to be positioned precisely at the centre of curvature of the LEED screen. In addition, both methods require that the incident electron beam be aligned with the centre of curvature of the LEED screen. Under special circumstances when the crystal is off-centre or the electron gun is misaligned, a more general method developed by A. C. Sobrero and W. H. Weinberg [3.15] should be used. This method can be used to check the alignment of the electron gun and the position of the crystal, in addition to determining the polar and azimuthal angles of incidence.

A similar method, also based on the knowledge of the beam indices, has been described by F. Sojka et al. [3.16; 3.17] to correct deviations from normal incidence for the precise determination of lattice constants with high-resolution LEED systems.

Further methods have been developed to determine the angles of incidence which are especially useful for studies with SPA-LEED when a second external electron gun and a large angle of incidence are used [3.18]. Such an arrangement allows the observation of morphological changes during homo- or hetero-epitaxial growth and the measurement of the strain state at different coverages. The precise knowledge of the incidence angle is essential in such studies. The authors describe four different cases where the polar angle ϑ of the incident beam can be determined. The first method is based on the fact that in the reciprocal net a constant distance exists between beams of a Laue zone while the incidence angle can be derived from the measurable distance on the screen (see Figure 3.9).

This method has the advantage that no knowledge about the lattice constants nor about the beam indices is required.

The second and third case apply to growth experiments and rough surfaces. The method utilises the fact that steps on surfaces cause an oscillation of the beam width and of the peak intensity of the (00) beam with energy. This effect is used to observe layer-by-layer growth in epitaxial growth but occurs on every surface with a non-zero step density. It is required that the interlayer distance be known. Maxima of the peak

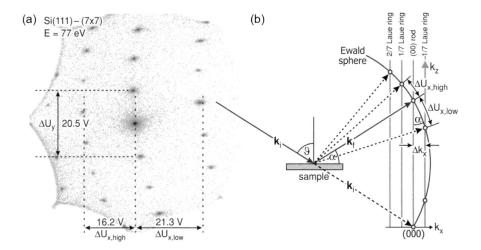

Figure 3.9 (a) LEED pattern for Si(111)–(7×7), recorded with the external electron gun at 77 eV. The intensity is displayed on an inverse logarithmic scale. The distances $\Delta U_{x,low}$, $\Delta U_{x,high}$ and ΔU_y between the Laue rings and the first 1/7 peaks are indicated. (b) Ewald construction for large angles of incidence for an external electron gun geometry. The scan is located on the Ewald sphere, where the y direction of the LEED picture is perpendicular to k_z and k_x. Reprinted from [3.18] C. Klein, T. Nabbefeld, H. Hattab, D. Meyer, G. Jnawali, M. Kammler, F.-J. Meyer zu Heringdorf, A. Golla-Franz, B. H. Müller, Th. Schmidt, M. Henzler and M. Horn-von Hoegen, *Rev. Sci. Instrum.*, vol. 82, p. 035111, 2011, with the permission of AIP Publishing.

intensity occur at the Bragg points, while minima occur between the Bragg points. The position of the maxima and minima on the energy scale depends on the normal component S of the momentum transfer:

$$S = 2d\left(k'_z - k^0_z\right) = 2d|\mathbf{k}|\cos\vartheta \tag{3.11}$$

where d is the normal distance between layers and $S = n2\pi$. At integer values of n, the peak intensity of the (00) beam has maxima, as illustrated in Figure 3.10. The angle of incidence ϑ can be derived from the fit of the energy dependence on the peak intensity. This is, by the way, not influenced by the inner potential by which the energy of the electron beam inside the crystal is increased, since the interference between the diffracted beams from different terraces occurs outside the crystal [3.19]. The fourth case applies to faceting of the surface after deposition of an adsorbate. From the

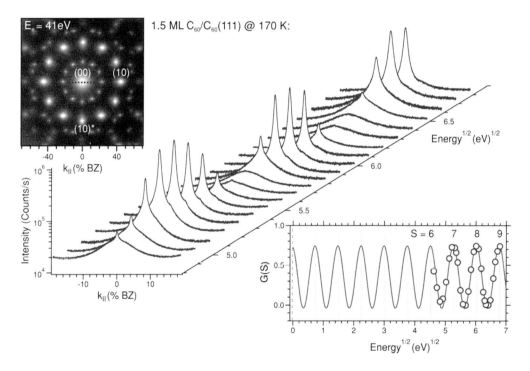

Figure 3.10 Upper left: LEED pattern of a rough $C_{60}(111)$ surface at 41 eV. Centre: (00) spot profiles as a function of the electron energy. The periodic variation between sharp spots at the 'in-phase' or Bragg condition and broad spots at the 'out-of-phase' or anti-Bragg condition is clearly visible. Bottom right: The ratio of the central spike intensity to the total intensity $G(S)$ is plotted as a function of the square root of the electron energy. The order of the Bragg conditions (integer scattering phase S) is determined from the oscillation period and the boundary condition at zero energy. Reprinted from [3.18] C. Klein, T. Nabbefeld, H. Hattab, D. Meyer, G. Jnawali, M. Kammler, F.-J. Meyer zu Heringdorf, A. Golla-Franz, B. H. Müller, Th. Schmidt, M. Henzler and M. Horn-von Hoegen, *Rev. Sci. Instrum.*, vol. 82, p. 035111, 2011, with the permission of AIP Publishing.

intersection of the (00) rods of two facets the orientation of the facet and the angle of incidence can be derived.

3.3.3   Measurement of LEED Intensities with a Video System

With a rear-view LEED system and a fluorescent screen, the diffracted intensities are frequently measured by using a video system. The resulting beam intensities are typically reported on an arbitrary scale but must be normalised to the intensity of the primary beam. The manufacturer usually provides a program to measure I(V) curves of selected beams. There is also similar software available from independent companies. The intensity is integrated over an area around the centre of the spot and the background is subtracted; this background is taken as the average intensity on the rim of an area surrounding the peaks, scaled by the area of the peak. The area of the peak can be set in the program. An improved procedure for the extraction of diffraction intensities from LEED images uses a 2-D fit of the peak with a Gaussian profile and interpolation of the background; see Figure 3.11, as described by K. M. Schindler et al. [3.20]. The improvements in the removal of background intensity and the tracking of spot positions are shown in Figure 3.12 together with the I(V) curve obtained by a simple background subtraction. The precision and accuracy of the LEED I(V) analysis are substantially increased using this evaluation of LEED images.

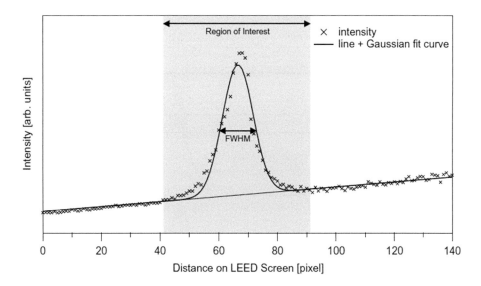

Figure 3.11 Line profile through a diffraction spot, with the fit curve (Gaussian on a line) and the derived region of interest (shaded band); the abscissa, representing angle of diffraction, has arbitrary units. Reprinted from [3.20] K. M. Schindler, M. Huth and W. Widdra, *Chem. Phys. Lett.*, vol. 532, pp. 116–118, 2012, with permission from Elsevier.

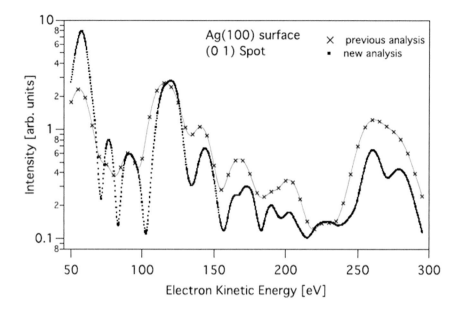

Figure 3.12 Experimental LEED I(V) curves of the clean Ag(001) surface for the (01) diffraction spot. Line marked with ×: simple background subtraction. Dotted line: fit background and subtraction showing a better resolution and a larger dynamic range. The new analysis has been performed with an improved integration of the spot intensity and subtraction of the background. Reprinted from [3.20] K. M. Schindler, M. Huth and W. Widdra, *Chem. Phys. Lett.*, vol. 532, pp. 116–118, 2012, with permission from Elsevier.

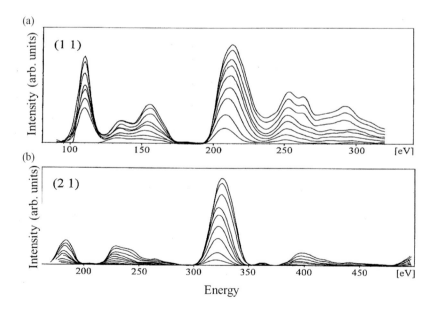

Figure 3.13 Experimental I(V) curves for two diffracted beams (upper and lower panels, respectively) from Cu(110) at various temperatures increasing from 110 K (highest curves) to 465 K (lowest curves). Reprinted from [3.21] W. Moritz, J. Landskron and M. Deschauer, *Surf. Sci.*, vol. 603, pp. 1306–1314, 2009, with permission from Elsevier.

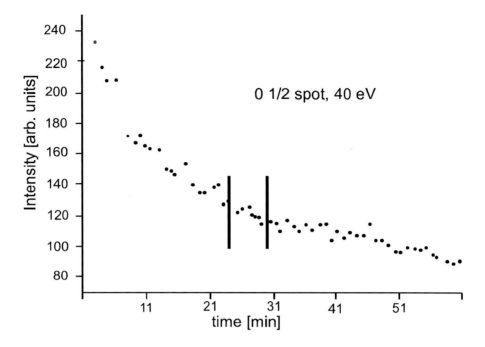

Figure 3.14 Intensity decay due to electron beam damage for thiouracil on Ag(111). The two vertical bars indicate the period chosen for the measurement. Reprinted from [3.21] W. Moritz, J. Landskron and M. Deschauer, *Surf. Sci.*, vol. 603, pp. 1306–1314, 2009. With permission from Elsevier.

The sample should be cooled to liquid nitrogen temperature or lower if possible, because LEED intensities are considerably weakened by thermal vibrations, especially at higher energies. As shown in Figure 3.13, the intensity decreases drastically with increasing temperature (while the background increases, not shown). The measurement at low temperatures exhibits more details in the I(V) curves and less uncertainty due to background subtraction. Also, even more importantly, the calculation of LEED intensities involves rough approximations for vibration amplitudes so that calculated intensities show worse agreement with experiment and higher error bars at higher temperatures.

A check should always be performed by repeating the measurement and evaluating the reproducibility. The sample alignment usually takes much more time than performing the measurement directly after preparation. Therefore, the sample should be prepared afresh and adjusted to the previously determined sample position such that the measurement can be started quickly. Figure 3.14 shows the intensity decay of a superlattice spot of a thiouracil monolayer on Ag(111) due to electron beam damage. The measurements were made after the initial rapid decay, so that during the time of the measurement only a relatively small intensity loss occurred [3.22].

The current in the diffracted beam is usually on the order of a fraction 10^{-2} to 10^{-4} of the primary beam and can be measured with a video camera or directly with a microchannel plate. Modern video techniques allow the measurement of a complete set of I(V) curves in less than 5 minutes.

3.4   Instrumental Response Function

The diffraction spots in the LEED pattern are broadened by two effects: imperfections in the surface and the instrumental response function; the latter concept was introduced by R. L. Park et al. [3.23] to describe the resolution of the LEED instrument. The instrumental response function has several contributions, among which are the energy spread and the angular width of the primary beam, as well as the finite resolution of the detector or fluorescent screen. The convolution of these different contributions leads to an instrumental broadening of the LEED beams which limits the range of defect distributions that can be detected by a beam profile analysis [3.24–3.26].

The instrumental response function represents the beam profile which would be measured from an infinite perfect crystal. The measured angular beam profile may be denoted by $J(\mathbf{q})$, where $\mathbf{q} = \mathbf{k}' - \mathbf{k}$ is the scattering vector. It is given as the convolution of two parts, the beam profile due to imperfections of the surface, $I(\mathbf{q})$, and the instrumental broadening $T(\mathbf{q})$:

$$J(\mathbf{q}) = T(\mathbf{q}) \otimes I(\mathbf{q}) \equiv \int T(\mathbf{q}')I(\mathbf{q}' - \mathbf{q})d\mathbf{q}'. \tag{3.12}$$

The Fourier transform of a convolution is given by the product of the Fourier transforms of the single components and therefore

$$F\{J(\mathbf{q})\} = F\{T(\mathbf{q})\}\Gamma\{I(\mathbf{q})\} = t(\mathbf{u})P(\mathbf{u}). \tag{3.13}$$

The quantity $t(\mathbf{u}) = F\{T(\mathbf{q})\}$ is defined as the transfer function. The full width at half maximum (FWHM) is defined as the transfer width [3.23]. The Fourier transform of the diffracted intensity $I(\mathbf{q})$ is the Patterson function $P(\mathbf{u})$, that is, the autocorrelation function of the crystal structure; $\mathbf{u}$ is the distance vector between two scatterers. The autocorrelation function $P(\mathbf{u})$ is a measure of the number of pairs of scatterers that are separated by a vector $\mathbf{u}$. It is a vector in real space; we use the symbol $\mathbf{u}$ to distinguish the distance vector from a position vector $\mathbf{r}$ of the atoms. From $P(\mathbf{u})$, information can be derived about the periodic structure and, in the case of disordered surfaces, about the defect distribution. In principle, $P(\mathbf{u})$ could be determined if $T(\mathbf{q})$ and $J(\mathbf{q})$ could be measured accurately. However, in practice, both $T(\mathbf{q})$ and $J(\mathbf{q})$ have uncertainties associated with them and the function $I(\mathbf{q})$ cannot be determined with a width less than the sum of the uncertainties in the widths of $T(\mathbf{q})$ and $J(\mathbf{q})$.

Equation (3.13) implies that the correlation between two scatterers cannot be detected in the range in which $t(\mathbf{u})$ is zero, normally for larger distances. Hence, the transfer width is regarded as the effective coherence length of the instrument.

However, it does not represent a limit on the region over which phase correlation exists or can be detected, as has been discussed by G. Comsa [3.27].

The angular divergence of the primary beam of most of the commercially available LEED systems is about 0.5°, together with a limited energy resolution of 0.1–0.2 eV. The transfer width determined by measuring the angular profile of a diffracted beam varies typically between 2 and 10 nm. This resolution is sufficient for structure analysis but is poor for the analysis of defect distributions. Therefore, the observation of an apparently 'good' LEED pattern in a conventional LEED system with sharp spots and low background does not guarantee a low concentration of defects or a surface free of contaminants. For the analysis of defect distributions, high-resolution LEED systems have been developed with a transfer width of up to 400 nm [3.4; 3.28; 3.29], as discussed in Section 3.6.1.

3.5   Brief Overview of Types of Available LEED Systems

LEED systems are commercially available, leading to their widespread use in surface science. A number of systems are offered by different companies; we describe only some types here. It is not our intention to review all available systems, nor to make recommendations.

There are mainly three different types of systems on offer. Widely used are display systems with a fluorescent screen; for example, Figure 3.15 shows a system from SPECS, which is also equipped with a shutter for protection of the screen. The system is based on a design developed by P. Heilmann et al. [3.3] in the group of K. Heinz and K. Müller, who also pioneered the development of video measurements of LEED intensities.

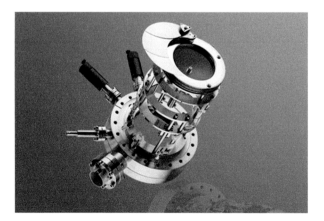

Figure 3.15 LEED system with a phosphorescent screen, ErLEED 100/150, Reverse View LEED SPECS. Reprinted from [3.30] www.specs.de, with permission from SPECS GmbH, Berlin, Germany.

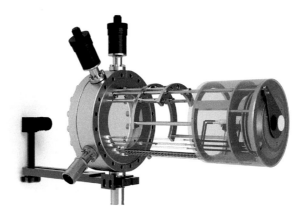

Figure 3.16 Microchannel plate system (MCP-LEED), LEED 800 MCP, LEED for Organic Films & Molecular Beam Epitaxy (Scienta Omicron, Uppsala, Sweden and Taunusstein, Germany). Reprinted from [3.31] www.scientaomicron.direct/96/electron-diffraction/femto, with permission.

The primary beam current used with the fluorescent screen is usually on the order of 100 nA, which corresponds to about 0.1 electron per surface atom per second. Materials which are sensitive to electrons and insulating surfaces require lower currents. For this purpose, a microchannel plate (MCP) for intensifying the diffracted beams can be used in front of a fluorescent screen. The primary current can be as low as 5 nA with a single microchannel plate and down to 100 pA with a double microchannel plate. The LEED pattern can then be viewed on a fluorescent screen behind the microchannel plate(s). The aberrations in the LEED pattern caused by the use of a planar screen and microchannel plates can be eliminated by correction field plates. An example of a microchannel plate system from Scienta Omicron is shown in Figure 3.16 [3.31].

A fluorescent screen is usually combined with a CCD camera to measure the LEED intensities from the image on the screen. To obtain high quality LEED I(V) data a highly-sensitive 12-bit digital camera is required. However, this is not necessary when a position sensitive detector is used behind the microchannel plate. The direct measurement with a delay line detector allows an even lower current down to the fA range, so the digital image can be directly stored and no camera is needed. The low current allows a small focus of the primary beam and the investigation of insulating surfaces. A model from OCI Vacuum Microengineering is shown in Figure 3.17 [3.32].

The use of microchannel plates and position sensitive detectors has the disadvantage that these components are usually sensitive to venting the UHV system, due to contamination by air, which is not the case for the fluorescent screen.

Systems have also been developed for special applications, for example a LEED system combined with evaporation sources to observe in situ the structures formed during adsorption; this capability is mainly applied to the study of organic molecules [3.33].

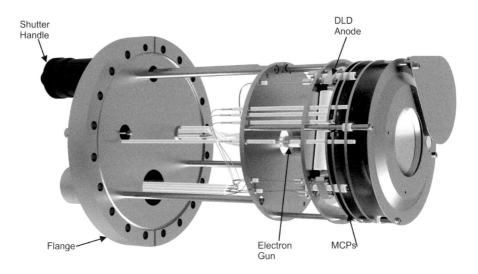

Shutter
Handle

DLD
Anode

Flange

Electron
Gun

MCPs

Figure 3.17 FemtoLEED spectrometer DLD-L800 with integral shutter (OCI Vacuum Microengineering). Reprinted from www.ocivm.com, with permission from OCI Vacuum Microengineering, Inc.

3.6   High-Resolution Instruments

As mentioned in Section 3.4, the transfer width of most LEED systems is in the range of about 2–10 nm. This means that such systems cannot detect spatial correlations between scatterers spaced farther apart than about 2–10 nm; thus, spatial ordering (in particular crystallinity, such as domain and island sizes) cannot be resolved over larger distances: this is considered to be low resolution. High-resolution instruments have transfer widths up to 400 nm or even more, corresponding to very narrow reflection profiles, provided the sample is perfect. Instruments producing a direct image of the surface are LEEM (low-energy electron microscope) and PEEM (photoemission electron microscope). We discuss LEEM in Section 3.6.2. With LEEM a resolution of 5 nm can be reached due to substantial improvements of the electron lenses since the first instruments were built. Even higher resolution may be possible in the future. The direct imaging of STM and AFM produces much higher spatial resolution (on the order of many microns) and is not considered here.

3.6.1   Spot Profile Analysis LEED (SPA-LEED)

To study the distribution of surface defects with high resolution requires specially designed electron guns and collimating systems known as high-resolution LEED (HR-LEED) or SPA-LEED. Spot profile analysis was developed by U. Scheithauer

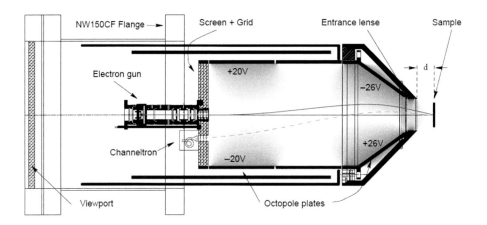

Figure 3.18 Horizontal cut through a third generation SPA-LEED with conical shape. Shown are the electron gun, the channeltron detector, the electrostatic deflection unit, the entrance lens and the sample in a position with distance d. The potential of the electrostatic deflection field is shown in a grey scale representation. The path for the incident electron beam and the path of those electrons which are recorded in the detector are shown. During scanning across the sample, the spot position on the sample stays constant. The scan is done by moving the focus at the detector. Republished with permission of Walter de Gruyter and Company, from [3.34] M. Horn-von Hoegen, *Z. Krist.*, vol. 214, pp. 591–629, 1999; permission conveyed through Copyright Clearance Center, Inc.

et al. [3.4] in the group of M. Henzler. The SPA-LEED instrument is a high-resolution LEED system with an octopole lens to collimate the primary and diffracted beams. The instrument and applications have been described in detail by M. Horn-von Hoegen [3.34]. The instrument has an internal electron gun, but an external gun can be applied to allow in-situ observation of growth processes. Commercially available is a system which reaches a transfer width of up to 400 nm. The dynamic range is 100-times higher than in conventional LEED systems. A schematic drawing of a cut through the instrument is shown in Figure 3.18 and a view of the new commercially available system is shown in Figure 3.19. Scienta-Omicron no longer produces the SPA-LEED system; however, a newly-designed system and new software are available from a new company [3.35].

The diffracted intensities are measured with a channeltron and the diffraction pattern is electronically stored. The spot profiles can be evaluated. An example of the diffraction pattern for the Si(111)-(7×7) reconstruction is shown in Figure 3.20, demonstrating the high resolution.

The diffraction intensities measured with the channeltron are not suited for I(V) analysis since the incidence angle is not constant but different for each spot. Some I(V) measurements are possible, nevertheless. The instrument has a phosphorescent screen, on which the diffracted intensities can be observed and measured with a video camera keeping the incidence angle constant; see Figure 3.21. The aperture of the

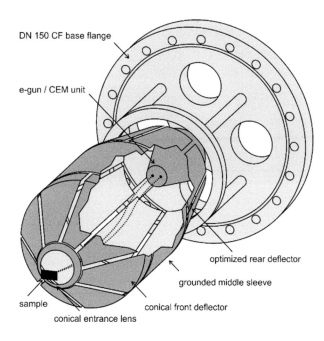

Figure 3.19 View of the new SPA-LEED system mounted on an 8″ flange. The fourth generation of the SPA-LEED system shows a highly symmetric design with round octopole deflectors and a compact and centred e-gun/CEM unit in order to minimise the image distortions and to increase the angular scan range. With permission from [3.35] Dr. Peter Kury, out-of-the-box systems GmbH.

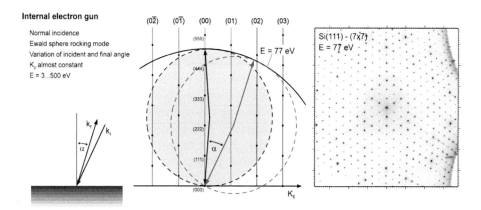

Figure 3.20 Diffraction geometry in reciprocal space for the SPA-LEED internal electron gun. The angles of the incident electron beam and of the diffraction pattern are both varied continuously in order to record the LEED pattern. The relative angle between incident and final scattering vectors stays constant. As a result, the recorded diffraction pattern is described by the envelope of the rocking Ewald sphere, that is, a sphere centred at the origin of the reciprocal space with a diameter twice the size of the Ewald sphere. The high-resolution pattern shown on the right is for Si(111)-(7×7) at 77 eV. Typical electron energies are 30–120 eV. Republished with permission of Walter de Gruyter and Company, from [3.34] M. Horn-von Hoegen, *Z. Krist.*, vol. 214, pp. 591–629, 1999; permission conveyed through Copyright Clearance Center, Inc.

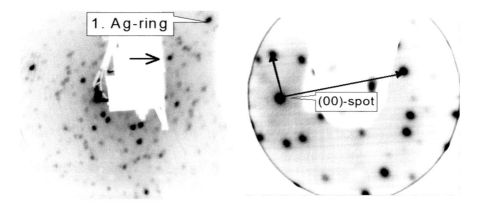

Figure 3.21 Diffraction patterns from thiouracil/Ag(111) from different LEED systems. Left panel: conventional LEED system; right panel: SPA-LEED system. Both images are taken from the same sample at 50 eV and 147 K. In the right panel, most of the first ring of 12 superlattice spots around the specular beam is visible on the left; in the left panel the specular beam and this ring are obscured by the specimen holder (bright object). The corresponding rings around three substrate reflections appear weakly in the left panel. The arrow marks the (0,1/2) spot. Reprinted from [3.21] W. Moritz, J. Landskron and M. Deschauer, *Surf. Sci.*, vol. 603, pp. 1306–1314, 2009. With permission from Elsevier.

screen is not as large as with conventional systems, but the high collimation of the primary beam allows measurements which cannot be resolved by a conventional LEED system.

3.6.2   Low-Energy Electron Microscopy (LEEM)

High resolution is also achieved by low-energy electron microscopy (LEEM), which allows the observation of the direct image as well as measurement of diffracted intensities; see Figure 3.22. The instrument was developed by W. Telieps and E. Bauer [3.5] in 1985. Several instruments have been built and are commercially available (see the link within [3.5] for further references and information about the capabilities of the LEEM instrument and its applications). The resolution achieved with LEEM can be as low as about 5 nm and allows the measurement of diffraction intensities from small areas and single domains, which is usually not possible with conventional LEED systems. Dark field images using the superlattice beams show the individual domains; an example is given in Figure 3.23 for Si(100) where the two domains with (2×1) and (1×2) reconstruction in different terraces are made visible.

The specular beam at normal incidence can be measured by the deflection of the primary and diffracted beam through the beam splitter. A further advantage is the fixed spot position on the screen with varying energy. Instruments equipped with additional experimental facilities are: SPELEEM (spectroscopic measurements),

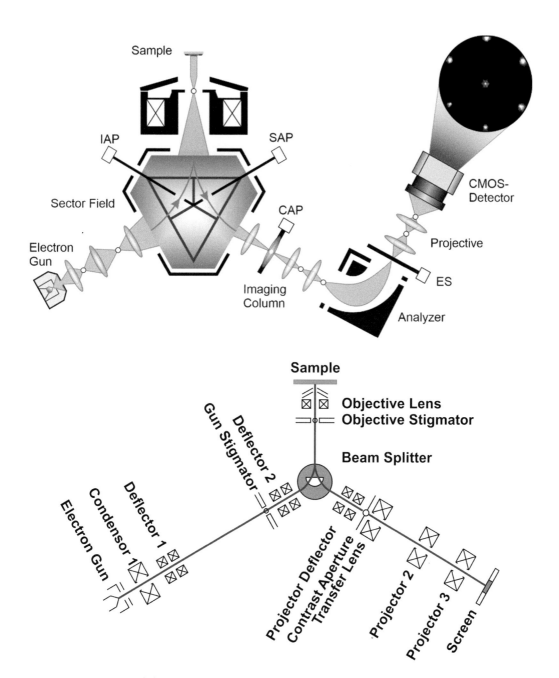

Figure 3.22 Schematic drawings of two LEEM instruments. The upper panel shows the beam path with an energy analyser in the diffracted beam. The diffraction pattern is from the reconstructed Au(111) surface and shows the satellite reflections around the (00) spot. Courtesy of F.-J. Meyer zu Heringsdorf, University Duisburg-Essen, with permission.

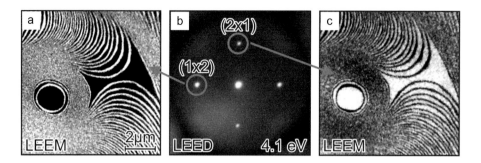

Figure 3.23 Dark field LEEM images from (2×1) and (1×2) domains of Si(100). The diffraction pattern shows the (00) spot and the $(0, \pm1/2)$ and $(\pm1/2, 0)$ spots of the two domains: the selected spot highlights its corresponding domain orientation. Reproduced with permission from [3.36] P. Kirschbaum, L. Brendel, K. R. Roos, M. Horn-von Hoegen and F.-J. Meyer zu Heringdorf, *Mater. Res. Express*, vol. 3, p. 085011, 2016 (Open Access).

SPLEEM (spin polarised), and instruments that allow both spectroscopic and spin resolved measurements.[1]

3.7   Some Developments for Special Applications

3.7.1   In-situ Observation of Adsorption Processes

Some conventional LEED instruments have been modified in the past two decades for special applications and improved measurement methods. A system allowing the study of sensitive surface structures with low electron doses using a slow scan CCD camera has been developed by G. Held et al. [3.37]. The observation of the diffraction pattern during deposition of adsorbates can be used to study phase transformations and phase transitions. An arrangement which allows real time monitoring of phase transitions of vacuum deposited organic films has been described by C. Seidel et al. [3.33]. Frequently, there is a need to prepare surfaces and monitor them by LEED, prior to further investigation in a different chamber without breaking the vacuum; this is achieved by a transport chamber for interconnection with other experiments [3.38].

3.7.2   Nanostructures

A compact UHV system has been developed in order to fabricate and analyse micro- and nanostructures on surfaces in situ [3.39]. The observation of nanostructures by LEED requires a LEED beam with a very small focus. The diameter of the primary beam usually depends on energy and beam intensity. It is difficult to focus a

[1] For commercially available instruments see: https://elmitec.de and www.specs-group.com/nc/specsgroup/knowledge/applications/productlines/detail/leempeem/, 2021.

low-energy electron beam because of the space charge effect which broadens the beam. The width usually varies from about 500 μm at low energies to 100 μm at higher energies and depends on the design of the electron gun. The system reported by Y. Kakefuda et al. [3.39] includes a low-energy electron gun which provides a minimum spot size of about 25 nm using electrostatic lenses, AES, EELS and a LEED optic.

3.7.3 Convergent Beam LEED

The transfer width of a conventional LEED system can be substantially improved from 10 nm to about 70 nm by a modulated beam combined with a time resolved collection of LEED images [3.40]. A convergent beam LEED (CBLEED) technique has been proposed [3.41] in analogy to the related technique in X-ray diffraction. The diffraction disks, replacing sharp spots, show the angular dependence of the diffraction intensity. The use of an STM tip as electron source could produce LEED patterns from areas as small as 50 nm.

3.7.4 Ultrafast Measurements

Recently RHEED has been used in combination with a pulsed photocathode to study the dynamics of ultrafast processes at surfaces [3.42]. The surface is excited by

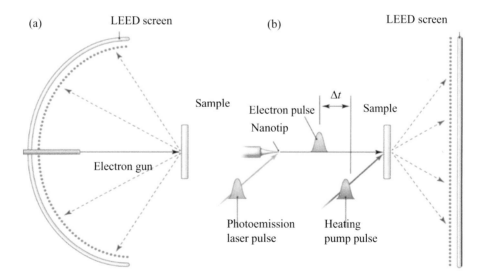

Figure 3.24 (a) Conventional LEED works by reflecting electrons back to a phosphor screen. (b) Transmission ultrafast LEED (T-ULEED). From [3.47] E. T. J. Nibbering, *Science*, vol. 345, pp. 137–138, 2014, https://doi.org/10.1126/science.1256199. Reprinted with permission from AAAS (online readers may view, browse and/or download material for temporary copying purposes only, provided these uses are for noncommercial personal purposes. Except as provided by law, this material may not be further reproduced, distributed, transmitted, modified, adapted, performed, displayed, published or sold in whole or in part, without prior written permission from the publisher).

femtosecond laser pulses and the transient changes of surface structure (e.g., upon excitation of a phase transition) is followed in a stroboscopic pump probe scheme [3.43; 3.44]. A temporal resolution of less than 330 fs has been achieved with a pulsed electron gun at high energies [3.45]. M. Gulde et al. [3.46] achieved time resolution in the picosecond regime with a pump-probe configuration in their transmission ultrafast LEED (T-ULEED) experiment with electron pulses lasting a few picoseconds; see Figure 3.24(b). Transmission electron diffraction patterns were obtained from bilayer graphene covered with PMMA (polymethylmethacrylate) at an electron energy of 1 keV [3.47].

The electron pulses are generated from a sharp tungsten tip (50 nm radius of curvature) by illumination with a second-harmonic pulse originating from the same laser output as the laser pump pulse [3.48]. This method ensures accurate time delay between the two pulses.

4 Interpretation of the Diffraction Pattern

LEED has found widespread application in surface science, since the LEED experiment can be performed in a small laboratory and LEED systems are commercially available. A main advantage compared to surface X-ray diffraction is that on the LEED screen most of the 2-D diffraction pattern is visible, thus allowing a quick and comprehensive overview of the symmetry and to some extent about the degree of ordering of the surface under examination. A LEED system is therefore included in most UHV chambers to control the quality of the surface preparation for a wide range of surface studies. A qualitative interpretation of the diffraction pattern is the most common use of LEED: it allows the identification of the surface unit cell, the estimation of the degree of ordering and the identification of different surface phases in adsorption systems (and thereby often a check on adsorbate coverage). The diffraction pattern thus reflects the translational symmetry and the crystalline order of the surface.

In most cases, the substrate lattice constants are precisely known (from bulk X-ray diffraction) and, for commensurate superlattices (due to adsorption or reconstruction), the lattice constants of the superlattice are well defined by the substrate lattice. From the diffraction pattern the size and orientation of the superlattice cell can be derived. However, the point symmetry (rotation axes, mirror planes, etc.) of the superlattice cell cannot always be uniquely determined due to the frequent co-existence of domains (domains on surfaces are finite 2-D 'crystallites' whose identical periodic lattices do not mutually match up). This point is discussed in Section 4.1.

In other cases, incommensurate superlattices may occur where a precise determination of the periodicity of the superlattice is desirable, for example, when epitaxial layers are thick enough such that the substrate is no longer visible in the diffraction pattern. These layers may be commensurate or incommensurate to the substrate or distorted due to strain in the layer. Here the indexing of the diffraction spots and a precise measurement of the reciprocal lattice constants may be necessary, which requires a precise determination of the incidence angle and a correction of distortions of the LEED pattern. Examples will be shown in Sections 4.2 and 4.3.

Real surfaces frequently exhibit one or more kinds of defects or disorder that can affect the diffraction pattern. More significantly, imperfections can be interesting in their own right, as well as important in terms of causing or strongly modifying useful physical, chemical and biological effects, from colour centres in materials science to catalysis in chemistry and biology. Such imperfections include: disordered overlayers (such as islanding, domains, lattice gas layers, phase transitions, randomly oriented or spinning molecules and nanocrystallites); point defects (including disordered vacancies and substitutional atoms); line defects (e.g., steps and grain boundaries); faceting (restructuring to expose multiple crystallographic orientations); quasicrystalline and modulated structures; amorphous surfaces (i.e., non-crystalline materials and melting surfaces); and thermal disorder (including atomic vibration and molecular libration).

Most types of disorder and defects generate diffuse intensity distributions in the diffraction pattern; the diffuse intensities often co-exist with sharp diffraction features due to ordered parts of the surface, for example, for a disordered overlayer on a crystalline substrate. Some of these cases have been subjected to LEED interpretation and even structural analysis; they are further discussed in this and other chapters as follows.

The effect on diffraction patterns of antiphase and out-of-phase boundaries between domains in overlayers with superlattices is treated in Section 4.6: periodic arrays of domain boundaries can be identified by a characteristic splitting of spots. Diffraction patterns due to twin domains (especially rotated domains) are considered in Section 4.1.2. Modulated structures, discussed in Section 4.7, generate distinctive diffraction patterns due to the superposition of commensurate or incommensurate lattices of adsorbate layers on substrates; a common example of such structures is a graphene layer on transition metals and other substrates [4.1]. The case of stepped surfaces, including kinked steps, is described in detail in Section 4.4: sufficiently well-ordered stepped and kinked surfaces can be reproducibly prepared and their diffraction patterns can be qualitatively interpreted; regular step and kink arrays lead to characteristic diffraction patterns. Section 4.5 discusses diffraction patterns of faceted surfaces: some metal surfaces decompose into different facets after adsorption of reactive gases; faceted surfaces then exhibit two or more (00) specular reflection spots.

Going beyond the diffraction pattern, multiple scattering in nanoparticles is described in Section 5.2.4. The theoretical treatment in LEED of disordered overlayers and point defects at crystalline surfaces is presented in Section 5.4.3. The LEED theory of various types of thermal effects is discussed in Section 5.5. The diffraction patterns and multiple scattering in quasicrystals and modulated structures are described in Sections 6.2 and 6.3, respectively, where further useful information about atomic positions is obtained.

4.1   Symmetry and Orientation of the Unit Cell

We first discuss the symmetry of the diffraction pattern. At normal incidence the diffraction pattern on the LEED screen gives directly the image of the 2-D reciprocal lattice and exhibits the point symmetry of the surface structure (rotation axes, mirror planes, etc.). We assume the screen to be spherical and the crystal to be positioned in the centre of the sphere (cf. Figure 4.1(a)).

The diffraction pattern may exhibit an apparent symmetry which is higher than that of the actual surface due to the existence of domains (cf. Figure 4.1(b)). The point

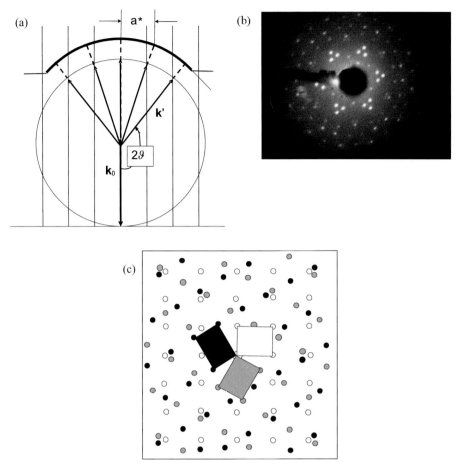

Figure 4.1 (a) Schematic drawing of the diffraction geometry at normal incidence. (b) LEED pattern of NTCDA (1,4,5,8-naphthalene-tetracarboxylic-dianhydride) on Ag(111) at 30 eV, 123 K. NTCDA forms a $\begin{pmatrix} 6 & 3 \\ 0 & 4 \end{pmatrix}$ superstructure; the symmetry is p1 [4.2]. (c) Diffraction patterns due to six domains with three 120° rotated orientations (shown in the figure) and their mirrored counterparts occur on the Ag(111) substrate which has symmetry p3m1.

symmetry of the diffraction pattern is that of the substrate if domains occur with overall equal areas; otherwise only the spot *positions* exhibit that symmetry, while the spot *intensities* do not. Two types of domains can be distinguished: antiphase domains and twin domains, defined as follows.

Antiphase domains are characterised by a shift vector between equivalent ordered domains and frequently occur when the superlattice unit cell is larger than that of the substrate: they are often called translational domains. Antiphase domains do not change the point symmetry of the diffraction pattern; however, short-range order of the antiphase domain boundaries may become visible in the diffraction pattern by characteristic broadening of some beams. Regular arrays of antiphase domains may become visible in the diffraction pattern as split spots, to be described in Section 4.6.

Twin domains also frequently occur in many structures that have a superlattice. If the point symmetry of the structure is lower than that of the substrate, then symmetrically equivalent domains can occur, called twin domains. Such domains are related by a rotation or a mirror operation, possibly in connection with a shift vector: these types of twin domains are often called rotational and mirrored domains, respectively. They occur in adsorbate layers as well as at ideally terminated clean surfaces, in the latter case due to the existence of steps if the bulk structure has glide vectors or screw axes normal to the surface.

Software is available to allow simulating experimentally observed LEED spot patterns and to help determining the corresponding surface unit cell for substrates and overlayers [4.3].

4.1.1   Domains due to Different Substrate Terminations

At clean unreconstructed surfaces, twin domains are the result of the symmetry properties of the 3-D substrate. Bulk rotation axes normal to the surface and bulk mirror planes with their normal parallel to the surface remain in the 2-D symmetry of the surface; bulk glide planes with both their normal and the glide vector in the surface plane remain as well. All other symmetry elements of the 3-D lattice vanish.

More generally, many crystals can have more than one surface termination. Examples include: hcp(0001) versus hcp(000$\bar{1}$) (cf. Figure 4.2), which have a bulk stacking sequence ...ABABAB... that can terminate either in an A layer or a B layer; and Si(100) (cf. Figure 4.3) in which alternate terminations are rotated 90° relative to each other; such terminations are identical to each other after rotations by 180° or 90°, respectively, and thus are examples of twin domains.

More generally, the surface frequently exhibits twin domains when screw axes or glide planes with a glide vector normal to the surface exist. A glide plane with a glide vector in the direction of the surface normal produces two terraces related by a mirror operation (cf. Figure 4.2). The diffraction pattern then exhibits a mirror plane due to the superposition of the diffraction patterns from two terraces. An example is the c-glide plane in the hexagonal close packed structure, space group P6$_3$/mmc. The two symmetrically equivalent domains result here from the 6$_3$ screw axes as well as from the c-glide plane. The point symmetry of the diffraction pattern is 6mm.

Other examples include the commonly observed 4-fold symmetry of the diffraction pattern of the Si(100) surface. The space group of silicon is Fd$\bar{3}$m (No. 227, Schoenflies symbol O$_h^7$); the full Hermann–Mauguin symbol is F4$_1$/d$\bar{3}$2/m showing that a 4$_1$ screw axis exists along the cubic translation vectors. The 4$_1$ screw axis would lead to four symmetrically equivalent terminations (due to cutting the bulk at inequivalent planes). Due to the p2mm symmetry of a single terrace only two distinct terraces exist, which are simply rotated by 90° relative to each other. The other two

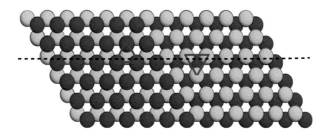

Figure 4.2 Two terraces (at left and right, respectively, and separated by a step) of the (0001) surface of the hcp lattice. The symmetry of the structure is p3m1. The two terraces are rotated by 60° relative to each other, as indicated by the two triangles of inter-atomic links. The c-glide plane is indicated by the dashed line. The symmetry of the diffraction pattern is p6mm if the two terraces occur with overall equal areas.

(a)

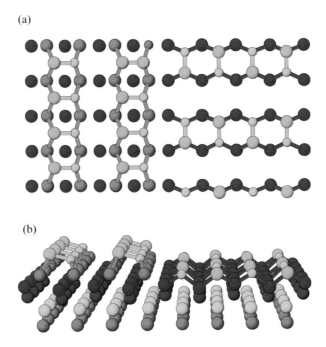

(b)

Figure 4.3 Model of two equivalent terraces (left and right, respectively) of the Si(100) surface exhibiting a (2×1) and a (1×2) reconstruction with asymmetric (i.e., tilted) dimers (green). The tilting can occur in two opposite directions; for example, on the left terrace, the higher silicon (shown larger and pointing up to the left) could equally point up to the right. The dimer axis is rotated by 90° between the two terraces. (a) and (b) give top-down and glancing-angle views, respectively. The four-layer sequence in Si(100) is indicated by different colours.

correspond to antiphase domains, assuming an unreconstructed bulk-like surface structure. The surface actually exhibits a reconstruction with asymmetric dimers: the asymmetry is due to a tilt of the dimer; this tilting can occur in two equivalent but opposite directions on the same terrace (cf. Figure 4.3(a)). The superposition of the diffraction pattern from the two terraces and both dimer orientations leads to an apparent pattern point symmetry 4mm, that is, having 4-fold rotational symmetry and two orthogonal mirror planes. The same applies, for example, to the (100) surface of magnetite, Fe_3O_4, with the same space group.

Another example is Al_2O_3 with the space group $R\bar{3}c$ (No. 167): the single domain has the 2-D symmetry p3m1. Due to the c-glide plane, two domains related by mirror operation exist and the point symmetry of the diffraction pattern is 6mm.

4.1.2   Twin Domains in Structures with Superlattices

Adsorbate structures and reconstructed surfaces that have a superlattice frequently form twin domains when their point symmetry is lower than that of the substrate. Some common superstructures and the number of possible twin domains are listed in Table 2.5. If the single domains are large enough to produce sufficiently sharp reflections, then the diffraction pattern shows the superposition of the patterns from all symmetrically equivalent domains. In Figure 4.4, two examples show the diffraction patterns of a (4×2) superlattice and of a c(4×2) superlattice. The two cases can be distinguished by the diffraction pattern.

It is not always possible to uniquely determine the unit cell of the superlattice from the diffraction pattern: the interpretation of the diffraction pattern may indeed be ambiguous. As an example, the (2×2) superlattice at a (111) surface may be the result of three rotated domains of a (2×1) superlattice or a single domain of a (2×2) superlattice (cf. Figure 4.5).

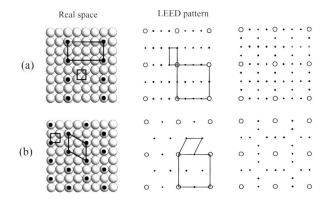

Figure 4.4 Superposition of diffraction patterns of twin domains. (a) (4×2) superstructure on a 4-fold substrate; (b) c(4×2) superstructure on a 4-fold substrate. Left column: real-space structure with adsorbates (black discs) on a square lattice. Middle and right columns: corresponding LEED patterns for a single domain and for a co-existence of two symmetrically equivalent domains, respectively. In the patterns, circles are (1×1) substrate spots, while dots are superlattice spots. The boxes show unit cells (in the left column) and reciprocal unit cells (in the middle column).

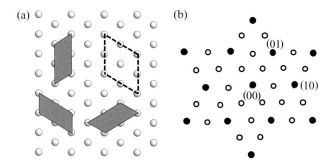

Figure 4.5 (a) Three symmetrically equivalent (2×1) unit cells (solid grey) on a 3-fold lattice and one (2×2) unit cell (dashed outline). (b) Corresponding diffraction pattern for both structures exhibiting in both cases an apparent (2×2) superlattice.

It is likewise not possible to unambiguously derive the point symmetry of the superstructure from the diffraction pattern, due to the mixing of domains in the pattern. The low-index surfaces of metals with fcc or hcp lattices exhibit a high symmetry. However, the diffraction pattern also exhibits this high symmetry even though the single domain of an adsorption layer may have a low symmetry or even no symmetry at all. These cases cannot be distinguished from the diffraction pattern alone; STM or AFM images could reveal the lower symmetry of the surface layer.

It is therefore incorrect in a detailed structural analysis, for example, with LEED I(V) calculations, to assume a high symmetry of the superstructure on the basis of an apparent high symmetry of the diffraction pattern: lower-symmetry superstructures must be allowed and tested, unless they can be a priori excluded by other data from different surface-sensitive techniques.

4.1.3 Glide Planes

At normal incidence, the symmetry of the diffraction pattern exhibits that of the surface structure (to the extent described in Section 4.1.2). At oblique (off-normal) incidence, a mirror plane or a glide plane can be observed in the diffraction pattern only if the incident beam is parallel to that plane. At general oblique incidence, no symmetry at all occurs in the diffraction pattern. In other words, the incident beam direction must remain invariant under the symmetry operation for this symmetry to be observable in the diffraction pattern. This is clearly seen in the structure factor of the reflection (hk) (cf. Eq. (2.49)). For an a-glide plane (glide vector in the a-direction) each atom at position (x_j, y_j) has a symmetrically equivalent atom at $(x_j + 1/2, -y_j)$. Only if the wave vectors of the incident and the diffracted beams, $\mathbf{k}'$ and $\mathbf{k}_0$, remain invariant under the mirror operation will the structure factor remain unchanged when the beam indices satisfy $h = $ odd and $k = 0$; otherwise the atomic scattering factors

$f_j(\mathbf{k'}, \mathbf{k}_0)$ become different. This can be seen in the following expression derived from Eq. (2.49):

$$F(\mathbf{k'}, \mathbf{k}_0) = \sum_{j=1}^{n} f_j(\mathbf{k'}, \mathbf{k}_0) \left[e^{2\pi i \left(hx_j + ky_j + lz_j \right)} + e^{2\pi i \left(h(x_j + 1/2) - ky_j + lz_j \right)} \right], \qquad (4.1)$$

where j now runs over pairs of glide-plane-related atoms.

This is different from X-ray diffraction where f_j depends only on $\mathbf{q} = \mathbf{k'} - \mathbf{k}_0$; in LEED, this results from multiple scattering. A glide plane causes extinction of odd-order reflections in a line of spots passing through the (00) spot and in the direction of the glide plane. Glide planes are observable in the four space groups pg, p2mg, p2gg and p4gm, which exhibit glide planes without parallel mirror planes (see Figure 2.12). The glide planes in other 2-D space groups, for example, in hexagonal lattices, exhibit mirror planes parallel to the glide planes, so that no extinction can be observed or the glide planes are implicitly generated by other symmetry elements.

Extinctions are systematic and occur at all energies. It is therefore useful to observe the diffraction pattern at various energies and in addition at oblique incidence to clearly identify the presence of a glide plane. At oblique incidence, at least some of the extinguished spots should become visible.

An example of a structure with symmetry p2gg on a substrate with p4mm symmetry is shown in Figure 4.6.

As mentioned in Section 4.1.2, differently oriented domains may occur due to different terminations of the substrate as well as due to a superstructure forming different domains on a single terrace of the substrate. The superposition of the

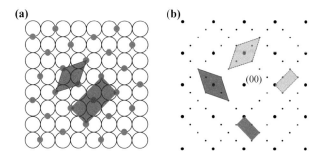

(a)   **(b)**

Figure 4.6 Diffraction pattern for a surface with two orthogonal glide planes. (a) Model of one rotational domain of Pd(100) (large circles) covered with half a monolayer of CO molecules (small circles) in 'bridge sites', with a $(2\sqrt{2} \times \sqrt{2})R45°$ structure, in matrix notation $\begin{pmatrix} 1 & -1 \\ 2 & 2 \end{pmatrix}$; the symmetry is p2gg; the rectangle is a unit cell and contains two CO molecules, while the rhombus is a unit cell of the CO layer alone, but not of the combined lattice, hence the diffraction pattern will show a rectangular array of spots. (b) Diffraction pattern at normal incidence for the coexistence of two domains rotated by 90° (the second shown in lighter grey); note the missing spots along the two main diagonals intersecting at the central (00) spot, due to glide plane symmetry within each domain and normal incidence of the electron beam.

diffraction pattern may therefore also hide an extinction due to a glide plane. It is thus indeed not always possible to determine the symmetry of the structure uniquely from the diffraction pattern.

4.2   Determination of the Lattice Constant

The experimental LEED spot patterns usually exhibit instrumental distortions from ideal reciprocal lattices, whether recorded by a CCD camera from a fluorescent screen (conventional LEED) or by a microchannel plate (MCP-LEED). These distortions do not allow a precise determination of the lattice constants and may also lead in some cases to an incorrect assignment of reflection indices. There are several causes for these deviations from the ideal imaging conditions, including: the position of the sample may deviate from the centre of the LEED screen; the incidence angle may deviate from normal incidence; the energy of the incident beam may deviate from the voltage displayed by the LEED power supply due to instrumental error or uncorrected differences in the work function of the filament, the filter grids or the sample; the diffraction pattern may also be distorted by magnetic fields arising from magnetic material in the chamber or insufficient μ-metal shielding of the Earth's magnetic field. These effects play a minor role in LEED I(V) analyses where the lattice constants of the substrate are known a priori with sufficient precision and the superstructure reflections can be safely indexed. Then, only the indices and the intensities are needed for the structure analysis, so that the reflection angles and the deviation from the ideal position need not be measured. For the direct determination of the lattice constant from the LEED pattern, however, the precise knowledge of the diffraction angles is essential.

There are also cases where the lattice constants are not known, for example for incommensurate epitaxial layers of such thickness that the substrate reflections are not visible and thus provide no reference scale to determine the relative lattice constants. In such cases a correction of the LEED images is required in order to derive lattice constants. An algorithm has been developed by F. Sojka et al. [4.4] to correct systematic distortions in the diffraction pattern and to determine the primary energy precisely. A relative accuracy of the lattice parameters of better than 1% has been achieved. The method is based on the use of a calibration sample of which the superlattice and the lattice constants are precisely known. As an example, Si(111) with a (7×7) reconstruction was chosen. The method produces a map of the systematic instrumental image distortions which is used to correct the images of other samples.

Care must be taken that the distance between the sample surface and the LEED screen or MCP system is retained for each subsequent measurement. The importance of the correction is shown in Figure 4.7 for Si(111)–(7×7) and three different LEED systems: a conventional LEED system (C-LEED), a microchannel plate LEED system (MCP-LEED) and a high-resolution spot profile analysis system (SPA-LEED). The corrected image shows the spot positions at the calculated positions.

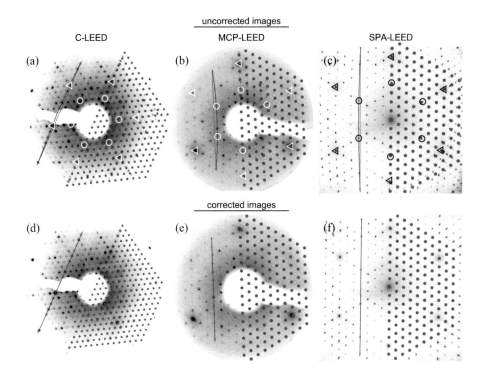

Figure 4.7 (a–c) Uncorrected and (d–f) corrected LEED images of Si(111)–(7×7) obtained with various devices: (a) and (d) C-LEED (conventional LEED system) at 74 eV, (b) and (e) MCP-LEED (microchannel plate LEED system) at 75 eV, (c) and (f) SPA-LEED (spot profile analysis LEED system) at 87 eV. For the sake of clarity, the ideal reciprocal lattice (red dots) is superimposed on one half of the respective patterns only. To highlight the distortions, some spots are linked by blue curves, which ideally should be straight lines. As an example of the size of the corrections at a given point in the reciprocal space for the given diffraction conditions, we choose the set of (4/7, 0) order spots of Si(111) (marked by white or black circles). If the k-space is scaled with respect to the (1, 0) order of Si(111) (marked by white or black triangles), without correction one would measure a reciprocal length of 10.5 nm^{-1} (1.05 Å^{-1}) in the case of C-LEED and SPA-LEED or 11.7 nm^{-1} (1.17 Å^{-1}) in the case of MCP-LEED, instead of the true value of 10.8 nm^{-1} (1.08 Å^{-1}). Republished with permission of the American Institute of Physics, from [4.4] F. Sojka, M. Meissner, Ch. Zwick, R. Forker and T. Fritz, *Rev. Sci. Instr.*, vol. 84, p. 015111, 2013; permission conveyed through Copyright Clearance Center, Inc.

4.3   Correction of the Incidence Angle

In many cases it is useful to apply oblique incidence to increase the amount of experimental data available for the LEED I(V) analysis. This is, for example, the case for structure analyses of large organic molecules where the large lattice constants do not allow the separate measurement of tightly spaced LEED spots at higher energies. The measurable energy range is then limited and the experimental database necessarily small, while the complexity of such structures usually requires relatively more data than for small unit cells. The database can be enhanced by measurement at oblique incidence, thus allowing the measurement of the (00) beam which at normal incidence is not accessible with a conventional LEED system or an MCP system; in addition, all other beams then become symmetrically inequivalent, greatly expanding the amount of measurable independent data. As mentioned in Chapter 3, the incidence angle is not defined precisely enough by mechanical adjustment with the manipulator, while it can be determined from the LEED pattern when some substrate reflections are visible for which the lattice constants are precisely known. A method to derive the incidence angle from the position of the reflections on the LEED screen has been described in Section 3.3.2. The method requires the knowledge of the lattice constants and the indices of the reflections.

Another method to correct the incidence angle has been developed by Sojka et al. [4.5] and tested with MCP-LEED and SPA-LEED. First the instrumental distortions must be removed in the way described in Sojka et al. [4.4] (see Figure 4.7). If the incidence angle deviates from normal incidence, then there remain distortions which are axially symmetric with respect to the tilt direction. An example is shown in Figure 4.8(c) and (d). The positions of all reflections are measured and fitted to calculated positions. Required is the knowledge of the substrate lattice, the indices of some substrate reflections and the superlattice.

An application of the method is shown in Figure 4.9. The diffraction pattern of PTCDA (3,4,9,10-perylenetetracarboxylic dianhydride) on Ag(111), with substrate symmetry p3m1, exhibits six domains as the adsorbate structure possesses no symmetry. For the measurement of I(V) curves and the analysis of the data it is essential to index the reflections correctly and assign them to a single domain. It is clearly seen in Figure 4.9 by comparing corrected and uncorrected images that some spots cannot be clearly indexed in the uncorrected image.

4.4   Stepped Surfaces

Defects at surfaces are thought to play an important role in many processes, one example of which is heterogeneous catalysis [4.7]. Steps and kinks are defects that can be created reproducibly and studied at surfaces. They have become models for the complex variety of structures found at more natural surfaces, such as those of nanoparticles. Not only are steps and kinks reproducible, but their nature and concentration can also be controlled by varying the cutting angle (i.e., the surface orientation) of a single crystal. Thus, we need to be able to determine accurately the structure of regularly stepped and kinked surfaces, collectively called high-Miller-index surfaces. The structural properties of stepped and kinked surfaces have been discussed in Section 2.1.12.

Regular step arrays exhibit characteristic diffraction patterns in which split spots occur at certain energies. When varying the energy of the primary beam the intensity switches between the neighbouring spots. This behaviour occurs systematically at all

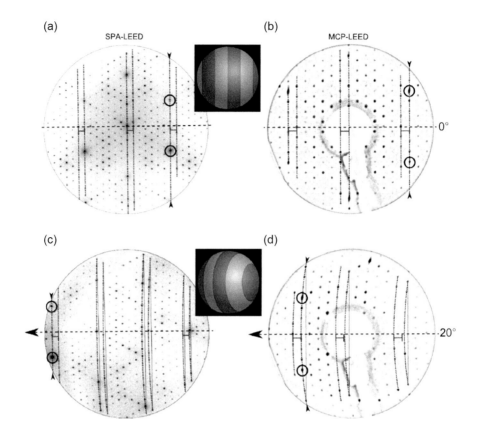

Figure 4.8 LEED patterns from Si(111)–(7×7): (a) and (b) with a non-tilted surface, and (c) and (d) with a surface tilted by 20°, obtained with two devices: (a) and (c) SPA-LEED at 159.6 eV; (b) and (d) MCP-LEED at 71.6 eV. The diameter of the depicted area is 488 pixels for the SPA-LEED images and 976 pixels for the MCP-LEED images. For the sake of clarity, reflections which should be on a straight line are linked by dotted black curves. Multiple equal distance markers (black) in each image demonstrate the equal ((a) and (b)) and varying ((c) and (d)) sizes of geometrically equivalent distances. A dashed arrow indicates the tilt direction. For each device separately, equivalent pairs of Si (111)–(1×1) spots are marked by black circles. All LEED images are contrast-processed and colour-inverted. Reprinted from [4.5] F. Sojka, M. Meissner, Ch. Zwick, R. Forker, M. Vyshnepolsky, C. Klein, M. Horn-von Hoegen and T. Fritz, *Ultramicroscopy*, vol. 133, pp. 35–40, 2013, https://doi.org/10.1016/j.ultramic.2013.04.005, with permission from Elsevier.

split spots and indicates the existence of steps on the surface. It can be easily explained by a kinematic consideration. A schematic model of a step is shown in Figure 4.10. For simplicity a structure factor f is assigned to the whole column of atoms in the surface slab. The columns at the step may exhibit different structure factors, due to different local relaxations and multiple scattering, which depend on the local surrounding structure.

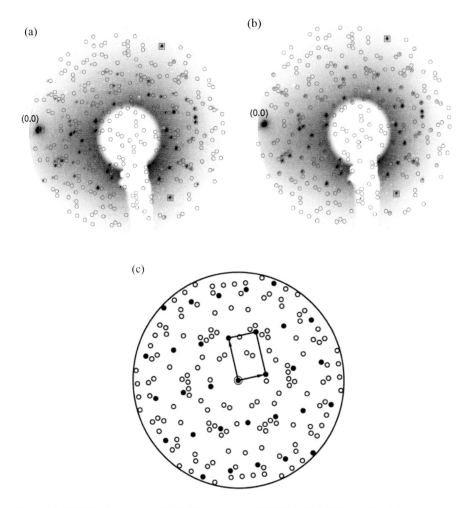

Figure 4.9 LEED patterns of a 13.9° tilted surface of PTCDA (3,4,9,10-perylenetetracarboxylic dianhydride) on Ag(111) taken at 43 eV using MCP-LEED. A correction of the distortion of the LEED image described in [4.4] is superimposed on the images. Green squares mark the positions of first order reflections of the substrate, while red circles mark the positions of the reflections of the PTCDA layer. (a) Original LEED image with strong axially symmetric distortion. (b) Corrected LEED image. In both cases, the simulation is oriented and scaled to the first order of the substrate reflections. (c) Kinematic construction of the LEED pattern at normal incidence with superposition of six domains. The unit cell of one domain is marked. Images (a) and (b) are reprinted from [4.5] F. Sojka, M. Meissner, Ch. Zwick, R. Forker, M. Vyshnepolsky, C. Klein, M. Horn-von Hoegen and T. Fritz, *Ultramicroscopy*, vol. 133, pp. 35–40, 2013, with permission from Elsevier. Panel (c) is reprinted from [4.6] K. Glöckler, C. Seidel, A. Soukopp, M. Sokolowski, E. Umbach, M. Böhringer, R. Berndt and W.-D. Schneider, *Surf. Sci.*, vol. 405, pp. 1–20, 1998, https://doi.org/10.1016/j.ultramic.2013.04.005, with permission from Elsevier.

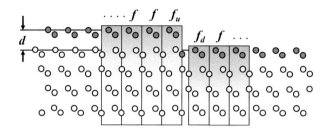

Figure 4.10 Schematic drawing of a step. At a step edge the structure factor f deviates from its average value; f_u and f_d may show different scattering amplitudes and scattering phases. Redrawn figure with permission of Walter de Gruyter and Company, from [4.7] M. Horn-von Hoegen, *Z. Krist.*, vol. 214, pp. 591–629, 1999; permission conveyed through Copyright Clearance Center, Inc.

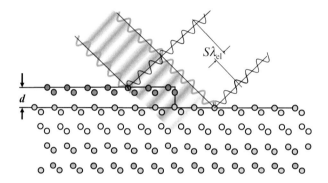

Figure 4.11 Phase contrast at a step edge: Electrons scattered from terraces separated by a single atomic step interfere with a phase difference $2\pi S$, due to a path difference given by the product of the phase S and the electron wavelength λ_{el}. The phase S is given by the product of the diffraction vector and the vector separating the terraces, $(\mathbf{k}' - \mathbf{k}_0)\cdot\mathbf{d}$. For half integer values of S, the electrons interfere destructively and the sharp LEED spot disappears. For integer values of S, the interference is constructive and the electrons are insensitive to surface roughness due to steps. Redrawn figure with permission of Walter de Gruyter and Company, from [4.7] M. Horn-von Hoegen, *Z. Krist.*, vol. 214, pp. 591–629, 1999; permission conveyed through Copyright Clearance Center, Inc.

According to Eq. (2.49), the intensity can be written as:

$$I(\mathbf{k}', \mathbf{k}_0) = |F(\mathbf{k}', \mathbf{k}_0)|^2 |G(h)|^2 |G(k)|^2$$

$$= \left| \sum_{j=1}^{n} f_j(\mathbf{k}', \mathbf{k}_0) e^{2\pi i \left(hx_j + ky_j + lz_j \right)} \right|^2 \delta(\mathbf{k}' - \mathbf{k}_0 + \mathbf{g}), \qquad (4.2)$$

where $|F(\mathbf{k}', \mathbf{k}_0)|^2$ represents the structure factor of one unit cell and $|G(h)|^2 |G(k)|^2$ represents the lattice factor. We can take a whole terrace as a unit cell and the lattice factor then describes a grating due to the regular array of steps. Even for narrow terraces the structure factor of a single terrace exhibits maxima near the Bragg points of the flat surface. These are the points where the phase difference between the terraces become a multiple of 2π. The phase difference between reflections from two terraces is illustrated in Figure 4.11 and the reciprocal lattice in Figure 4.12.

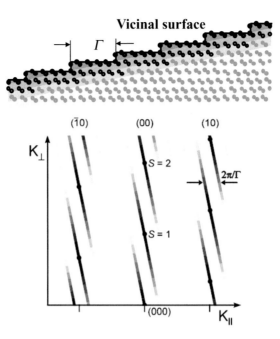

Figure 4.12 Diffraction pattern from a stepped (or vicinal) surface of Si(111) with a regular step array. Top panel: profile of surface with step width Γ. Bottom panel: reciprocal lattice in a plane perpendicular to the step edges through the specular reflection marked by (000), where $S = 0$, with beam intensities shown by grey levels; dots represent substrate Bragg reflection conditions, where no splitting occurs. All split LEED spots show a splitting $2\pi/\Gamma$ between Bragg conditions, due to the periodic step sequence. Redrawn with permission of Walter de Gruyter and Company, from [4.7] M. Horn-von Hoegen, *Z. Krist.*, vol. 214, pp. 591–629, 1999; permission conveyed through Copyright Clearance Center, Inc.

When the Ewald sphere intersects a Bragg point, then only one spot is visible, as can be seen in Figure 4.12. Between the Bragg points split spots become visible. The step edge is normal to the direction of the splitting. The diffraction pattern is schematically shown in Figure 4.13.

Increasing the energy of the primary beam results in a switch of the intensity to the spot with the larger diffraction angle. The inclination of the surface, that is, the polar angle of the inclination, points into the opposite direction.

It should be noted that the occurrence of split spots is typical for stepped surfaces with identical terraces separated by a constant step vector. If the terraces are not equivalent, as for example at the stepped Si(100) surface, see Figure 4.3 (where two differently oriented terraces alternate), then the step periodicity is doubled and a third spot appears between the split spots.

4.5   Faceted Surfaces

Some surfaces which form a clean unreconstructed surface exhibit facets after adsorption of reactive gases [4.8; 4.9]. Faceting has also been observed at metal

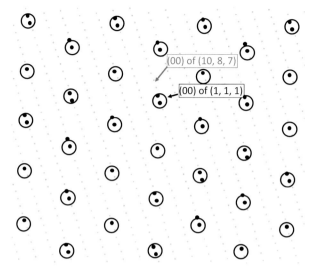

Figure 4.13 LEED pattern for a fcc(10, 8, 7) kinked surface consisting of beams (shown as large dots) determined by the conjunction of the circles [which represent broadened beams due to the finite width of the (1, 1, 1) terraces] and the small grey dots (which represent the reciprocal periodicity of the steps). As a function of increasing energy, the grey dots converge toward the immobile (00) specular beam of the (10, 8, 7) surface. At the same time, the circles, which define which beams are intense, converge toward the immobile (00) specular beam position of the (1, 1, 1) terraces. As a result of these two motions, different beams in the pattern appear alternately split or not as a function of energy.

surfaces after coverage with other metal layers [4.10]. The initially flat surface (*hkl*) then breaks up into separate nanoscale surfaces with typically two or three different orientations; the orientation of the overall surface plane remains. Its Miller indices can be decomposed into the Miller indices of the individual facets according to the vectorial rule

$$(h, k, l) = \sum_i a_i (h_i, k_i, l_i),$$ (4.3)

where a_i represents the relative sizes of the facets (h_i, k_i, l_i), following K. Hermann [4.11]. For example, the (115) surface of Cu forms facets (401) + (−401) + (113) after oxygen adsorption [4.12] (cf. Figure 4.14): the STM image shows clearly that the oxygen covered surface is facetted; the orientation of the facets is determined from the diffraction pattern.

Faceting causes an increase of the surface area. This must obviously be compensated by the smaller surface energy of the adsorbate covered facets. A schematic drawing of the reciprocal lattices of multiple facets is shown in Figure 4.15.

When the facets are large enough to diffract independently (i.e., incoherently), faceting is observed in the diffraction pattern by a superposition of the independent patterns from the different facets and is identified by the energy-independent positions

(4 0 1)   ($\bar{4}$ 0 1)   (1 1 3)

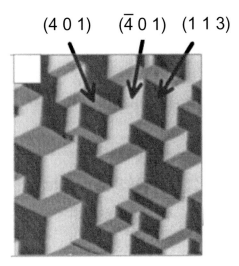

Figure 4.14 Faceting of the Cu(115) surface after oxygen adsorption at $p_{O_2} = 5 \times 10^{-8}$ mbar at 617 K observed by STM; image size: 250×250 nm. The Miller indices are modified from the original image to be consistent with the facet formation described in the text. Reprinted from [4.12] N. Reinecke and E. Taglauer, *Surf. Sci.*, vol. 454–456, pp. 94–100, 2000, with permission from Elsevier.

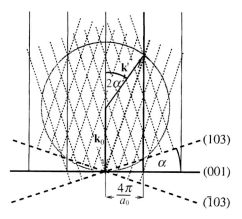

Figure 4.15 Reciprocal lattices for two facets (dashed lines) inclined from the overall surface plane (horizontal full line)

of the separate specular (00) spots of the individual facet orientations. The spots belonging to a single facet converge towards its (00) spot with increasing energy. The only spot from that single facet which is not moving is its (00) spot. By contrast, the flat, non-faceted surface exhibits a single (00) spot and all reflections converge towards this specular spot with increasing energy. Typical for faceted surfaces therefore are neighbouring spots moving in non-convergent directions when the energy is varied.

The indices of the facets can be determined from the diffraction angles of the (00) spots. At incidence normal to the overall surface plane, half of the polar angle of the (00) spot defines the inclination of the corresponding facet with respect to the overall surface plane. There may be ambiguities in the determination of the indices. These can be overcome in most cases by considering the condition in Eq. (4.3) and the identification of all (00) spots.

4.6   Antiphase Boundaries

Antiphase domains frequently occur in adsorbate layers, particularly when their coverage deviates from the ideal value for a superlattice. In these cases, the vacant or excess sites frequently order as line defects that form dislocations between out-of-phase domains. The expression 'antiphase' domain is used for all domains that are connected by a shift vector. The dislocations are called domain walls.

The domain walls are frequently periodically arranged for energetic reasons, forming what one might call a 'super-superstructure', and this becomes visible in a splitting of the superlattice reflections. This case is similar to that of stepped surfaces, see Figure 4.13. The difference between antiphase domains in an adsorbate structure and a stepped surface is that in the latter case the shift vector has a component normal to the surface while in an adsorbate structure the shift vector is parallel to the surface. If a periodic array of domain boundaries exists, that is, if a superlattice could be defined, then spot splitting occurs. In the case of irregular domain boundaries, the spots are diffuse, that is, streaked in the direction of the spot splitting.

There are some general rules to interpret diffraction patterns with split spots: (i) If the splitting is energy dependent, that is, if there exist energies where the spots appear sharp while they are split or diffuse at other energies, then steps are involved, see Figure 4.12. (ii) The direction of the spot splitting is normal to the direction of the domain boundaries. (iii) The shift vector between the domains determines which spots are split or diffuse. If the shift vector is ½ of a lattice vector, then the odd order reflections in the corresponding direction in the reciprocal lattice are split. For example, if the shift vector is ¼ of a lattice vector then each fourth reflection remains without satellites. These relations are schematically illustrated in Figure 4.16 for the case of antiphase domains with shift vectors of ½ of the lattice vectors $\mathbf{a}$ and $\mathbf{b}$.

For antiphase domains in adsorbate layers, we can distinguish between heavy (high-density) and light (low-density) walls. An example is shown in Figure 4.17. We may consider the simplest case of a linear array of antiphase boundaries in a c(2×2) superlattice, or $\begin{pmatrix} 1 & 1 \\ 1 & -1 \end{pmatrix}$ in matrix notation, on a square or rectangular lattice, assuming a periodic sequence. The domains are shifted by $(\mathbf{a}' + \mathbf{b}')/2$, where $\mathbf{a}'$ and $\mathbf{b}'$ are the lattice vectors of the c(2×2) superlattice, cf. Figure 4.17. We can define an enlarged supercell consisting of two domains and a long period $M\mathbf{a}' = M(\mathbf{a} + \mathbf{b})$. Here M is always an odd integer because the domains are shifted by half a unit cell in the case of heavy walls; for light walls the domains are also shifted by half a unit cell,

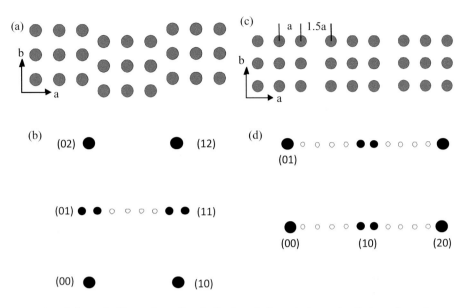

Figure 4.16 Schematic illustration of periodic domain boundaries and splitting of certain spots. (a, b): Shift vector **b**/2 between domains, so spot splitting occurs in (hk) reflections with k = odd. (c, d): Shift vector **a**/2, so spot splitting occurs in reflections (hk) with h = odd.

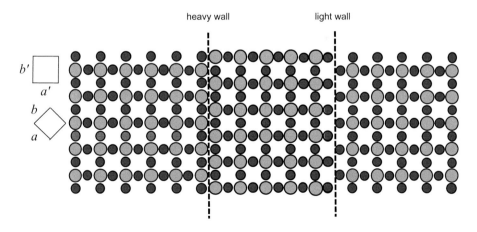

Figure 4.17 Schematic illustration of heavy and light domain walls in a c(2×2) superstructure on a square substrate lattice: dark grey atoms are the substrate, while light grey atoms are adatoms. Heavy walls (at left) have excess occupied sites, while light walls (at right) have vacant sites. In the heavy walls the adsorbate atoms may be displaced from the ideal position due to repulsion between adsorbate atoms, while relaxations are also possible in the light domain walls. The domain boundary is normal to the [11] direction of the substrate lattice. The substrate unit cell and the c(2×2) unit cell are indicated on the left.

but the adsorption sites at the domain boundaries are empty. This represents the idealized case, as in reality the individual domains have variable domain sizes and the average long period can thus take any real value. In the idealized case, the domain size including the boundary is $N = M/2$ which implies that N is not an integer. Counting the number N_d of occupied c(2×2) cells in a single domain, N_d becomes $(M - 1)/2$ for light and heavy walls. For heavy walls the boundary sites are occupied, which means that there are $N_d + 1$ occupied sites per single domain. The boundary sites are empty for the light walls. The short period would be perpendicular to the long period, with length $(\mathbf{a} - \mathbf{b})/2$.

We can define a structure factor $F_s(\mathbf{k}', \mathbf{k}_0)$ for the domain sequence with the long period $M\mathbf{a}'$ including the domain boundaries:

$$F_s(\mathbf{k}', \mathbf{k}_0) = \sum_{j=1}^{M} f_j(\mathbf{k}', \mathbf{k}_0) e^{2\pi i \left(h_s x_j + k_s y_j + l z_j\right)}, \tag{4.4}$$

where f_j represents the structure factor of a c(2×2) cell consisting of the adsorbate atom and the substrate atoms within the domain, or a vacant or excess site at the domain wall. This structure factor also allows us to include relaxation at the domain walls if required. The indices h_s, k_s and the coordinates x_j, y_j, z_j refer to the period of the domain sequence. The structure factor for the supercell can be calculated in the same way for other sequences of antiphase domains, for example for domains in the $(\sqrt{3} \times \sqrt{3})$R30° superstructure, where three out-of-phase domains exist.

In the example considered here (cf. Figure 4.17), we have two domains with the period $\mathbf{a}_s = (2N_d + 1)\mathbf{a}' = (2N_d + 1)(\mathbf{a} + \mathbf{b})$ and $\mathbf{b}_s = \mathbf{b}'$, while $\mathbf{a}' = (\mathbf{a} + \mathbf{b})$ and $\mathbf{b}' = (\mathbf{a} - \mathbf{b})$ are the translation vectors of the c(2×2) superlattice; $\mathbf{a}$ and $\mathbf{b}$ are the translation vectors of the substrate lattice and the domain boundary is normal to the [11] direction. The shift vector between the two domains is $\mathbf{d} = \frac{1}{2}(\mathbf{a}' + \mathbf{b}') = \mathbf{a}$. The phase factor between the two domains becomes $e^{2\pi i(\mathbf{k}' - \mathbf{k}_0) \cdot \mathbf{d}} = e^{\pi i(h_s + k_s)}$, because the indices h_s, k_s are related here to the translation vectors of the domain sequence. We define a structure factor $F_d(\mathbf{k}', \mathbf{k}_0)$ for the single domain with the period $M\mathbf{a}'/2$ including the domain boundaries. There are N_d occupied c(2×2) cells and one boundary cell with factor f_{bnd} in the single domain, yielding:

$$F_d(\mathbf{k}', \mathbf{k}_0) = \sum_{j=1}^{N_d} f_j(\mathbf{k}', \mathbf{k}_0) e^{2\pi i \left(h_s x_j + k_s y_j + l z_j\right)}$$
$$+ f_{bnd}(\mathbf{k}', \mathbf{k}_0) e^{2\pi i \left(h_s x_{N_d+1/2} + k_s y_{N_d+1/2} + l z_{N_d+1/2}\right)}. \tag{4.5}$$

In the kinematic limit with no relaxations, for heavy boundaries f_{bnd} is ½ of an occupied c(2×2) cell while for light boundaries it is ½ of an empty c(2×2) cell. The structure factor for the complete supercell then becomes

$$F_s(\mathbf{k}', \mathbf{k}_0) = F_d(\mathbf{k}', \mathbf{k}_0) \left[1 + e^{i\pi(h_s + k_s)}\right]. \tag{4.6}$$

The lattice factor gives a grid of reflections (h_s, k_s) with $h_s + k_s$ = even. The relation to the reciprocal lattice of the substrate is

$$h = \frac{1}{2}\left(\frac{h_s}{(2N_d + 1)} + k_s\right) \text{ and } k = \frac{1}{2}\left(\frac{h_s}{(2N_d + 1)} - k_s\right), \tag{4.7}$$

where N_d is the number of c(2×2) cells in a single domain without the boundary. The structure factor vanishes for the odd order reflections of the supercell. The structure factor of a single domain $F_d(\mathbf{k'}, \mathbf{k}_0)$ is nearly that of the perfect c(2×2) structure, except for the missing or excess atoms in the domain boundary, and it exhibits maxima at the positions of the substrate spots, while the superlattice spots are split. It should be noted that heavy and light walls cannot be distinguished from the diffraction pattern alone: a quantitative intensity analysis would be needed to reveal the actual type of the walls. The domain structure factor, shown in Figure 4.18(a), implies that only split spots occur near the (1/2, 1/2) position, which is itself extinguished. The other reflections rapidly become too weak to be observable even for small numbers of $N_d > 3$. The diffraction pattern is schematically drawn in Figure 4.18(b). The domain walls are always normal to the splitting of the superlattice spots.

The diffraction pattern of the 1-D domain superlattices shown in Figure 4.18(c) can be distinguished from 2-D superlattices corresponding to a regular arrangement of islands shown schematically in Figure 4.19(a) for c(2×2) islands on a square substrate lattice and the corresponding case on a hexagonal lattice in Figure 4.19 (c). The quartet of split spots, Figure 4.19(b), is rotated compared to that shown in Figure 4.18(c). There are no satellites around the integer order spots because the shift vectors between the domains are always ½ of a lattice vector of the c(2×2) lattice. This causes the odd order reflections to split while no splitting occurs at the even order reflections. The indices in Figures 4.18 and 4.19 refer to the substrate lattice and the split spots are the half-order reflections generated by the c(2×2) structure. In Figure 4.19(c) a regular array of ($\sqrt{3} \times \sqrt{3}$)R30° islands on a hexagonal substrate is shown. Some ($\sqrt{3} \times \sqrt{3}$)R30° unit cells are included in the figure to illustrate that three different shift vectors occur for the three antiphase domains. A light wall is assumed and with the domain size drawn in the figure the superlattice of domains would have a (24×24) periodicity, referred to the substrate lattice. The shift vectors between the domains are lattice vectors of the substrate lattice but not simply ½ of a lattice vector of the ($\sqrt{3} \times \sqrt{3}$)R30° lattice, so that satellites occur around the substrate reflections, also due to the empty sites in the domain walls.

The formal kinematic calculation described above can be extended to other superstructures and domain boundaries in all directions. If the domain boundaries do not exhibit a regular spacing, then elongated streaked superlattice spots occur instead of spot splitting. Multiple scattering plays an important role for LEED and the observed diffraction pattern deviates from those calculated by the kinematic theory, i.e., more satellite intensities are visible, the intensity relations are different and they depend on energy. Nevertheless, the characteristic features of the

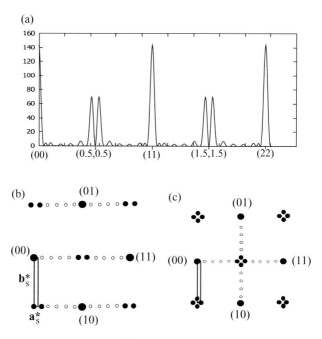

Figure 4.18 (a) Structure factor $|F_s(\mathbf{k}', \mathbf{k}_0)|^2$ of a single domain of the structure shown in Figure 4.17, showing that only the spots near the half-order positions exhibit measurable intensity. (b) Reciprocal lattice of a supercell consisting of two antiphase domains, each containing five c(2×2) cells on a square lattice. The lattice vectors of the c(2×2) superlattice are $\mathbf{a}'$, $\mathbf{b}'$ (cf. Figure 4.17) and the supercell has the lattice vectors $\mathbf{a}_s = (2 \times 5 + 1)\, \mathbf{a}'$, $\mathbf{b}_s = \mathbf{b}'$. The reciprocal lattice vectors $\mathbf{a}_s^*$ and $\mathbf{b}_s^*$ are shown. The open circles represent reflection positions where the intensity is usually too weak to be observed. (c) Diffraction pattern of a surface where two mutually rotated 1-D domain superlattices exist. The two domain systems are symmetrically equivalent on the square substrate lattice and the diffraction pattern shows the superposition of both superlattices.

diffraction pattern remain and the kinematic calculation is useful to obtain a qualitative picture of the diffraction pattern.

In the c(2×2) superstructure on a square lattice shown in Figures 4.17 and 4.18 as an example, it is unlikely that in reality straight domain boundaries would exist, so that diffuse half-order spots would be observed. It is also likely that relaxation at the domain boundaries occurs. Furthermore, there is a continuous transition from domain structures to modulated structures up to completely relaxed adsorbate layers which then correspond to commensurate or incommensurate superposition of rigid lattices. The latter case occurs, for example, for the adsorption of noble gases on metal surfaces. The change of intensities of superlattice reflections with increasing relaxation at domain boundaries has been investigated by P. Zeppenfeld et al. [4.13] for the case of Xe/Pt(111). The effect of domain boundaries on the diffraction pattern of 1-D domain systems has also been investigated using a kinematic analysis by F. Timmer and J. Wollschläger [4.14] for the case of rare earth elements on Si(111).

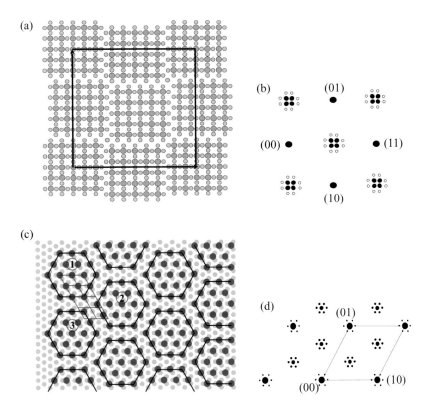

Figure 4.19 (a) Schematic drawing of a 2-D arrangement of domain boundaries in a c(2×2) structure on a square substrate lattice, with (b) the corresponding diffraction pattern showing a quartet of split spots around the position of the half order reflections. (c) 2-D arrangement of ($\sqrt{3} \times \sqrt{3}$)R30° domains on a hexagonal lattice, with (d) the corresponding diffraction pattern. Satellite reflections occur at the positions of a superlattice formed by the domains. The domain sizes and the positions of the split spots are not drawn to scale in either case.

4.7   Modulated Layers

Modulated structures exhibit periodic or aperiodic displacements of the atoms from their mean position. The structure of the mean positions is assumed to be periodic and is defined here as averaged positions within one unit cell of the unmodulated structure. On surfaces a 2-D lattice mismatch between a periodic adsorbate layer and a substrate may lead to lattice modulations. Common examples are graphene monolayers on metal surfaces, for example graphene on Ir(111) [4.15]. Modulated surfaces exhibit characteristic diffraction patterns where each reflection of the substrate lattice is surrounded by a set of satellite reflections. One of the satellite reflections is the main reflection of the mean graphene lattice. As example the diffraction pattern of graphene on Ir(111) is shown in Figure 4.20.

The superposition of two rigid lattices with a lattice mismatch generates a moiré pattern (which is a regular interference pattern between the two lattices); in the kinematic theory the diffraction pattern would then be just the superposition of two

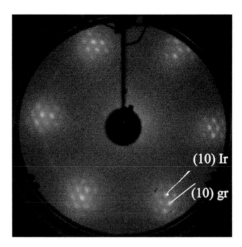

Figure 4.20 Diffraction pattern of a monolayer graphene on Ir(111) at 78 eV. The (00) spot is hidden behind the specimen holder. Reprinted from [4.15] S. K. Hämäläinen, M. P. Boneschanscher, P. H. Jacobse, I. Swart, K. Pussi, W. Moritz, J. Lahtinen, P. Liljeroth and J. Sainio, *Phys. Rev. B*, vol. 88, p. 201406(R), 2013, https://journals.aps.org/prb/abstract/10 .1103/PhysRevB.88.201406, with permission from APS.

reciprocal lattices. The lattices are in reality not rigid: the interaction between the layers in competition with the bonds within the layers generates lattice modulations, namely regular patterns of atomic relaxations. The lattice mismatch between adsorbate and substrate is not the only origin of a lattice modulation. In 3-D crystals charge density waves have been observed with X-ray diffraction. Charge density waves are also considered to be lattice modulations, independent from the fact that a modulation of the electron density is always accompanied by a displacement of the atom core, even though this may be very small. Charge density waves have also been found on surfaces [4.16; 4.17].

For a more detailed description, we may consider a periodic 1-D modulation within a single layer. The atom positions are described by $\mathbf{r}_{n,m} = \mathbf{r}_{n,m}^0 + \Delta\mathbf{r}_n$. where $\mathbf{r}_{n,m}^0$ is the mean position in the rigid lattice without modulation. We consider as the simplest example a mono-atomic lattice with a 1-D modulation $\Delta\mathbf{r}_n$ in the direction of the **a**-vector with a period of N unit cells. The displacement $\Delta\mathbf{r}_n$ can always be described by a Fourier series and we consider here only a simple sinusoidal displacement function:

$$\mathbf{r}_{n,m} = \left[n + A_x \sin\left(2\pi \frac{n}{N-1} \right) \right] \mathbf{a} + m\mathbf{b}$$
$$= (n + \Delta_n)\mathbf{a} + m\mathbf{b}. \tag{4.8}$$

Here A_x is the amplitude of the modulation. Generally, three different amplitudes in three directions can be considered: two for transverse modulations and one for a longitudinal modulation. For simplicity we assume here only a longitudinal modulation in the **a**-direction. In the equation for the structure factor, the sine function appears in the exponent describing the phase factor:

$$F(\mathbf{k}', \mathbf{k}_0) = \sum_{n,m=-\infty}^{\infty} f_n(\mathbf{k}', \mathbf{k}_0) e^{2\pi i((\mathbf{k}'-\mathbf{k}_0)\cdot \mathbf{r}_{n,m})}. \qquad (4.9)$$

With the condition for a substrate reflection

$$(\mathbf{k}' - \mathbf{k}_0)\cdot \mathbf{a} = h \text{ and } (\mathbf{k}' - \mathbf{k}_0)\cdot \mathbf{b} = k, \qquad (4.10)$$

and assuming constant form factors f_n we obtain for the structure factor:

$$F(\mathbf{k}', \mathbf{k}_0) = f(\mathbf{k}', \mathbf{k}_0) \sum_{n,m=-\infty}^{\infty} e^{2\pi i h n} e^{2\pi i h m} e^{2\pi i A_x h \sin\left(2\pi \frac{n}{N-1}\right)}. \qquad (4.11)$$

We can make use of the expansion

$$e^{iy\sin\vartheta} = \sum_{p=-\infty}^{\infty} J_p(y) e^{ip\vartheta} \qquad (4.12)$$

for the third exponential in Eq. (4.11). $J_p(y)$ is the Bessel function of p-th order and argument y. Inserting Eq. (4.12) into Eq. (4.11) we obtain

$$F(h, k) = f(h, k) \sum_{p=-\infty}^{\infty} J_p(2\pi h A_x)\delta\left(h + \frac{p}{N-1}\right)\delta(k). \qquad (4.13)$$

We see that each main reflection (h, k) with integer values h, k is accompanied by an infinite set of satellite reflections; h takes integer values for the main reflections and fractional values for the satellite reflections. The Bessel functions decay rapidly with order p so that for small modulations only few satellite reflections become visible. $J_0(2\pi h A_x)$ describes the intensity of the main reflection.

The simplified kinematic theory described in this section is applicable in X-ray diffraction, but does not describe the intensities correctly in LEED, because satellite reflections generated by the modulation function cannot be distinguished from satellites generated by multiple scattering. Because of multiple scattering, more satellite reflections occur in LEED than in X-ray diffraction and the intensity relations between the intensities of main reflections and satellites cannot be evaluated directly to derive modulation amplitudes. It remains nevertheless true that the general features of the diffraction image as described by the kinematic theory are preserved in the multiple scattering theory, namely that with increasing order of the main reflections the satellite intensities tend to increase while the intensities of the main reflections tend to decrease. A more detailed description is given in Sections 6.3.7 and 6.3.8.

5 LEED Theory
Basic Formalisms

As mentioned in Section 2.2, a kinematic, that is, single-scattering, theory of LEED cannot describe experimental intensities with an accuracy that is sufficient to determine atomic positions and other non-structural information about surfaces. This degree of accuracy requires the inclusion of multiple scattering at a level of sophistication that is similar to that of electronic band structure calculations; in fact, some early versions of LEED theory employed methods of 3-D band structure theory, such as Bloch waves and pseudopotentials. However, the goal of surface structure determination by iterative optimisation of atomic positions with lower-dimensional periodicity and sometimes large 2-D unit cells requires very efficient calculational schemes of the multiple scattering of electrons.

This chapter reviews the main methods of LEED theory in recent use. More complete descriptions of the variety of basic methods developed for LEED can be found in earlier publications [5.1–5.3]. After basic geometrical considerations in Section 5.1, our account describes low-energy electron scattering by a single atom in terms of phase shifts in Section 5.2.1. The treatment of multiple scattering using plane waves and spherical waves is addressed in Section 5.2.2. This approach is then applied to the scattering by clusters of atoms in Section 5.2.3, and in particular to nanoparticles (NanoLEED) in Section 5.2.4. Electron scattering by 2-D infinite monolayers and multilayers of atoms is described in Section 5.2.5, with both exact and iterative schemes. This is followed in Section 5.2.6 by 2-D layers of finite or infinite thickness, representing the surface of bulk crystals.

Symmetry plays an important role, not only in terms of understanding and classifying structures, but also in accelerating the computations and reducing computer memory usage and is, therefore, extensively treated in Section 5.3. Approximations can also greatly reduce computational efforts and are described in Section 5.4 for tensor LEED (TLEED), frozen LEED and diffuse LEED (DLEED). Section 5.5 discusses thermal effects and, to some extent, structural disorder, which must be taken into account for accurate structure determination. Chapter 6 continues the treatment of LEED theory for special categories of materials and structures, including quasicrystals and modulated structures. Chapter 6 also addresses the practical issues involved in actual quantitative structural determination.

5.1   Diffraction Geometry

The most common experimental approach to LEED consists of a well-collimated (i.e., 'parallel') beam of near-mono-energetic electrons incident on a flat surface. If the surface is periodic, as is often the case with the surface of a single crystal, such geometry creates a series of well-collimated diffracted beams yielding a pattern of sharp spots on a display. If such a surface has long-range disorder, as in a poorly ordered overlayer of added atoms or molecules, the diffraction pattern will have diffuse intensity distributed between any spots due to ordered parts of the surface. The case of a quasicrystal surface produces a pattern with sharp spots.

For studying nanostructures it may also be useful to consider a convergent beam that is focused on a particular nanoscale region of a surface, for example a nanoparticle supported by a surface. One may also direct a collimated beam onto a single nanostructure and observe the diffusely scattered electrons. We shall in this chapter describe methods for calculating LEED intensities for all these different kinds of geometry.

For a collimated beam incident on a planar surface, the beam is normally represented in the vacuum outside the surface as a plane wave $e^{i\mathbf{k}_0 \mathbf{r}}$ with wave vector $\mathbf{k}_0$, see Figure 5.1.

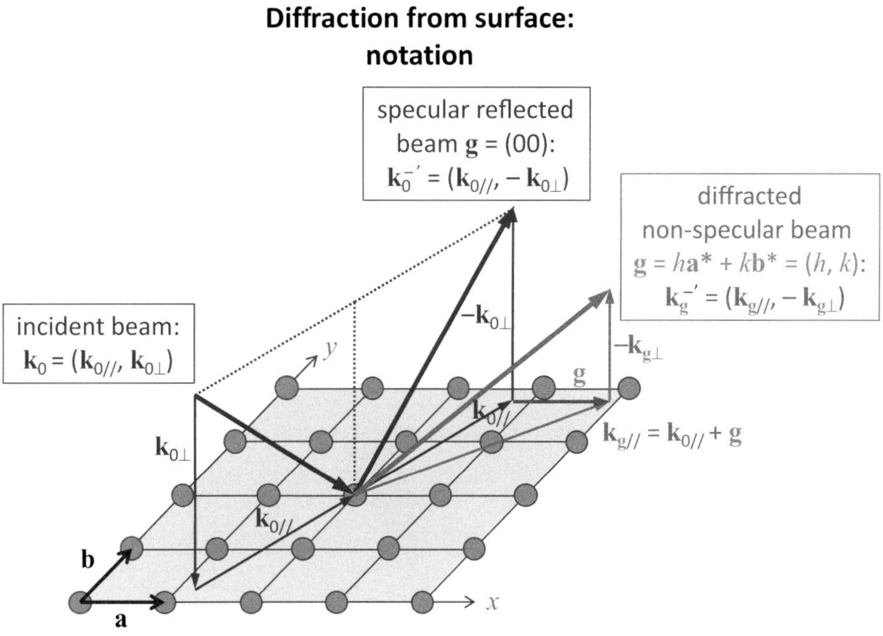

Diffraction from surface: notation

specular reflected beam $\mathbf{g} = (00)$:
$$\mathbf{k}_0^{-\prime} = (\mathbf{k}_{0//}, -\mathbf{k}_{0\perp})$$

diffracted non-specular beam $\mathbf{g} = h\mathbf{a}^* + k\mathbf{b}^* = (h, k)$:
$$\mathbf{k}_g^{-\prime} = (\mathbf{k}_{g//}, -\mathbf{k}_{g\perp})$$

incident beam:
$$\mathbf{k}_0 = (\mathbf{k}_{0//}, \mathbf{k}_{0\perp})$$

$-\mathbf{k}_{0\perp}$

$-\mathbf{k}_{g\perp}$

$\mathbf{k}_{0\perp}$

$\mathbf{k}_{0//}$

$\mathbf{g}$

$\mathbf{k}_{g//} = \mathbf{k}_{0//} + \mathbf{g}$

$\mathbf{k}_{0//}$

b

a

x

y

Figure 5.1 Diffraction from a periodic *surface* with 2-D lattice vectors **a** and **b** (grey). The beam incident $\mathbf{k}_0$ on the surface (red) is specularly reflected into the (00) beam (blue); it also diffracts into non-specular beams $\mathbf{k}_g^-$ (green), where **g** (orange) is any 2-D reciprocal lattice vector (h,k).

We have

$$|\mathbf{k}_0| = \frac{\sqrt{2m_e E}}{\hbar} = \frac{2\pi}{\lambda}, \quad \mathbf{k}_0 = (k_{0x}, k_{0y}, k_{0z}),$$

$$k_{0x} = |\mathbf{k}_0| \cos\varphi \sin\vartheta, \quad k_{0y} = |\mathbf{k}_0| \sin\varphi \sin\vartheta, \quad k_{0z} = |\mathbf{k}_0| \cos\vartheta, \tag{5.1}$$

where E and m_e are the kinetic energy and mass of the electron, respectively, while ϑ and φ are the polar and azimuthal angles of $\mathbf{k}_0$ with respect to the surface normal and the x-axis, respectively.

As explained in Chapter 2, a 2-D periodic surface structure with a 2-D reciprocal lattice $(\mathbf{a}^*, \mathbf{b}^*)$ causes diffracted plane waves (beams) $A_\mathbf{g} e^{i\mathbf{k}_\mathbf{g} \mathbf{r}}$ with wave vectors $\mathbf{k}_\mathbf{g}$, such that $|\mathbf{k}_\mathbf{g}| = |\mathbf{k}_0|$ (for energy conservation), and $\mathbf{k}_\mathbf{g} = \mathbf{k}_0 + \mathbf{g} = (k_{0x} + g_x, k_{0y} + g_y, k_{\mathbf{g}z})$ (due to parallel momentum exchange with the surface), where $\mathbf{g} = 2\pi(h\mathbf{a}^* + k\mathbf{b}^*)$ (to satisfy Laue conditions in 2-D), see Figure 5.2. The indices h and k are often used to label the diffracted beams, for example, the beam (h, k); h and k are often integers, for example, beams $(0, 0)$ or $(2, 1)$, and can be fractional in the presence of superlattices, for example, $(1/2, 0)$ or $(1/4, 3/4)$. As a result of energy conservation, we have in the vacuum outside the surface:

$$k_{\mathbf{g}z}^{\text{outside}} = -\sqrt{\left(\frac{2\pi}{\lambda}\right)^2 - (k_{0x} + g_x)^2 - (k_{0y} + g_y)^2}$$

$$= -\sqrt{\frac{2m_e E}{\hbar^2} - (k_{0x} + g_x)^2 - (k_{0y} + g_y)^2}. \tag{5.2}$$

Diffraction from atomic layer: notation

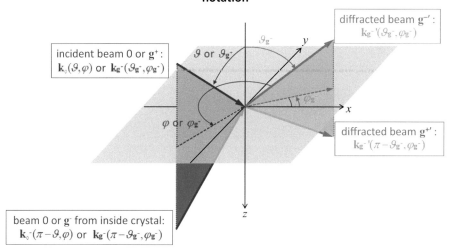

Figure 5.2 Diffraction from a 2-D periodic *layer* (grey), showing typical nomenclature for beams and angles. The + and − signs refer to beam propagation into versus out of the crystal, respectively; primes indicate scattered (reflected or diffracted) beams. The incident beam (red) can arrive directly from the electron source ($\mathbf{k}_0$) or may have already been diffracted by an atomic layer above it ($\mathbf{k}_\mathbf{g}^+$). The beam incident from inside the crystal (blue) may have been specularly reflected from a deeper layer ($\mathbf{k}_0^-$) or may have been otherwise diffracted from a deeper layer ($\mathbf{k}_\mathbf{g}^-$).

The negative sign in front of the square root signifies a wave travelling away from the surface into vacuum. If the argument of the root is positive, we obtain a propagating wave that can reach a distant external detector. However, if the argument is negative, due to large $|g_x|$ or $|g_y|$, the wave becomes evanescent (exponentially decaying): it travels parallel to the surface and cannot reach a distant detector.

Inside the surface, the LEED electrons feel a different energy due to any electric dipole layer present at the surface (this dipole layer is related to the work function that keeps the material's electrons from spilling out of a surface): this manifests itself as an upward energy shift by the 'inner potential', V_0, relative to vacuum. This inner potential may depend on the electron energy.

In addition, LEED electrons can suffer energy losses to the material; those electrons that lose energy are removed from consideration, since we assume measurement at the same energy as the nominal incident beam's energy, that is, we only follow electrons that undergo elastic scattering. This loss of electrons to inelastic processes is commonly represented by a damping of the LEED wave amplitudes through an imaginary part of the inner potential, iV_{0i}, or, equivalently, by an inelastic mean free path (IMFP), which severely limits the penetration depth of electrons into the surface. The imaginary part of the inner potential usually depends on the electron energy, for example rising with energy to represent a decreasing inelastic mean free path. The imaginary part of the inner potential is the main contributor to the energy width of peaks in I(V) curves (other contributions come from the energy spread of the electron source and the energy resolving grids, as well as simple overlap of neighbouring peaks).

As a result, the LEED electron's energy, which is E in vacuum, becomes complex inside the surface: $E + V_0 + iV_{0i}$. We obtain, inside the material, a wave vector with unchanged components parallel to the surface $(k_{0x} + g_x, k_{0y} + g_y)$ and a modified complex component perpendicular to the surface:

$$k_{gz}^{\text{inside}} = -\sqrt{\frac{2m_e}{\hbar^2}(E + V_0 + iV_{0i}) - \left(k_{0x} + g_x\right)^2 - \left(k_{0y} + g_y\right)^2}. \tag{5.3}$$

Inside the surface all 'plane' waves decay exponentially in the directions perpendicular to the surface, due to damping, in addition to the exponential decay whenever $|g_x|$ or $|g_y|$ is large. All of these waves can potentially reach and transmit electrons from one atomic layer to the next. However, the larger $|g_x|$ and $|g_y|$ are, the stronger the exponential decay will be: as a result, we can truncate the number of plane waves used, especially when layer spacings are large. For smaller layer spacings, plane waves with larger $|g_x|$ and $|g_y|$ can reach from layer to layer and must therefore be included; in fact, this growth in number of plane waves rapidly becomes computationally unstable, so that one must avoid using plane waves with small layer spacings. The solution is to use spherical waves in small layer spacings, that is, to combine such closely-spaced atomic planes into a layer with two or more atomic planes, within which a spherical-wave treatment is performed.

Consequently, most LEED formalisms represent a surface as one or more 2-D infinite slabs of finite thickness; these slabs are separated by layer spacings of at least about 0.1 nm (1 Å). Within slabs, spherical waves are used, while between and outside slabs plane waves are used. We will describe both cases in the following.

5.2   Scattering Theory

We shall stepwise build a scattering theory by starting with the simplest scatterer, a single atom. We then consider electron multiple scattering from a cluster of several atoms: this will lead in particular to diffraction from nanoparticles. In this section, we will discuss the case of a 2-D periodic layer of atoms, an important ingredient of extended surfaces. We shall then assemble such atomic layers into a surface that is thick enough to include the penetration depth of LEED electrons.

We shall assume that a collimated incident electron beam can be represented by a plane wave and that the scattering of LEED electrons is elastic. The case of a convergent LEED beam will be allowed and represented as a linear combination of a continuous set of plane waves; this case will be discussed in the context of diffraction from nanoparticles.

We shall ignore electron spin and magnetic effects. These assumptions have proved adequate to match theory to experiment for the vast majority of surfaces studied.

We also adopt the successful muffin-tin model of the scattering by a solid, chosen for its computational efficiency. This model represents each atom core as a spherically symmetric potential, as sketched in Figure 5.3, which also shows common terminology. The potential within the 'muffin-tin radius' of each atom is frequently derived from atomic or band structure calculations, as described briefly in Section 2.2.3. Section 6.1.6 gives methods to calculate the 'muffin-tin potential' for the individual

Conventional muffin-tin potential model

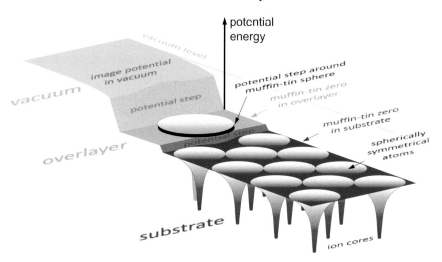

Figure 5.3 Schematic drawing of the LEED muffin-tin potential, indicating terminology

atoms from the free atom potentials and the surrounding potentials in the surface slab; there we also describe a new method that avoids artificial potential steps (Section 6.1.6.3), such as the potential step shown in Figure 5.3 surrounding the adatom in the overlayer (such steps are also frequent in more complex substrates, e.g., covalent, compound and ionic materials, but are unusual in close-packed metallic substrates, as illustrated in Figure 5.3).

Usually, the radii of the muffin-tin spheres have been chosen so that they just touch each other, but overlapping spheres are helpful, as discussed in Section 6.1.6.3. Between these atomic spheres are regions of constant potential: usually a single constant potential value, called muffin-tin zero or muffin-tin constant, or also inner potential, suffices for the entire material, especially in compact materials; in more complex cases different values can be given to different layers parallel to the surface, for example in overlayers of adatoms or admolecules, leading to potential steps between layers. This muffin-tin constant ends with a more or less abrupt potential step outside the surface, where the potential rises gradually to the vacuum level in the form of the image potential. While these potential steps outside the surface and between layers produce both reflection and refraction (change of direction), the reflection is relatively weak and usually neglected, except in very low energy electron diffraction (VLEED).

5.2.1 Atomic Scattering: Phase Shifts

It is convenient to use the phase shift formalism to describe plane-wave scattering by the spherical atomic scattering potential of the muffin-tin model, as follows. A plane wave $e^{i\mathbf{kr}}$ is scattered by a single atom into an outgoing wave $f(\vartheta)\frac{e^{ikr}}{r}$, centred on that atom, in the asymptotic limit of large distances r from the scatterer; here ϑ is the scattering angle. The scattering factor $f(\vartheta)$ for a spherical potential can be expressed in terms of a series of 'spherical' waves (more properly called 'partial' waves, centred on the atom) with amplitudes t_l given by the phase shifts δ_l according to:

$$f(\vartheta) = -4\pi\sum_{l=0}^{\infty}(2l+1)t_lP_l(\cos\vartheta), \quad t_l = -\frac{\hbar^2}{2m_e}\frac{1}{2ik}\left(e^{2i\delta_l}-1\right) = -\frac{\hbar^2}{2m_e}\frac{1}{k}\sin\delta_le^{i\delta_l},$$

(5.4)

where P_l are Legendre polynomials. The amplitudes t_l are elements of the so-called t-matrix **t**. In practice, the series over the angular quantum number l can be truncated at $l = l_{max} \sim 5$ to 10 for common LEED energies up to about 400 eV. Such phase shifts depend on the atomic element (and also on its ionic charge), as well as on the electron energy. The phase shifts themselves are obtained by numerical integration of the radial Schrödinger equation within the spherical atomic potential from the origin at $r = 0$ to the muffin-tin radius r_m, where the radial wave function has the resulting value $R_l(r_m)$ and radial derivative $R_l'(r_m)$:

$$e^{2i\delta_l} = \frac{L_l\mathrm{h}_l^{(2)}(kr_m) - \mathrm{h}_l^{(2)\prime}(kr_m)}{\mathrm{h}_l^{(1)\prime}(kr_m) - L_l\mathrm{h}_l^{(1)}(kr_m)}, \quad L_l = \frac{R_l'(r_m)}{R_l(r_m)}.$$

(5.5)

Here $h_l^{(1)}$ and $h_l^{(2)}$ are spherical Hankel functions of the first and second kinds, respectively; their radial derivatives $h_l^{(1)\prime}$ and $h_l^{(2)\prime}$ are also used.

For a spherical atomic potential, the matrix **t** is diagonal, with elements $t_{lm,l'm'} = t_l \delta_{ll'}$ (using the Kronecker symbol, with value 1 when $l = l'$ or 0 otherwise, and the magnetic quantum number $m = -l, \ldots, l$). The spherical atomic potential needed for the integration of the radial Schrödinger equation is typically obtained from solutions of the electronic structure of atoms in solids, for which many schemes are available and are often provided together with LEED codes. In some cases, phase shifts must be pre-calculated, stored and input to the LEED code, while, in other cases, the phase shifts are calculated within the LEED code only when needed. In many cases the spherical atomic potential is obtained from a superposition of free atom potentials.

An essential point is here the choice of the muffin-tin radius. For clean metal surfaces with close packed lattices this works well; the muffin-tin radius is chosen as half of the nearest neighbour distance in the bulk. For ionic or covalent compounds the choice of the muffin-tin radii is less obvious: choosing half of the inter-atomic distance in the elemental or compound bulk structures as the muffin-tin radius leads to large discontinuities of the individual potentials to the interstitial muffin-tin constant; with covalent bonds this generates a large interstitial volume which is poorly described by a constant potential and thus uniform electron density. Such potential discontinuities do not appear explicitly in the LEED multiple scattering formalism; nevertheless, the choice of the muffin-tin radius does affect the calculated LEED intensities, in particular in the lower energy range below about 50–80 eV.

Until recently, the muffin-tin model was restricted to non-overlapping spheres. It has been shown that the use of overlapping spheres leads to the same secular equations as non-overlapping spheres and the same formalism can be used with overlapping spheres provided that the muffin-tin radii do not exceed the centres of the neighbouring potentials [5.4]. The use of overlapping spheres allows minimising the potential steps between the muffin-tin spheres. This method has been developed by J. Rundgren [5.5] and applied to calculating phase shifts for surface structures where the differing environment of each atom in the surface slab can be taken into account. This method led in a number of cases to a substantial improvement of the LEED analysis compared to separately calculated phase shifts from distinct elemental bulk structures.

For a more realistic non-spherical atomic potential (often called a full potential), the matrix **t** becomes non-diagonal: all its elements can be non-zero, making their computation and the calculation of multiple scattering more complex. Some attempts have been made to include full potentials in the multiple scattering formalism [5.6–5.9], but no LEED code currently in use for structure determination includes such full potentials, to our knowledge.

Thermal atomic vibrations are normally and conveniently treated through 'temperature-dependent' phase shifts of the individual atoms, based on the Debye–Waller formalism of X-ray diffraction. This assumes isotropic and uncorrelated vibrations, thus neglecting correlations of vibrations between neighbouring atoms, as are present in phonons. Thermal vibrations tend to decrease the intensity of diffracted beams, as they redirect intensity to a more diffuse background in all scattering directions. With the Debye model of thermal vibrations, the intensity scattered by a crystal atom in the single-scattering (kinematic) limit is multiplied by the Debye–Waller factor:

$$e^{-2M} = e^{-\frac{3\hbar^2 |s|^2 T}{m_a k_B \Theta_D^2}}, \tag{5.6}$$

where s is the momentum transfer, T the sample temperature, m_a the atomic mass of the surface atoms, k_B the Boltzmann constant and Θ_D the Debye temperature of the material. For the multiple-scattering (dynamical) case of LEED, one assumes that the scattering amplitude $f(\vartheta)$ of each atom is multiplied by the Debye–Waller factor to yield an effective scattering amplitude $e^{-M} f(\vartheta)$, expressed in terms of effective 'temperature-dependent' phase shifts, after which multiple scattering is calculated as in the case of no thermal effect. This temperature-dependent scattering amplitude can be simply obtained by replacing the previously discussed phase shifts δ_l by the following modified temperature-dependent phase shifts (with m_e the electron mass):

$$\delta_l(T) = \frac{1}{2i} \ln \left[1 - \frac{4kim_e}{\hbar^2} t_l(T) \right]. \tag{5.7}$$

This approximation is widely used: it assumes isotropic thermal motion while correlation effects are ignored; the thermal motion of each atom is assumed to be independent from its neighbours. A more detailed discussion of thermal effects is given in Section 5.5.

5.2.2 Multiple Scattering: Plane Waves and Spherical Waves

To efficiently treat the multiple scattering of electrons by an assembly of atoms, we will select simple wave types to represent the full LEED wave function. Specifically, as discussed in Section 5.2, we will express the full electron wave function as a linear combination of either plane waves or 'spherical' waves (more accurately, if less evocatively, called 'partial' waves). The plane waves are a natural choice for a well-collimated incident beam and well-collimated diffracted beams due to a 2-D periodic surface and because the muffin-tin model contains regions of constant potential. The spherical waves are convenient in view of the assumed spherical shape of the atomic scattering potentials. Using spherical waves centred on each atom is called a 'multi-centre expansion'.

An alternative approach is the 'one-centre expansion', in which all spherical waves are centred on the same point (not necessarily an atomic nucleus): its main disadvantage is the need for larger angular momenta l, roughly proportional to the outer radius of the cluster of atoms that it describes, instead of the atomic radius in the multi-centre expansion: this increases the computational time considerably. Hence the one-centre expansion is rarely used in LEED.

Most LEED theories use both plane waves and spherical waves in different parts of the calculations: as mentioned, plane waves are often used outside and between slabs of atoms, while spherical waves are often used within slabs of atoms, as well as within clusters of atoms; this mixed approach is sometimes called 'combined space method'. Spherical waves are thus also convenient within nanoparticles.

The spherical waves can be limited in number by the size of the phase shifts: as mentioned, this number is often truncated at $l = l_{max} \sim 5$ to 10 for common LEED energies. For each value of the angular quantum number l, there are $2l + 1$ allowed values of the magnetic quantum number m. This results in a total of $(l_{max} + 1)^2$ partial waves for a given value of l_{max}.

The plane waves are given by the 2-D reciprocal lattice of a periodic surface: $\mathbf{k_g}^\pm = \mathbf{k_0}^\pm + \mathbf{g} = (k_{0x} + g_x, k_{0y} + g_y, \pm k_{gz})$, where $\mathbf{g} = 2\pi(h\mathbf{a^*} + k\mathbf{b^*})$. Here we have added signs $\pm$ to represent two sets of plane waves: those moving into a surface $(+)$ and those moving out of a surface $(-)$. The number of plane waves used is limited by the fact that for large $|g_x|$ or $|g_y|$ the waves become strongly evanescent on both sides of an atomic layer and therefore do not carry electrons from one atomic layer to another.

5.2.3   Multiple Scattering in a Cluster of Atoms

The scattering of electrons by an assembly of atoms can, in principle, be solved self-consistently and relatively efficiently through a multiple-scattering approach. Multiple scattering means that we allow an initially simple electron wave to scatter from individual atoms as many times as needed; normally, each subsequent scattering contributes a decreasing correction to the resulting total wave. Self-consistency is obtained when the multiple scattering converges.

However, scattering in LEED can be strong, depending on the atoms, scattering angles and electron energies involved and can prevent such convergence. In that case, it is still often possible to obtain a self-consistent solution, at a higher computational cost, with a closed formula. Typically, multiple scattering can be written as an infinite series of terms corresponding to increasing numbers of scatterings (e.g., a non-scattered term plus a single-scattering term plus a double-scattering term plus a triple-scattering term, etc.), as symbolically illustrated with the series $1 + x + x^2 + x^3 \ldots$. This series can usually be summed into a closed formula, here $(1 - x)^{-1}$, which normally involves an inversion: since the mathematical quantities, like x here, involved in LEED are matrices, we perform matrix inversion in closed form. The computational expense of matrix inversion is often much larger than for a convergent series expansion, but it gives an exact and self-consistent solution of the multiple scattering problem. More effective and common than full matrix inversion is the method of Gaussian elimination and back substitution, which is also exact and self-consistent.

Much work has identified other efficient convergent multiple-scattering LEED schemes. We here describe the dominant approach, which is based on the Korringa–Kohn–Rostoker (KKR) methodology. It uses the atom scattering matrix $\mathbf{t}$ and propagates spherical waves $h_l^{(1)}(kr)Y_{lm}(\vartheta, \varphi)$ through regions of constant potential between atoms, in accordance with the muffin-tin model.

We start by describing the multiple scattering between a pair of atoms: this will then be extended to multiple scattering within a cluster of N atoms, before considering one periodic plane of atoms, several periodic planes of atoms, and ultimately an infinitely deep surface of atoms.

In this process, we in effect propagate simple spherical waves all the way from one atom *nucleus* to another, while the scattering itself is completely contained in the t-matrix: it is as if we had point scatterers representing dimensionless atoms but having the scattering properties of real extended atoms. As a result, the argument r in $h_l^{(1)}(kr)Y_{lm}(\vartheta, \varphi)$ will be taken as the distance between two atomic nuclei.

Suppose that a wave has been scattered from one atom in a pair of atoms: leaving that first atom is a wave that can be expressed as a linear combination of spherical waves. To include the effect of a subsequent scattering by the other atom, we must then express how a spherical wave of given angular momentum $L' = (l', m')$ centred on the first atom at $\mathbf{r}_1$ propagates to the second one, and how it decomposes into spherical waves centred and incident on that second atom at $\mathbf{r}_2$. The result is given by a Green's function, which, for a constant potential between the atoms, takes the following form:

$$\overline{G}_{LL'}^{21} = -4\pi i \frac{2m_e}{\hbar^2} k \sum_{L_1} i^{l_1'} a(L, L', L_1) h_{l_1}^{(1)}(k|\mathbf{r}_2 - \mathbf{r}_1|) Y_{L_1}(\mathbf{r}_2 - \mathbf{r}_1). \tag{5.8}$$

Here, $k = |\mathbf{k_g}^{\pm}|$, while $L_1 = (l_1, m_1)$ runs over all values of l_1 and m_1 compatible with $L = (l, m)$ and $L' = (l', m')$, namely $|l - l'| \leq l_1 \leq l + l'$ and $m + m' = m_1$ (this summation will be assumed implicitly throughout this book, unless otherwise stated). This compatibility corresponds to non-zero values of the Clebsch–Gordan (or Gaunt) coefficients

$$a(L, L', L_1) = \int Y_L^*(\Omega) Y_{L'}(\Omega) Y_{L_1}^*(\Omega) d\Omega, \tag{5.9}$$

where Ω ranges over all values of solid angle.

Equation (5.8) gives the amplitude of the spherical wave $L = (l, m)$ centred on and arriving at the second atom at $\mathbf{r}_2$ due to a spherical wave $L' = (l', m')$ centred on and leaving the first atom at $\mathbf{r}_1$. The Hankel function $h_{l_1}^{(1)}$ and the spherical harmonic Y_{L_1} describe the propagation, while the Clebsch–Gordan coefficients are due to the expansion of a spherical wave centred on one atom into spherical waves centred on the second atom.

We now have the tools to express an arbitrary scattering path involving the given pair of atoms 1 and 2. Consider the path drawn in Figure 5.4: the matrix expression $\mathbf{t}^2 \overline{\mathbf{G}}^{21} \mathbf{t}^1 \overline{\mathbf{G}}^{12} \mathbf{t}^2 \overline{\mathbf{G}}^{21} \mathbf{t}^1$ should be read, from right to left, as first a scattering $\mathbf{t}^1$ at atom 1 (labelled by the superscript 1), followed by propagation $\overline{\mathbf{G}}^{21}$ from atom 1 to atom 2 (superscripts '21'), then a scattering $\mathbf{t}^2$ at atom 2, followed by propagation $\overline{\mathbf{G}}^{12}$ back to atom 1, scattering there again and finally propagating again with $\overline{\mathbf{G}}^{21}$ to atom 2 for a final scattering $\mathbf{t}^2$ there. This particular multiple-scattering path involves four scatterings and three inter-atomic propagations. It can be extended to arbitrary length. The matrix notation implies summation over intermediate spherical waves of all possible angular momenta $L = (l, m)$.

Multiple scattering by two atoms

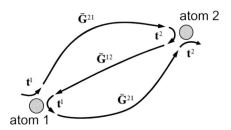

Figure 5.4 A multiple scattering path for a pair of atoms, showing nomenclature

We can now express and sum up all possible scattering paths, of any length, within the pair of atoms 1 and 2. We classify these paths according to the terminal atom, that is, whether each path ends at atom 1 or at atom 2, resulting in two series $\mathbf{T}^1$ and $\mathbf{T}^2$, respectively:

$$\mathbf{T}^1 = \mathbf{t}^1 + \mathbf{t}^1\overline{\mathbf{G}}^{12}\mathbf{t}^2 + \mathbf{t}^1\overline{\mathbf{G}}^{12}\mathbf{t}^2\overline{\mathbf{G}}^{21}\mathbf{t}^1 + \mathbf{t}^1\overline{\mathbf{G}}^{12}\mathbf{t}^2\overline{\mathbf{G}}^{21}\mathbf{t}^1\overline{\mathbf{G}}^{12}\mathbf{t}^2 + \cdots, \tag{5.10}$$

$$\mathbf{T}^2 = \mathbf{t}^2 + \mathbf{t}^2\overline{\mathbf{G}}^{21}\mathbf{t}^1 + \mathbf{t}^2\overline{\mathbf{G}}^{21}\mathbf{t}^1\overline{\mathbf{G}}^{12}\mathbf{t}^2 + \mathbf{t}^2\overline{\mathbf{G}}^{21}\mathbf{t}^1\overline{\mathbf{G}}^{12}\mathbf{t}^2\overline{\mathbf{G}}^{21}\mathbf{t}^1 + \cdots. \tag{5.11}$$

In these two infinite series, we first have the single scattering from atom 1 ($\mathbf{t}^1$) or atom 2 ($\mathbf{t}^2$), respectively. Next comes the double-scattering term, followed by triple scattering and then quadruple scattering, etc. All paths in Eq. (5.10) end at atom 1, while all paths in Eq. (5.11) end at atom 2. The meaning of $\mathbf{T}^1$ and $\mathbf{T}^2$ can be easily understood by neglecting all multiple scattering, such that $\mathbf{T}^1 = \mathbf{t}^1$ and $\mathbf{T}^2 = \mathbf{t}^2$: we obtain the scattering matrix of the single atoms individually; the full expressions for $\mathbf{T}^1$ and $\mathbf{T}^2$ simply add to that all the multiple-scattering paths between the two atoms.

The two series Eqs. (5.10) and (5.11) can be solved self-consistently in closed form, once we specify the proper phases of the incident wave. If the incident wave is $e^{i\mathbf{k}_{\mathrm{in}}\mathbf{r}}$, there is an amplitude ratio of $e^{i\mathbf{k}_{\mathrm{in}}(\mathbf{r}_1-\mathbf{r}_2)}$ between the amplitudes of this wave arriving from the outside at atoms 1 and 2. We shall insert this phase factor within the definition of $\overline{\mathbf{G}}^{21}$, which now becomes

$$\overline{G}^{21}_{LL'} = -4\pi i\frac{2m_e}{\hbar^2}k\sum_{L_1}i^{l_1}a(L,L',L_1)h^{(1)}_{l_1}(k|\mathbf{r}_2-\mathbf{r}_1|)Y_{L_1}(\mathbf{r}_2-\mathbf{r}_1)e^{i\mathbf{k}_{\mathrm{in}}(\mathbf{r}_1-\mathbf{r}_2)}. \tag{5.12}$$

The same expression with $\mathbf{r}_1$ and $\mathbf{r}_2$ exchanged gives $\overline{G}^{12}_{LL'}$. Inspection shows that Eqs. (5.10) and (5.11) are very closely linked, namely:

$$\mathbf{T}^1 = \mathbf{t}^1 + \mathbf{t}^1\overline{\mathbf{G}}^{12}\mathbf{T}^2, \tag{5.13}$$

$$\mathbf{T}^2 = \mathbf{t}^2 + \mathbf{t}^2\overline{\mathbf{G}}^{21}\mathbf{T}^1. \tag{5.14}$$

Now, Eqs. (5.13) and (5.14) constitute a self-consistent set of equations that can be solved in closed form for $\mathbf{T}^1$ and $\mathbf{T}^2$. Using matrix notation, we get

$$\begin{pmatrix} \mathbf{T}^1 \\ \mathbf{T}^2 \end{pmatrix} = \begin{pmatrix} \mathbf{I} & -\mathbf{t}^1 \overline{\mathbf{G}}^{12} \\ -\mathbf{t}^2 \overline{\mathbf{G}}^{21} & \mathbf{I} \end{pmatrix}^{-1} \begin{pmatrix} \mathbf{t}^1 \\ \mathbf{t}^2 \end{pmatrix}, \tag{5.15}$$

where $\mathbf{I}$ is a unit matrix and everything is known on the right-hand side. The dimension of the matrices $\mathbf{I}$, $\mathbf{G}$, $\mathbf{t}$, and $\mathbf{T}$ is determined by the maximum number of phase shifts required to describe the atomic scattering: if $l \leq l_{max}$, and all values of m such that $-l \leq m \leq l$ are included, then this dimension is $(l_{max} + 1)^2$ (the larger matrix to be inverted will have double that dimension). We note that the geometrical series obtained by expanding the large-matrix inversion of Eq. (5.15) reproduces Eqs. (5.10) and (5.11), similar to $(1 - x)^{-1} = 1 + x + x^2 + \dots$. Equation (5.15) is rigorously valid even if Eqs. (5.10) and (5.11) do not converge well or do not converge at all, because Eqs. (5.13) and (5.14) are rigorous and do not depend on convergence of the series.

We can now generalise these results from a pair of atoms to a cluster of N atoms at positions $\mathbf{r}_1$ through $\mathbf{r}_N$. Equation (5.15) becomes

$$\begin{pmatrix} \mathbf{T}^1 \\ \mathbf{T}^2 \\ \mathbf{T}^3 \\ \vdots \\ \mathbf{T}^N \end{pmatrix} = \begin{pmatrix} \mathbf{I} & -\mathbf{t}^1 \overline{\mathbf{G}}^{12} & -\mathbf{t}^1 \overline{\mathbf{G}}^{13} & \cdots & -\mathbf{t}^1 \overline{\mathbf{G}}^{1N} \\ -\mathbf{t}^2 \overline{\mathbf{G}}^{21} & \mathbf{I} & -\mathbf{t}^2 \overline{\mathbf{G}}^{23} & \cdots & -\mathbf{t}^2 \overline{\mathbf{G}}^{2N} \\ -\mathbf{t}^3 \overline{\mathbf{G}}^{31} & -\mathbf{t}^3 \overline{\mathbf{G}}^{32} & \mathbf{I} & \cdots & -\mathbf{t}^3 \overline{\mathbf{G}}^{3N} \\ \vdots & \vdots & \vdots & \ddots & \vdots \\ -\mathbf{t}^N \overline{\mathbf{G}}^{N1} & -\mathbf{t}^N \overline{\mathbf{G}}^{N2} & -\mathbf{t}^N \overline{\mathbf{G}}^{N3} & \cdots & \mathbf{I} \end{pmatrix}^{-1} \begin{pmatrix} \mathbf{t}^1 \\ \mathbf{t}^2 \\ \mathbf{t}^3 \\ \vdots \\ \mathbf{t}^N \end{pmatrix}. \tag{5.16}$$

The matrix to be inverted now has dimension $N(l_{max} + 1)^2$, proportional to the number of atoms in the cluster (again, computationally this matrix is actually not inverted: more efficient standard methods are available to solve this equation). Solving Eq. (5.16) produces the amplitudes $\mathbf{T}^i$ of outgoing waves at each atom $i = 1, \dots, N$, due to the incident plane wave $e^{i\mathbf{k}_{in}\mathbf{r}}$. To obtain the amplitude of a total outgoing plane wave $e^{i\mathbf{k}_{out}\mathbf{r}}$, we must combine the waves leaving the different atoms with appropriate phases:

$$\mathbf{T} = \sum_{i=1}^{N} e^{i(\mathbf{k}_{in} - \mathbf{k}_{out})\mathbf{r}_i} \mathbf{T}^i. \tag{5.17}$$

We next generalise this result to the case of a non-planar incident wave [5.10], such as a convergent spherical wave with an arbitrary angular profile; it will result in diffraction in all directions instead of diffraction into specific beams. We can represent a general incident wave as a continuous superposition of plane waves $\varphi(\mathbf{k}_{in}; \mathbf{r}) = (2\pi)^{-\frac{3}{2}} e^{i\mathbf{k}_{in}\mathbf{r}}$ with amplitudes $A(\mathbf{k}_{in})$, so that the total incident electron wave function can be written as an integral over all incident directions contained within a suitable solid angle Ω_{in}:

$$\Psi_{in}(\mathbf{r}) = \int_{\Omega_{in}} A(\mathbf{k}_{in}) \varphi(\mathbf{k}_{in}; \mathbf{r}) d\mathbf{k}_{in} = (2\pi)^{-\frac{3}{2}} \int_{\Omega_{in}} A(\mathbf{k}_{in}) \exp(i\mathbf{k}_{in}\mathbf{r}) d\mathbf{k}_{in}. \tag{5.18}$$

At a position $\mathbf{R}$ far from the scattering cluster, the wave scattered by the cluster of N atoms at positions $\mathbf{r}_i$ becomes:

$$\Psi_{sc}(\mathbf{R}) = -\sqrt{\frac{\pi}{2}} \frac{1}{k^2} \frac{e^{ikR}}{R} \sum_{i,L} e^{-ik\hat{\mathbf{R}}\mathbf{r}_i} Y_L'(\hat{\mathbf{R}}) X_L^i(k, \mathbf{r}_i, \Omega_{in}), \text{ with} \tag{5.19}$$

$$X_L^i(k, \mathbf{r}_i, \Omega_{in}) = \sum_{L'} \int_{\Omega_{in}} A(\mathbf{k}_{in}) e^{i\mathbf{k}_{in}\mathbf{r}_i} Y_{L'}^*(\hat{\mathbf{k}}_{in}) d\hat{\mathbf{k}}_{in} T_{LL'}^i, \tag{5.20}$$

using $T_{LL'}^i$ from Eq. (5.16) and the notation $\hat{\mathbf{v}}$ for the unit vector in the direction of any vector $\mathbf{v}$.

We can write Eq. (5.19) as $\Psi_{sc}(\mathbf{R}) = \frac{e^{ikR}}{R} f(\theta_{sc}, \varphi_{sc})$ to highlight the similarity with the atomic scattering factor $f(\theta_{sc})$, which has no azimuthal dependence. This results in the following differential cross-section into solid angle Ω_{sc}:

$$\frac{d\sigma}{d\Omega}(\Omega_{sc}) = (2\pi)^3 \frac{|f(\Omega_{sc})|^2}{\int_{\Omega_{in}} |A(\hat{\mathbf{k}}_{in})|^2 d\hat{\mathbf{k}}_{in}}. \tag{5.21}$$

For the special case of the collimated electron beam of normal LEED, with unique incident direction $\mathbf{k}_{in}$, we obtain:

$$\frac{d\sigma}{d\Omega}(\Omega_{sc}) = (2\pi)^3 |f(\Omega_{sc})|^2, \tag{5.22}$$

with the changed definition

$$X_L^i(k, \mathbf{r}_i, \Omega_{in}) \rightarrow X_L^i(k, \mathbf{r}_i, \mathbf{k}_{in}) = \sum_{L'} e^{i\mathbf{k}_{in}\mathbf{r}_i} Y_{L'}^*(\hat{\mathbf{k}}_{in}) T_{LL'}^i.$$

5.2.4   Multiple Scattering in Nanoparticles: NanoLEED

Nanoparticles contain at least tens to thousands of atoms. Their finite structure prevents the computational efficiencies provided by the 2-D periodicity of surfaces.

Equation (5.16) solves self-consistently the multiple scattering in a cluster of N atoms through a matrix inversion (or a related method such as Gaussian elimination/ back substitution). Due to its matrix dimension $D = N(l_{max} + 1)^2$, and the fact that the computational cost of matrix inversion scales with the cube (or the square for some iterative schemes) of the cluster size, the cost rapidly becomes unaffordable. A modest cluster of 10 atoms and an average value of $l_{max} = 6$ already gives a matrix dimension of $D = 490$. For nanoparticles of hundreds or thousands of atoms, this approach is prohibitive. Even the series expansion, as in Eqs. (5.10) and (5.11), quickly becomes very demanding.

For that reason, more efficient approximate methods have been adapted for larger clusters, under the common name NanoLEED [5.10–5.13], namely: the sparse matrix canonical grid (SMCG) method and the UV method, described in more detail in

Sections 5.2.4.1–5.2.4.4. Their common feature is to more efficiently calculate $\mathbf{x} = \mathbf{A}^{-1}\mathbf{b}$, as in Eq. (5.16), which is the solution of the matrix-vector equation $\mathbf{A}\mathbf{x} = \mathbf{b}$. These two methods scale computationally as $D\log D$, instead of the cube of the matrix dimension D. These methods have distinct capabilities and are best used in parallel with each other in a single code, to solve each part of the multiple-scattering problem with the most effective method.

The SMCG and UV methods are not limited to nanoparticles: they can be applied equally well in complex periodic surface layers, for example, for a surface with a large bulk unit cell or large surface-unit-cell reconstruction, or in an overlayer of complex molecules such as C_{60}. Disordered surfaces can also be handled by NanoLEED [5.10], since the method does not require periodicity. The NanoLEED approach can also be combined with tensor LEED (TLEED) and other approximations to enable automated structural optimisation, cf. Section 5.4 [5.13].

In addition, convergent beams (such as the incident convergent wave shown in Eq. (5.18)) can also be accommodated, in addition to the traditional collimated beams of LEED. This allows studying individual nanostructures or nanoscale areas of a surface, since a convergent electron beam can sample just the small structure of interest, as proposed under the name convergent-beam LEED [5.14]; it may even be possible to focus a beam down to the nanoscale [5.15]. This enables the use of highly focused low-energy electron microscopy (LEEM), for example [5.16; 5.17]. Another experimental method to expose a small surface area uses a nearby STM tip to create a narrow beam of electrons aimed at a surface [5.18]. Such a tip-emitted beam expands to some extent with an angular spread of perhaps 5°, so it is only approximately collimated. While probe diameters of the order of 40 μm have already been achieved experimentally by this method, the potential exists of exposing regions smaller than about 40 nm, close to the size of many nanostructures.

5.2.4.1   The Sparse-Matrix Canonical Grid Method

In the sparse-matrix canonical grid (SMCG) method [5.12; 5.19], the scaling of the computational cost is improved by the fast Fourier transform (FFT). This is made possible by changing matrix $\mathbf{A}$ to be strictly periodic (as for a periodic structure), that is, $A_{n,m} = A_{n-m}$, even though the nanoparticle structure may not be periodic at all. (The matrix A is called a Toeplitz matrix, and contains only $2D - 1$ independent elements, instead of D^2). The major step of the inversion of such a matrix can be performed by fast Fourier transform (FFT), which scales as $D\log D$. However, the use of FFT imposes a constraint on D, which must be an integer power of 2.

For LEED, this method would require that the atoms occupy a periodic, rectangular spatial grid. However, with an arbitrary non-periodic structure, including any nanostructure, we can construct a regular grid of points and refer each atom to its nearest grid point, as illustrated in Figure 5.5; then the propagation of an electron from an atom i to an atom j proceeds via the grid points P and Q nearest to atoms i and j, respectively, that is, along the path $i \rightarrow P \rightarrow Q \rightarrow j$.

This SMCG method is exact if a sufficient number of partial waves $(l_{max} + 1)^2$ is used, so as to enable the accurate propagation of electrons between each atom and its

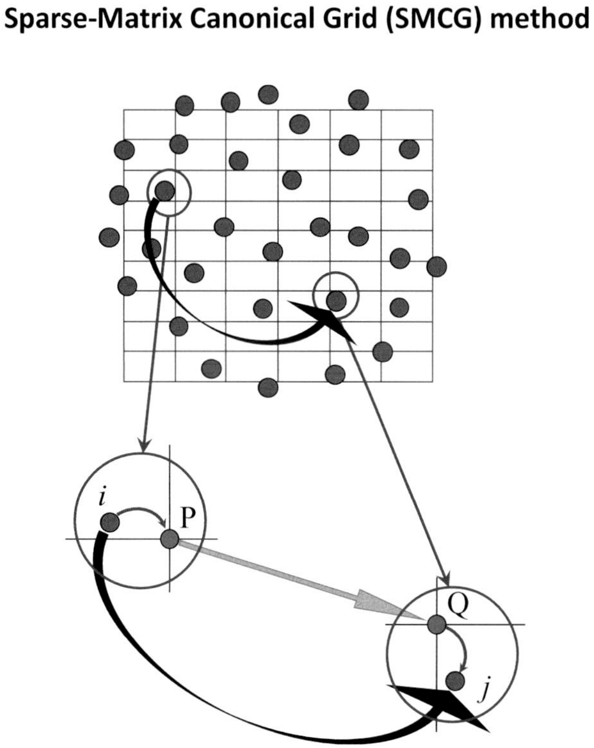

Sparse-Matrix Canonical Grid (SMCG) method

Figure 5.5 Principle of the sparse-matrix canonical grid (SMCG) method. The scattering path from any atom i to any atom j is made to pass through their nearest grid points P and Q, respectively, of a periodic grid overlain on the nanoparticle. The bottom panel zooms in on points i and j. Figure redrawn with permission from [5.12] G. M. Gavaza, Z. X. Yu, L. Tsang, C. H. Chan, S. Y. Tong and M. A. Van Hove, *Phys. Rev. B*, vol. 75, p. 014114, 2007. http://dx.doi.org/10.1103/PhysRevB.75.014114. © (2007) by the American Physical Society.

nearest grid point; this number of partial waves increases with the distance between atom and grid point (because l_{max} should be roughly proportional to that distance) and is reduced by using a denser grid. The major part of the computation, however, turns out to be the scattering between the regular grid points, which is solved with FFT in times proportional to $N_g \log N_g$, where N_g is the number of grid points used (in each dimension, FFT requires a number of grid points that is a power of 2): the number N_g is approximately related to the number of atoms. Thus, a compromise is needed between grid density (which reduces the distance between atoms and grid points and thus l_{max}) and the total number of grid points (which affects the cost of FFT).

5.2.4.2   The UV Method

An alternative method to SMCG uses singular value decomposition (SVD) [5.12; 5.20]. If the rank of matrix **A** (i.e., the number of its non-zero eigenvalues) is $r < D$, where D

is the matrix dimension, $\mathbf{A}$ can be factored into a product of three matrices, $\mathbf{A}_{D \times D} = \mathbf{U}_{D \times r} \mathbf{R}_{r \times r} \mathbf{V}_{r \times D}$ (the subscripts here indicate the matrix dimensions), where the smaller matrix $\mathbf{R}$ is diagonal with dimension r and contains the r non-zero eigenvalues of $\mathbf{A}$, while $\mathbf{U}$ and $\mathbf{V}$ are rectangular. LEED does not produce vanishing eigenvalues, but this approach can still be used approximately by equating small eigenvalues to zero: we can then replace $\mathbf{A}$ by $\mathbf{URV}$. Thereby, the smaller is the rank r, the faster is the computation. To efficiently find the rank and the singular values of a matrix, the so-called UV method can be used for the SVD decomposition. This approach also leads to computation times proportional to $D \log D$.

The UV method is an approximation exploiting prior knowledge of or prediction for the value of the rank r. If one can predict or estimate a certain value r_e for the rank of the matrix $\mathbf{A}$, then the decomposition can be performed using not the entire matrix $\mathbf{A}$, but only linearly independent rows and columns *sampled* from the matrix $\mathbf{A}$. This way, the decomposition complexity is lowered to Dr_e^2 and so its computational weight is drastically reduced, since the bulk of the calculation is spent on the matrix-vector multiplication.

'Sampling' is done with an algorithm that can select from the rows and columns of the matrix $\mathbf{A}$ a set of r_e linearly independent rows and r_e linearly independent columns. To ensure the accuracy of the method, one must overestimate the rank r of $\mathbf{A}$ when predicting the value of r_e ($r_e > r$) and, in order to ensure a high method efficiency, the difference between estimated and real rank should be as low as possible ($r_e \cong r$). This is achieved in practice by using empirical values obtained from prior studies on the system of interest. The resulting accuracy depends on the choice of the sampled rows and columns. In LEED practice a rank of ~15 is found to be adequate for neighbouring atoms, declining rapidly to ~3 for atoms distant by 1.5 nm (15 Å) or more.

Figure 5.6 illustrates the application of the SMCG and UV methods to an ordered (4×4) layer of C_{60} molecules on a Cu(111) surface [5.12]: they are visually almost indistinguishable from an exact calculation performed with the conjugate gradient (CG) method, which is an iterative method that scales as D^2 and converges to the result of exact matrix inversion. In the combined SMCG+UV+CG method, the relative compute times at 100 eV were in the proportions SMCG : UV : CG ~ 100 : 16.1 : 4.4, while the scaling was confirmed to be very close to $D \log D$.

5.2.4.3   NanoLEED for Nanoparticles

The SCMG and UV methods are most efficient in different circumstances. SMCG outperforms UV for large numbers of atoms and large inter-atomic distances. Also, the more conventional CG method is more efficient than either SMCG or UV for small numbers of atoms and small inter-atomic distances [5.12]. It is therefore useful to combine these three methods into a single 'NanoLEED' code which selects the most efficient approach at each stage: a single structure often includes near, intermediate and distant neighbours, so different methods are applied to different pairs of atoms in the same structure.

Figure 5.7 illustrates the application of NanoLEED to nanowires of different sizes. The LEED pattern becomes diffuse due to the finite size of the nanoparticles, such that

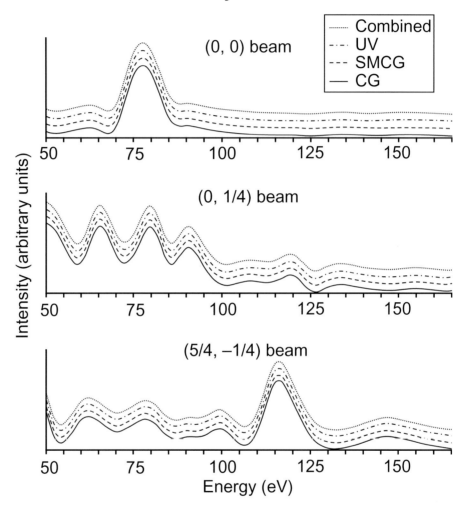

Figure 5.6 LEED intensities calculated for three beams diffracted from Cu(111)+(4×4)–C_{60} by the conjugate gradient method (CG), the sparse-matrix canonical grid method (SMCG), the UV method and their combined use (the latter approach is described in Section 5.2.4.3). Figure redrawn with permission from [5.12] G. M. Gavaza, Z. X. Yu, L. Tsang, C. H. Chan, S. Y. Tong, and M. A. Van Hove, *Phys. Rev. B*, vol. 75, p. 014114, 2007. http://dx.doi.org/10.1103/ PhysRevB.75.014114. © (2007) by the American Physical Society.

the inherent periodicity of the upper surface (exposed to a collimated LEED beam) produces a pattern of broadened spots that correspond directly to the conventional sharp LEED spot pattern of the infinite surface.

Figure 5.8 illustrates the structural sensitivity of LEED intensities from a single surface layer of nanowires. It was found that 'circular' detector scans around the

(a) Short & thick Si nanowire

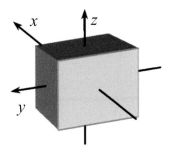

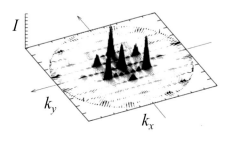

(b) Long & thin Si nanowire

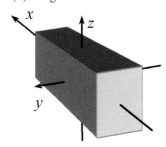

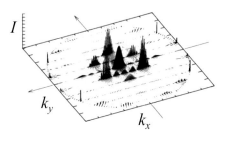

Figure 5.7 LEED patterns calculated by NanoLEED for Si nanowires of two different lateral sizes sketched on the left: (a) $(x, y, z) = (4.9, 1.2, 0.7)$ nm and (b) $(x, y, z) = (10.2, 0.8, 0.7)$ nm. The z-axis corresponds to the surface normal in usual LEED experiments from surfaces; the upper z-face exposes an ideal bulk-like Si(111) surface; normal incidence of a collimated LEED beam directed toward $-z$ is assumed. The $\langle 1, 1, -2 \rangle$ Si crystallographic direction points along x. The LEED patterns (on the right) show bulk-induced maxima in directions implied by the Si (111) surface structure; these maxima are broadened in inverse proportion to the different x- and y-dimensions of the nanoparticles. For clarity, the plots only show intensities along radial lines of the patterns, causing artificial fine structure. Figure redrawn with permission from [5.10] G. M. Gavaza, Z. X. Yu, L. Tsang, C. H. Chan, S. Y. Tong, and M. A. Van Hove, *Phys. Rev. B*, vol. 75, p. 235403, 2007. http://dx.doi.org/10.1103/PhysRevB.75.235403. © (2007) by the American Physical Society.

nanowire axis show sensitivity to even minor structural changes. In this example of a silicon nanowire, the upper Si surface monolayer (saturated with hydrogen which can produce an outward expansion of the outermost layer spacing) is allowed to 'warp', being more expanded near the centre than near the ends of the nanowire, by a difference of only 0.004 nm (0.04 Å). The effect on LEED intensities of such small structural changes can be appreciable under suitable diffraction geometries.

5.2.4.4   NanoLEED with Matrix Inversion in Subclusters

One finds that strong multiple scattering may occasionally cause poor convergence in NanoLEED (e.g., within a SiH_3 group, surprisingly, meaning that strong multiple

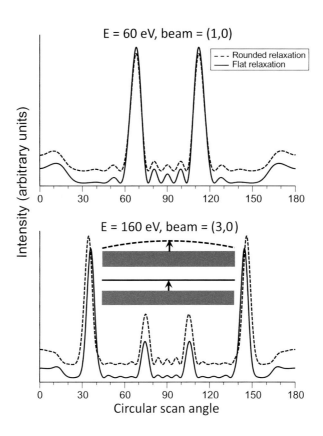

Figure 5.8 Effect on calculated LEED intensities of 'rounding' the relaxation of the topmost SiH$_3$ layer of a silicon nanowire of dimensions 10.2 × 1.8 × 0.7 nm exposing a H-saturated Si (111) surface to the incident collimated beam: the nanowire is shown as two insets at the bottom, with length to left and right, and with height up and down; the lower inset shows a uniform outward relaxation of the outer monolayer (full line); the upper inset shows a rounded expansion. The rounding consists in expanding the outer Si monolayer (decorated with H$_3$) more at the centre and less at the ends of the nanowire (cf. lower nanowire with curved outer monolayer shown as dashed line): the centre of the monolayer is expanded outward from the bulk by 0.012 nm (0.12 Å), while the far ends of the monolayer are only expanded outward by 0.008 nm (0.08 Å): the difference of 0.004 nm (0.04 Å) induces noticeable changes of the relative peak heights or shifts of the major peaks in azimuthal direction. Shown are 'circular' scans halfway around the long axis of the nanowire, with 90° being along the vertical axis of the nanowires and marking specular reflection from the top (relaxed) facet of the nanowire. Figure redrawn with permission from [5.10] G. M. Gavaza, Z. X. Yu, L. Tsang, C. H. Chan, S. Y. Tong, and M. A. Van Hove, Phys. Rev. B, vol. 75, p. 235403, 2007. http://dx.doi.org/10.1103/ PhysRevB.75.235403. © (2007) by the American Physical Society.

scattering can build up between the Si and H atoms). To overcome this limitation, such problematic small subclusters of atoms can be treated with complete matrix inversion to guarantee full and accurate inclusion of multiple scattering within them: the result is a (non-diagonal) scattering matrix describing the exact and complete scattering by that subcluster, which can then be included as a non-spherical 'pseudo-atom' in the NanoLEED method [5.13]. This approach uses the idea of one-centre expansion applied earlier by D. K. Saldin and J. B. Pendry to LEED [5.21] and follows the cluster approach to LEED developed in the 1980s [5.22].

5.2.5 Multiple Scattering in Atomic Layers

5.2.5.1 A Periodic Plane of Atoms

We here return to 2-D periodic arrays of atoms present in a surface. In Eq. (5.16), the dimension of the central matrix to be inverted is $N(l_{max} + 1)^2$, which rapidly becomes prohibitive for clusters with many atoms N. For a periodic layer of atoms, however, as in a surface, we can exploit the periodicity very effectively. With a Bravais lattice (i.e., a single atom in each unit cell), all atoms are equivalent, so that the total wave leaving each atom is the same (apart from a simple phase factor as in Eq. (5.17) and further explained in Sections 5.2.5.2 and 5.2.5.4):

$$\mathbf{T}^1 = \mathbf{T}^2 = \ldots = \mathbf{T}^N \equiv \boldsymbol{\tau}.$$

We shall henceforth reserve the symbol $\boldsymbol{\tau}$ for the scattering by a Bravais lattice plane of atoms. In analogy with Eqs. (5.13) and (5.14), we can now write:

$$\boldsymbol{\tau} = \mathbf{t} + \mathbf{t}\left(\sum_n{}' \overline{\mathbf{G}}^{in}\right)\boldsymbol{\tau}, \tag{5.23}$$

where n extends over all atoms other than the atom i under consideration (the exclusion of atom i is indicated by the prime on the summation symbol). The index i can be dropped from the result since the outcome is the same for all atoms i. We define the new 'planar' Green's function

$$G_{LL'}^{ii} = \sum_n{}' \overline{G}_{LL'}^{in} = -4\pi i \frac{2m_e}{\hbar^2} k \sum_{L_1} \sum_{\mathbf{P}}{}' i^{l_1} a(L, L', L_1) \, h_{l_1}^{(1)}(k|\mathbf{P}|) Y_{L_1}(\mathbf{P}) e^{-i\mathbf{k}_{in}\mathbf{P}}, \tag{5.24}$$

which includes a sum over all atoms at positions $\mathbf{P}$, except at the origin $\mathbf{P} = \mathbf{0}$. (This exclusion is again denoted by the prime on the summation symbol.) The superscripts of $G_{LL'}^{ii}$ have become meaningless since the right-hand side of Eq. (5.24) no longer depends on the atom index i, but we shall retain them for later generalisation. Now, Eq. (5.24) implies that

$$\boldsymbol{\tau} = \left(1 - \mathbf{t}G^{ii}\right)^{-1}\mathbf{t} = \mathbf{t}\left(1 - G^{ii}\mathbf{t}\right)^{-1}. \tag{5.25}$$

Equation (5.25) essentially solves the multiple scattering in a Bravais-lattice plane. Although it contains an infinite number of atoms, the matrices have a relatively small dimension, namely $(l_{max} + 1)^2$, and we have a self-consistent solution valid whatever the amount of multiple scattering may be. We are still employing the spherical-wave representation, and, consequently, $\boldsymbol{\tau}$ has matrix elements $\tau_{LL'}$. The quantity $\tau_{LL'}$ indicates that an incident spherical wave L' of amplitude 1 centred on any of the atoms of the layer eventually produces an amplitude $\tau_{LL'}$ in a departing spherical wave L centred on the same atom.

The lattice sum in Eq. (5.24) is expected to converge only if sufficient damping is present in the form of the imaginary part of k in the Hankel function $h^{(1)}$. Thus, when inelastic effects are weak, such as at kinetic energies near or below the Fermi level, a different summation procedure is required. K. Kambe has formulated an alternative method based on an Ewald summation scheme [5.23–5.25]. Basically, it consists of a Fourier summation over the distant part of the atomic lattice and a direct summation over the near part of the atomic lattice. Even for normal LEED cases where the kinetic energy is far above the Fermi level, the Kambe approach is computationally slightly faster than the direct-space summation of Eq. (5.24), but at the expense of more complicated programming.

5.2.5.2   Several Periodic Planes of Atoms

Just as we assembled several atoms into clusters, we can stack several periodic layers into a thicker slab of atoms: this slab can represent a surface of finite thickness. If there is enough damping, this may even suffice to represent an infinitely thick surface: this case is often called the 'giant-matrix method'. There are also more effective ways to create an infinitely thick surface by stacking individual layers and/or slabs, as we will discuss in Section 5.2.6.

Following a derivation that is parallel to the case of assembling individual atoms, we can stack N individual periodic atomic layers, each having the same Bravais lattice with an atom at location $\mathbf{r}_i$, $i = 1, \ldots, N$. In very close analogy with Eq. (5.16) we obtain

$$
\begin{pmatrix} \mathbf{T}^1 \\ \mathbf{T}^2 \\ \mathbf{T}^3 \\ \vdots \\ \mathbf{T}^N \end{pmatrix} = \begin{pmatrix} \mathbf{I} & -\boldsymbol{\tau}^1 \mathbf{G}^{12} & -\boldsymbol{\tau}^1 \mathbf{G}^{13} & \cdots & -\boldsymbol{\tau}^1 \mathbf{G}^{1N} \\ -\boldsymbol{\tau}^2 \mathbf{G}^{21} & \mathbf{I} & -\boldsymbol{\tau}^2 \mathbf{G}^{23} & \cdots & -\boldsymbol{\tau}^2 \mathbf{G}^{2N} \\ -\boldsymbol{\tau}^3 \mathbf{G}^{31} & -\boldsymbol{\tau}^3 \mathbf{G}^{32} & \mathbf{I} & \cdots & -\boldsymbol{\tau}^3 \mathbf{G}^{3N} \\ \vdots & \vdots & \vdots & \ddots & \vdots \\ -\boldsymbol{\tau}^N \mathbf{G}^{N1} & -\boldsymbol{\tau}^N \mathbf{G}^{N2} & -\boldsymbol{\tau}^N \mathbf{G}^{N3} & \cdots & \mathbf{I} \end{pmatrix}^{-1} \begin{pmatrix} \boldsymbol{\tau}^1 \\ \boldsymbol{\tau}^2 \\ \boldsymbol{\tau}^3 \\ \vdots \\ \boldsymbol{\tau}^N \end{pmatrix}. \quad (5.26)
$$

Comparing Eq. (5.26) with Eq. (5.16), the single-atom matrix $\mathbf{t}$ is replaced by the single-Bravais-plane matrix $\boldsymbol{\tau}$, while $\overline{\mathbf{G}}^{ij}$ is replaced by a similar $\mathbf{G}^{ij}$, where

$$
G_{LL'}^{ij} = -4\pi i \frac{2m_e}{\hbar^2} k \sum_{L_1} \sum_{\mathbf{P}} {}' i^{l_1} a(L, L', L_1) h_{l_1}^{(1)} \left(k |\mathbf{r}_j - \mathbf{r}_i + \mathbf{P}| \right) Y_{L_1} (\mathbf{r}_j - \mathbf{r}_i + \mathbf{P}) e^{-i\mathbf{k}_{in}(\mathbf{r}_j - \mathbf{r}_i + \mathbf{P})}.
$$

$$(5.27)$$

Here, $\mathbf{P}$ extends over the lattice points of any of the planes, except the point $\mathbf{r}_j - \mathbf{r}_i + \mathbf{P} = 0$. The amplitude of the total plane wave scattered from this stack of N layers can again be obtained from Eq. (5.17), by simply inserting there the result of Eq. (5.26).

5.2.5.3   Diffraction Matrices for a Bravais-Lattice Layer

A periodic surface scatters an incoming electron beam into a set of well-defined departing electron beams: to obtain the intensity of these outgoing beams, we first need a relation between an arbitrary general t-matrix $t_{LL'}$, giving the diffraction amplitude between two spherical waves L and L', and a plane-wave diffraction amplitude $M_{\text{out,in}}$ between two plane waves $\mathbf{k}_{\text{in}}$ and $\mathbf{k}_{\text{out}}$. For a lattice with 2-D periodicity, such a relation is given by

$$M_{\text{out,in}} = -\frac{8\pi^2 i}{A k_{\text{outz}}} \frac{2m_e}{\hbar^2} \sum_{LL'} Y_L(\mathbf{k}_{\text{out}}) t_{LL'} Y_{L'}^*(\mathbf{k}_{\text{in}}), \tag{5.28}$$

where A is the area of the 2-D unit cell ($1/A$ is the number density of atoms in each atomic plane). In particular, for the incident beam $\mathbf{k}_0$ and the outgoing beam $\mathbf{k}_{\mathbf{g}}^-$, we obtain, by inserting Eq. (5.17) into Eq. (5.28), the reflected amplitude

$$M_{\mathbf{g},0}^{-+} = -\frac{16\pi^2 i m_e}{A k_{\mathbf{g}z}^- \hbar^2} \sum_{LL'} Y_L(\mathbf{k}_{\mathbf{g}}^-) T_{LL'} Y_{L'}^*(\mathbf{k}_0), \tag{5.29}$$

where $T_{LL'}$ is calculated from the results of Eq. (5.26), via Eq. (5.17).

A surface will often be represented by a stack of layers (whether single Bravais-lattice planes or slabs of such planes). Each of these layers will diffract the plane waves according to the Laue conditions in 2-D. Therefore, starting with the incident beam $\mathbf{k}_0$, all beams $\mathbf{k}_{\mathbf{g}}^{\pm}$ can be generated and each of these can be incident on another layer, from one side or the other (into the surface or outward from the surface). We therefore need the amplitude of any of the diffracted plane waves $\mathbf{k}_{\mathbf{g}'}^{\pm}$ (where $\mathbf{g}'$ is in general different from $\mathbf{g}$) due to such an incident wave.

For a layer containing only one atomic plane (we call it a Bravais-lattice layer), we get

$$M_{\mathbf{g}'\mathbf{g}}^{\pm\pm} = -\frac{16\pi^2 i m_e}{A k_{\mathbf{g}z}^{\pm} \hbar^2} \sum_{LL'} Y_L\left(\mathbf{k}_{\mathbf{g}'}^{\pm}\right) \tau_{LL'} Y_{L'}^*\left(\mathbf{k}_{\mathbf{g}}^{\pm}\right) + \delta_{\mathbf{g}'\mathbf{g}}\delta_{\pm\pm}. \tag{5.30}$$

The two Kronecker symbols (equal to unity when $\mathbf{g}' = \mathbf{g}$ and when the signs are the same, respectively) in this expression represent the unscattered plane wave of amplitude 1 which is transmitted without change of direction through the layer (the transmitted wave is modified by the wave which is scattered in the forward direction). Note the right-to-left logical order of the indices of $M_{\mathbf{g}'\mathbf{g}}^{\pm\pm}$.

The following form equivalent to Eq. (5.29) was derived by J. B. Pendry [5.1]:

$$M_{\mathbf{g}'\mathbf{g}}^{\pm\pm} = \frac{8\pi^2 i m_e}{A k k_{\mathbf{g}'z}^{+} \hbar^2} \sum_{lml'm'} \left[i^l(-1)^m Y_{l-m}\left(\mathbf{k}_{\mathbf{g}}^{\pm}\right)\right] (1-X)_{lm,l'm'}^{-1} \left[i^{-l'} Y_{l'm'}\left(\mathbf{k}_{\mathbf{g}'}^{\pm}\right)\right] \sin\delta_{l'} e^{i\delta_{l'}}$$

$$+ \delta_{\mathbf{g}'\mathbf{g}}\delta_{\pm\pm}, \tag{5.31}$$

with

$$X_{lm,l'm'} = \sum_{l'+m'=\text{even}} C^l(l'm', l'', m'') F_{l'm'} \sin\delta_{l'} e^{i\delta_{l'}}, \tag{5.32}$$

$$C^l(l'm',l'',m'')=4\pi(-1)^{(l-l'-l'')/2}(-1)^{m'+m''}Y_{l'-m'}(\theta=\pi/2,\varphi=0)\int Y_{lm}Y_{l'm'}Y_{l''-m''}d\Omega,$$

(5.33)

$$F_{l'm'}=\sideset{}{'}\sum_{\mathbf{P}}e^{i\mathbf{k}\mathbf{P}}h_{l'}^{(1)}(k|\mathbf{P}|)(-1)^{m'}e^{-im'\varphi(\mathbf{P})}.$$

(5.34)

Here, $C^l(l'm',l'',m'')$ is a set of Clebsch–Gordan coefficients that is different from the set $a(lm,l'm',l_1m_1)$ of Eq. (5.9). However, these two sets are equivalent under suitable permutations, apart from different prefactors. In Eq. (5.34), $\varphi(\mathbf{P})$ is the azimuth in the surface plane of the lattice vector $\mathbf{P}\neq\mathbf{0}$. (One minor discrepancy occasionally appears in some versions of these formulae: the complex wave number k in the denominator of Eq. (5.31) is sometimes replaced by its absolute value or by its real part.)

5.2.5.4 Diffraction Matrices for a Layer with N Periodic Atomic Planes

Equation (5.31) can be generalised to the case of a layer containing multiple periodic atomic planes, namely a composite layer, N:

$$M_{\mathbf{g}'\mathbf{g}}^{\pm\pm}=-\frac{16\pi^2 im_e}{Ak_{\mathbf{g}z}^+\hbar^2}\sum_{LL'}Y_L(\mathbf{k}_{\mathbf{g}}^{\pm})\sum_{i=1}^N\left\{e^{i\left(\pm\mathbf{k}_{\mathbf{g}}^{\pm}\mp\mathbf{k}_{\mathbf{g}'}^{\pm}\right)\cdot\mathbf{r}_i}T_{LL'}^i\right\}Y_{L'}^*(\mathbf{k}_{\mathbf{g}}^{\pm})+\delta_{\mathbf{g}'\mathbf{g}}\delta_{\pm\pm},\quad(5.35)$$

using $T_{LL'}^i$ from Eq. (5.26). One small modification is needed here in the Green's function of Eq. (5.27): the incident wave is now $\mathbf{k}_{\mathbf{g}}^{\pm}$ rather than $\mathbf{k}_{\text{in}}$. Simultaneously with the final form of $G_{LL'}^{ji}$ we now give an alternate form derived by S. Y. Tong [5.2], based on a reciprocal-lattice summation rather than a direct-lattice summation:

$$G_{LL'}^{ji}=e^{-i\mathbf{k}_{\mathbf{g}}^{\pm}\cdot(\mathbf{r}_j-\mathbf{r}_i)}\hat{G}_{LL'}^{ji},\text{ with}$$

(5.36)

$$\hat{G}_{LL'}^{ji}=-\frac{8\pi ikm_e}{\hbar^2}\sum_{L_1}\sideset{}{'}\sum_{\mathbf{P}}i^{l_1}a(L,L',L_1)h_{l_1}^{(1)}(k|\mathbf{r}_j-\mathbf{r}_i+\mathbf{P}|)Y_{L_1}(\mathbf{r}_j-\mathbf{r}_i+\mathbf{P})e^{-i\mathbf{k}\mathbf{P}}$$

(5.37)

$$\text{or }\hat{G}_{LL'}^{ji}=-\frac{16\pi^2 im_e}{A\hbar^2}\sum_{\mathbf{g}_1}\frac{e^{i\mathbf{k}_{\mathbf{g}_1}^+\cdot(\mathbf{r}_j-\mathbf{r}_i)}}{k_{\mathbf{g}_1z}^+}Y_L^*(\mathbf{k}_{\mathbf{g}_1}^{\pm})Y_{L'}(\mathbf{k}_{\mathbf{g}_1}^{\pm}).$$

(5.38)

Equation (5.38) has computational advantages over Eq. (5.37) under certain circumstances, namely when the atomic planes of the layer are not spaced too closely: the sum over $\mathbf{g}_1$ in principle runs over the infinite 2-D reciprocal lattice, but is limited by the evanescent character of waves with large $|\mathbf{g}_1|$ for larger spacings between atomic planes.

We note that Eqs. (5.27) and (5.29) assume a reference point (origin of coordinates) at the centre of an atom in the layer under consideration: $M_{\mathbf{g}'\mathbf{g}}^{\pm\pm}$ is the ratio of outgoing to incident plane-wave amplitudes measured in that atomic centre (or at least extrapolated to that atomic centre as if we had point scatterers). In Eq. (5.35) the corresponding reference point is the origin of the vectors $\mathbf{r}_i$ and thus can be an arbitrary point. However, it is important to be consistent in the use of a particular reference point in each layer when carrying out a calculation, especially in connection with the utilisation of symmetries.

5.2.5.5   Layer Reflection and Transmission Matrices

We now introduce a notation that will be convenient when we treat the stacking of layers in a surface. We distinguish between reflection and transmission at a layer: transmission occurs when electrons emerge on the side of the layer opposite to that from which they impinge on it, while reflection occurs when they emerge on the same side, with an appropriate generalisation in the case of evanescent waves. This definition implies, for example, that grazing specular diffraction is treated as a reflection rather than as a transmission, even though the scattering angle may be close to $0°$. Thus, we define reflection and transmission matrices

$$\mathbf{r}^{+-} = \mathbf{M}^{+-}, \quad \mathbf{r}^{+-} = \mathbf{M}^{+-}, \quad \mathbf{t}^{++} = \mathbf{M}^{++}, \quad \mathbf{t}^{--} = \mathbf{M}^{--}, \tag{5.39}$$

where $\mathbf{M}^{\pm\pm}$ comes from Eqs. (5.29), (5.30) or (5.35).

5.2.6   Layer Stacking

We have obtained the scattering properties of 2-D periodic atomic layers of finite thickness, whether of mono-atomic thickness or of multi-atomic thickness. These layers can next be stacked onto each other to form thicker slabs of layers, including a semi-infinite slab. We start with the latter, solving the complete semi-infinite surface in a closed and self-consistent form, using Bloch waves.

5.2.6.1   Bloch Wave and Transfer Matrix Methods

The Bloch wave method and its close relative, the transfer matrix method, together with the matrix-inversion method (Section 5.2.5), were the earliest solutions of the LEED problem. The Bloch wave and transfer matrix methods have the virtues of providing a connection with the familiar band structure theory of bulk materials and of giving a simple interpretation of the electron diffraction process at elementary surfaces. However, these advantages have not been particularly useful in treating more complex surfaces involving adsorbates or reconstructions.

In this approach, one applies the Bloch condition to the electronic wave functions in the direction perpendicular to the surface for the bulk layers that have a well-defined periodicity perpendicular to the surface. An eigenvalue problem then provides the Bloch functions, the relative amplitudes of which are determined by a matching across the surface to the plane waves in the vacuum.

At a point midway between layers i and $i + 1$ in the bulk (away from any structural and electronic deviations due to the surface), we decompose a Bloch wave into plane waves defined by the 2-D periodicity of the surface:

$$\varphi_i(\mathbf{r}) = \sum_{\mathbf{g}} \left(b_{ig}^+ e^{i\mathbf{k}_g^+ \cdot \mathbf{r}} + b_{ig}^- e^{i\mathbf{k}_g^- \cdot \mathbf{r}} \right). \tag{5.40}$$

We include both penetrating and emerging plane waves in this expansion: these plane waves are represented in Figure 5.9 as trios of arrows pointing to the right (into the surface) and to the left (toward the vacuum), respectively.

The Bloch condition, due to the periodicity perpendicular to the surface, implies that the relations

$$b_{i+1}^+ = b_i^+ e^{i\mathbf{k}_B \mathbf{a}} \tag{5.41}$$

and

$$b_{i+1}^- = b_i^- e^{i\mathbf{k}_B \mathbf{a}} \tag{5.42}$$

hold for each plane-wave amplitude of the Bloch wave of Eq. (5.40), if $\mathbf{a}$ is the repeat vector that relates one layer to the next ($\mathbf{a}$ need not be perpendicular to the surface). Here $e^{i\mathbf{k}_B \mathbf{a}}$ is an eigenvalue to be determined, which will then specify the value of the Bloch wave vector $\mathbf{k}_B$, familiar in band structures.

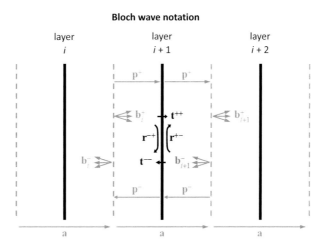

Figure 5.9 Bloch wave notation, where $\mathbf{b}_i^+$ and $\mathbf{b}_i^-$ are plane-wave amplitudes of a Bloch wave travelling to the right between layers i and $i + 1$ of a periodic sequence of identical layers (see Eq. (5.40)). Three atomic layers are shown as thick black vertical lines. Each atomic plane has reflection and transmission matrices for plane waves given by $\mathbf{r}^{+-}, \mathbf{r}^{-+}, \mathbf{t}^{++}$, and $\mathbf{t}^{--}$, according to Eq. (5.39). Shown in grey are midway planes (dashed lines), the layer repetition vector $\mathbf{a}$ and the plane-wave propagators $\mathbf{p}^+$ and $\mathbf{p}^-$ for propagation by $+\mathbf{a}/2$ and $-\mathbf{a}/2$, respectively. Adapted with permission from Springer Nature Customer Service Centre GmbH: Springer, *Low-Energy Electron Diffraction: Experiment, Theory and Structural Determination*, by M. A. Van Hove, W. H. Weinberg and C.-M. Chan, © (1986) [5.3].

Unlike in bulk band structures, however, which normally assume no electron damping (no inelastic effects, especially below the Fermi level), in the LEED case we do have damping (given by an inelastic mean free path): as a result, the Bloch wave vector $\mathbf{k}_B$ is always a complex quantity in LEED, representing the damping of waves from one layer to the next: one then speaks of a complex band structure.

In Eq. (5.40), the expansion coefficients depend on the index i because each layer intermixes the plane waves that are scattered by it. We shall need this mixing information and use it to relate the components of the Bloch waves φ_i, and φ_{i+1}, on either side of layer $i + 1$. This mixing is described by the layer-diffraction matrices $M_{\mathbf{g'g}}^{\pm\pm}$ of Eqs. (5.16) and (5.21), which we shall write here in terms of reflection and transmission matrices similar to those given in Eq. (5.39),

$$
\begin{aligned}
R_{\mathbf{g'g}}^{-+} &= p_{\mathbf{g'}}^- r_{\mathbf{g'g}}^{-+} p_{\mathbf{g}}^+ = p_{\mathbf{g'}}^- M_{\mathbf{g'g}}^{-+} p_{\mathbf{g}}^+, \\
R_{\mathbf{g'g}}^{+-} &= p_{\mathbf{g'}}^+ r_{\mathbf{g'g}}^{+-} p_{\mathbf{g}}^- = p_{\mathbf{g'}}^+ M_{\mathbf{g'g}}^{+-} p_{\mathbf{g}}^-, \\
T_{\mathbf{g'g}}^{++} &= p_{\mathbf{g'}}^+ t_{\mathbf{g'g}}^{++} p_{\mathbf{g}}^+ = p_{\mathbf{g'}}^+ M_{\mathbf{g'g}}^{++} p_{\mathbf{g}}^+, \\
T_{\mathbf{g'g}}^{--} &= p_{\mathbf{g'}}^- t_{\mathbf{g'g}}^{--} p_{\mathbf{g}}^- = p_{\mathbf{g'}}^- M_{\mathbf{g'g}}^{--} p_{\mathbf{g}}^-,
\end{aligned}
\tag{5.43}
$$

where

$$
p_{\mathbf{g}}^{\pm} = e^{\pm i \mathbf{k}_{\mathbf{g}}^{\pm} \mathbf{a}/2}.
\tag{5.44}
$$

Many of these quantities are shown in Figure 5.9.

Now we can relate the components of the Bloch waves φ_i and φ_{i+1}:

$$
\begin{aligned}
\mathbf{b}_{i+1}^+ &= \mathbf{T}^{++}\mathbf{b}_i^+ + \mathbf{R}^{+-}\mathbf{b}_{i+1}^-, \\
\mathbf{b}_i^- &= \mathbf{T}^{--}\mathbf{b}_{i+1}^- + \mathbf{R}^{-+}\mathbf{b}_i^+,
\end{aligned}
\tag{5.45}
$$

using matrix/vector notation. Equations (5.41), (5.42) and (5.45) can be combined, using double-length vectors, to yield the following eigenvalue equation:

$$
\begin{pmatrix} \mathbf{T}^{++} & \mathbf{R}^{+-} \\ (\mathbf{T}^{--})^{-1}\mathbf{R}^{-+}\mathbf{T}^{++} & -(\mathbf{T}^{--})^{-1}\mathbf{R}^{-+}\mathbf{R}^{+-} + (\mathbf{T}^{--})^{-1} \end{pmatrix} \begin{pmatrix} \mathbf{b}_i^+ \\ \mathbf{b}_{i+1}^- \end{pmatrix} = e^{i\mathbf{k}_B \mathbf{a}} \begin{pmatrix} \mathbf{b}_i^+ \\ \mathbf{b}_{i+1}^- \end{pmatrix}.
\tag{5.46}
$$

Solving this equation yields eigenvectors which give the Bloch waves φ_i that we need, as well as the eigenvalues giving the Bloch wave vector $\mathbf{k}_B$ for the electron energy and incidence direction under consideration. The double matrix dimension of Eq. (5.46) corresponds to the presence of Bloch waves that travel toward the bulk as well as Bloch waves that travel toward the vacuum: only the former are relevant for LEED, as those Bloch waves travelling toward the vacuum would require an electron source of the same energy E deep inside the surface.

The LEED wave functions in the bulk can also be obtained with an approach equivalent to the Bloch wave method, namely the transfer matrix method. The transfer matrix $\mathbf{S}$ relates the plane waves from one layer to the next through

$$\begin{pmatrix} \mathbf{b}_{i+1}^+ \\ \mathbf{b}_{i+1}^- \end{pmatrix} = \mathbf{S} \begin{pmatrix} \mathbf{b}_i^+ \\ \mathbf{b}_i^- \end{pmatrix}, \quad \text{with } \mathbf{S} = \begin{pmatrix} \mathbf{T}^{++} - \mathbf{R}^{+-}(\mathbf{T}^{--})^{-1}\mathbf{R}^{-+} & \mathbf{R}^{+-}(\mathbf{T}^{--})^{-1} \\ -(\mathbf{T}^{--})^{-1}\mathbf{R}^{-+} & (\mathbf{T}^{--})^{-1} \end{pmatrix},$$

(5.47)

which follows from our earlier relations. Diagonalising $\mathbf{S}$ also yields the Bloch wave functions and eigenvalues.

We still need to match the Bloch waves to the plane waves in vacuum, which consist of one incident wave of given amplitude and a set of diffracted waves with unknown amplitudes: the matching will fix those unknown amplitudes. The simplest case is that of an abrupt termination of the bulk structure at the surface, such that the Bloch waves can be equated there to the vacuum waves: that equation solves the unknown amplitudes. However, in general, the surface structure deviates from the bulk structure, whether through different layer spacings, overlayers, 2-D superlattices, etc. In the general case, we have to match the plane waves between each pair of surface layers as well as between them and the vacuum waves, and between them and the bulk layers (where the Bloch waves exist): this provides enough conditions to solve the problem but is clearly more involved than for the case of an abrupt bulk termination.

The computational cost of the Bloch wave and transfer matrix approaches scales with the number of plane waves g used as $(2g)^3 + (2gs)^3$, if s is the number of surface matching planes. We shall describe more efficient schemes in Sections 5.2.6.2–5.2.6.5.

5.2.6.2 Layer-by-Layer Stacking

The multiple scattering by a pair of layers can be expressed as a series similar to that between two atoms (cf. Eqs. (5.10) and (5.11)) and can also be summed up exactly as in Eq. (5.46), using a closed matrix inversion. Referring to Figure 5.10, the reflection from the vacuum side of the double layer is the sum of reflection from layer A, reflection from layer B (involving transmission twice through layer A), double reflection from layer B, triple reflection from layer B, etc.:

$$\begin{aligned} \mathbf{R}^{-+} &= \mathbf{P}_1^- \mathbf{r}_A^{-+} \mathbf{P}_1^+ \\ &\quad + \mathbf{P}_1^- \mathbf{t}_A^{--} \mathbf{P}^- \mathbf{r}_B^{-+} \mathbf{P}^+ \mathbf{t}_A^{++} \mathbf{P}_1^+ \\ &\quad + \mathbf{P}_1^- \mathbf{t}_A^{--} \mathbf{P}^- \mathbf{r}_B^{-+} \mathbf{P}^+ \mathbf{r}_A^{+-} \mathbf{P}^- \mathbf{r}_B^{-+} \mathbf{P}^+ \mathbf{t}_A^{++} \mathbf{P}_1^+ + \cdots \\ &= \mathbf{P}_1^- \left[\mathbf{r}_A^{-+} + \mathbf{t}_A^{--} \mathbf{P}^- \mathbf{r}_B^{-+} \mathbf{P}^+ \left(\mathbf{I} - \mathbf{r}_A^{+-} \mathbf{P}^- \mathbf{r}_B^{-+} \mathbf{P}^+ \right)^{-1} \mathbf{t}_A^{++} \right] \mathbf{P}_1^+ . \end{aligned}$$

(5.48)

The various plane-wave propagators $\mathbf{P}$ are similar to those of Eq. (5.44), with $\mathbf{a}/2$ replaced by the appropriate displacement vectors. Note that we here include transmission ($\mathbf{t}_A^{++}$ and $\mathbf{t}_A^{--}$) through the intervening layer A: this transmission is a strong perturbation of the waves and cannot be treated as a weak effect.

Similarly, the reflection from the opposite side of the layer pair and the transmissions from both sides can be summed up to a closed form (for simplicity we now make planes 1 and A coincide, as well as planes 2 and B coincide):

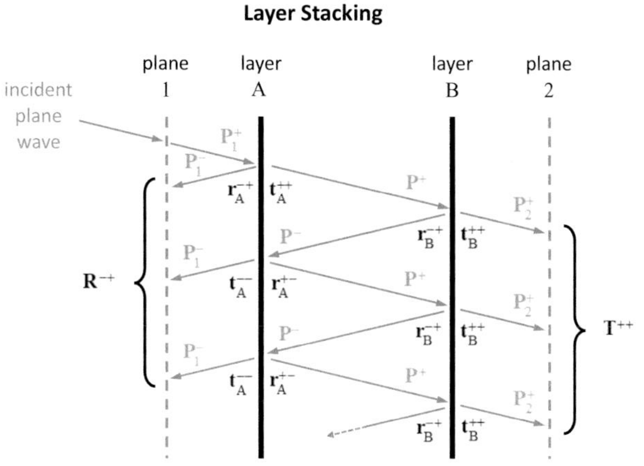

Layer Stacking

Figure 5.10 Layer stacking scheme for two layers. Two atomic layers A and B are shown as thick black vertical lines. Shown in grey are external reference planes 1 and 2 (dashed lines) and the plane-wave propagators $\mathbf{P}^+$ and $\mathbf{P}^-$ (for propagation from one layer to the next) as well as the plane-wave propagators $\mathbf{P}_1{}^+$, $\mathbf{P}_1{}^-$, $\mathbf{P}_2{}^+$, $\mathbf{P}_2{}^-$ (for propagation from plane 1 to layer A and back, and from layer B to plane 2 and back, respectively). The internal reflections between layers A and B can repeat as often as needed. Each arrow (except the incident plane wave) represents all possible diffracted waves $\mathbf{g}$. The exiting waves are summed to yield the converged reflection matrices $\mathbf{R}^{-+}$ and transmission matrices $\mathbf{T}^{++}$ for the pair of layers. The scheme is repeated for incidence from the right, unless left/right symmetry yields directly: $\mathbf{R}^{+-} = \mathbf{R}^{-+}$ and $\mathbf{T}^{--} = \mathbf{T}^{++}$. Adapted with permission from Springer Nature Customer Service Centre GmbH: Springer, *Low-Energy Electron Diffraction: Experiment, Theory and Structural Determination*, by M. A. Van Hove, W. H. Weinberg and C.-M. Chan, © (1986) [5.3].

$$\mathbf{R}^{-+} = \mathbf{r}_A^{-+} + \mathbf{t}_A^{--}\mathbf{P}^-\mathbf{r}_B^{-+}\mathbf{P}^+\left(\mathbf{I} - \mathbf{r}_A^{+-}\mathbf{P}^-\mathbf{r}_B^{-+}\mathbf{P}^+\right)^{-1}\mathbf{t}_A^{++},$$

$$\mathbf{R}^{+-} = \mathbf{r}_B^{+-} + \mathbf{t}_B^{++}\mathbf{P}^+\mathbf{r}_A^{+-}\mathbf{P}^-\left(\mathbf{I} - \mathbf{r}_B^{-+}\mathbf{P}^+\mathbf{r}_A^{+-}\mathbf{P}^-\right)^{-1}\mathbf{t}_B^{--},$$

$$\mathbf{T}^{++} = \mathbf{t}_B^{++}\mathbf{P}^+\left(\mathbf{I} - \mathbf{r}_A^{+-}\mathbf{P}^-\mathbf{r}_B^{-+}\mathbf{P}^+\right)^{-1}\mathbf{t}_A^{++},$$

$$\mathbf{T}^{--} = \mathbf{t}_A^{--}\mathbf{P}^-\left(\mathbf{I} - \mathbf{r}_B^{-+}\mathbf{P}^+\mathbf{r}_A^{+-}\mathbf{P}^-\right)^{-1}\mathbf{t}_B^{--},$$

$$(5.49)$$

with $P_\mathbf{g}^\pm = e^{\pm i\mathbf{k}_\mathbf{g}^+\mathbf{r}_{BA}}$, where $\mathbf{r}_{BA}$ links a point in layer A with a point in layer B (specifically, these are the points with respect to which the layer reflection and transmission matrices are defined).

To stack more than two layers, we can add single atomic layers to a growing stack by repeating the steps of Eqs. (5.49), using reflection and transmission matrices calculated in the previous stacking step. This should be repeated until the stack thickness makes the reflection matrix elements $R_{\mathbf{g}'0}^{-+}$ (giving the amplitudes of the diffracted plane waves) converge due to the inelastic mean free path, which typically happens after about 10 mono-atomic layers. The computational cost of layer stacking scales as g^3N, for g plane waves and N layers.

5.2.6.3   Layer Doubling

In the bulk, layer stacking can be considerably accelerated by so-called layer doubling, illustrated in Figure 5.11. The bulk periodicity perpendicular to the surface (also used in the Bloch wave approach) allows doubling the layer thickness at each step: the first step combines two layers into one double layer, as discussed in Section 5.2.6.2. The second step would combine two such double layers into one quadruple layer. The third step would combine two quadruple layers into one octuple layer. A fourth step produces a stack of 16 layers, which is usually sufficient for convergence of the reflection by the stack.

The layer doubling method yields a full reflection matrix for the stack of equal layers. Other layers different from the bulk layers can then be added to the surface at will, using the individual layer stacking approach described in Section 5.2.6.2. For efficiency, the bulk reflection can be stored and used repeatedly for many different

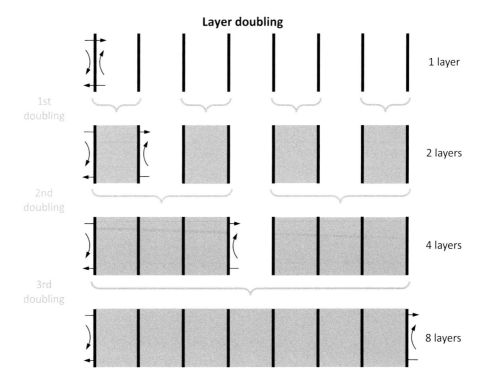

Figure 5.11 Layer doubling scheme, with surface at left, assuming identical atomic layers. First, the reflection and transmission matrices of a single layer (represented by arrows at top left) are used with the layer stacking scheme to pair up the left-most two layers, producing new reflection and transmission matrices for a pair of layers (arrows in second row). The process is repeated to grow the layer stack to 4, 8, 16, ... layers until convergence of the reflection matrices at the surface. Adapted with permission from Springer Nature Customer Service Centre GmbH: Springer, *Low-Energy Electron Diffraction: Experiment, Theory and Structural Determination*, by M. A. Van Hove, W. H. Weinberg and C.-M. Chan, © (1986) [5.3].

configurations of surface layers (such as different overlayer spacings and registries), a convenient feature in a surface structure search. The computational cost of layer doubling scales as $g^3 \ln N$, for g plane waves and N layers (N is now a power of 2).

5.2.6.4   Renormalised Forward Scattering (RFS)

An important aspect of the layer stacking described in Section 5.2.6.2 is the inclusion of transmission matrices in all the multiple scattering paths that are transmitted through layers. The reason is that forward transmission affects the waves strongly, both by redistributing amplitudes among plane waves and by modifying their phases significantly. By contrast, the backward reflection by an atomic layer can more justifiably be treated as a perturbation. That is the idea behind the renormalised forward scattering (RFS) method.

In RFS, the reflection by any layer is considered to be weak, and therefore the perturbation is based on an expansion of the total reflectivity of the surface in terms of the number of reflections. The first order contains all paths that have been reflected only once but transmitted any number of times; the second order contains only triple-reflection paths (an odd number of reflections is needed to bring electrons back out of the surface), and so on.

Figure 5.12 shows the time sequence of computations. The incident beam is weakly reflected (**r**) and also strongly transmitted (**t**) by the first layer; its

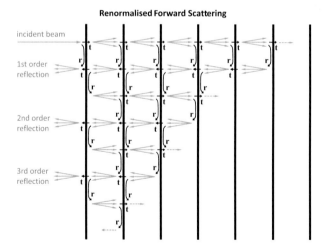

Figure 5.12 Renormalised forward scattering scheme. Each layer (thick black lines) may be different from the others, including different layer spacings and 2-D superlattices (which are normally assumed commensurate). Grey arrows indicate plane-wave propagation into and out of the surface, while black arrows indicate reflection (**r**) and transmission (**t**) at each layer (the matrices **r** and **t** may differ from layer to layer and may have left/right asymmetry). Each order of back scattering can penetrate to any depth required for convergence. Adapted with permission from Springer Nature Customer Service Centre GmbH: Springer, *Low-Energy Electron Diffraction: Experiment, Theory and Structural Determination*, by M. A. Van Hove, W. H. Weinberg and C.-M. Chan, © (1986) [5.3].

transmission causes a 'shower' of plane waves to hit the second layer, where they are reflected as well as transmitted; this transmission reaches the third layer and is then transmitted to the fourth layer, etc., until inelastic effects have weakened the penetrating waves sufficiently. Then the process is reversed: the reflected waves from the deepest layer are propagated outward, causing at each subsequent layer a weak reflection (into the surface) and a strong transmission (out of the surface); at each outward transmission, the waves reflected in the previous stage (during pene- tration of the incident waves) are picked up and added to the outward flux of waves; this results in the first-order exit from the surface. For the second order, the inward reflections from the first order process are propagated inward (again causing weak reflection and strong transmission at each layer) until they die out with depth, after which the new outward reflections are picked up and propagated outward into vacuum: there they are added to the previously obtained reflections from the first order. Additional orders are obtained by iteration of this scheme, until convergence of the total reflected amplitudes.

Typically, a dozen atomic layers and a few orders of reflection suffice for convergence. Occasionally, RFS does not converge well: then one may alternatively apply layer doubling or else combine layers into slabs within which strong multiple scattering can first be solved by matrix inversion (Eq. (5.16)) or by layer-by-layer stacking (Eqs. (5.49)) before handling by RFS. The computational effort of RFS scales as g^2N, for g plane waves and N layers, making it the most efficient method available for LEED. It is also very flexible in terms of surface deviations from the bulk (overlayers, reconstructions, superlattices, etc.).

5.2.6.5   The Case of Stepped Surfaces

Stepped and kinked surfaces (characterised by high Miller indices, i.e., vicinal orien- tations) present a special challenge to the multiple scattering theory of LEED. These surfaces have a relatively large area A of their 2-D unit cell (which stretches from one step edge to the next), a relatively small interlayer spacing d (measured perpendicular to the macroscopic surface) and relatively low symmetry (at most a mirror plane), at all depths below the surface. The number of plane waves (including evanescent waves) needed in LEED theory scales roughly as A/d, which rapidly becomes prohibitively large for stepped surfaces with wider terraces. Worse, in the plane wave representation, layer stacking fails to converge for interlayer spacings d below approximately 0.1 nm (1 Å), thus preventing structure determination for most stepped surfaces.

The giant-matrix method (cf. Section 5.2.5.2) is more effective. It uses spherical waves within a composite layer that should include enough atomic layers to match the penetration depth of electrons into the substrate; there is one layer for each periodic- ally inequivalent atom in a terrace as well as in the buried continuation of the terrace down to the electron penetration depth. This approach requires matrix dimensions that increase in proportion to the number of included atoms (i.e., layers), which grows roughly as $1/d$ instead of A/d. Other advantages of using the giant-matrix method include: the calculation of matrix elements is quite fast and the solution of the system of linear equations is only required for the measured beams (not for the full beam set required by plane-wave methods) and only for one incident beam.

Several other schemes have been proposed to deal with stepped surfaces. One approach is the use bundles of chains of atoms parallel to the steps, using cylindrical waves within chains [5.26]; it allows disordered arrays of steps and thus diffuse LEED, but it does not benefit from the periodicity of ordered steps. Two other approaches combine the plane-wave and spherical-wave expansions in such a way as to generate effectively larger interlayer spacings between groups of layers, bundled together to form a step: a variation of the transfer-matrix method [5.27] and a scheme based on a three-centre layer doubling algorithm [5.28]; these roughly halve the lower limit on d to about 0.05 nm (0.5 Å), but do not solve the problem in general.

We next briefly describe a method that applies to any stepped or kinked surface, at the cost of requiring higher accuracy in some parts of the calculation. It was first developed by X.-G. Zhang and A. Gonis [5.29; 5.30] for electronic band-structure problems. This real-space multiple-scattering theory (RS-MST) [5.31; 5.32] uses the spherical wave representation, like the giant-matrix method, but adds two further techniques: Fourier transformation to convert the surface problem into a one-dimensional problem; and the concept of 'removal invariance'. The latter technique states that removing one layer from a semi-infinite periodic stack of layers does not change the surface properties, such as the surface reflectivity: this leads to a self-consistent equation for the scattering matrix from that semi-infinite stack. To allow surface relaxations, namely changes in interlayer spacings and layer registries parallel to the surface, tensor LEED (cf. Section 5.4.1) is applied very effectively. Adsorbates, reconstructions, etc., could also be treated explicitly by adding atomic layers on top of the semi-infinite substrate, for example in a way similar to the calculation of multiple scattering in a cluster of atoms (cf. Section 5.2.3).

The RS-MST method was tested against layer doubling for Cu(311) with a bulk interlayer spacing d of 0.109 nm (1.09 Å) and Cu(331) with a bulk interlayer spacing d of 0.083 nm (0.83 Å) [5.32]: the agreement was very good wherever layer doubling itself converged and for energies below about 130 eV, above which some numerical instabilities occurred in RS-MST; these instabilities should be eliminated by increased accuracy in calculating interplanar propagators and in performing matrix inversion. The method was also applied to determine, from experiment, surface relaxations down to the fourth layer in Pt(210) [5.31], which has a bulk interlayer spacing d of 0.08765 nm (0.8765 Å): in particular, an outermost inward relaxation by $23 \pm 4\%$ was found, giving an interlayer spacing of 0.06749 ± 0.00351 nm (0.6749 ± 0.0351 Å), in qualitative agreement with other results.

5.2.6.6   Beam Subsets Independent in the Bulk

Usually the surface layers exhibit a superlattice in comparison with the 2-D lattice of the substrate layers and not all beams are generated by the substrate layers; that is, if the beam indices (hk) are related to the substrate lattice, then only the beams with integer (hk) are generated by the substrate and all beams with fractional indices have their origin in the surface layers. The total beam set generated by the superlattice can

Beam subsets for a (2x2) superlattice

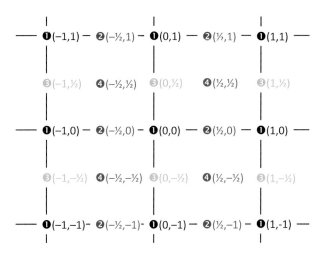

Figure 5.13 Beam subsets for a (2×2) superlattice on a rectangular substrate lattice. The 2-D reciprocal lattice of the substrate is drawn black with symbols ❶ and integer beam labels (h, k). The half-order beam (½, 0) gives rise to a shifted subset of beams (½ + h, k) marked in blue as ❷. The half-order beam (0, ½) likewise gives rise to beam subset (h, ½ + k) marked in green as ❸, while (½, ½) generates the beam subset (½ + h, ½ + k) marked in red as ❹. Adapted with permission from Springer Nature Customer Service Centre GmbH: Springer, *Low-Energy Electron Diffraction: Experiment, Theory and Structural Determination*, by M. A. Van Hove, W. H. Weinberg and C.-M. Chan, © (1986) [5.3].

be divided into subsets, as illustrated in Figure 5.13. One of the beams of each subset acts as a primary beam in the substrate layers exciting only beams belonging to this subset. The layer scattering matrices of substrate layers can therefore be block-diagonalised for subsets of beams, as illustrated in Figure 5.14. This block-diagonalisation can be used to save considerable computing effort in the various layer stacking schemes. For example, with layer doubling, the calculation time is substantially reduced if the backscattering matrices from the substrate are calculated separately for each subset.

5.3   Symmetry in Calculations

The time required for the calculation of the layer scattering matrices $M_{g'g}^{\pm\pm}$ (Eq. (5.35)) increases rapidly with the size of the surface unit cell and the number of atoms in it. The size of the matrices to be inverted can be reduced considerably by making use of symmetry relations. Symmetries are nearly always present when adsorbate structures are investigated on low-index surfaces of highly symmetric crystals like metals or semiconductors. The adsorbate structure itself may have a lower symmetry than the

**Block diagonalisation of a
substrate-layer diffraction matrix
with beam subsets for a (2x2) superlattice**

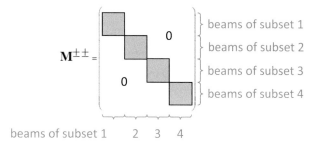

Figure 5.14 Block diagonalisation of a substrate-layer diffraction matrix with beam subsets for a (2×2) superlattice. The superlattice may be caused by an overlayer (e.g., ¼-monolayer adsorption) or by a reconstruction (e.g., missing atoms or commensurate modulation). Adapted with permission from Springer Nature Customer Service Centre GmbH: Springer, *Low-Energy Electron Diffraction: Experiment, Theory and Structural Determination*, by M. A. Van Hove, W. H. Weinberg and C.-M. Chan, © (1986) [5.3].

clean surface but usually at least some of the symmetry elements of the bulk structure are retained. As mentioned in Section 2.1.6, the existence of the surface reduces the 3-D space group of the substrate crystal to one of the 17 2-D space groups where only rotational axes normal to the surface and mirror planes with their normal parallel to the surface remain. Of the glide planes, only those with the normal in the surface plane and an in-plane glide vector remain. The symmetry of the LEED pattern, including its spot intensities, is that of the point symmetry of the structure combined with the incident beam. The incident beam must be invariant with respect to all symmetry operations. This is a result of multiple scattering and would not be the case for X-ray diffraction where the structure factor depends on the diffraction vector $\mathbf{q} = \mathbf{k}' - \mathbf{k}$ and not on the individual vectors $\mathbf{k}'$ and $\mathbf{k}$. As a result, the full 2-D symmetry can only be used at normal incidence, while at oblique incidence only a mirror or glide plane can be used, assuming that the incident beam lies in that plane.

Most frequently, experiments are performed at normal incidence and at incidence within a mirror plane, since the orientation of the probe is then easily controlled by the symmetry of the diffraction spot intensities. The calculation of LEED intensities is usually divided into two parts. One part solves the multiple-scattering problem within a single atomic layer, mostly a composite layer with several atoms per unit cell. The second part uses a plane-wave expansion and combines the layer scattering matrices by using layer doubling, the RFS scheme or other equivalent schemes as described in Section 5.2. Symmetry operations can be used in both steps. We first describe the use of symmetry operations in the calculation of layer scattering matrices. There are several ways to solve the multiple scattering equations (Eqs. (5.16) and (5.29)). The most general is the self-consistent solution by matrix inversion which will be considered in Sections 5.3.1 and 5.3.2; symmetry can be used similarly within other methods.

5.3.1   Symmetry in Reciprocal Space

We start with Eq. (5.35) for the reflection and transmission matrices for a multi-atomic layer where we explicitly write here the indices (l, m) instead of the abbreviation $L \equiv (l, m)$, because we need the single indices when describing the symmetry operators in angular momentum space.

$$M^{\pm\pm}_{\mathbf{g}'\mathbf{g}} = -\frac{16\pi^2 i m_e}{Ak^+_{\mathbf{g}z}\hbar^2} \sum_{lm,l'm'} Y_{lm}\left(\mathbf{k}^{\pm}_{\mathbf{g}'}\right) \sum_{i=1}^{N} \left\{ e^{i\left(\pm\mathbf{k}^{\pm}_{\mathbf{g}}\mp\mathbf{k}^{\pm}_{\mathbf{g}'}\right)\mathbf{r}_i} T^{i}_{lm,l'm'} \right\} (-1)^{m'} Y_{l'-m'}\left(\mathbf{k}^{\pm}_{\mathbf{g}}\right) + \delta_{\mathbf{g}'\mathbf{g}}\delta_{\pm\pm}.$$

(5.50)

Here we have used $(-1)^{m}Y_{l,-m} = Y^{*}_{L}$. The sum over j runs over all atoms in the unit cell, which means over all subplanes of the layer. The vector $\mathbf{r}_j$ gives the position of the atom j with respect to the origin of the unit cell. It is not required that an atom sits in the origin of the unit cell. The directions of the incoming and outgoing waves are indicated by the superscript $\pm$ of the wave vectors and define whether reflection or transmission matrices are calculated. With the $+$ sign indicating the incidence from the vacuum side, we have $\mathbf{T}^{++} = \mathbf{M}^{++}$, $\mathbf{R}^{+-} = \mathbf{M}^{+-}$ and the transmission and reflection from the crystal side are given by $\mathbf{T}^{--} = \mathbf{M}^{--}$ and $\mathbf{R}^{-+} = \mathbf{M}^{-+}$. The indices $\pm$ are left out for convenience in the following equations, keeping in mind that the matrices for incidence from the crystal side must be calculated with the beam set $\mathbf{k}^{-}_{\mathbf{g}}$.

The matrices $\mathbf{T}^{j}$ describe the scattered wave around the j-th atom. They are the solution of a set of linear equations:

$$T^{j}_{lm,l'm'}\left(\mathbf{k}_{\mathbf{g}}\right) = t^{j}(E) + t^{j}(E)\sum_{l''m''}\sum_{\mu=1}^{N} G^{j\mu}_{lm,l''m''}(\mathbf{k}_0) \cdot e^{i\mathbf{k}_{\mathbf{g}}\left(\mathbf{r}_j-\mathbf{r}_\mu\right)} \cdot T^{\mu}_{l''m'',l'm'}\left(\mathbf{k}_{\mathbf{g}}\right), \quad (5.51)$$

$$G^{j\mu}_{lm,l'm'}(\mathbf{k}_0) = -4\pi i\frac{2m_e}{\hbar^2}k\sum_{l_1m_1}\sum_{\mathbf{P}}{}' i^{l_1}a(L,L',L_1)h^{(1)}_{l_1}\left(k\left|\mathbf{r}_j-\mathbf{r}_\mu+\mathbf{P}\right|\right)Y_{l_1m_1}\left(\mathbf{r}_j-\mathbf{r}_\mu+\mathbf{P}\right)e^{-i\mathbf{k}_0\mathbf{P}}.$$

(5.52)

The propagator matrices $\mathbf{G}^{j\mu}(\mathbf{k}_0)$ describe the transport of a spherical wave from point μ to point j, and are here defined without the phase factors $\exp[i\mathbf{k}_{\mathbf{g}}(\mathbf{r}_j - \mathbf{r}_\mu)]$ which have been shifted into Eq. (5.51); they depend therefore on the incoming wave $\mathbf{k}_0$ only. The sum over $\mathbf{P}$ runs over all lattice points within a limiting radius determined by the energy and the damping parameter. The quantities $t^{j}(E)$ are diagonal matrices describing a single scattering event at atom j. The dimension of the matrix which must be inverted to solve Eq. (5.51) is $N(l_{\max} + 1)^2$, where N is the number of atoms in the unit cell and $l_{\max} + 1$ is the number of angular momentum components used. The matrix $\mathbf{M_{g'g}}$ must be invariant under any symmetry operation that leaves unchanged both the crystal and the incident beam. Unfortunately, this symmetry property cannot be used in the present form of Eqs. (5.50) and (5.51), as the distance vector $\mathbf{r}_j - \mathbf{r}_\mu$ is in general not invariant under a symmetry operation. When there is only a single atom in the unit cell the origin can be chosen such that the phase factors $\exp[i(\mathbf{k_g} - \mathbf{k_{g'}})\mathbf{r}_j]$ in Eq. (5.51) vanish. It then becomes immediately clear that symmetry-adapted functions can be used. This is still possible when the origin of the layer is chosen to be at special points in the unit cell that have the full symmetry. A symmetry operation transforms point j into point j', which must also be a lattice point in that case. The phase factors then remain unchanged:

$$e^{i\mathbf{k_g}\mathbf{r}_j} = e^{i\mathbf{k_g}\mathbf{r}_{j'}} \text{ when } \mathbf{r}_j - \mathbf{r}_{j'} = \mathbf{r}_n,$$

where $\mathbf{r}_n$ is a translation vector and $\mathbf{k_g}$ is a vector of the reciprocal net. In the following it is assumed that all the vectors $\mathbf{k_g}$ and $\mathbf{k_{g'}}$ belong to the reciprocal net of the surface lattice.

When there are atoms at general positions in the unit cell the phase factors are no longer invariant under a symmetry operation and the matrices $\mathbf{T}$ no longer contain symmetries. Also, the propagator matrices $\mathbf{G}^{j\mu}(\mathbf{k}_0)$ are in general not invariant under a symmetry operation. This means that for an atom in a general position no symmetries can be used. The local point symmetry of that atom is 1 and the wave field around this atom also has no symmetry. Nevertheless, each general point in the unit cell is associated with one or more symmetrically equivalent points, the number of which depends on the position of that point and the space group. The matrices $\mathbf{T}^j(\mathbf{k_g})$ can be transformed into each other by simple symmetry operators.

It is important to note that the sum over equivalent positions can be done prior to inversion of the matrix. To do that it is necessary to first perform the sum over equivalent beams in Eq. (5.50). There are n_g equivalent beams $\mathbf{k_g}$ generated by all symmetry operations acting on $\mathbf{k_g}$: this is usually called the beam star of $\mathbf{k_g}$. Here and in the remainder of this section, $\mathbf{k_g}$ denotes this beam star, that is, the set of symmetrically equivalent beams, saving a further subscript. Only for one beam of the star are the reflection and transmission matrix elements needed: by symmetry, the other corresponding matrix elements are equal. It is convenient to define the following quantities in angular momentum space [5.33]:

$$a^j_{lm}(\mathbf{k_{g'}}) = (n_{g'})^{1/2} \sum_{s=1}^{n_{g'}} e^{\left(-i\mathbf{k_{g_s'}}\mathbf{r}_j\right)} Y_{lm}(\mathbf{k_{g_s}}), \tag{5.53}$$

$$b^j_{lm}(\mathbf{k_g}) = (n_g)^{1/2} t^j_l(E) \sum_{s=1}^{n_g} e^{\left(i\mathbf{k_{g_s}}\mathbf{r}_j\right)} (-1)^m Y_{l-m}(\mathbf{k_{g_s}}). \tag{5.54}$$

The matrices $\mathbf{T}^j(\mathbf{k_g})$ are not explicitly needed to calculate $\mathbf{M_{gg'}}$ and Eq. (5.51) is solved directly for the vectors $\mathbf{Z}^j(\mathbf{k_g})$:

$$\mathbf{Z}^j\left(\mathbf{k_g}\right) = \mathbf{b}^j\left(\mathbf{k_g}\right) + \mathbf{t}^j(E)\sum_{\mu}\mathbf{G}^{j\mu}(\mathbf{k_0})\mathbf{Z}^\mu\left(\mathbf{k_g}\right), \text{where}$$

$$Z^j_{lm}\left(\mathbf{k_g}\right) = b^j_{lm}\left(\mathbf{k_g}\right) + t^j_{lm}(E)\sum_{l'm'}\sum_{\mu=1}^{N} G^{j\mu}_{lm,\,l'm'}(\mathbf{k_0})Z^\mu_{l'm'}\left(\mathbf{k_g}\right) \qquad (5.55)$$

$$= \sum_{l'm'}(1-\mathbf{X})^{-1}_{\nu lm,\,\mu\,l'm'}b^\mu_{l'm'}\left(\mathbf{k_g}\right).$$

The quantities $\mathbf{a}^j(\mathbf{k_g})$ and $\mathbf{b}^j(\mathbf{k_g})$ are vectors in angular momentum space and $\mathbf{X}$ is a short notation for a large matrix with the propagator matrices $\mathbf{G}^{j\mu}$ as submatrices. They are not invariant under a symmetry operation but simple transformation rules exist. A symmetry operation of a space group may be denoted by $(\mathbf{W}, \mathbf{w})$, see Section 2.1.8. $\mathbf{W}$ is a point group operation and $\mathbf{w}$ is a glide vector in a 2-D group. The operation acting on a vector $\mathbf{r}^j$ transforms it into $\mathbf{r}^{j'}$. This is equivalent to a symmetry operation of the wave field incident on atom j, which is kept fixed. When a symmetry operation is applied to the beam star the vectors $\mathbf{a}^j(\mathbf{k_g})$ and $\mathbf{b}^j(\mathbf{k_g})$, Eqs. (5.53) and (5.54), remain invariant. The symmetry operation here acts only on the wave field and since $\mathbf{a}^j(\mathbf{k_g})$ depends on the beam star $\mathbf{k_g}$ it is invariant under a symmetry operation:

$$(\mathbf{W}, \mathbf{w})\cdot\mathbf{a}^j\left(\mathbf{k_g}\right) = \mathbf{a}^{j'}\left(\mathbf{k_g}\right). \qquad (5.56)$$

We use the notation $\mathbf{D}_{jj'}$ for the matrix transforming a vector $\mathbf{r}^j$ to a vector $\mathbf{r}^{j'}$ or transforming the whole wave field. From the definitions given in Eqs. (5.53) and (5.54) and the properties of the spherical harmonics, it follows directly that a rotation about an n-fold axis with rotation angle $\varphi_n = 2\pi/n$ implies:

$$a^{j'}_{lm}\left(\mathbf{k_g}\right) = a^{j}_{lm}\left(\mathbf{k_g}\right)e^{im\varphi_n}. \qquad (5.57)$$

The rotation of a position vector is equivalent to a rotation of the beam star in the opposite direction. A mirror plane at an angle γ to the x-axis leads to:

$$a^{j'}_{lm}\left(\mathbf{k_g}\right) = a^{j}_{l-m}\left(\mathbf{k_g}\right)(-1)^m e^{im2\gamma}. \qquad (5.58)$$

A glide plane is connected with a phase factor in the reflected amplitude due to the glide vector $\mathbf{w}$ and therefore

$$a^{j'}_{lm}\left(\mathbf{k_g}\right) = a^{j}_{l-m}\left(\mathbf{k_g}\right)(-1)^m e^{im2\gamma}e^{i\mathbf{k_g}\mathbf{w}}. \qquad (5.59)$$

The transformation properties can be formally written as:

$$\mathbf{a}^{j'}\left(\mathbf{k_g}\right) = \mathbf{a}^j\left(\mathbf{k_g}\right)\mathbf{D}^{jj'}. \qquad (5.60)$$

For the vectors $\mathbf{b}^j(\mathbf{k_g})$ the same relations apply for the inverse matrices:

$$\mathbf{b}^{j'}\left(\mathbf{k_g}\right) = \left(\mathbf{D}^{jj'}\right)^{-1}\mathbf{b}^j\left(\mathbf{k_g}\right). \qquad (5.61)$$

The matrices $\mathbf{D}^{jj'}$ are unitary matrices in angular momentum space. They are diagonal for a rotation and change the sign of the m indices for a mirror or glide plane. For the point groups they are independent of the wave vectors $\mathbf{k_g}$ or $\mathbf{k_{g'}}$, while for groups with a glide plane they contain a phase factor $u_g = \exp(i\mathbf{k_g}\mathbf{w})$. In the 2-D symmetry groups only values $u_g = \pm 1$ occur. In Appendix G the index selection rules for all special positions in the 2-D symmetry groups are listed. The possible symmetry operations and glide vectors for all 17 2-D space groups are listed in Table 2.4.

The reflection and transmission matrices must remain unchanged under a symmetry operation of the crystal, and this leads to equivalent relations for the vectors $\mathbf{Z}^j(\mathbf{k_g})$:

$$\mathbf{Z}^{j'}(\mathbf{k_g}) = \left(\mathbf{D}^{jj'}\right)^{-1}\mathbf{Z}^j(\mathbf{k_g}). \tag{5.62}$$

These relations enable us to perform the sum over equivalent positions $\mathbf{r}_i$ in Eq. (5.50). For this purpose, it is necessary to split the sum over subplanes or over all atoms in the unit cell into two parts. Summation indices (j, μ) refer to all atoms in the unit cell, where (j_0, μ_0) designate only symmetrically independent positions. Finally (j', μ') designate the set of equivalent positions generated by the symmetry operations. A summation index j' includes the position j_0 unless it is explicitly indicated otherwise under the summation sign. It should be kept in mind that at oblique incidence all rotation axes are lost and only mirror or glide planes coinciding with the plane of incidence are retained. The symmetry of the unit cell which can be used in the multiple scattering calculation is then only one of the groups pm, cm and pg. With these definitions one obtains from Eq. (5.55):

$$Z_{lm}^{j_0}(\mathbf{k_g}) = b_{lm}^{j_0}(\mathbf{k_g}) + t_l^{j_0}(E)\sum_{\mu_0}\sum_{\mu'\neq\mu_0}\sum_{l'm'}G_{lm,l'm'}^{j_0\mu'}(\mathbf{k_0})\sum_{l''m''}\left(\mathbf{D}^{\mu_0\mu'}\right)_{l'm',l''m''}^{-1}Z_{l''m''}^{\mu_0}(\mathbf{k_g}), \tag{5.63}$$

where use has been made of Eq. (5.62). We define new symmetrised interlayer propagators $\mathbf{G}^S$

$$\mathbf{G}^{S,j_0\mu_0}(\mathbf{k_0}) = \sum_{\mu'}\mathbf{G}^{j_0\mu'}(\mathbf{k_0})\left(\mathbf{D}^{\mu_0\mu'}\right)^{-1}, \tag{5.64}$$

and Eq. (5.63) can then be written as

$$\mathbf{Z}^{j_0}(\mathbf{k_g}) = \mathbf{b}^{j_0}(\mathbf{k_g}) + \mathbf{t}^{j_0}(E)\sum_{\mu_0}\mathbf{G}^{S,j_0\mu_0}(\mathbf{k_0})\mathbf{Z}^{\mu_0}(\mathbf{k_g}). \tag{5.65}$$

This equation is completely equivalent to Eq. (5.55); the only difference is that the interlayer propagators $\mathbf{G}^S$ describe the propagation of a spherical wave from a subplane j_0 to a subplane μ_0, where the latter contains all symmetrically equivalent positions and is no longer a subplane within the former definition. The index j_0 in Eq. (5.65) now runs over all points in the asymmetric unit of the surface unit cell, that is, the scattering matrices $\mathbf{T}^j$ or the vectors $\mathbf{Z}^j$ must be calculated only for these points. Now the reflection and transmission matrices are simply given by

$$M_{\mathbf{g'g}}^{\pm\pm} = \sum_j \mathbf{a}^j\left(\mathbf{k_{g'}^{\pm}}\right)\mathbf{Z}^j\left(\mathbf{k_g^{\pm}}\right) = \sum_{j_0}n^{j_0}\mathbf{a}^{j_0}\left(\mathbf{k_{g'}^{\pm}}\right)\mathbf{Z}^{j_0}\left(\mathbf{k_g^{\pm}}\right) + \delta_{\mathbf{g'g}}\delta_{\pm\pm}, \tag{5.66}$$

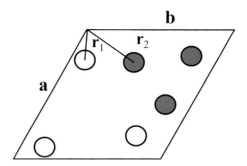

Figure 5.15 Example of a unit cell with symmetry p3 and two symmetrically inequivalent atoms in the unit cell with the local symmetry p1. Using the symmetry in Eqs. (5.51) and (5.52), the symmetrised subplane j_0 in Eq. (5.55) now contains all three symmetrically equivalent atoms (open circles) and subplane μ_0 contains the other three symmetrically atoms (shaded circles). Only two propagator matrices $\mathbf{G}^{S,j_0\mu_0}(\mathbf{k}_0)$ are needed with $j_0 = 1, 2$ and $\mu_0 = 1, 2$.

where n^{j_0} is the multiplicity of point j_0. This is illustrated in Figure 5.15, where a hexagonal cell is shown with six atoms in the unit cell, two of which are independent with local symmetry p1. Without symmetry the whole plane is subdivided into six subplanes; if the symmetry p3 is used, we have two subplanes and the matrix to be inverted has dimension $2(l_{\max} + 1)^2$. In the calculation of the symmetrised propagator matrices $\mathbf{G}^{S,j_0\mu_0}(\mathbf{k}_0)$, Eq. (5.64), we must of course include all single propagator matrices according to all interatomic vectors within the unit cell.

So far, the size of the matrix to be inverted has been reduced because only symmetrically independent atoms need to be explicitly included. But this is not the only consequence of symmetry that can be used. Equation (5.50) has now been brought into a form suited for the introduction of symmetry-adapted functions. The symmetrisation in angular momentum space is coupled to that in $\mathbf{k}$-space in the sense that symmetries in angular momentum space can only be used when the corresponding sum over equivalent beams and the sum over equivalent atomic positions are also performed. That means that Eq. (5.65) must be solved instead of Eq. (5.55); otherwise the symmetry in angular momentum space is destroyed.

5.3.2   Symmetry in Angular Momentum Space

The appropriate way to take into account symmetries in connection with spherical harmonics is to use symmetry-adapted functions. These are linear combinations of spherical harmonics that are either invariant or have the required transformation properties under symmetry operations. The scattered wave from each atom can be expanded in terms of symmetry-adapted functions where the local point symmetry of the atom applies. It is necessary to introduce symmetries into the lattice sum given in Eq. (5.27) or (5.52). The first step is to perform the sum over equivalent atomic sites as well as the sum over equivalent beams in the system of equations defining the matrices $\mathbf{T}$ and the reflection and transmission matrices.

Symmetry-adapted functions for all crystallographic point groups have been tabulated [5.34]; the irreducible representations for the point symmetries and the character tables can be found there. However, only the 17 2-D space groups are needed here and we shall summarise the index selection rules in the following. The 10 2-D point groups are considered first, since for these groups the transformation matrices do not depend on the wave vector $\mathbf{k_g}$ or $\mathbf{k_{g'}}$. Consequently, only the unit representation is needed, provided that the wave vectors $\mathbf{k_g}$ belong to the reciprocal net of the unit cell. When the point j is on an n-fold axis, Eq. (5.57) implies that

$$a_{lm}^{j'}(\mathbf{k_g}) = a_{lm}^{j}(\mathbf{k_g})e^{im2\pi/n}, \tag{5.67}$$

and m should satisfy the condition $m = 0 \pmod{n}$. It is unimportant whether point j is at the origin or not, provided that the origin is chosen properly at the principal axis as usual; then the condition for m holds true for all other axes. It should be noted that this is only true for point groups and symmorphic groups (cm and c2mm): a glide operation can add a phase factor and m can take other values.

When point j lies on a mirror plane, Eq. (5.58) must be applied, where γ is the angle between the x-axis and the mirror plane. For all atoms on a mirror plane the spherical harmonics involving the vectors $\mathbf{a}^j$, $\mathbf{b}^V$ and $\mathbf{Z}^V$ can be replaced by the symmetry-adapted function

$$Y_{lm}^S = \frac{1}{\sqrt{2}}\{Y_{lm} + (-1)^m e^{im2\gamma}Y_{l-m}\}, \tag{5.68}$$

and only positive values of m are needed. These selection rules remain unaltered for the vectors $\mathbf{Z}^V$ and consequently also for the propagator matrices $\mathbf{G}^{S,j\mu}$. From the properties of the transformation matrices it follows that for an atom j_0 on an n-fold axis

$$G_{lm,\,l'm'(\mathbf{k_0})}^{S,J_0,\mu_0} = 0, \quad \text{unless } m = 0 \bmod (n). \tag{5.69}$$

This is a consequence of the fact that $\mathbf{G}^S$ is a propagator from a subplane containing all symmetrically equivalent atoms. It is easy to show that for the indices $l'm'$ on the left-hand side of Eq. (5.69) the local point symmetry of atom μ_0 has to be taken. For an atom in a general non-symmetrical position, all spherical waves are needed and no reduction is possible. Here the reduction results from the fact that the sum over equivalent positions can be performed before inverting the matrix. For each atom, a different set of indices lm must be defined, according to its local symmetry, in such a way that the matrix to be inverted is reduced to its minimum size. The interlayer propagator matrices $G_{lm,\,l'm'}^{S,\,V_0\mu_0}$ are now rectangular matrices, where the indices lm take values belonging to the local symmetry of point V_0 and the indices $l'm'$ refer to the local symmetry of point μ_0. A scheme of the symmetrised matrix $\mathbf{G}^S$ is shown in Figure 5.16.

The index selection rules are described in detail in Appendix G, see Table G.1. They apply for the point groups and symmorphic groups, which are here the groups cm and cmm: the propagators for the centred lattices can be calculated by choosing the primitive unit cell and retaining the mirror plane. For the other groups with glide

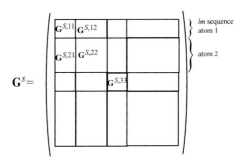

Figure 5.16 Scheme of the symmetrised propagator matrix $\mathbf{G}^S$. The submatrices are rectangular because some angular momentum components can be omitted as described by Eqs. (5.67–5.69).

planes (pg, p2gg, p2mg and p4gm) the calculation becomes more complicated. As mentioned in Appendix G, a glide operation changes the sign of the reflected amplitudes for those beams having an odd-order index parallel to the glide plane. The consequence is that different representations are required for different beams. A glide plane parallel to the x-axis with a glide vector of half of the translation vector implies that

$$a_{v,lm}(\mathbf{k_g}) = a_{v,l-m}(\mathbf{k_g}) \cdot (-1)^m e^{i\pi h}, \tag{5.70}$$

where $\mathbf{k_g} = \mathbf{k_0} + h\mathbf{a}^* + k\mathbf{b}^*$ and h, k are indices of the beam $\mathbf{k_g}$, while $\mathbf{a}^*$ and $\mathbf{b}^*$ are the basis vectors of the 2-D reciprocal net. The superlattice translation vectors $\mathbf{a}$ and $\mathbf{b}$ must be used with integer indices h and k. Due to the shift of the atom a phase factor occurs in the symmetry operator and the propagator matrices have to be calculated twice for the two beam sets (i.e., the set with even index h and the set with odd index h), each having a different set of symmetry-adapted functions. This could be applied only to atoms placed directly on a glide plane. This position, however, is not a special position, that is, it has the local point symmetry 1 and a shift of the atom off the glide plane does not produce split positions, unlike positions on mirror planes. We therefore do not recommend using this possibility to reduce the size of the matrices. A detailed description of the index selection rules for the special positions in all 2-D point groups is given in Appendix G.

The full use of symmetries still allows optimising atomic positions, as long as those positions respect the overall symmetry as well as the local symmetry of each atom. The local symmetry of an atom cannot be changed in the optimisation procedure. For instance, if an atom is located on a mirror plane, it can be moved only within that mirror plane; that atom cannot be allowed to move out of the mirror plane, because the single atom would become two atoms on either side of that mirror plane, thus changing the number of atoms, changing the stoichiometry and potentially placing atoms too close together.

If it is desired to let atoms break the initial symmetry by moving away from a mirror or glide plane or from a rotational axis, then a new symmetry (or no symmetry) should be chosen for the optimised structure. Before starting the calculation of interlayer propagators, the local symmetry of each atom needs to be labelled and the allowed shifts of the atomic positions have to be determined. Which coordinates are free is given by the Wyckoff positions.

The local symmetry fixes the *lm* sequence of the spherical harmonics for all atoms and the sequence can be stored in an index array. The sequence of atoms does not change during optimisation of the atomic parameters. All further referencing to the indices *l* and *m* is done via this index image. The subsequent calculation is then independent of the actual coordinates of each atom as long as its local symmetry and multiplicity are not changed. The ten point groups and the groups cm and cmm can be handled that way. For the remaining four groups containing glide planes the procedure requires more effort since an additional loop must be incorporated. The set of beams must be decomposed into two or more groups for which different symmetry-adapted functions are required. The propagator matrices must be calculated again, but the sum over lattice points need not be repeated, since it can be stored and used again. There is still an important gain in computing time and memory space compared with the calculation made without using symmetries. All 17 2-D space groups can be handled with the same program. Details of the calculation of interlayer propagator matrices and the lattice sum are given in Appendix G.

5.4 Approximations

The computational effort increases drastically with the number of symmetrically independent atoms in the unit cell. A number of approximations have been proposed to treat structures with large unit cells. Some of the methods turned out to be insufficiently precise for general applications and are not used any more. We consider here only three methods: tensor LEED, frozen LEED and diffuse LEED. Tensor LEED is widely used in connection with optimisation methods and reduces the computation time substantially, especially in its symmetrised version. Frozen LEED is less used but should provide similar improvement as tensor LEED. Diffuse LEED applies to crystalline surfaces with some disorder that produces diffuse LEED patterns.

Methods applied to nanostructures with a very large number of atoms and without requiring translation symmetry have been described in Section 5.2.4. NanoLEED is not considered to be an approximation because the numerical precision can be chosen, for example by the grid size in the sparse-matrix canonical grid (SMCG) method (see Section 5.2.4.1) and leads to a correct solution within numerical errors. Approximation methods, on the other hand, contain systematic errors as some parts of the multiple scattering equations are omitted. The errors, nevertheless, are small and tolerable if the limits for parameter variation are set sufficiently small.

5.4.1 Tensor LEED Approximation

Tensor LEED (TLEED) is an efficient method for calculating the change in the scattering amplitude as a function of the shift of an atom. It is used in combination with optimisation procedures where many I(V) calculations are required for the variation of atomic parameters. A full dynamical calculation is performed for a reference structure and the change in diffraction intensities is approximately calculated from the shift of atomic parameters, most often shifted atomic positions. It is valid in a limited range of about 0.04 nm (0.4 Å) for coordinate shifts [5.35]. Variations of chemical element, Debye–Waller factors and site occupation factors are possible as well. The approximate part of the tensor LEED calculation is relatively very fast, enabling extensive exploration of many structures deviating from a single reference structure. This exploration is particularly useful for steepest descent schemes to optimise atomic positions and other parameters (steepest descent leads to a new reference structure, but this iterative step is usually not necessary, due to the good quality of the tensor LEED approximation).

The theory and applications of tensor LEED are described in a number of publications by P. J. Rous and J. B. Pendry, in particular [5.36; 5.37]. The formalism described there refers to the application in the RFS scheme, see Section 5.2.6.4. We describe here the tensor LEED formalism for use with the layer doubling method. That means we need to calculate the change of all matrix elements of the reflection and transmission matrices for all composite layers in which atom parameters are optimised (composite layers contain more than one atom per 2-D unit cell). Then those layers must be combined with the other layers by the layer doubling scheme. This is the most general application. Tensor LEED requires much storage space in this case. Alternatively, all atoms in the surface slab can be put in a single layer of sufficient thickness so that backscattering from deeper layers can be ignored: this approach is often called the giant-matrix method, cf. Section 5.2.5.2. Although the dimension of the matrix to be inverted becomes very large, the method is quite efficient as only one vector of the reflection matrix $\mathbf{R}_{\mathbf{g'g}}$ need be calculated, for $\mathbf{g} = 0$, that is, the incident beam. There is only one incident beam $\mathbf{k}_0$ and only the reflection amplitudes for those beams which are measured need to be calculated. It should be checked before starting an analysis whether the giant-matrix method or the layer doubling method is more efficient or whether the RFS scheme is applicable when small layer distances occur. If layer distances are large enough, larger than about 0.1 nm, the RFS scheme is certainly the most efficient for stacking layers.

We start again with Eq. (5.35), slightly adapted, which describes the scattering amplitude for an incoming wave $\mathbf{k}_{\mathbf{g}}$ into the direction $\mathbf{k}_{\mathbf{g'}}$:

$$M_{\mathbf{g'g}}^{\pm\pm} = -\frac{16\pi^2 i m_e}{A k_{\mathbf{g}z}^+ \hbar^2} \sum_{LL'} Y_L\left(\mathbf{k}_{\mathbf{g'}}^{\pm}\right) \sum_{i=1}^{N} \left\{ e^{i\left(\pm\mathbf{k}_{\mathbf{g}}^{+}\mp\mathbf{k}_{\mathbf{g'}}^{\pm}\right)\mathbf{r}_i} T_{LL'}^i \right\} Y_{L'}^*\left(\mathbf{k}_{\mathbf{g}}^{\pm}\right) + \delta_{\mathbf{g'g}}\delta_{\pm\pm},$$

(5.72)

where N is the number of atoms in the unit cell. The directions of incoming and outgoing waves are omitted in the following equations for convenience and will be inserted in the final equations to distinguish reflection and transmission matrices, as in Section 5.2.5. Here, $\mathbf{k}_{\mathbf{g}}^{+}$ is the incoming wave from the vacuum side and $\mathbf{k}_{\mathbf{g}}^{-} = -\mathbf{k}_{\mathbf{g}}^{+}$ is the incoming wave from the crystal side. The signs of the wave vectors must be chosen according to the matrices that are calculated. The vectors $\mathbf{r}_j$ are the position vectors of all atoms in the unit cell with respect to the origin of the composite layer. The use of symmetries and the reduction to symmetrically independent atoms in the unit cell are described in Appendix H.

The scattering matrices $\mathbf{T}^j$ describe the wave leaving the atom at point $\mathbf{r}_j$ without further scattering. $\mathbf{T}^j$ is calculated in the angular momentum expansion:

$$T^j_{lm,\,l'm'}(\mathbf{k_g}) = t^j_l(E) + t^j_l(E) \sum_{\mu=1}^{N} \sum_{l''m''} G^{j\mu}_{lm,\,l''m''}(\mathbf{k}_0) e^{i\mathbf{k_g}(\mathbf{r}_j - \mathbf{r}_\mu)} T^{\mu}_{l''m'',\,l'm'}(\mathbf{k_g}). \tag{5.73}$$

One can define the following vectors in angular momentum space:

$$a^j_{lm}(\mathbf{k}_{g'}) = e^{-i\mathbf{k}_{g'}\mathbf{r}_j} Y_{lm}(\mathbf{k}_{g'}), \tag{5.74}$$

$$b^j_{lm}(\mathbf{k_g}) = e^{i\mathbf{k_g}\mathbf{r}_j}(-1)^m Y_{l-m}(\mathbf{k_g}), \tag{5.75}$$

and

$$\begin{aligned} Z^j_{lm}(\mathbf{k_g}) &= t^j_l(E) b^j_{lm}(\mathbf{k_g}) + t^j_l(E) \sum_{l'm'} \sum_{\mu=1}^{N} G^{j\mu}_{lm,\,l'm'}(\mathbf{k}_0) Z^{\mu}_{l'm'}(\mathbf{k_g}) \\ &= t^j_l(E) \sum_{l'm'} \sum_{\mu=1}^{N} \left[(1 - X^{-1}) \right]^{j\mu}_{lm,\,l'm'} b^{\mu}_{l'm'}(\mathbf{k_g}). \end{aligned} \tag{5.76}$$

With these vectors, Eq. (5.72) can be written as

$$M^{\pm\pm}_{\mathbf{g'g}} = -\frac{16\pi^2 i m_e}{A k^+_{gz} \hbar^2} \sum_{lm} \sum_{j=1}^{N} a^j_{lm}(\mathbf{k}^{\pm}_{g'}) t^j_l Z^j_{lm}(\mathbf{k}^{\pm}_g) + \delta_{\mathbf{g'g}} \delta_{\pm\pm}. \tag{5.77}$$

Here $\mathbf{Z}^j(\mathbf{k_g})$ represents the incoming spherical waves from all surrounding atoms, and $\mathbf{a}^j(\mathbf{k}_{g'})$ is the diffracted wave, leaving atom j without further scattering. In Eq. (5.77) the single scattering matrix $\mathbf{t}^j$ occurs now between the vectors $\mathbf{a}^j(\mathbf{k}_{g'})$ and $\mathbf{Z}^j(\mathbf{k_g})$. This allows us to modify $\mathbf{t}^j$ and leave the incoming waves from the surroundings unchanged. In the tensor LEED approximation the atom at point $\mathbf{r}^j$ is shifted from the position in the reference structure, while the incoming wave field is transported to the new position, as is the outgoing wave field, but the propagator matrices $\mathbf{G}^{j\mu}(\mathbf{k}_0)$ in the X matrix (Eq. (5.76)) remain unchanged. An illustration of the method is shown in Figure 5.17. The tensor LEED approximation can be described as a 'frozen X matrix' approximation.

A shift of atom j by $\delta\mathbf{r}^j$ changes the diffracted amplitude, that is, the matrix element $M_{\mathbf{g'g}}$ of the reference structure, to $M'_{\mathbf{g'g}}$:

$$M'_{\mathbf{g'g}} = M_{\mathbf{g'g}} + \delta M_{\mathbf{g'g}}. \tag{5.78}$$

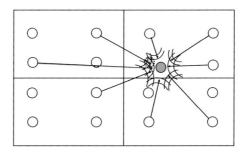

Figure 5.17 Illustration of the tensor LEED approximation. The shaded atom is shifted from its original position and the field of spherical waves coming from the surrounding atoms remains unchanged but is propagated to the new position.

We approximate $\delta M_{\mathbf{g'g}}$ by shifting only the atomic t-matrix $t^j(E)$ in Eq. (5.77) and leaving the propagator matrices occurring in Eq. (5.76) unchanged. We thus replace $t_l^j(E)$ by:

$$t_{lm,l'm'}^{'j} = t_{lm,l'm'}^{j}\delta_{ll'}\delta_{mm'} + \delta t_{lm,l'm'}^{j}, \tag{5.79}$$

which becomes a non-diagonal matrix. To obtain the angular momentum expansion from the phase factors in real space we can make use of the spherical wave expansion of a plane wave:

$$e^{i\mathbf{k}\mathbf{r}} = \sum_{lm} 4\pi i^l j_l(kr) Y_{lm}(\Omega_r)(-1)^m Y_{l-m}(\Omega_k). \tag{5.80}$$

With

$$\mathbf{t}^{'j}(\mathbf{k_g'},\mathbf{k_g}) = e^{i\mathbf{k_g'}\delta\mathbf{r}^j} t^j(\mathbf{k_g'},\mathbf{k_g}) e^{-i\mathbf{k_g}\delta\mathbf{r}^j}, \tag{5.81}$$

we obtain

$$\delta\mathbf{t}^j(\mathbf{k_g'},\mathbf{k_g}) = e^{-i\mathbf{k_g'}\delta\mathbf{r}^j}\mathbf{t}^{'j}(k_g',\mathbf{k_g})e^{i\mathbf{k_g}\delta\mathbf{r}^j} - \mathbf{t}^j(\mathbf{k_g'},\mathbf{k_g}). \tag{5.82}$$

In Eq. (5.82) the change of the t-matrix is defined as a function of the wave vectors; however, we need the expansion in spherical harmonics. The transformation of a t-matrix from spherical waves to a function of wave vectors is given by

$$t(-\mathbf{k}',\mathbf{k}) = 8\pi^2\sum_{lm}\sum_{l'm'}(-1)^\ell Y_{lm}(\mathbf{k}')t_{lm,l'm'}(-1)^{m'}Y_{l'-m'}(\mathbf{k}), \tag{5.83}$$

and the reciprocal transformation gives

$$t_{lm,l'm'} = \frac{1}{8\pi^2}\int\int(-1)^{l+m}Y_{l-m}(\mathbf{k}')t(\mathbf{k}',\mathbf{k})Y_{l'm'}(\mathbf{k})d\Omega_k d\Omega_{k'}. \tag{5.84}$$

Inserting Eq. (5.82) into Eq. (5.84) we obtain

$$
\delta t^{j}_{lm,\,l'm'} = \sum_{l_2=0,\,m_2}^{l_s} 4\pi i^{l_2} j_{l_2}\left(k\delta r^{j}\right) Y_{l_2 m_2}\left(\delta \mathbf{r}^{j}\right)
$$

$$
\cdot \sum_{l_3=0,\,m_3}^{l_{max}} \sum_{l_5=0,\,m_5}^{l_s} (-1)^{m_2+m}\, a(l_2-m_2, l_3 m_3, l-m) t^{j}_{l_3}(-1)^{m_3+m_5}
$$

$$
\cdot a(l_3-m_3, l_5-m_5, l'm') 4\pi i^{l_5} j_{l_5}\left(k\delta r^{j}\right) Y_{l_5 m_5}\left(\delta \mathbf{r}^{j}\right) - t^{j}_{lm\,l'm'}\delta_{ll'}\delta_{mm'},
$$

(5.85)

where summation over m_2 is implied from $-l_2$ to $+l_2$, and similarly for m_3 and m_5. The upper limit l_s for l_2 and l_5 is set by the value where the spherical Bessel function $j_l(k\delta r^{j})$ becomes negligibly small, usually $l_s \leq 3$ or 4. The quantities $a(LM, L'M', lm)$ are Clebsch–Gordan-like coefficients or Gaunt coefficients (Eq. (5.9)):

$$
a(LM, L'M', lm) = \int (-1)^{M} Y_{L-M}(\Omega) Y_{L'M'}(\Omega)(-1)^{m} Y_{l-m}(\Omega) d\Omega,
$$

(5.86)

with the conditions

$$
M - M' + m = 0,\ |L' - l| \leq L \leq |L' + l|,\ L + L' + l = \text{even}.
$$

(5.87)

We obtain, finally, for the change in the diffraction amplitudes:

$$
\delta M_{\mathbf{g'g}} = -\frac{16\pi^2 i m_e}{A k^{+}_{gz}\hbar^2} \sum_{j} \sum_{l_2 m_2} 4\pi i^{l_2} j_{l_2}\left(k\delta r^{j}\right) Y_{l_2 m_2}\left(\delta \mathbf{r}^{j}\right)
$$

$$
\cdot \sum_{lm} \sum_{l'm'} \sum_{l_3 m_3} \sum_{l_5 m_5} a^{j}_{lm}(\mathbf{k'_g})\Big\{ (-1)^{m_2+m} a(l_2-m_2, l_3 m_3, l-m) t^{j}_{l_3}
$$

$$
\cdot (-1)^{m_3+m_5} a(l_3-m_3, l_5-m_5, l'm') 4\pi i^{l_5} j_{l_5}\left(k\delta r^{j}\right) Y_{l_5 m_5}\left(\delta \mathbf{r}^{j}\right) - t^{j}_{l}\Big\} Z^{j}_{l'm'}(\mathbf{k_g}).
$$

(5.88)

It is convenient to define the vectors

$$
S^{a,j}_{l_2 m_2} = 4\pi i^{l_2} j_{l_2}\left(k\delta r^{j}\right) Y_{l_2 m_2}\left(\delta \mathbf{r}^{j}\right)
$$
$$
S^{b,j}_{l_5 m_5} = 4\pi i^{l_5} j_{l_5}\left(k\delta r^{j}\right) Y_{l_5 m_5}\left(\delta \mathbf{r}^{j}\right)
$$

(5.89)

and a tensor:

$$
F^{j}_{l_2 m_2, l_5 m_5}(\mathbf{k_{g'}}, \mathbf{k_g}) = \sum_{lm} \sum_{l'm'} \sum_{l_3 m_3} a^{j}_{lm}(\mathbf{k'_g}) \Big[(-1)^{m_2+m} a(l_2-m_2, l_3 m_3, l-m) t^{j}_{l_3}
$$

$$
\cdot (-1)^{m_3+m_5} a(l_3-m_3, l_5-m_5, l'm') - t^{j}_{l}\Big] Z^{j}_{l'm'}(\mathbf{k_g}).
$$

(5.90)

The change of the scattering amplitude is then obtained as:

$$
\delta M_{\mathbf{g'g}} = -\frac{16\pi^2 i m_e}{A k^{+}_{gz}\hbar^2} \sum_{l_2 m_2} \sum_{l_5 m_5} \sum_{j=1}^{N} S^{a,j}_{l_2 m_2} F^{j}_{l_2 m_2, l_5 m_5} S^{b,j}_{l_5 m_5}.
$$

(5.91)

The tensor $F^j_{l_2 m_2, l_5 m_5}(\mathbf{k_{g'}}, \mathbf{k_g})$ can be split into two independent parts, one for the incident beam set $\mathbf{k_g}$ and a second part for the diffracted beam set $\mathbf{k_{g'}}$, which allows the computational storage space to be reduced. We now include the direction of the wave vectors, which is needed to differentiate the reflection and transmission matrices:

$$
f^{a,j,\pm}_{l_2 m_2, l_3 m_3} = \sum_{lm} a^j_{lm}\left(\mathbf{k'}^{\pm}_g\right)\left[(-1)^{m_2+m} a(l_2 - m_2, l_3 m_3, l - m)t^j_{l_3} - t^j_l\right],
$$

$$
f^{b,j,\pm}_{l_3 m_3, l_5 m_5} = \sum_{l'm'} (-1)^{m_3 + m_5} a(l_3 - m_3, l_5 - m_5, l'm')Z^j_{l'm'}\left(\mathbf{k}^{\pm}_g\right).
$$

$$(5.92)$$

Now the scattering matrix is obtained as the product of two independent parts:

$$
\delta M^{\pm\pm}_{\mathbf{g'g}} = -\frac{16\pi^2 i m_e}{A k^+_{gz}\hbar^2}\sum_{j=1}^{N}\sum_{l_2 m_2}\sum_{l_3 m_3} S^{a,j}_{l_2 m_2} f^{a,j,\pm}_{l_2 m_2, l_3 m_3}\sum_{l_5 m_5} f^{b,j,\pm}_{l_3 m_3, l_5 m_5} S^{b,j}_{l_5 m_5}.
$$

$$(5.93)$$

Equation (5.93) is the final solution. The tensor $F^j_{l_2 m_2, l_5 m_5}(\mathbf{k_{g'}}, \mathbf{k_g})$ and its two parts defined in Eq. (5.92) depend only on the reference structure and need to be calculated only once, while the shift of atom j only affects the vectors $\mathbf{S}^a$ and $\mathbf{S}^b$. The change in scattering amplitude is thus obtained by a multiplication of a vector with a matrix instead of a new matrix inversion, saving considerable computing time. The Bessel functions j_l occurring in the shift vectors also need to be calculated only once for a grid of shifts and can be interpolated to intermediate values.

The tensor LEED approximation allows a very fast calculation of I(V) curves for structures in the vicinity of a reference structure. It does require a large amount of storage space depending on the size of the unit cell and the number of phase shifts used. The tensor $F^j_{l_2 m_2, l_5 m_5}(\mathbf{k_{g'}}, \mathbf{k_g})$ depends on the wave vectors of the incoming beams $\mathbf{k_g}$ and the diffracted beams $\mathbf{k_{g'}}$, while the two parts f^a and f^b have to be stored for all beams and all energies. As mentioned at the beginning of this section, the tensor LEED approximation is particularly useful in cases where the RFS scheme can be used or all atoms are combined in a single layer, that is, where the giant-matrix method is used. Then only one incoming wave exists and the second part of the tensor only needs to be stored for the measured beams, not for the whole beam set required in the coupling of layers. The coupling of layers with the layer doubling method is then avoided as well. In many cases this largely overcomes the larger computational effort needed in the multiple scattering calculation for the reference structure.

Tensor LEED can be symmetrised in the same way as a full dynamical calculation. When the structure exhibits a symmetry, this must not be changed in the optimisation process, as mentioned in Section 5.3. The use of symmetries reduces the size of the tensor substantially, as only selected indices are required in the (l, m) sequence of the spherical harmonics. The symmetrisation of the tensor is described in Appendix H.

5.4.2   Frozen LEED Approximation

A further approximation, frozen LEED, has been proposed by Z. X. Yu and S. Y. Tong [5.38]. Frozen LEED has close conceptual similarities to tensor LEED, including the use of a reference structure for which a full dynamical calculation is performed. However, frozen LEED exhibits a larger radius of convergence of about 0.08 nm (0.8 Å): this larger radius allows exploring other nearby minima in a structural search, rather than only the nearest minimum, without needing a new full dynamical calculation for a new reference structure, for example, by use of simulated annealing.

Instead of calculating the change of the diffraction amplitude as a function of a shift of a single atom, as is done in tensor LEED, frozen LEED considers a shift of a whole subplane. For that purpose, we must slightly rewrite Eq. (5.73), which defines the matrices **T** for all atoms in the unit cell while the propagator matrices contain the sum over all inter-atomic distances including the translation vectors, by adding a sum over translationally equivalent atoms **P**:

$$T^j_{LL'}(\mathbf{k_g}) = t^j_l(E) + t^j_l(E) \sum_{\mathbf{P}} \sum_{\mu=1}^{N} \sum_{L''} G^{j\mu}_{LL''}(\mathbf{k_0}) e^{i\mathbf{g}(\mathbf{r_\mu} - \mathbf{r}_j + \mathbf{P})} T^\mu_{L''L'}(\mathbf{k_g}), \quad (5.94)$$

where the sum over inter-atomic distances has been extracted from the propagator matrices **G**$^{j\mu}$ which are now defined as

$$G^{j\mu}_{LL'}(\mathbf{k_0}) = -4\pi i \frac{2m_e}{\hbar^2} k \sum_{\mathbf{P}} \sum_{L_1} i^{l_1} a(L, L', L_1) h^{(1)}_{l_1} (k|\mathbf{P} + \mathbf{r}_j - \mathbf{r_\mu}|) Y_{L_1}(\mathbf{P} + \mathbf{r}_j - \mathbf{r_\mu}) e^{-i\mathbf{k_0}\mathbf{P}}.$$
$$(5.95)$$

For the case $j = \mu$ (i.e., for one subplane) the sum over lattice points reduces to the sum over lattice vectors. This is the same for all subplanes:

$$G^P_{LL'}(\mathbf{k_g}) = -4\pi i \frac{2m_e}{\hbar^2} k \sum_{\mathbf{P}\neq 0} \sum_{l_1 m_1} i^{l_1} a(L, L', L_1) h^{(1)}_{l_1}(k|\mathbf{P}|) Y_{L_1}(\mathbf{P}) e^{i\mathbf{k_g}\mathbf{P}}. \quad (5.96)$$

Equation (5.94) becomes:

$$T^j_{LL'}(\mathbf{k_g}) = t^j_l(E) + t^j_l(E) G^P_{LL'}(\mathbf{k_g}) + t^j_l(E) \sum_{\mathbf{P},\mu\neq j} \sum_{L''} G^{j\mu}_{LL''}(\mathbf{k_0}) \cdot e^{i\mathbf{k_g}(\mathbf{r}_j - \mathbf{r_\mu} + \mathbf{P})} T^\mu_{L''L'}(\mathbf{k_g}). \quad (5.97)$$

We can define subplane t-matrices τ^j (as already defined in Eq. (5.23)) which are calculated from a much smaller set of linear equations:

$$\tau^j_{LL'}(\mathbf{k_g}) = t^j_l(E) + t^j_l(E) \sum_{L''} G^P_{LL''}(\mathbf{k_g}) \tau^j_{L''L'}(\mathbf{k_g}). \quad (5.98)$$

We insert these into Eq. (5.97). After some rearrangement of the matrix multiplications and making use of the matrix equation $(\mathbf{A} + \mathbf{B})^{-1} = (\mathbf{1} + \mathbf{A}^{-1}\mathbf{B})^{-1}\mathbf{A}^{-1}$ we obtain (analogously to Eq. (5.26)):

$$T^j_{LL'}(\mathbf{k_g}) = \tau^j_{LL'}(\mathbf{k_g}) + \tau^j_{LL'}(\mathbf{k_g}) \sum_{\mathbf{P},\mu\neq j} \sum_{L''} G^{j\mu}_{LL''}(\mathbf{k_0}) e^{i\mathbf{k_g}(\mathbf{r_\nu} - \mathbf{r_\mu})} T^\mu_{L''}(\mathbf{k_g}). \quad (5.99)$$

We use the vectors $\mathbf{a}^j$ and $\mathbf{b}^j$ in angular momentum space, defined previously in Eqs. (5.73) and (5.74), and an additional vector $\mathbf{c}^j$:

$$a_L^j(\mathbf{k}_{\mathbf{g}'}) = e^{-i\mathbf{k}_{\mathbf{g}'}\mathbf{r}_j} Y_L(\mathbf{k}_{\mathbf{g}'}), \tag{5.100}$$

$$b_L^j(\mathbf{k}_{\mathbf{g}}) = e^{i\mathbf{k}_{\mathbf{g}}\mathbf{r}_j}(-1)^m Y_{l-m}(\mathbf{k}_{\mathbf{g}}), \tag{5.101}$$

$$c_L^j(\mathbf{k}_{\mathbf{g}}) = \sum_{L'} \tau_{LL'}^j(\mathbf{k}_{\mathbf{g}}) b_{L'}^j(\mathbf{k}_{\mathbf{g}}). \tag{5.102}$$

The matrices $\mathbf{T}^j(\mathbf{k}_{\mathbf{g}})$ need not be calculated explicitly; similar to Eq. (5.76), we solve the matrix equation of Eq. (5.99) for vectors $\mathbf{Z}^j(\mathbf{k}_{\mathbf{g}})$:

$$
\begin{aligned}
Z_{lm}^j(\mathbf{k}_{\mathbf{g}}) &= \sum_{l'm'} \tau_{lm,l'm'}^j(\mathbf{k}_{\mathbf{g}}) b_{l'm'}^j(\mathbf{k}_{\mathbf{g}}) + \sum_{l''m''} \tau_{lm,l''m''}^j(\mathbf{k}_{\mathbf{g}}) \sum_{l'm'} \sum_{\mu=1,\mu\neq j}^{N} G_{l''m'',l'm'}^{j\mu}(\mathbf{k}_0) Z_{l'm'}^\mu(\mathbf{k}_{\mathbf{g}}) \\
&= c_L^j(\mathbf{k}_{\mathbf{g}}) + \sum_{L''} \tau_{LL''}^j(\mathbf{k}_{\mathbf{g}}) \sum_{L'} \sum_{\mu=1,\mu\neq j}^{N} G_{L'',L'}^{j\mu}(\mathbf{k}_0) Z_{L'}^\mu(\mathbf{k}_{\mathbf{g}}) \\
&= \sum_{L''} \tau_{LL''}^j(\mathbf{k}_{\mathbf{g}}) \sum_{L'} \sum_{\mu=1,\mu\neq j}^{N} \left[(1-X)^{-1}\right]_{L''L'}^{j\mu} b_{L'}^\mu(\mathbf{k}_{\mathbf{g}}).
\end{aligned}
\tag{5.103}
$$

The vectors $\mathbf{Z}^j(\mathbf{k}_{\mathbf{g}})$ are now defined in terms of the subplane scattering matrices $\boldsymbol{\tau}$ instead of atomic $\mathbf{t}^j$ matrices, as in Eq. (5.76). The solution vectors $\mathbf{Z}^j(\mathbf{k}_{\mathbf{g}})$ can be written as:

$$Z_L^j(\mathbf{k}_{\mathbf{g}}) = \sum_{L'} T_{LL'}^j(\mathbf{k}_{\mathbf{g}}) b_{L'}^J(\mathbf{k}_{\mathbf{g}}). \tag{5.104}$$

The scattering amplitudes are then obtained as:

$$M_{\mathbf{g}'\mathbf{g}}^{\pm\pm} = -\frac{16\pi^2 i m_e}{A k_{\mathbf{g}z}^+ \hbar^2} \sum_L \sum_j a_L^j(\mathbf{k}_{\mathbf{g}'}^{\pm}) Z_L^j(\mathbf{k}_{\mathbf{g}}^{\pm}) + \delta_{\mathbf{g}'\mathbf{g}}\delta_{\pm\pm}. \tag{5.105}$$

We can now shift the whole subplane by $\delta\mathbf{r}$ and calculate the change in $M_{\mathbf{g}'\mathbf{g}}$:

$$\delta M_{\mathbf{g}'\mathbf{g}}^{\pm\pm} = -\frac{16\pi^2 i m_e}{A k_{\mathbf{g}z}^+ \hbar^2} \sum_L \sum_j a_L'^j(\mathbf{k}_{\mathbf{g}'}^{\pm}) Z_L'^j(\mathbf{k}_{\mathbf{g}}^{\pm}) - M_{\mathbf{g}'\mathbf{g}}^{\pm\pm}. \tag{5.106}$$

In Eq. (5.106) $\mathbf{a}'^j(\mathbf{k}_{\mathbf{g}})$ and $\mathbf{Z}'^j(\mathbf{k}_{\mathbf{g}})$ are vectors for the shifted structure, while in Eq. (5.105) $\mathbf{a}^j(\mathbf{k}_{\mathbf{g}})$ and $\mathbf{Z}^j(\mathbf{k}_{\mathbf{g}})$ are for the reference structure. With layer doubling or the RFS scheme we have only one incident beam $\mathbf{k}_0$. At normal incidence the new vector $\mathbf{Z}'^j$ is then calculated from:

$$Z_L'^j(\mathbf{k}_0) = c_L'^j(\mathbf{k}_0) + \left[Z_L^{0j}(\mathbf{k}_0) - c_L^{0j}(\mathbf{k}_0)\right] e^{i\mathbf{k}_{0z}\delta\mathbf{r}_{jz}}. \tag{5.107}$$

The quantities $\mathbf{a}'^j(\mathbf{k}_{\mathbf{g}})$, $\mathbf{c}'^j(\mathbf{k}_0)$ and $\mathbf{Z}'^j(\mathbf{k}_{\mathbf{g}})$ must be recalculated for each new structure. The method is illustrated in Figure 5.18.

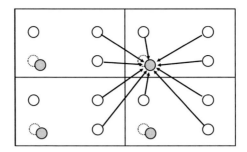

Figure 5.18 Illustration of the frozen LEED approximation. Shown is a section of the plane with four unit cells. One of the four atoms in the unit cell is shifted together with the translationally equivalent ones, that is, the whole periodic subplane is shifted. The wave field from the surrounding atoms remains unchanged but is transported to the shifted subplane positions. This is possible because the subplane has been extracted from the matrix to be inverted, see Eq. (5.99). For the wave field the plane wave expansion is used, in contrast to tensor LEED where the spherical wave expansion is used.

When the method is applied to one composite layer which must be combined with other layers using layer doubling or related methods, then all beams of the whole set of diffracted beams appear as incident beams and $\mathbf{Z}^{i,j}(\mathbf{k_g})$ must be calculated and stored for all beams. The calculation of new $\mathbf{a}^{ij}$, $\mathbf{c}^{ij}$ and $\mathbf{Z}^{ij}$ does not cause any problems.

Symmetries can be used by summation over symmetry equivalent beams in Eqs. (5.100–5.102). Symmetry in angular momentum components can be used as outlined in Section 5.3. The method is very fast and does not require much storage space when the whole surface slab is treated as a single layer. In the case of layer stacking, the required space is multiplied by the number of beams; therefore, the required space for storage can become very large.

5.4.3   Diffuse LEED for Disordered Surfaces

Disorder causes diffuse diffraction into all directions. In general, this applies to structures without long-range order, such as crystalline surfaces with randomly positioned defects (e.g., atomic vacancies or substitutions) or with irregularly placed adatoms or admolecules, which will be discussed in this section. Other candidates are randomly oriented or spinning molecules and disordered steps on a crystalline surface, as well as amorphous materials: such cases have not been addressed so far with diffuse LEED. The special case of finite particles is treated in Section 5.2.4. On the other hand, ideal quasicrystals (see Section 6.2) and modulated structures (see Section 6.3) have sufficient long-range order to prevent diffuse diffraction.

We here discuss the case of lattice gas disorder of adsorbates on crystalline surfaces, first proposed by J. B. Pendry and D. K. Saldin [5.39]. In this situation, adsorbates randomly occupy identical sites on the periodic substrate surface, while themselves lacking 2-D periodicity. Because the local electron scattering is very similar to the periodic case, the diffuse LEED (DLEED) intensity contains the same local structural information as conventional LEED for ordered overlayers, namely adsorption site, bond lengths, etc. It must be noted that the diffuse intensity also contains contributions from any other defects, such as scattering from vacancies, disordered steps, impurities and phonons: extracting adsorbate structure therefore requires good sample preparation and low temperatures.

We shall in this section briefly review the experimental and theoretical approaches to extract the local structural information of disordered adsorbates. Reviews were published by M. A. Van Hove [5.40] and U. Starke et al. [5.41]. DLEED for other forms of disorder, such as vacancies and impurities, has not yet been fully developed, but may use similar methods, as supported by a study of disordered Pd substitution in the outermost layer of Cu(100) [5.42].

In the presence of an overlayer with lattice gas disorder, the substrate continues to produce a (1×1) diffraction pattern. In fact, the integer-order (1×1) spots also contain useful structural information about the adsorbate, as has been shown for methanol on Pd(111) [5.43].

DLEED intensities between (1×1) spots provide additional data but are inherently weak compared to the spot intensities [5.44]. A high-sensitivity 'digital LEED' detector using a wedge-and-strip anode was specifically developed to measure weak DLEED intensities [5.45]. Furthermore, disorder is not the only cause of diffuse diffraction. Another source is thermal diffuse scattering due to phonons, including molecular vibrations. Disorder and phonons may contribute similar amounts; the thermal diffuse scattering can easily exceed the scattering due to disorder, especially if the adsorbate density is low. Since thermal diffuse scattering is mostly caused by the substrate in the case of adsorbates on solid surfaces [5.46], it is customary to subtract the diffuse scattering due to the clean surface from that due to the adsorbate-covered surface. Such subtraction can also help reduce the effect of impurities and defects such as steps and vacancies, assuming that their presence and structure are not much modified by adsorption.

Diffuse LEED intensities were initially measured as angular maps at a few fixed energies. However, the polar and azimuthal angular dependence of DLEED intensities also depends on any 2-D correlation within the disordered adsorbate layer through an autocorrelation function; some correlation must be expected, as a sufficient density of adsorbates is needed for the DLEED measurement and mutual repulsion between adsorbates then causes a degree of local ordering. To remove such correlation effects, one can measure DLEED intensities $I(E)$ at pairs of nearby energies E (e.g., 4 eV apart) and take the logarithmic derivative with respect to energy: $L = (\partial I / \partial E)/I$. In practice, one normally uses the function $Y = L/\left(1 + V_{0i}^2 L^2\right)$ to avoid singularities in L (here V_{0i} is the imaginary part of the inner potential; a corresponding R-factor was devised for angular Y maps [5.39]; the same function Y is also used in defining the Pendry R-factor for I(V) curves, see Section 6.1.2) [5.47].

More recently, diffuse LEED intensities have been measured as a function of energy in a way similar to I(V) curves for spots [5.41; 5.43]: then the abovementioned autocorrelation function is independent of energy and can be ignored. This requires tracking diffuse intensity at constant parallel momentum transfer, as if tracking spots (such tracking can be done by pre-recording spot movements for an ordered structure, or by interpolation between the positions of substrate-induced integer-order spots which are always present).

The theory of diffuse LEED was developed based on the concept that the electronic inelastic mean free path limits the distance that electrons travel within the surface. In the case of disordered adsorbate layers, individual electrons are not likely to scatter from a second adsorbate after scattering from a first adsorbate; this is especially true near normal incidence, where scattering by $90°$ from the incident direction toward a second adsorbate is unlikely, due to the angular dependence of typical atomic scattering factors. Therefore, in DLEED theory one may assume the limit of large inter-adsorbate distances, that is, low adsorbate density: the problem is thereby reduced to that of a single adsorbate on a crystalline substrate. This assumption is supported by experimental and theoretical LEED results for O adsorbates on Ni(100) [5.41] and Pd substitutions in Cu(100) [5.42], which show that the structure of diffuse or fractional-order I(V) curves depends little on adatom density (apart from an overall intensity increase with density).

With a single adsorbate, the electron scattering can be subdivided into three steps, as shown in Figure 5.19(a) [5.39; 5.41; 5.47]. Step 1 follows the incident beam $\mathbf{k}_{in}$ and its diffraction by the substrate until the electrons reach the adsorbate. In this process, the electrons can follow any possible path, including a direct approach to the adsorbate and all paths diffracted by the (1×1) substrate lattice, that is, all integer-order beams $\mathbf{k}_{in} + \mathbf{g}_{1 \times 1}$. Step 2 follows the electrons as they *first* scatter from the adsorbate until they *last* scatter from the adsorbate: this includes direct scattering by the adsorbate and all possible paths from the adsorbate to the substrate and back to the same adsorbate; this also includes all multiple scattering within the adsorbate alone and within the substrate alone, as well as single and multiple scattering by the substrate back to the adsorbate. Step 3 takes the electrons that leave the adsorbate for the last time and follows them to the detector. For step 3, we must realise that the outgoing electrons will be detected in one specific direction $\mathbf{k}_{out}$ determined by the detector location. If we then backtrack from the detector to the adsorbate, we have a set of possible paths that are directly analogous to those in step 1, but with reversed direction. This implies, in the forward direction, following the direct path from adsorbate to detector and all integer-order beams $\mathbf{k}_{out} + \mathbf{g}_{1 \times 1}$ within the (1×1) substrate.

In this scheme, steps 1 and 3 are familiar multiple scattering processes in LEED from ordered surfaces, efficiently done with plane waves, as described in Section 5.2.6. Step 2 is better handled as a cluster problem with spherical waves, as described in Section 5.2.3.

The preceding theoretical model of DLEED is exact in the limit of a low adsorbate density. Its most time-consuming part is step 2, which is best calculated with spherical waves. Fortunately, most of the scattering in step 2 is relatively weak. The strongest scattering in step 2 is normally the direct scattering of the incident beam by the adsorbate to the detector: this cannot be neglected and may include strong intra-molecular multiple scattering. All other scattering paths in step 2 involving the substrate are of third or higher order in terms of multiple scattering: for instance, an

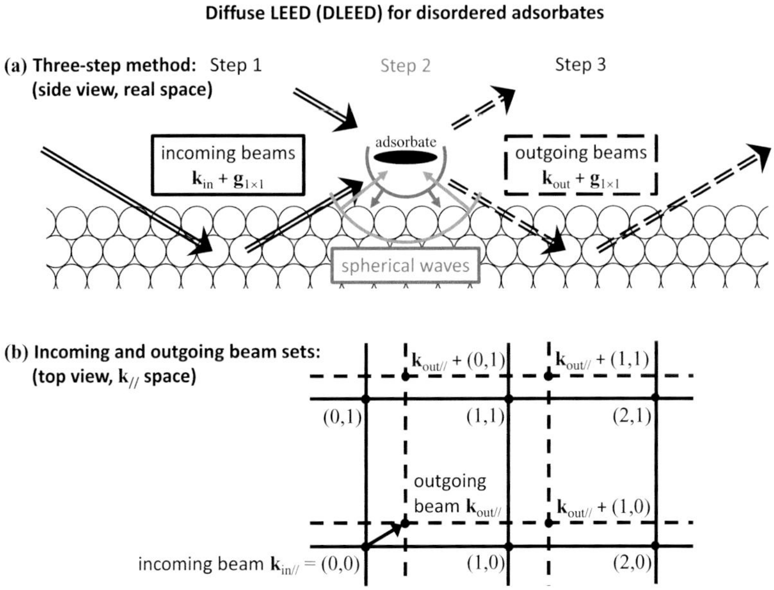

Diffuse LEED (DLEED) for disordered adsorbates

Figure 5.19 Multipe scattering model for diffuse LEED from a disordered adsorbate layer on a crystalline substrate: (a) three-step model; (b) relevant beam sets

electron could follow the third order path from adsorbate to substrate to adsorbate, or more complex paths. This suggests neglecting in step 2 all but the direct scattering by the adsorbate (which is very simple for an atomic adsorbate, and still allows including multiple scattering within the adsorbate if it is multi-atomic like a molecule, for example as a composite layer, cf. Section 5.2.5.4): calculations show this to be a very good approximation [5.40; 5.41]. This approximation is also computationally very beneficial: the single adsorbate can now be treated as a conventional layer with scattering matrices between the two beam sets $\mathbf{k}_{in} + \mathbf{g}_{1\times1}$ and $\mathbf{k}_{out} + \mathbf{g}_{1\times1}$. (As seen in Eqs. (5.28) and (5.35), such scattering matrices contain a pre-factor $1/A$ that specifies the density of scatterers in the layer; for DLEED that factor can be selected to match the actual density of the disordered layer.) These two beam sets are shown in 2-D reciprocal space in Figure 5.19(b): the outgoing set $\mathbf{k}_{out} + \mathbf{g}_{1\times1}$ is simply a shifted copy of the incoming set $\mathbf{k}_{in} + \mathbf{g}_{1\times1}$.

The above argument leads to a simple model: perform a conventional LEED calculation based on plane waves, using just the two plane-wave sets $\mathbf{k}_{in} + \mathbf{g}_{1\times1}$ and $\mathbf{k}_{out} + \mathbf{g}_{1\times1}$, with a conventional layer (of appropriate density) for the adsorbate. (This is an example of the application of the beam set neglect method of calculating LEED intensities [5.40; 5.48].)

A further simplification is possible: if $\mathbf{k}_{out}$ is a fractional-order beam of the substrate, then standard LEED codes can be used virtually unchanged for DLEED by assuming a fictitious superlattice [5.40; 5.47]. For example, if $\mathbf{k}_{out}$ corresponds to a (0.5, 0.5) or (0.25, 0.25) beam relative to the substrate's (1×1) pattern, then a standard LEED code can be applied by using a fictitious c(2×2) or p(2×2) superlattice. This approach was used, for example, to successfully study the disordered substitutional adsorption of Pd in the outermost layer of Cu(100) [5.42].

DLEED can also be viewed as an electron hologram, and this technique is described in Section 6.1.1.2 (see also [5.49–5.54], and for positrons instead of electrons see [5.55]).

DLEED was first used to determine the structure of disordered O on W(100) [5.56]: O was found to adsorb in hollow sites; a later refinement (using tensor LEED combined with DLEED) found lateral W relaxations towards the O position [5.57]. DLEED was subsequently applied to study the initial stages of adsorption and the changes in adsorption site or bond lengths with increasing coverage for O and S on Ni(100) [5.58], Cl on Ti(0001) [5.59; 5.60], K on Co(10-10) [5.61] and K on Ni(100) [5.62]. Partial occupation of hcp and fcc sites of NO was found on Ni(111) [5.63], and for I on Rh(111) (also using tensor LEED) [5.64]. The early stages of surface alloy formation have been studied for Pd/Cu(100) [5.42]. A detailed analysis of the short-range order with mixed adsorption sites and comparison with full dynamic LEED calculations was performed for O on Ni(111) [5.65]. The DLEED calculation has been extended to spin dependent scattering and applied to CO on Pt(111) [5.66; 5.67], where the incoherent magnetic asymmetry parameters proved to be very sensitive to the adsorbate-substrate distance [5.68]. Molecular adsorption has been studied with DLEED in cases where no ordered structures could be obtained: Pt(111) + C_6H_6 [5.69], Pt(111) + C_2H_4 [5.70], Pt (111) + H_2O [5.71; 5.72], and methanol on Pd(111) [5.43].

5.5   Thermal Effects

Vibration amplitudes at surfaces are usually assumed to be larger than in the bulk [5.73]. There is also evidence that the vibration of surface atoms exhibits enhanced anisotropic and anharmonic effects. A detailed knowledge of the anisotropy and anharmonicity of thermal vibrations at surfaces would allow a better understanding of numerous surface properties like phase transitions, surface reconstructions, adsorption and desorption phenomena and growth processes. LEED studies of thermal vibrations on surfaces are rare, even though the surface sensitivity should make LEED an appropriate tool to study these effects. The reasons are experimental errors due to the influence of defects and the difficult analysis due to the multiple scattering. Both difficulties could be overcome in principle. The experimental control afforded by STM, AFM or SPA-LEED allows in many cases the preparation of very well ordered surfaces with low defect density; also the LEED programs could be developed further, such that a non-expert could analyse the data, as is the case, for example, with X-ray diffraction.

In the X-ray diffraction literature, the expression 'thermal vibration' is normally not used in connection with the analysis of diffraction data because the effects of static displacements (defects) and dynamic displacements (vibrations) cannot be distinguished experimentally. It has therefore been proposed to use the term 'atomic displacement parameter' (ADP) [5.74] and that has become the common terminology. The term 'Debye–Waller factor' is also commonly used. The determination of the thermal contribution to the displacement parameter usually requires temperature-dependent measurements. Neutron diffraction measurements, where energy resolution allows the measurement of phonons, are in general more reliable than X-ray data. Electron diffraction has the same drawback as X-ray diffraction: static and dynamic displacements cannot be distinguished.

In most LEED surface structure analyses a Debye temperature is used which assumes isotropic displacements. The fitting of Debye temperatures to experimental I(V) curves cannot be considered adequate for determining vibration amplitudes, unless careful temperature dependent measurements are made and the static contribution can be separated. The expression 'atomic displacement parameter' or ADP is therefore used in the following and includes both static and dynamic contributions. The more common expression 'thermal vibration' is nevertheless also used in connection with temperature-dependent measurements.

The main experimental method to investigate thermal vibration at surfaces is high-resolution electron energy loss spectroscopy (HREELS) [5.75], which allows measuring the frequencies of surface vibrations. In structure analyses with diffraction techniques it is the amplitudes of the vibrational displacements that are measured. Vibrational frequencies and amplitudes are related and it would be useful to combine and compare results from LEED and HREELS, but this is rarely done. The vibration amplitudes play an important role in structure determination. With both LEED and X-ray diffraction, a high accuracy of the structural parameters can only be reached by including atomic displacement parameters in the structure refinement procedure. For example, LEED I(V) analyses of coadsorbed CO and oxygen on Ru(0001) led to a substantial improvement of the fit of the experimental data when anisotropic vibrations were included [5.76]. A further reason to study vibration amplitudes with diffraction methods is the access it provides to understanding molecular motion and diffusion paths at surfaces. With X-ray diffraction, an anharmonic motion of Cs adsorbed on Cu(100) has been determined [5.77]. Although this has not yet been done with LEED, the method is certainly capable of studying anharmonic motion: it is a promising future application for LEED.

LEED is sensitive enough to determine anisotropic displacement parameters even at clean metal surfaces; an example for Cu(110) is shown in Section 5.5.8. This should also be possible in adsorbate layers. The point of interest is the thermally induced change of bond lengths in the adsorbed state and the interaction with the substrate. Here the temperature dependence and the anisotropy of the thermal parameters play an important role. In many cases details should become observable by LEED studies which would be difficult to detect with other methods. Nevertheless, a combination of two or more methods is in any case advantageous to reach a conclusive interpretation of the results.

5.5.1   Thermal Vibration in the Kinematic Theory of Diffraction

We here briefly reiterate the treatment of thermal vibrations in the kinematic theory in order to elucidate the approximations made in the case of multiple scattering.

A detailed description of the kinematic theory can be found in B. T. M. Willis and A. W. Pryor [5.78], while a review of theoretical and experimental methods to analyse atomic displacements has been given by F. W. Kuhs [5.74]. The commonly used atomic displacement parameters have been described by K. N. Trueblood et al. [5.79]. In the kinematic theory the diffracted intensity is given by

$$I(\mathbf{q}) = \sum_{i,j} f_i(\mathbf{q}) \cdot f_j^*(\mathbf{q}) \cdot e^{i\mathbf{q}(\mathbf{r}_i - \mathbf{r}_j)} \langle e^{i\mathbf{q}(\mathbf{u}_i - \mathbf{u}_j)} \rangle, \qquad (5.108)$$

where $\mathbf{r}_i$, $\mathbf{r}_j$ are the equilibrium positions of atoms i and j, the sum is taken over all atoms of the crystal, $\mathbf{u}_i$ and $\mathbf{u}_j$ are the corresponding displacements at a given time, $\mathbf{q} = \mathbf{k}' - \mathbf{k}$ is the scattering vector and $f_i(\mathbf{q})$ are the atomic scattering factors. The pointed brackets denote the average over time. In this description the displacements include static displacements due to defects like vacancies, interstitial atoms, etc., as well as thermal vibrations. Assuming a Gaussian distribution of displacements $\mathbf{u}_i$ and $\mathbf{u}_j$, the average is given by:

$$\langle \exp[i\,\mathbf{q} \cdot (\mathbf{u}_i - \mathbf{u}_j)] \rangle = \exp\left\{ -\frac{1}{2} \langle [\mathbf{q} \cdot (\mathbf{u}_i - \mathbf{u}_j)]^2 \rangle \right\}$$

$$= \exp\left\{ -\frac{1}{2} \left[\langle (\mathbf{q} \cdot \mathbf{u}_i)^2 \rangle + \langle (\mathbf{q} \cdot \mathbf{u}_j)^2 \rangle \right] + \langle (\mathbf{q} \cdot \mathbf{u}_i)(\mathbf{q} \cdot \mathbf{u}_j) \rangle \right\}.$$

$$(5.109)$$

It should be pointed out here that the assumption of a Gaussian distribution is an approximation, which is not justified in all cases, certainly not for static displacements around defects where the displacement depends on the distance from the defect [5.80]. A better description of the distribution may be even more important for LEED because, due to multiple scattering, resonances may occur along the scattering paths between nearest neighbours. For the same reason, a more detailed description of correlated motion should be included. Nevertheless, we here use this approximation of a Gaussian distribution to derive the expression for the kinematic Debye–Waller factor. In the following, we may assume for simplicity that all atoms are equal and then obtain for the average:

$$\langle (\mathbf{q} \cdot \mathbf{u}_i)^2 \rangle = \langle (\mathbf{q} \cdot \mathbf{u}_j)^2 \rangle = \langle (\mathbf{q} \cdot \mathbf{u})^2 \rangle. \qquad (5.110)$$

The intensity can be written in a form that yields a first-order term and higher-order terms:

$$I(\mathbf{q}) \propto |f(\mathbf{q})|^2 \sum_{ij} e^{i\mathbf{q}(\mathbf{r}_i - \mathbf{r}_j)} e^{-\langle (\mathbf{q} \cdot \mathbf{u})^2 \rangle} \cdot \{ 1 + \langle (\mathbf{q} \cdot \mathbf{u}_i)(\mathbf{q} \cdot \mathbf{u}_j) \rangle + \cdots \}. \qquad (5.111)$$

The first term in Eq. (5.111) is a sharp but weakened reflection I_{Bragg}, while the second and higher order terms constitute the thermal diffuse scattering (TDS) I_{TDS}:

$$I = I_{\text{Bragg}} + I_{\text{TDS}}.$$

The atomic displacements result in a reduction of the elastically scattered Bragg intensity by a factor $\exp(-2M)$:

$$
\begin{aligned}
I_{\text{Bragg}} &= |F(\mathbf{q})|^2 |T(\mathbf{q})|^2 \delta(\mathbf{q} - \mathbf{g}) \\
&= |F(\mathbf{q})|^2 \exp(-2M) \delta(\mathbf{q} - \mathbf{g}).
\end{aligned}
\tag{5.112}
$$

Here $\mathbf{g}$ is a vector of the 2-D reciprocal net and

$$
M = \frac{1}{2} \left\langle (\mathbf{q} \cdot \mathbf{u})^2 \right\rangle.
\tag{5.113}
$$

The intensity that is missing from the Bragg peak is the thermal diffuse scattering. Correlated displacements occurring in the term $\left\langle (\mathbf{q} \cdot \mathbf{u}_i)(\mathbf{q} \cdot \mathbf{u}_j) \right\rangle$ in Eq. (5.111) cause the one-phonon peak at the Bragg energies, while uncorrelated displacements lead to a constant background distributed over all directions. Although the TDS cannot be neglected in LEED experiments, its treatment in the multiple scattering theory has not been solved. Experimentally it has been found that the one-phonon part is distributed as $1/q$ away from the Bragg peaks, while the multiphonon part is more equally distributed in reciprocal space, corresponding to the result of the kinematic theory. Experimental studies of the TDS in LEED are rare. The first measurements were made in the early days of LEED by M. B. Webb and coworkers [5.81]. In later investigations by M. Henzler and coworkers using a SPA-LEED system with high energy and angle resolution, the single- and multi-phonon parts could be separated [5.82]. It was found that kinematic description of the one-phonon peak which causes a Lorentzian foot of the Bragg peak is also valid for LEED. No indication of multiple scattering effects in the beam profile were found. In LEED systems with lower resolution the one-phonon peak is usually included in beam profile and I(V) measurements. As the intensity of the TDS follows the Bragg intensity, only small influences on the analysis of the beam profiles or I(V) curves can be expected. Also, in X-ray diffraction the TDS is usually neglected; the multiple scattering theory described in Section 5.5.3 neglects the TDS as well.

In the following, only the Bragg intensity is described. We first mention that the conventional derivation of the Debye–Waller factor is obtained when the average of the left-hand side of Eq. (5.109) is calculated from the distribution of atomic displacements, namely the probability density function (PDF) $p(\mathbf{u})$. The Debye–Waller factor $T(\mathbf{q})$ is given as the Fourier transform of the PDF:

$$
T(\mathbf{q}) = \int p(\mathbf{u}) \exp(i\mathbf{q} \cdot \mathbf{u}) d\mathbf{q},
\tag{5.114}
$$

which is equivalent to

$$
T(\mathbf{q}) = \exp\left(-\frac{1}{2} q^2 \langle u^2 \rangle \right),
\tag{5.115}
$$

if the PDF is a Gaussian distribution; the PDF in general can be different from a Gaussian distribution. When the Debye–Waller factor is experimentally determined, the corresponding PDF is obtained by the reverse Fourier transform of the Debye–Waller factor:

$$p(\mathbf{u}) = (2\pi)^{-3} \int T(\mathbf{q}) \, \exp\left(-i\mathbf{q}\cdot\mathbf{u}\right) d\mathbf{q}. \tag{5.116}$$

5.5.2   Thermal Parameters

The atomic displacement parameter (ADP) enters the intensity calculations in one of several possible forms. In X-ray structure analysis for *isotropic* displacements, usually the so-called B-factor is used, which is defined as $B = 8\pi^2 \langle u^2 \rangle$. B is frequently called the temperature factor, although it is well known that the displacements include defects and lattice distortions around defects; u is the mean square displacement from the equilibrium position, given in nm or Å when q is given in nm^{-1} or Å^{-1}, respectively. In the case of isotropic displacements, u is the radius of the spherically symmetric PDF. It is usually defined as the radius where the integral of the probability density reaches ½. The Debye–Waller factor is a function of the mean square of the scalar product $\mathbf{q} \cdot \mathbf{u}$:

$$\left\langle (\mathbf{q} \cdot \mathbf{u})^2 \right\rangle = \left\langle (q_x u_x)^2 \right\rangle + \left\langle (q_y u_y)^2 \right\rangle + \left\langle (q_z u_z)^2 \right\rangle = \frac{3}{2} |q|^2 |u|^2, \tag{5.117}$$

where u is the radius of the probability density sphere; the B-factor uses the mean square value $\langle u^2 \rangle = 3/2 \, |u|^2$, independent of lattice and orientation of the diffraction vector.

For *anisotropic* atomic displacement parameters (ADPs), the average of the mean square of $\mathbf{q} \cdot \mathbf{u}$ depends on the diffraction vector. Six independent parameters are used to describe the ellipsoid of the probability density assuming a Gaussian distribution of the displacements. Frequently used are β_{kl} with $k, l = 1, 2, 3$. The quantities β_{kl} represent a symmetric quadratic form with six parameters $\beta_{11}, \beta_{22}, \beta_{33}$ and $2\beta_{12}, 2\beta_{13}, 2\beta_{23}$, which can be used to describe an ellipsoid. The lengths of the semi-axes of the ellipsoid cannot be directly determined from the parameters β_{11}, β_{22} and β_{33}; the determination of the size and orientation of the thermal ellipsoid from the β_{kl} is described in Appendix I. In X-ray diffraction the parameters are usually related to the lattice vectors, which in general form a non-orthonormal coordinate system. The parameters β_{kl} are dimensionless and are related to the lengths of the lattice vectors. To compare the values to the isotropic case a transformation to absolute values in nm^2 or Å^2 is required. There are symmetry restrictions on the parameters, for special positions in the unit cell; the orientation of the ellipsoid and their semi-axes have to be consistent with the site symmetry.

Another set of six parameters, usually called U_{kl}, is also frequently used. They are also related to the translation lattice but are not dimensionless. The use of U_{kl} instead of β_{kl} has the advantage that U_{11}, U_{22} and U_{33} give directly the semi-axis sizes of the ellipsoid in nm^2 or Å^2 and can be compared with the values obtained with the assumption of isotropic displacements. For the definition of U_{kl}, see Appendix I.

By contrast, in LEED calculations it is common to use not the displacement parameter u but the Debye temperature Θ_D. In the high temperature limit the relation to the displacement parameter is given by:

$$\langle u^2 \rangle = \frac{3\hbar^2 T}{2 m_a k_B \Theta_D^2}, \tag{5.118}$$

with m_a the mass of the atom and k_B the Boltzmann constant. Through this relation, the Debye temperature gives the mean square displacement as a function of temperature: it has the advantage that only one parameter is needed in temperature dependent measurements. The Debye temperature has been derived theoretically in the Debye model mainly for mono-atomic structures: in practice, the Debye model is not used for compounds. In LEED studies of compounds (including adsorption of foreign elements, for example), one may optimise Debye temperatures for different constituent elements or atoms in inequivalent positions, for example, in different layers, at a fixed temperature: this then corresponds to the determination of separate isotropic mean square displacement parameters for the different elements or inequivalent atoms and is not related to the Debye model.

In analogy to X-ray diffraction, there have been early attempts in LEED studies to derive an average temperature factor from the decrease of the intensity with increasing energy. However, this approach has not been successful in LEED and leads to different factors depending on which diffracted beam was used. It is not possible in general to derive an overall temperature factor from LEED I(V) curves, due to multiple scattering effects. In X-ray diffraction an overall temperature factor is used for the normalisation of structure factors in direct methods, but not in structure refinement.

5.5.3   Harmonic Vibrations

The distribution of atomic displacements is described by the probability density function (PDF), $p(u)$. For harmonic vibrations in the high temperature limit it can be assumed to be a Gaussian distribution. First, we consider isotropic displacements. The distribution is then defined by a single parameter, the mean square displacement $\langle u^2 \rangle$:

$$p(u) = \frac{1}{(2\pi)^{3/2} \langle u^2 \rangle^{1/2}} \exp\left(-\frac{u^2}{2\langle u^2 \rangle}\right). \tag{5.119}$$

The atomic Debye–Waller factor is given by the Fourier transform of the PDF,

$$T(q) = \exp\left(-\frac{1}{2} q^2 \cdot \langle u^2 \rangle\right), \tag{5.120}$$

and is independent of the crystal system and of the direction of the scattering vector $\mathbf{q}$, but strongly dependent on the scattering angle.

In the case of anisotropic vibrations, assuming mean square displacements $\langle u_x^2 \rangle, \langle u_y^2 \rangle, \langle u_z^2 \rangle$ in three orthogonal directions, the atomic Debye–Waller factor is given by

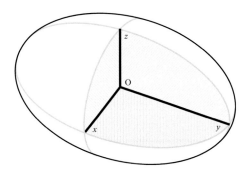

Figure 5.20 Ellipsoid of the probability density. The semi-axis lengths Ox, Oy and Oz are usually chosen such that the atom has 50% probability of being inside the ellipsoid; the semi-axis lengths are proportional to the respective mean square displacements.

$$T(\mathbf{q}) = \exp\left(-\frac{1}{2}\cdot\mathbf{q}^{\mathrm{T}}\langle\mathbf{u}^{\mathrm{T}}\mathbf{u}\rangle\mathbf{q}\right), \tag{5.121}$$

using the conventional crystallographic formulation, where $\mathbf{u}^{\mathrm{T}}$ denotes the transpose of vector $\mathbf{u}$. The probability density function is then given by

$$p(\mathbf{u}) = \left(\frac{\det\mathbf{B}^{-1}}{8\pi^3}\right)^{1/2}\exp\left(-\frac{1}{2}\,\mathbf{u}^{\mathrm{T}}\,\mathbf{B}^{-1}\,\mathbf{u}\right), \tag{5.122}$$

with a symmetric matrix

$$\mathbf{B} = \begin{pmatrix} \langle u_x^2\rangle & \langle u_x u_y\rangle & \langle u_x u_z\rangle \\ \langle u_x u_y\rangle & \langle u_y^2\rangle & \langle u_y u_z\rangle \\ \langle u_x u_z\rangle & \langle u_y u_z\rangle & \langle u_z^2\rangle \end{pmatrix}. \tag{5.123}$$

The exponent $\mathbf{u}^{\mathrm{T}}\,\mathbf{B}^{-1}\,\mathbf{u}$ is a quadratic function of the atomic displacements, so that $\mathbf{u}^{\mathrm{T}}\,\mathbf{B}^{-1}\,\mathbf{u} = const.$ defines an ellipsoid, cf. Figure 5.20. Conventionally the size of the ellipsoid is chosen such that the integral over its volume is 0.5. The six independent matrix elements describe the lengths of the semi-axes and the orientation of the ellipsoid. The mean square displacements u_x, u_y and u_z refer to an orthonormal coordinate system. The description in general oblique coordinate systems and the relation to the parameters β_{kl} mentioned in Section 5.5.2 are given in Appendix I.

5.5.4   Anharmonic Vibrations

In surface structure determinations by LEED, anharmonic thermal vibrations have received little attention. On the other hand, physical properties like thermal expansion, temperature dependence of elastic constants or adsorption and desorption kinetics can be explained only by anharmonic effects [5.83]. The influence of anharmonic corrections on the inter-atomic force constants on the adsorption kinetics has been theoretically investigated [5.83]; experimental results, however, are rare [5.84]. It is therefore desirable to include anharmonic effects in the multiple scattering theory in addition to anisotropy, in the hope of gaining more insight into the anharmonicity of the inter-atomic forces from diffraction experiments. To that end, this section gives a brief overview of anharmonicity in materials, followed by a description of how anharmonic effects can be included in the multiple scattering LEED theory in Section 5.5.5.

We need to first discuss how anharmonicity can be handled within a single scattering theory of diffraction. There exist various possibilities to include anharmonic terms in the generalised Debye–Waller factor, both in the static contribution and in the dynamic contribution. In the single-scattering context of X-ray diffraction, methods have been reviewed by F. W. Kuhs [5.74] and a detailed description is given by B. T. M. Willis and A. W. Pryor [5.78]. One approach is the expansion of the Debye–Waller factor in moments:

$$\langle \exp(i\,\mathbf{q}\cdot\mathbf{u})\rangle = \sum_{N=0}^{\infty}(1/N!)\,\langle(\mathbf{q}\cdot\mathbf{u})^{N}\rangle. \tag{5.124}$$

Another option is through cumulants where the series expansion appears in the exponent:

$$\langle \exp(i\,\mathbf{q}\cdot\mathbf{u})\rangle = \exp\left(\sum_{N=0}^{\infty}(1/N!)\,\langle(\mathbf{q}\cdot\mathbf{u})^{N}\rangle_{cum}\right). \tag{5.125}$$

The cumulant expansion has the advantage that in the harmonic case all cumulants of order higher than 2 vanish, which enables identifying and separating anharmonic effects. Experiments on surfaces are rare; one example is the motion of Cs on Cu(001) determined by X-ray diffraction where the cumulant expansion was used [5.85].

The atomic displacements consist of two contributions: the static part arising from defects and the dynamic (thermal) part, both governed by the inter-atomic forces. The static part depends strongly on the defect concentration and the nature of the defects; there is no reason for the static displacements at defect sites to exhibit a Gaussian distribution, as has been pointed out by M. A. Krivoglaz [5.80]. The dynamic part is also affected by the static displacements near a defect site but the vibration amplitude is governed by the inter-atomic force constants which are usually calculated assuming a harmonic crystal potential. Calculations using a full crystal potential including all atomic interactions would be extremely complicated. As an approximation, anharmonic terms in the crystal potential mostly assume a one-particle potential $V(\mathbf{u})$, in which one particle is assumed to move in the fixed potential caused by the rigid lattice of surrounding atoms.

For a theoretical calculation of anharmonic lattice vibrations, the partition function of an atom in a one-particle potential $V(\mathbf{u})$ is required. The dynamic part of the Debye–Waller factor then becomes, at temperature T:

$$\langle \exp(i\,\mathbf{q}\cdot\mathbf{u}) \rangle = \frac{\int \exp(i\,\mathbf{q}\cdot\mathbf{u})\,\exp(-V(\mathbf{u})/k_BT)\,d\mathbf{u}}{\int \exp(-V(\mathbf{u})/k_BT)\,d\mathbf{u}}. \tag{5.126}$$

The one particle potential is expanded in a Taylor series by atomic displacements:

$$\begin{aligned}
V(\mathbf{u}) &= V_0 + V_1 + V_2 + V_3 + V_4 \ldots \\
&= const + \sum_{a,l}\left(\frac{dV}{d\mathbf{u}_{a,l}}\right)\mathbf{u}_{a,l} + \frac{1}{2!}\sum_{a,l}\sum_{a',l'}\left(\frac{dV}{d\mathbf{u}_{a,l}}\right)\left(\frac{dV}{d\mathbf{u}_{a',l'}}\right)\mathbf{u}_{a,l}\mathbf{u}_{a',l'} \\
&\quad + \frac{1}{3!}\sum_{a,l'}\sum_{a',l'}\sum_{a'',l''}\cdots
\end{aligned} \tag{5.127}$$

where α, α', α'', etc. label the atoms and l, l', l'', etc. label the unit cells. The derivatives are taken at the equilibrium position so that V_1 vanishes. $V(\mathbf{u}) = V_0 + V_2$ represents the harmonic approximation. The third and fourth order terms in the anharmonic expansion, Eq. (5.127), are usually considered to be sufficient for cases where a perturbation treatment is adequate. Higher order terms, of course, become important for larger displacements. The anharmonic terms are mainly dominated by the short-range repulsion between nearest neighbours.

The explicit calculation of the dynamic part of the Debye–Waller factor using the expansion in Eq. (5.127) is complicated and has been done only for simple crystal structures. Most evaluations of experimental results use an expansion of the displacements. This includes then also the static part of the Debye–Waller factor, but is of course not suited to draw a precise conclusion on the form of the inter-atomic pair potentials. The most convenient approach is the expansion of $\mathbf{q}\cdot\mathbf{u}$ by moments or cumulants, as described in Eqs. (5.124) and (5.125). The cumulant expansion is preferred as the harmonic and anharmonic terms can be separated [5.74]. Because effects are small, usually only the third and fourth order terms are included: the result then shows the difference from the harmonic approximation.

Anharmonicity also occurs in such special cases as libration and free rotation of molecules. Libration occurs when a cluster of atoms vibrates relatively rigidly against its surroundings, for example, when the internal displacements within a molecule or a part of a molecule are much smaller than the displacements of that atom cluster as a whole. The atomic displacements are then calculated as a function of the movement of a rigid body. For example, C_{60} molecules in some circumstances can rotate around an axis normal to a substrate surface. Such a movement cannot be treated as independent thermal motions of individual atoms. In X-ray diffraction, such cases can be handled with the kinematic theory, see for example [5.86].

In the case of LEED, the full multiple scattering theory for libration modes has not been worked out, while an approximation has been used to study the bending mode of CO on Ru(0001) [5.87]. In this study the bending mode of the CO molecule was simulated by coherent superposition of scattered waves from different inclined positions of the molecule, as illustrated in Figure 5.21.

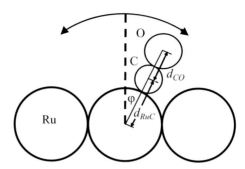

Figure 5.21 Example of a libration mode: the bending mode vibration of a CO molecule which is assumed to be internally rigid. Redrawn from [5.87] M. Gierer, H. Bludau, H. Over, and G. Ertl, *Surf. Sci.*, vol. 346, pp. 64–72, 1996, with permission from Elsevier.

5.5.5   Multiple Scattering Theory of Thermal Effects

In the multiple scattering theory, the diffracted intensity is formally described in the same way as in the kinematic theory by summing the scattering amplitudes of all atoms in the unit cell (see introductory remarks in Section 2.2.1 and Eq. (2.49)):

$$I(\mathbf{k}, \mathbf{k}') = \left| \sum_{v} F_v(\mathbf{k}, \mathbf{k}') e^{i(\mathbf{k}'-\mathbf{k})\mathbf{d}_v} \right|^2. \tag{5.128}$$

The only difference is that atomic scattering factors $F_v(\mathbf{k}, \mathbf{k}')$ are now generalised scattering factors which represent all multiple scattering paths ending in atom v (this form can also be seen in Eq. (5.17), where $\mathbf{T}^i$ represents all multiple scattering paths ending in atom i); these scattering factors depend explicitly on the wave vectors $\mathbf{k}$ and $\mathbf{k}'$, that is, on both the incoming and the diffracted beams. The generalised scattering factor depends also on the crystal structure surrounding that atom, in contrast to the kinematic theory where the atomic scattering factors are assumed to depend only on the scattering vector $\mathbf{q} = \mathbf{k}' - \mathbf{k}$ and are independent of the crystal structure. The generalised scattering factors are usually calculated in a spherical wave expansion as matrices $\mathbf{T}$ and are given by Eq. (5.29), which is written here in slightly different form and without pre-factors:

$$F_v(\mathbf{k}, \mathbf{k}') = \sum_{lm, l'm'} (-1)^m Y_{l-m}(\vartheta_k, \varphi_k) T_{v,lm,l'm'}(\mathbf{k}) Y_{l'm'}(\vartheta_{k'}, \varphi_{k'}). \tag{5.129}$$

The matrices $\mathbf{T}$ result from the solution of self-consistent equations which take into account the multiple scattering processes (see Eq. (5.29) and its derivation).

Introducing the atomic displacement $\mathbf{u}_v$ of atom v, the average diffracted intensity is now given by:

$$I(\mathbf{k}, \mathbf{k}') = \left\langle \left| \sum_{v} F_v(\mathbf{k}, \mathbf{k}', \mathbf{u}_v) e^{i(\mathbf{k}'-\mathbf{k})(\mathbf{d}_v+\mathbf{u}_v)} \right|^2 \right\rangle_T. \tag{5.130}$$

Because of the multiple scattering, the generalised scattering factors themselves now depend on the displacements and therefore the average cannot be restricted to the phase factors. We neglect correlations and the thermal diffuse scattering and average the scattering amplitudes. This corresponds to the approximation made in the kinematic theory when only the first term of Eq. (5.111) is included. We obtain for the intensity of the Bragg scattering:

$$
I(\mathbf{k}, \mathbf{k}') = \left| \sum_\nu \langle F_\nu(\mathbf{k}, \mathbf{k}', \mathbf{u}_\nu) e^{i(\mathbf{k}'-\mathbf{k})(\mathbf{d}_\nu+\mathbf{u}_\nu)} \rangle_T \right|^2
$$

$$
= \left| \sum_\nu \langle F_\nu(\mathbf{k}, \mathbf{k}') \rangle_T e^{i(\mathbf{k}'-\mathbf{k})\mathbf{d}_\nu} \right|^2, \tag{5.131}
$$

with thermally averaged scattering factors:

$$
\langle F_\nu(\mathbf{k}, \mathbf{k}') \rangle_T = \sum_{lm, l'm'} (-1)^m Y_{l-m}(\Omega_k) \left\langle \sum_\nu T_\nu(\mathbf{k}, \mathbf{u}_\nu) e^{i(\mathbf{k}'-\mathbf{k})\mathbf{u}_\nu} \right\rangle_T Y_{l'm'}(\Omega_k). \tag{5.132}
$$

The average on the right-hand side of Eq. (5.132) can be defined as an averaged matrix **T**. It was shown in [5.1; 5.88] that, in the harmonic approximation, the thermally averaged matrices **T** are given by solving the multiple scattering equations using the thermally averaged single scattering matrices **t**. Correlations between atomic displacements as well as correlations in the multiple scattering series are neglected here:

$$
T_\nu(\mathbf{k}, \langle \mathbf{u}_\nu \rangle_T) = t_\nu(E, \langle \mathbf{u}_\nu \rangle_T) + t_\nu(E, \langle \mathbf{u}_\nu \rangle_T) \sum_{\mu \neq \nu} G_{\nu\mu}(\mathbf{k}, \mathbf{d}_\mu - \mathbf{d}_\nu) T_\mu(\mathbf{k}, \langle \mathbf{u}_\mu \rangle_T),
$$

$$
\tag{5.133}
$$

with

$$
t_\nu(E, \langle \mathbf{u}_\nu \rangle_T) = t_\nu(E) \langle e^{i(\mathbf{k}'-\mathbf{k})\cdot\mathbf{u}_\nu} \rangle_T. \tag{5.134}
$$

Equation (5.133) means that the multiple scattering equations are solved with the averaged atoms which are held immobile in a rigid but distorted lattice, thus excluding correlations in the thermal motion; in other words, the electron is scattered by spherically symmetric atomic potentials which are displaced by variable vectors $\mathbf{u}_\nu$, causing a fluctuating phase factor that is averaged in Eq. (5.134). Deformation of the electron density or the potential is not considered. In the harmonic approximation and assuming a Gaussian PDF, the average in Eq. (5.134) can be replaced by

$$
\langle e^{i(\mathbf{k}'-\mathbf{k})\cdot\mathbf{u}_\nu} \rangle_T = e^{-\frac{1}{2}\langle [(\mathbf{k}'-\mathbf{k})\cdot\mathbf{u}_\nu]^2 \rangle_T}. \tag{5.135}
$$

This expression can be expanded in spherical harmonics, yielding diagonal matrices **t**, the diagonal elements of which lead to complex temperature dependent phase shifts, resulting in Eq. (5.154) further below. This formulation makes the calculation of thermal effects especially easy.

If we want to introduce anisotropic and anharmonic vibrations, we can start from the PDF $p(\mathbf{u})$:

$$\left\langle e^{i(\mathbf{k}'-\mathbf{k})\cdot\mathbf{u}} \right\rangle_T = \int p(\mathbf{u})\cdot e^{i(\mathbf{k}'-\mathbf{k})\cdot\mathbf{u}}\cdot d\mathbf{u}. \tag{5.136}$$

We can use a multipole expansion of the PDF:

$$p(\mathbf{u}) = \sum_{n=1}^{\infty} \sum_{l_c=0,m_c}^{l_{c,\max}} R_n(u)c_{n,l_c m_c} Y_{l_c m_c}(\vartheta_u, \varphi_u), \tag{5.137}$$

where $c_{n,l_c m_c}$ are appropriate multipole expansion coefficients and $R_n(u)$ radial functions of $u = |\mathbf{u}|$. In principle the sum over n is unlimited, but in practice only one or two radial functions are needed. Gaussian functions are used here to match the Gaussian shape of the thermal ellipsoid. It is necessary in practice to derive the expansion coefficients and radial functions from the mean square displacements in different directions assuming Gaussian distributions; otherwise an intractable number of fit parameters would remain in structure refinement [5.89–5.91]. The calculation procedure is described in Section 5.5.6, see also the detailed description in [5.92].

The multipole expansion of the PDF is not restricted to harmonic vibrations and allows the inclusion of anharmonic terms as well. This case requires special care to select the appropriate parameters; an example will be discussed in Section 5.5.7. The thermally averaged atomic scattering factor for single scattering is now given by:

$$t(T,\mathbf{k},\mathbf{k}') = t(0,\mathbf{k},\mathbf{k}') \int p(\mathbf{u})e^{i(\mathbf{k}-\mathbf{k}')\mathbf{u}}d\mathbf{u}, \tag{5.138}$$

where the atomic scattering factor at temperature $T = 0$ is

$$t(0,\mathbf{k},\mathbf{k}') = \frac{-2\pi i}{k} \sum_{l} (2l+1)e^{i\delta_l} \sin(\delta_l)P_l(\cos\vartheta_{kk'}), \tag{5.139}$$

and δ_l are the phase shifts. The scattering factors can be written as

$$t(0,\mathbf{k},\mathbf{k}') = \frac{-8\pi^2}{k} \sum_{lm,\,l'm'} e^{i\delta_l} \sin(\delta_l)\delta_{ll'}\delta_{mm'} Y_{lm}(\vartheta_{k'},\varphi_{k'})Y_{l-m}(\vartheta_k,\varphi_k)(-1)^m, \tag{5.140}$$

and a spherical wave expansion can be used for the phase factors:

$$e^{i\mathbf{k}\cdot\mathbf{r}} = \sum_{lm} 4\pi i^l j_l(kr)Y_{lm}(\Omega_r)(-1)^m Y_{l-m}(\Omega_k). \tag{5.141}$$

Following the calculation for the isotropic case as given by J. B. Pendry [5.1] and using the same notation, we define expansion coefficients for the Fourier transform of the probability density function:

$$\int p(\mathbf{u})e^{i(\mathbf{k}'-\mathbf{k})\mathbf{u}}d\mathbf{u} = \sum_{l_1,m_1 l_2,m_2} W_{l_1 m_1,l_2 m_2} Y_{l_1 m_1}(\vartheta_k,\varphi_k)Y_{l_2-m_2}(\vartheta_{k'},\varphi_{k'})(-1)^{m_2}. \tag{5.142}$$

After multiplying both sides with the spherical harmonics and integrating over the angles we obtain:

$$W_{LM,L'M'} = 16\pi^2 \sum_{n=1}^{\infty} \sum_{l_c m_c} B_{l_1 m_1, l_2 m_2, l_c m_c} C_{n, l_c m_c}$$

$$\cdot \int R_n(u)(-1)^{l_1+l_2} j_{l_1}(ku) j_{l_2}(k'u) u^2 du. \tag{5.143}$$

We can now define temperature dependent expansion coefficients of the scattering factors:

$$t(T, \mathbf{k}, \mathbf{k}') = 8\pi^2 \sum_{lm} \sum_{l'm'} t_{lm,l'm'}(T) Y_{lm}(\Omega_k)(-1)^m Y_{l'-m'}(\Omega_{k'}), \tag{5.144}$$

with the coefficients of the angular momentum expansion

$$t_{lm,l'm'}(T) = \frac{1}{8\pi^2} \iint t(T, \mathbf{k}, \mathbf{k}')(-1)^m Y_{l-m}(\vartheta_k, \varphi_k) Y_{l'm'}(\vartheta_{k'}, \varphi_{k'}) d\Omega_k d\Omega_{k'}, \tag{5.145}$$

where $d\Omega_k = \sin^2 \vartheta_k d\vartheta_k d\phi_k$ and similarly for $d\Omega_{k'}$. Inserting Eq. (5.142) into Eq. (5.138) and using the expansions in Eqs. (5.141) and (5.144), we obtain

$$t_{lm,l'm'}(T) = \sum_{l_1 m_1} \sum_{l_2 m_2} W_{l_1 m_1, l_2 m_2} D_{l_1 m_1, l_2 m_2, lm, l'm'}, \tag{5.146}$$

where

$$D_{l_1 m_1, l_2 m_2, lm, l'm'} = \frac{2\pi}{k} \sum_{l''m''} e^{i\delta_{l''}} \sin \delta_{l''} B_{l_1 m_1, lm, l''m''} B_{l_2 m_2, l'm', l''m''} \tag{5.147}$$

are temperature independent factors and

$$B_{l_1 m_1, l_2 m_2, lm} = \int Y_{l_1 m_1}(\Omega) Y_{l_2 - m_2}(\Omega) Y_{lm}(\Omega) d\Omega, \tag{5.148}$$

with the conditions

$$m_1 - m_2 + m = 0,$$
$$|l_2 - l| \le l_1 \le |l_2 + l|, \tag{5.149}$$
$$l_1 + l_2 + l = \text{even}$$

for non-vanishing coefficients. Equations (5.149) and (5.150) define a temperature-dependent atomic scattering matrix which now contains non-zero off-diagonal terms, instead of the diagonal matrix for the case of isotropic displacements. These matrices can be inserted in conventional LEED programs requiring only slight changes to take care of the non-diagonal form.

The maximum number of phase shifts used is $l_{max} + 1$. This sets the upper limit for the indices lm, $l'm'$ and $l''m''$, that is, $l'' \le l_{max}$ in Eq. (5.147). The limits set in Eq. (5.149) lead to

$$l_1 \le 2l_{max} \text{ and } l_2 \le 2l_{max}. \tag{5.150}$$

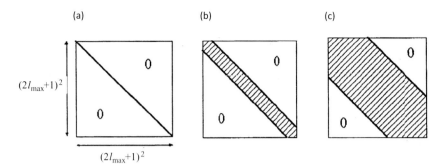

Figure 5.22 Off-diagonal terms in the matrix of the expansion coefficients $W_{l_1 m_1, l_2 m_2}$, which describe the thermal ellipsoid in angular momentum space (Eq. (5.142)); '0' indicates vanishing matrix elements. (a) isotropic case, (b) small anisotropy with $u_x/u_y \leq 1.5$ and (c) larger anisotropy. Adapted from [5.91] W. Moritz and J. Landskron, "Thermal Vibrations at Surfaces Analysed with LEED", in *Solid-State Photoemission and Related Methods*, W. Schattke and M. A. Van Hove, eds., Wiley-VCH, pp. 433–459, 2003, with permission from Wiley.

Thus, in the calculation of the expansion coefficients $W_{l_1 m_1, l_2 m_2}$ twice the number of angular momentum components are required compared to the partial wave expansion of the scattering amplitude, in which $l_{max} + 1$ phase shifts are used. Asymmetry of the thermal ellipsoid produces non-zero off-diagonal terms in **W**. Care must be taken to include a sufficient number of phase shifts if large anisotropies occur. In Figure 5.22 the off-diagonal terms in the matrix **W** are schematically shown for three cases. A diagonal matrix occurs for the isotropic case, a narrow band of off-diagonal terms occurs for small anisotropy on the order of $u_x/u_y \leq 1.5$ and a broad band occurs for large anisotropy. If l_{max} is chosen too small, the width of the band of off-diagonal terms exceeds $2l_{max}$ and the thermal ellipsoid cannot be adequately described.

In the case of isotropic and harmonic vibrations, the result must be consistent with the conventional derivation leading to a diagonal matrix. For spherically symmetric vibrations all coefficients $c_{n,l_c m_c}$ vanish unless $(l_c, m_c) = (0, 0)$ which leads to $l_1 = l_2$ and $m_1 = m_2$ in Eq. (5.146) and thus to a diagonal form of the matrix of expansion coefficients (see Figure 5.22).

With

$$p(u) = (2\pi)^{-3/2} \langle u^2 \rangle^{-1/2} \exp\left(-\frac{u^2}{2\langle u^2 \rangle}\right), \tag{5.151}$$

$W_{l_1 m_1, l_2 m_2}$ becomes

$$W_{l_1 m_1, l_2 m_2} = 16\pi^2 \delta_{l_1 l_2} \delta_{m_1 m_2} \int R_n\left(\langle u^2 \rangle\right) (-1)^{l_1+l_2} j_{l_1}\left(k \cdot \langle u^2 \rangle\right) j_{l_2}\left(k' \cdot \langle u^2 \rangle\right) u^2 du \tag{5.152}$$

and

$$D_{l_1 m_1, l_2 m_2, lm, l'm'} = \frac{2\pi}{k} \sum_{l''m''} e^{i\delta_{l''}} \exp\left(i\delta_{l''}\right) \sin\left(\delta_{l''}\right) \delta_{l_1 m_1, lm} \delta_{l_2 m_2, l'm'}. \tag{5.153}$$

This leads to complex temperature-dependent phase shifts $\delta_l(T)$:

$$t_{lm,l'm'}(T) = -\frac{1}{k}\sum_{l_1 l''} i^{l_1} \exp\left(-2\langle u^2\rangle k^2\right) j_{l_1}\left(-2i\langle u^2\rangle k^2\right) \exp\left(i\delta_{l''}(T)\right) \sin\left(\delta_{l''}(T)\right)$$

$$\cdot \left(\frac{4\pi(2l_1+1)(2l''+1)}{(2l+1)}\right)^{1/2} B(l'', l_1, l),$$

(5.154)

which is the result derived by J. B. Pendry [5.1]; the imaginary part of $\delta_l(T)$ corresponds to the loss of coherent electron scattering due to the thermal disorder. The derivation is more convenient when the Fourier transform of the PDF is performed prior to the expansion in spherical harmonics, which immediately leads to the same result.

5.5.6 Multipole Expansion Coefficients for Harmonic Vibrations

It is neither convenient nor possible in practice to optimise the expansion coefficients $c_{n,lm}$ in Eq. (5.137) directly by fitting them to the experimental data. The appropriate approach is to use the conventional parameters of the thermal ellipsoid as free parameters in the analysis. This requires calculating the expansion coefficients as a function of the vibration amplitudes, which reduces the number of free parameters as well. For anisotropic harmonic vibrations there is a maximum number of six independent parameters which define the mean square displacements in three directions and give the orientation of the ellipsoid. The multipole expansion in Eq. (5.137), on the other hand, requires the radial functions and a large number of expansion coefficients; their number depends on the ratio of the displacement parameters.

We start from the isotropic case where only one parameter exists, the mean square displacement, while the distribution function is Gaussian. In this case,

$$p(\mathbf{u}) \propto \exp\left(-\frac{u^2}{\langle u^2\rangle}\right),$$

(5.155)

and all the expansion coefficients but $c_{0,00}$ in Eq. (5.137) vanish. If we now assume anisotropy with

$$p(\mathbf{u}) \propto \exp\left\{-\left(\frac{u_x^2}{\langle u_x^2\rangle} + \frac{u_y^2}{\langle u_y^2\rangle} + \frac{u_z^2}{\langle u_z^2\rangle}\right)\right\},$$

(5.156)

the expansion of the ellipsoid in general requires higher order coefficients. In the expansion

$$p(\mathbf{u}) = \sum_{n,lm} c_{n,lm} R_n(u) Y_{lm}(\Omega_u)$$

(5.157)

all odd-order terms in l vanish because of the centrosymmetry. As the probability distribution is a real quantity, there are some restrictions for the complex coefficients $c_{n,lm}$:

$$c_{n,l_0} = \text{real},$$

$$c_{n,l-|m|} = \begin{cases} c^*_{n,l|m|} & \text{if } m = \text{even} \\ -c^*_{n,l|m|} & \text{if } m = \text{odd}. \end{cases} \tag{5.158}$$

To derive the expansion coefficients from the given parameters of the ellipsoid, we use the deviation from the isotropic average (neglecting prefactors):

$$p'(\mathbf{u}) = p(\mathbf{u}) - \exp\left(-\frac{u^2}{\langle u^2 \rangle}\right), \tag{5.159}$$

and calculate

$$c_{n,lm}R_{n,lm}(\langle u \rangle) = \int p'(\mathbf{u}) Y_{lm}(\Omega_u) d\mathbf{u}. \tag{5.160}$$

The probability density function $P(\mathbf{u})$ for harmonic vibrations is given by [5.78]:

$$P(\mathbf{u}) = \frac{1}{\sqrt{8\pi^3 \langle u_x \rangle^2 \langle u_y \rangle^2 \langle u_z \rangle^2}} \exp\left(-\frac{1}{2}\left[\frac{u_x^2}{\langle u_x^2 \rangle} + \frac{u_y^2}{\langle u_y^2 \rangle} + \frac{u_z^2}{\langle u_z^2 \rangle}\right]\right), \tag{5.161}$$

where $\langle u_x \rangle^2$, $\langle u_z \rangle^2$ and $\langle u_z \rangle^2$ are the mean square displacements in the x-, y- and z-directions. If the atoms sit in surface sites with 3-, 4- or 6-fold symmetry, only two parameters remain, frequently denoted by $u_\parallel$ and $u_\perp$.

The appropriate radial functions for a Gaussian distribution are

$$R_0(u) = \frac{1}{\left(2\pi \sigma_0^2\right)^{3/2}} \exp\left(-\frac{u^2}{2\sigma_0^2}\right), \tag{5.162}$$

$$R_n(u) = \frac{u^2}{\left(2\pi \sigma_n^2\right)^{3/2}} \exp\left(-\frac{u^2}{2\sigma_n^2}\right), n \geq 1. \tag{5.163}$$

The expansion coefficients $c_{n,lm}$ and the widths of the radial functions σ_n can be determined from the inverse relation

$$\sum_{n,lm} c_{n,lm} R_n(u) = \int d\Omega_u P(\mathbf{u}) Y_{lm}(\Omega_u). \tag{5.164}$$

The parameters of the vibration ellipsoid, u_x, u_y and u_z, are fit parameters in the LEED calculations; both the expansion coefficients and the radial functions are calculated in each iteration step from these parameters. The number of expansion coefficients needed to match the Gaussian shape of the PDF depends on the anisotropy of the vibration ellipsoid and on the electron energy through the Bessel functions in Eq. (5.141). In the calculations for CO/Ru(0001) shown in Section 5.5.9, two radial functions R_0 and R_1 with $l_{\max} = 12$ are found to be sufficient.

The orientation of the ellipsoid of the thermal vibration provides another three degrees of freedom. The rotation of the ellipsoid can be performed by rotator functions for the atomic t-matrix [5.89]:

$$t^{rot}_{l_1 m_1, l_2 m_2} = \sum_{m'm''} D^{(l_1)}_{m_1 m'}(\alpha\beta\gamma)\, t_{l_1 m', l_2 m''}\, D^{(l_2)*}_{m_2 m''}(\alpha\beta\gamma), \tag{5.165}$$

where α, β, γ are the Euler angles of the rotation. The calculation of the rotation operators $D^{(l)}_{mm'}(\alpha\beta\gamma)$ can be found in textbooks on quantum mechanics (e.g., [5.93]); a computer program for the rotation of t-matrices has been published by M. Blanco-Rey et al. [5.94; 5.95].

When an ellipsoid with large anisotropy must be represented in the multipole expansion, terms with higher order l, m are required. When a limited order of coefficients is given it is also convenient to obtain the expansion coefficients by an optimisation procedure to approximate the expanded distribution to a Gaussian with given parameters.

The shape of the PDFs which can be obtained with a given number of angular momentum components is illustrated in Figure 5.23. We see that a small number of angular momentum components is totally inappropriate to describe a PDF with high anisotropy. This circumstance is not specific to the multipole expansion and cannot be overcome by alternative formulations using a spherical wave expansion.

5.5.7 Multipole Expansion Coefficients for Anharmonic Vibrations

The expansion in Eq. (5.137) allows using non-Gaussian PDFs, so that one can easily treat anharmonic vibrations as well. It also allows describing static displacements by choosing the appropriate expansion coefficients in accordance with the local symmetry of the site. A priori there are no assumptions whether the vibrating atoms move in a harmonic potential or not. As the PDF is determined by the atomic potential within which the atom vibrates, anharmonic contributions to the one-particle potential can be analysed in terms of the resulting PDF itself. Anharmonic vibrations produce a deviation of the PDF from a Gaussian. Two examples are shown in Figure 5.24. As the parity of the spherical harmonics is $(-1)^l$, that is, $Y_{lm}(\pi - \vartheta, \varphi + \pi) = (-1)^l Y_{lm}(\vartheta, \varphi)$, a generalised, non-Gaussian PDF can be expanded as a sum of centrosymmetric and antisymmetric components (the radial functions are always centrosymmetric):

$$p(\mathbf{u}) = \sum_n R_n(u)\left(\sum_{lm} c_{n,lm} Y_{lm}(\vartheta, \varphi) + \sum_{l'm'} c_{n,l'm'} Y_{l'm'}(\vartheta, \varphi) \right), \tag{5.166}$$

with even l and odd l'. The antisymmetric contributions represent an anharmonic distortion of the Gaussian, while they shift the first moment (mean) of the PDF which corresponds to the averaged position of the vibrating atom.

Anharmonic deviations caused by the centrosymmetric components of the distribution function leave the atomic position unchanged but affect the even moments of the distribution and therefore the mean square amplitude in a given direction. In this way the shape of the ellipsoid of thermal vibrations is modified, for example, with lobes constrained by local symmetry of the surface atom, cf. Figure 5.24. By

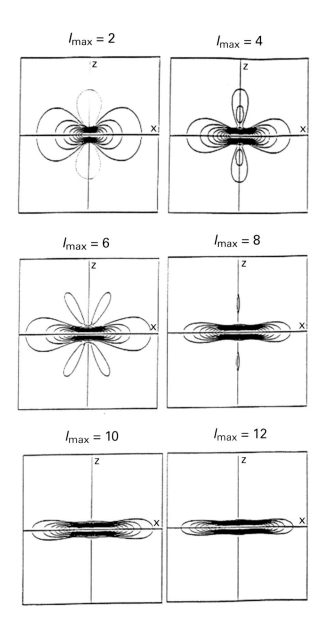

Figure 5.23 Illustration of the influence of the maximum number of angular momentum components of the shape of the PDF; $l_{max} = 12$ in the expansion, Eq. (5.157), is the minimum number required to approximate an ellipsoid with an axis ratio of 1:10. Reprinted from [5.91] W. Moritz and J. Landskron, "Thermal Vibrations at Surfaces Analysed with LEED", in *Solid-State Photoemission and Related Methods*, W. Schattke and M. A. Van Hove, eds., Wiley-VCH, pp. 433–459, 2003. With permission from Wiley.

(a)   (b)

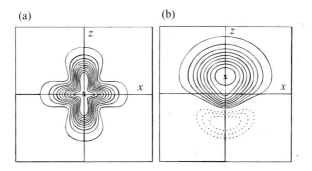

Figure 5.24 Examples for PDFs resulting from an anharmonic one-particle potential, presented as sections of the PDF in the x–z plane. (a) A PDF obtained from even order coefficients up to quadrupole terms in Eq. (5.166). The l = even coefficients produce a centrosymmetric distribution. (b) The shift of the mean position due an odd-order coefficient $c_{0,10} \neq 0$. The dotted lines correspond to negative PDF. Reprinted from [5.91] W. Moritz and J. Landskron, "Thermal Vibrations at Surfaces Analysed with LEED", in *Solid-State Photoemission and Related Methods*, W. Schattke and M. A. Van Hove, eds., Wiley-VCH, pp. 433–459, 2003. With permission from Wiley.

considering anharmonic effects, the number of non-vanishing off-diagonal elements in the atomic t-matrix increases in comparison to the harmonic case.

5.5.8   Example: Cu(110)

The multipole expansion of the probability density function is the appropriate way to introduce anisotropic and anharmonic thermal vibration into the multiple scattering formalism. The averaged non-diagonal scattering matrices can be used in conventional LEED codes without large changes, as well as for XPD, NEXAFS, or other spectroscopic methods. The computational effort can be reduced by using symmetries in real space and separate calculation of structural and thermal parameters, as both sets of parameters can be separately optimised in most cases.

As an example, anisotropy of the thermal motion was determined from temperature dependent LEED I(V) measurements of the clean Cu(110) surface, cf. Figure 5.25.

In the first step of the analysis, isotropic vibrations were assumed. The result for the rms displacements in the first three layers is shown in Figure 5.26(a).

The analysis optimised the four outermost layer distances, as well as rms displacements in the outermost three layers. In the second step anisotropic displacements were allowed only in the top layer, cf. Figure 5.26(b). Isotropic displacements could in principle be determined using Pendry's R-factor in which the scale factors for the intensities disappear, since the logarithmic derivative of the intensities is used in the Y-function. However, in the analysis of anisotropic displacements it is necessary to use a common scale factor for all I(V) curves, so the R-factor R_2 was chosen since it directly compares the intensities.

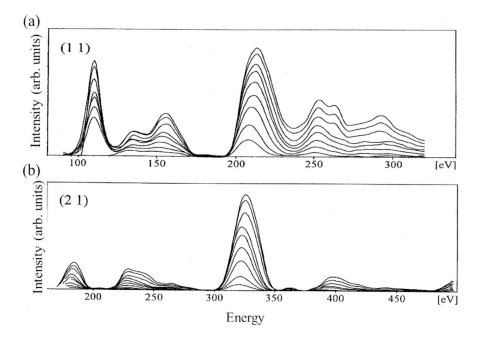

Figure 5.25 Experimental I(V) curves from Cu(110) for (a) the (11) beam and (b) the (21) beam at nine temperatures from 110 K (uppermost curves) to 465 K (lowest curves). Reprinted from [5.91] W. Moritz and J. Landskron, "Thermal Vibrations at Surfaces Analysed with LEED", in *Solid-State Photoemission and Related Methods*, W. Schattke and M. A. Van Hove, eds., Wiley-VCH, pp. 433–459, 2003. With permission from Wiley.

The results show clearly an enhancement of the displacement amplitudes in the [100] direction, cf. Figure 5.27, from about 0.011 nm (0.11 Å) in the $[1,\bar{1},0]$ direction to about 0.014 nm (0.14 Å) in the [100] direction, in agreement with theoretical calculations and with ion scattering results. The amplitudes normal to the surface are approximately the same as determined assuming isotropic vibrations.

5.5.9   Example: Ru(0001)+($\sqrt{3}\times\sqrt{3}$)R30°–CO

As an example of anisotropic thermal motion in adsorbate layers, the structure of CO molecules on Ru(0001) has been investigated by temperature-dependent LEED measurements [5.76]. In its ($\sqrt{3}\times\sqrt{3}$)R30° structure, CO stands perpendicular to the Ru (0001) surface, bonding through C to a top site; it occupies one third of the top sites and thus induces a slight out-of-plane buckling in the outermost Ru layer. The structure model and the notation of the parameters used in the analysis of rms displacements are shown in Figure 5.28.

The I(V) analysis was performed with both isotropic and anisotropic displacements. The geometrical results are practically identical. The possibility of anisotropy in the vibration of the topmost Ru atoms has been checked as well, although it seems *a priori* unlikely. No indications were found for such anisotropy, so that for all substrate

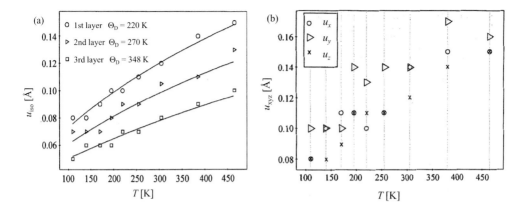

Figure 5.26 (a) Results of the vibration analysis of Cu(110) with isotropic displacements, showing rms displacements in the three outermost layers as a function of temperature; the values of Θ_D have been fitted for the top three layers and the corresponding u_{iso} are obtained through Eq. (5.118). (b) Results of the analysis with anisotropic displacements in the outermost layer (but isotropic displacements in the second and third layers); u_x and u_y are parallel and normal, respectively, to the ridges in the surface, while u_z is normal to the surface. Reprinted from [5.91] W. Moritz and J. Landskron, "Thermal Vibrations at Surfaces Analysed with LEED", in *Solid-State Photoemission and Related Methods*, W. Schattke and M. A. Van Hove, eds., Wiley-VCH, pp. 433–459, 2003. With permission from Wiley.

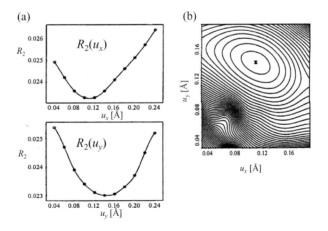

Figure 5.27 R-factor R_2 as a function of the lateral displacements of the Cu atoms in the top layer of Cu(110). (a) In the directions $[1\bar{1}0]$ (u_x) and [100] (u_y). (b) Contour plot of the R-factor, where × marks a minimum. Reprinted from [5.91] W. Moritz and J. Landskron, "Thermal Vibrations at Surfaces Analysed with LEED", in *Solid-State Photoemission and Related Methods*, W. Schattke and M. A. Van Hove, eds., Wiley-VCH, pp. 433–459, 2003. With permission from Wiley.

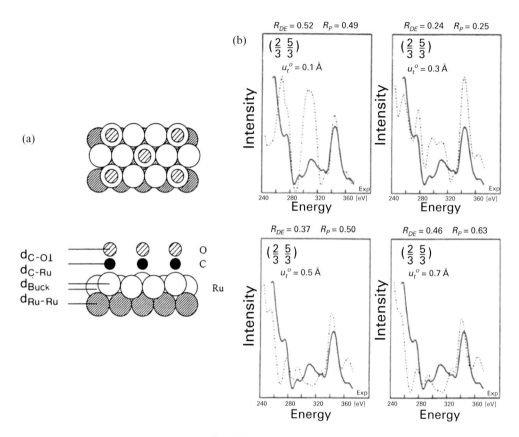

Figure 5.28 (a) Model of the $(\sqrt{3}\times\sqrt{3})R30°$ structure of CO on Ru(0001), with top view in the upper panel and side view in the lower panel showing layer spacings. (b) I(V) curves of the (2/3, 5/3) beam illustrating the influence of the rms displacements of oxygen parallel to the surface: solid lines are experimental, dashed lines theoretical best fits. Shown are four calculated I(V) curves where the parallel vibration amplitude of oxygen is kept fixed at values between 0.1 and 0.7 Å, as indicated in the figure. The peak positions fit best at 0.3 Å. All other structural parameters are optimised and all data were used in the analysis. Reprinted from [5.76] J. Landskron, W. Moritz, B. Narloch, G. Held, and D. Menzel, *Surf. Sci.*, vol. 441, pp. 91–106, 1999, with permission from Elsevier.

layers isotropic vibrations were used in the final analysis. The best-fit Debye temperatures were 310 K in the top substrate layer and 410 K in the second and subsequent layers. For each temperature, the perpendicular and parallel components of the displacement parameters for C and O were optimised. Note that in this treatment the motions of C and O are independent of each other, which is probably not quite realistic, as C and O likely have correlated motions due to their strong mutual bonding. Two R-factors were used to compare theory to experiment, R_1 and R_P (defined in Section 6.1.2), both leading to practically the same result. R_1 is the average of the absolute values of the difference to the experimental data. The R-factors show a noticeably better agreement between experimental and calculated data with anisotropic rather than isotropic vibrations. With isotropic vibrations the R-factor is $R_P = 0.29$ at 25 K and increases to 0.39 at 350 K, while with anisotropic vibrations the optimum values for R_P are 0.26 at 25 K and 0.34 at 350 K, showing significant improvement.

The increase of the R-factors with temperature could be the result of anharmonicity, that is, a non-Gaussian probability density or a dependence of the lateral vibrational amplitude on the azimuthal direction in-plane. A 3-fold shape of the PDF in-plane would not be allowed in the harmonic approximation and for an atom in a 3-fold coordinated site. By inclusion of higher-order terms in the multipole expansion it was tested whether a deviation from the lateral isotropy in the PDF could be detected, but no significant improvement was found. From previous ESDIAD results [5.96] it is indeed not expected that preferential directions exist for the vibration parallel to the surface. Nevertheless, it is conceivable that the actual correlated motion of the C and O atoms is important, although this was not considered in this LEED analysis. The remaining deviation from ideal fit could also result from defects, whose number increases with temperature. Some discussion about correlated motion and estimates of errors is given by E. Prince [5.85] which refers to the kinematic theory. The applicability to the multiple scattering theory in LEED has not yet been investigated.

The results for the temperature dependence of the vibration amplitudes are shown in Figure 5.29. All amplitudes increase with temperature. The parallel rms displacement for the oxygen atom is noticeably larger than that for the carbon atom except for the lowest temperature where both amplitudes are equal. Both parallel components are larger than the perpendicular components, as one would expect. After optimising the thermal vibrations, the structural parameters were optimised again, but no deviations were detected within the error limits. To show how sensitive the superlattice intensities are to the thermal parameters, I(V) curves for the (2/3, 5/3) beam are shown in Figure 5.28(b), together with the corresponding R-factors for this particular beam. The displacements do not reach the zero-point motion at low temperatures, which indicates the presence of static contributions to the displacements, probably resulting from point defects and steps.

5.6 From Calculated Amplitudes to Intensities

We connect here the wave amplitude to the diffracted intensity, that is, calculation to experiment. Equation (2.49) provided this link for the kinematic theory, where structure factor and lattice factor can be decoupled for identical atoms. We must generalise that result to the case of multiple scattering in the dynamical theory.

Assume an incident plane wave $A_{in}e^{i\mathbf{k}_{in}\mathbf{r}}$ and a diffracted plane wave $A_{out}e^{i\mathbf{k}_{out}\mathbf{r}}$, where A_{in} is usually set to unity, while A_{out} can come from $\mathbf{T}$ of Eq. (5.17) (in the case

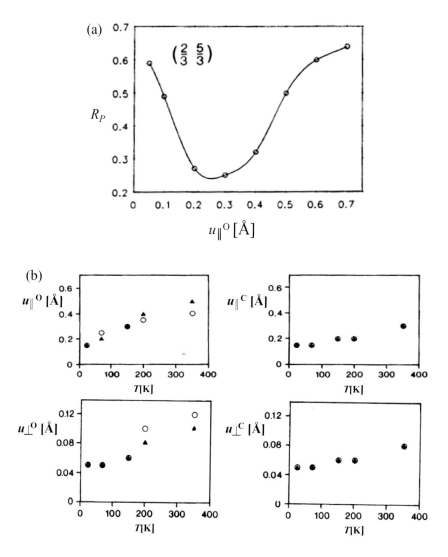

Figure 5.29 (a) Pendry R-factor of the (2/3, 5/3) beam as function of the parallel displacement of the O atoms in Ru(0001)+($\sqrt{3}\times\sqrt{3}$)R30°–CO. (b) Displacements of O and C parallel and normal to the surface as a function of temperature. All experimentally available beams were used in the analysis. Reprinted from [5.76] J. Landskron, W. Moritz, B. Narloch, G. Held, and D. Menzel, *Surf. Sci.*, vol. 441, pp. 91–106, 1999, with permission from Elsevier.

of divergent outgoing waves, as for diffraction from a nanocluster, $\mathbf{k}_{out}$ gives any chosen outgoing direction). For plane waves, the quantities $k|A_{in}|^2$ and $k|A_{out}|^2$, where $k = |\mathbf{k}_{in}| = |\mathbf{k}_{out}|$ are the fluxes in the incident and diffracted beams (or directions), namely the number of electrons per unit time crossing through a unit area oriented perpendicular to the direction of electron motion. For plane waves and a planar surface, the fluxes integrated over the beam cross-sections are then, in the theory, $Ck|A_{in}|^2$ and $Ck|A_{out}|^2(\cos \vartheta_{out}/\cos \vartheta_{in})$, respectively, where C is the cross-section of the incident beam and $C(\cos \vartheta_{out}/\cos \vartheta_{in})$ the corresponding cross-section of the diffracted beam, given the polar angles of incidence ϑ_{in} and of emergence ϑ_{out} relative to the surface normal. The theoretical reflectivity R^{calc} of the surface thus becomes

$$R^{calc} = \frac{\cos \vartheta_{out}}{\cos \vartheta_{in}} \frac{|A_{out}|^2}{|A_{in}|^2}. \tag{5.167}$$

The equivalent experimental reflectivity, to which the theoretical reflectivity of Eq. (5.167) should be compared, is the ratio of the measured intensity (current) I_{out} of a diffracted beam to that of the incident beam, I_{in}:

$$R^{exp} = \frac{I_{out}}{I_{in}}. \tag{5.168}$$

6 LEED Theory

Applications

We discuss here the methods of quantitative LEED I(V) analysis and their application to relatively complex types of surface structures: quasicrystalline and modulated surfaces.

We first give a short review of the methods that are used to derive a model of the surface structure from the LEED I(V) curves, in Section 6.1. Finding the correct model of the surface structure – as opposed to precise coordinates – is one of the major challenges in LEED. A variety of different approaches have been used, from direct methods to structure search methods. A generally applicable method has still not been found and there is continuing interest in the development of direct methods. We next outline the present status of the field. We also review briefly the phase shift calculation and the development of methods to overcome problems in the muffin-tin potential model.

Sections 6.2 and 6.3 discuss two types of surfaces, quasicrystalline and modulated surfaces, where quantitative analysis is possible but rarely used. With the increasing speed of computers in combination with the calculation methods of NanoLEED, many more surface modulations could be quantitatively analysed by LEED. Regarding quasicrystals we think that it should be possible to identify the composition of clusters by means of LEED, which is still a major problem in X-ray diffraction. To obtain quantitative results in these areas would be impossible or very difficult with methods other than LEED.

6.1 Quantitative Structure Analysis

The structure determination method which has been applied most is the trial and error approach coupled with structure refinement by various optimisation procedures. Numerous surface structures were analysed by LEED: some of these can be found, for example, in the Surface Structure Database (SSD) [6.1; 6.2] and in a review in Landolt-Börnstein, Vol. 45 [6.3].

One of the most common applications of LEED is certainly the observation of the diffraction pattern with the purpose to control the preparation of the surface and to check whether the surface is ordered, disordered or faceted, as well as to identify the size and orientation of superlattices in adsorbate layers and surface reconstructions.

A short description of this qualitative interpretation of the LEED pattern without quantitative multiple scattering calculations is given in Chapter 4.

The surface structure analysis by LEED consists of two steps, similar to the analysis by X-rays. In the first step, a structure model must be found; in the second step, that structure model must be refined by optimising all its parameters. In many cases more than one plausible model must be investigated.

In 3-D X-ray analysis, including in powder diffraction, the first step is rather well developed and direct methods can be applied, such that the analysis has become routine, at least for structures of low and moderate complexity. In X-ray diffraction the term 'direct methods' is solely used for methods to determine the phases of the complex structure factors. These methods are based on the fact that the electron density is positive and on the relation between intensities and phases [6.4]. If the phases of the structure factors are known, an image of the electron density is obtained via the Fourier transform of the structure factors. The model obtained in that way must still be refined by optimisation methods.

Direct methods are not, or not yet, applicable in LEED. Although many approaches have been developed with partial success, as will be discussed in this section, no generally applicable method to determine a structure model directly from the LEED intensity data is available. Multiple scattering severely limits the applicability of the kinematic theory: this implies that all standard X-ray methods of deriving models from intensity data, such as Patterson search methods, difference Fourier analysis, phase determination by the heavy atom method and related methods (see, e.g. [6.5]), are not reliably applicable. Also, widely used methods, including the so-called pixel methods, such as the 'charge flip' method by which the electron density is determined in an iteration process [6.6], cannot be applied in LEED. The phases of the atomic diffraction amplitudes cannot be determined by the atom positions alone, as is the case in the kinematic theory of diffraction, but by complex scattering factors and the influence of the surroundings of the atom through multiple scattering. We discuss in the next paragraphs the various approaches to removing or at least reducing multiple scattering effects and the special cases where limited success has been reached. Whether a generally applicable direct method can be developed for LEED remains an open question.

Consequently, to start the structural analysis with LEED, one or more suitable models for the surface structure must be proposed, based on the diffraction pattern, specifying size, orientation and symmetry of the surface unit cell and the chemical composition of the surface. In most cases, this information is not sufficient to propose a unique model; then more information can be used from further experimental methods like STM, AFM and various spectroscopic measurements, in order to constrain the surface composition, bond lengths, bond energies, coordination, etc. In most cases, a series of guessed structures are then tested against LEED experiment. Often, the correct structure model was found earlier by DFT calculations (and occasionally by other experimental techniques); in most modern surface structure determinations by LEED, the result is in fact simultaneously or subsequently confirmed by DFT calculations [6.7; 6.8].

With LEED, surface structures are usually determined by using the 'trial and error' approach. This means that the experimental I(V) curves are compared with calculated ones for various reasonable models and, once a promising model has been found, the parameters of the model are refined with optimisation procedures to improve the fit to experiment. Usually several promising models are found and locally optimised; normally no 'global' optimisation is used to explore other potential models. This procedure is lengthy; the main problem is to find the right initial model or models. The trial and error method always presents the danger that the correct model has not been thought of or some details of the model have been overlooked. Once some agreement between experimental and calculated curves is found for a specific model, the parameters of this model are locally varied until an optimum agreement has been reached. But the result depends on the correctness or completeness of the initial model. Details not included in the model are not found by local optimisation.

To overcome these problems, several approaches have been proposed to use global optimisation, in which many more possible combinations of all parameters are automatically explored. The method aims to avoid getting trapped in local minima of the R-factor, which are always present in the parameter space. This is a fairly general approach, but it is not completely model-free. For example, a chemical element that is missing in the model will not be found by global optimisation. Therefore, the term 'direct method' should not be used or confused with global optimisation. The various methods to derive models and to optimise structural parameters have been reviewed by E. A. Soares et al. [6.9].

In Section 6.1.1, we give a short overview of approaches that attempt to derive structural information directly from LEED intensity data without multiple scattering calculations; Sections 6.1.2–6.1.4 address several more specific aspects. These approaches are based on the Patterson function, holography, or thermal diffuse scattering. The resulting preferred models need to be subsequently refined by optimisation methods, which will be summarised in Section 6.1.5. The influence of non-structural parameters on such structure determination will be addressed in Section 6.1.6.

6.1.1 Approaches to Obtain Structural Information Directly from LEED Intensities

6.1.1.1 Averaging Methods and the Patterson Function

The strong interaction of electrons with the surface atoms and the resulting multiple scattering effects impede the use of the kinematic theory. The first attempt to eliminate multiple scattering effects was the constant momentum transfer averaging (CMTA), proposed by M. G. Lagally, M. B. Webb and co-workers [6.10], in which LEED intensities are averaged that were obtained at one point in reciprocal space but at different energies and angles of incidence, both polar angle and azimuth. By repeating this at other points, a range of averaged data is produced across a part of reciprocal space, for example, over a range of momentum transfer values that emphasises kinematic effects (such as Bragg peaks) and thereby suppresses multiple scattering effects. The averaged intensities are then analysed by a kinematic calculation. The fact that averaging of intensities does not really reduce multiple scattering effects to a negligible background has prevented a successful application of the method for general cases.

The CMTA method was not applied much until it was revived to construct a Patterson function which is submitted to a kinematic analysis. The Patterson function is obtained as the Fourier transform of the intensities and in the kinematic theory it shows all interatomic distances occurring in the surface structure. The first application of the Patterson function to derive interlayer distances from the (00) beam intensities was to analyse some clean metal surfaces by D. L. Adams and U. Landman, with limited success [6.11]. Due to multiple scattering effects, additional peaks occur in the Patterson function which cannot be assigned to interatomic distances.

To overcome this problem, H. Wu and S. Y. Tong [6.12; 6.13] proposed to apply the CMTA method to *averaged* LEED I(V) curves from which they derived the Patterson function. They showed that, by including a broad range of incidence angles and energies, the peaks in the Patterson function could be assigned to interatomic distances. They applied the method successfully to the Si(111)+(4×1)–In surface [6.14]; they could use only the superstructure I(V) curves due to the superposition of three domains and assumed the Si atom positions in the substrate to be known. The same structure was investigated later by T. Abukawa et al. [6.15] using a single domain surface and an extended data set which allowed determining the Si positions in the substrate from the Patterson function. The Patterson function was also used to select possible models in the more complicated system Si(111)+(1×1)–YSi$_2$ [6.16] and to also distinguish proposed models for the Si(111)+(1×1)–Fe surface [6.17]. A slightly different method to analyse the Patterson function, which was called 'integral-energy phase-summing method' [6.18], was applied to reconstructed Si(111) surfaces, namely Si(111)–(7×7) and Si(111)–(3×2). Multiple scattering effects are relatively weak in silicon and the kinematical analysis in the abovementioned studies on Si surfaces probably worked relatively well for that reason. The application to other surfaces has been rare: for the GaN(0001)–($\sqrt{3} \times \sqrt{3}$)R30° system the site of the Ga adatom could be identified with the Patterson function [6.19]; and a similar study of the GaN(0001)–(1×1) surface exhibited a Ga adlayer on a Ga-terminated surface [6.20].

The determination of a Patterson function from LEED I(V) data requires a very large data set, as has been shown by T. Abukawa et al. [6.15]. The derivation of structure models from the Patterson function, which only shows the lengths and direction of interatomic distances, requires a sufficient amount of data such that structure search methods can be applied. With the development of fast experimental techniques and automated evaluation of diffraction images, the method probably could be further developed. However, the applicability to adsorbed layers on metal substrates or systems with strong multiple scattering effects has not yet been proven.

6.1.1.2   Holography

Further methods to obtain structural information from the intensities have been developed by applying ideas of optical holography. The principle of holography, as originally proposed by D. Gabor [6.21], is to reconstruct the image of an object from the interference pattern (called hologram) due to the wave scattered by the object interfering with the unscattered portion of the primary incident wave (e.g., from a laser): their interference contains the structural details of the object. The reconstruction of the image is typically done by shining a wave identical to the incident wave through the hologram, especially in the optical case with a laser: this produces an image of the original object by simple interference. The idea can be used with electrons: in this case the reconstruction is done computationally, since it is difficult to reproduce the incident electron wave and shine it onto a hologram.

The hologram, whether produced with photon or electron waves, can be written as the interference of object wave $O(\mathbf{k}', \mathbf{k})$ due to the object with reference wave $R(\mathbf{k}', \mathbf{k})$, which is the unscattered incident wave:

$$I(\mathbf{k}', \mathbf{k}) = |R(\mathbf{k}', \mathbf{k}) + O(\mathbf{k}', \mathbf{k})|^2$$

$$= |R(\mathbf{k}', \mathbf{k})|^2 + |O(\mathbf{k}', \mathbf{k})|^2 + |R^*(\mathbf{k}', \mathbf{k})O(\mathbf{k}', \mathbf{k}) + R(\mathbf{k}', \mathbf{k})O^*(\mathbf{k}', \mathbf{k})|. \quad (6.1)$$

Here $\mathbf{k}$ is the incident wave vector and $\mathbf{k}'$ the wave vector of the scattered wave. The assumption in optical holography is that the reference wave is much stronger than the object wave, such that the second order terms $|O(\mathbf{k}', \mathbf{k})|^2$ can be omitted. The second assumption is that the reference wave does not depend on the direction of $\mathbf{k}'$, that is, the reference wave is isotropic. These assumptions are not valid for LEED, as has been pointed out by S. Y. Tong [6.22] and others, and the method must be modified for application in LEED. The computational image reconstruction process is ideally a simple Fourier transformation; however, the measurement of intensities rather than wave amplitudes introduces the complication of unknown wave phases and adds a twin image.

In the following, we give a brief description of the approach to interpret a hologram and produce an image from LEED, following the review of K. Heinz et al. [6.23]. The diffraction geometry is illustrated in Figure 6.1. The method requires a reference atom acting as a beam splitter that produces both the reference wave R and the object wave

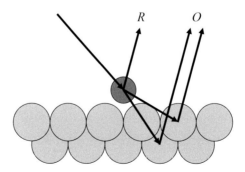

Figure 6.1 An adsorbate atom (darker grey) on a surface splits the incoming wave into reference wave R and object wave O and serves as a reference atom for the surface structure (schematic)

O (this wave *R* plays the role of the original 'unscattered' reference wave, while the wave *O* plays the role of the 'scattered' object wave).

The most obvious application where the object wave and the reference wave can be at least partially separated is for disordered adsorbate layers: for disordered surfaces the diffuse scattering between sharp spots necessarily involves a scattering event from the disordered layer: this event then defines the reference wave *R*. The case of ordered adsorbate layers is slightly less favourable, as the superstructure intensities occur only partly from single scattering in the superstructure, while scattering from the substrate also contributes through multiple scattering, so that no pure reference wave *R* exists. The method is not limited to overlayers, however: any atom in the surface can be viewed as a beam splitter; the drawback is then that multiple atoms in the surface qualify as beam splitters, so that the hologram represents the superposition of their multiple neighbourhoods, further complicating the resulting image.

In its simplest form the reference wave scattered from the beam splitter at position **r** can be written as

$$R(\mathbf{k}, \mathbf{k}') = F_r(\mathbf{k}, \mathbf{k}') \exp(i\mathbf{k}'\mathbf{r}), \tag{6.2}$$

while the scattered amplitude from the object is given by:

$$O(\mathbf{k}, \mathbf{k}') = \sum_j F_j(\mathbf{k}, \mathbf{k}') \exp\left(i\mathbf{k}'(\mathbf{r} - \mathbf{r}_j)\right). \tag{6.3}$$

Here F_r and F_j are the multiple scattering amplitudes due to the beam splitter atom and the object atoms at positions *j*, respectively. As F_r is assumed to be nearly structureless, that is, independent of $\mathbf{k}'$, the multiple scattering amplitude F_r is replaced by the single scattering formula. If $|O(\mathbf{k}', \mathbf{k})|^2$ is omitted, the Fourier transform (or more rigorously the Helmholtz–Kirchhoff integral for electrons) of the interference term gives

$$A(\mathbf{r}) = \int (RO^* - OR^*) \exp(-i\mathbf{k}'\mathbf{r}) d^2\mathbf{k}'_\parallel, \tag{6.4}$$

where $\mathbf{k}'_\parallel$ is the parallel component of $\mathbf{k}'$. At a fixed energy, the parallel components are the sampling points of the diffuse intensity or of the superstructure spots in the case of ordered structures. The image $A(\mathbf{r})$ consists of a superposition of two terms, called the real image and the twin image. By Eq. (6.1), the interference term $(RO^* - OR^*)$ can be replaced by $I(\mathbf{k}, \mathbf{k}') - |R(\mathbf{k}, \mathbf{k}')|^2$, if $|O(\mathbf{k}', \mathbf{k})|^2$ is assumed to be negligible; $|R(\mathbf{k}, \mathbf{k}')|^2$ can also be omitted, since it only refers to the beam splitter atom and contains no information on other atoms (the beam splitter atom will in any case be at the origin of the resulting image). The image is then approximated as

$$A(\mathbf{r}) \approx \int I(\mathbf{k}, \mathbf{k}') \exp(i\mathbf{k}'\mathbf{r}) d^2\mathbf{k}'_\parallel. \tag{6.5}$$

Here, $I(\mathbf{k}, \mathbf{k}')$ is the diffuse intensity or the intensity of the superstructure spots without the integer order spots. This simple evaluation does not work very well due to the neglect of multiple scattering effects and the assumption that $|O(\mathbf{k}', \mathbf{k})|^2$ is small.

Holography was first applied to surfaces in the context of photoelectron diffraction [6.24; 6.25], in which electron scattering is very similar to that in LEED: such 'photoelectron holography' (PEH) fits the original optical principle of holography well, since a single adatom can serve as a photoelectron emitter and thus as a reference atom (cf. Figure 6.1). While considerable progress was made with PEH, the relatively complex electron scattering, including multiple scattering and in particular strong forward focusing at higher energies, has limited its applicability in terms of the achievable precision of determining atomic positions [6.26]; nevertheless, useful qualitative structural information is accessible in many cases. For disordered surfaces, photoelectron diffraction (PED) is the method of choice. The reference wave is here generated by a photon which has the advantage to be element-specific and the analysis does not have the problem that the object wave is strong. Nevertheless, further improvements of holographic methods have been proposed and applied for PED, for example, so-called differential photoelectron holography [6.27; 6.28], and may become applicable in future developments in holography by LEED.

The first application of holography in LEED was for diffuse LEED (DLEED) to study the adsorption site of disordered adatoms [6.29–6.31]. As illustrated in Figure 6.1, now the reference wave is the single scattered wave from the adatom, while the object wave consists of the backscattered waves from all other surrounding atoms after being scattered first at the adatom. The strongly scattering adatom is considered as a beam splitter, as shown in Figure 6.1. The 2-D diffuse intensity was inverted by a Fourier transform to reconstruct a real-space image around the adsorption site. This unfortunately gives poor images and spurious peaks arising from multiple scattering effects.

A number of approaches were proposed to overcome these problems (e.g., [6.32; 6.33]). A better 3-D image of the surface was obtained for K/Ni(100) from a set of diffuse LEED patterns, by compensating the anisotropy of the reference wave by an appropriate scattered-wave kernel [6.34; 6.35]; this was also applied to SiC(111)–(3×3) [6.36] and 6H-SiC(0001)–(2×2) [6.37]. A summary is given in U. Starke et al. [6.38]. To reduce disturbances, an iterative procedure was proposed [6.39]; a further approach separates the multiple scattering paths into one part solely taking place in the known substrate and a second part involving the unknown surface [6.40]; thereby simultaneous occupation of different adsorption sites was identified for O/Ni(111) [6.41]. H. Wu et al. [6.42] introduced the selective holographic atomic-reference-pair (SHARP) transformation to directly reconstruct 3D atomic images from LEED I(V) curves using a multiple-incident angle and multiple-energy integral: they applied it to Si(111)+$(\sqrt{3}\times\sqrt{3})$R30°–Ga. The LEED holographic method has been reviewed by K. Heinz et al. [6.23].

Although in several cases a direct image of the adsorption site could be obtained, the holographic method is still not generally applicable, in particular for more complicated adsorption structures and especially for materials with strong multiple scattering, such as metals. Holography by LEED is thus not established as a routine method for surface structure analysis.

6.1.1.3 Further Approaches

The strong correlation of vibrations between near-neighbour atoms induces broad features in the thermal diffuse scattering, called correlated thermal diffuse scattering (CTDS), which can be analysed with a kinematic theory. This has been applied to the higher energies of 500–2,000 eV in medium energy electron diffraction (MEED) [6.43].

6.1.2 Comparison of Measured and Calculated I(V) Curves

The main methods of structure analysis by LEED rely on the fitting of theoretically calculated I(V) curves to experimental curves. This requires criteria to quantify the agreement between experimental and calculated intensities: these criteria are usually captured in so-called R-factors (a name abbreviated from reliability factors or residual factors).

For structure analysis by LEED, one normally measures continuous I(V) curves (here we shall use the equivalent quantity $I(E)$ in formulas, while in the LEED literature the designations I(V), I–V, IV or I/V are customary in text). The measurement is usually done with sufficiently small energy steps so that interpolation between the points is possible, a continuous curve can be drawn and the first derivative dI/dE – or even the second derivative – can be calculated.

In close analogy with X-ray crystallography (where individual Bragg reflection intensities rather than continuous curves are measured), we can define R-factors that quantify the closeness of theoretical and experimental data: this will both allow estimating the validity of a given structural model and enable a refinement of the structural parameters. The following simple R-factors have been proposed (summation rather than integration applies to larger energy intervals):

$$R_1 = A_1 \int |I_{ex} - cI_{th}| dE, \tag{6.6}$$

$$R_2 = A_2 \int |I_{ex} - cI_{th}|^2 dE. \tag{6.7}$$

Here the subscripts ex and th stand for experiment and theory, respectively, and the integration (or summation) ranges over the energy intervals common to experiment and theory in each diffracted beam considered. In practice, with these R-factors as well as with other R-factors defined later in this section, one calculates separate R-factors for different beams and then averages these beam-specific R-factors with weights proportional to the energy intervals of the different beams. The constant c is inserted to account for the usually unknown relative intensity scales in the experiment and in the theory, and it serves to normalise the curves to each other (usually for each beam separately), for example through:

$$c = \int I_{ex} dE \bigg/ \int I_{th} dE. \tag{6.8}$$

The prefactors A_1 and A_2 in Eqs. (6.6) and (6.7) are designed to render the R-factor dimensionless and to provide normalisation. The usual choice is

$$A_1 = 1 \bigg/ \int I_{ex} dE, \tag{6.9}$$

$$A_2 = 1 \bigg/ \int I_{ex}^2 dE. \tag{6.10}$$

The R-factor R_2 is used only occasionally, mainly in refinement of parameters when the structure model has been determined. R_2 compares the square of the difference between experimental and calculated intensities, therefore it emphasises the strong intensities, while R_1 compares the linear difference between intensities giving all peaks the same weight. R_1 is rarely used in LEED I(V) analyses but the analogous R-factor is frequently applied in X-ray analyses, with the difference that there X-ray structure factors are compared instead of intensities. Neither of these R-factors requires continuous I(V) curves, so they can be defined for discrete energies allowing large energy steps in the calculation, thus saving computing time but losing details of the intensity functions. R_1 has been called R_{DE} in these cases and has been found to be quite reliable up to steps of 15 eV on the energy scale when the I(V) curves contain enough points [6.44].

The comparison of continuous I(E) curves makes the LEED analysis more closely related to X-ray *powder* diffraction than to X-ray *single crystal* analysis. There is, nevertheless, a difference with respect to powder diffraction, as the comparison of peak positions and their intensities presents some problems in LEED. The peak positions cannot be indexed as is the case in X-ray powder diffraction. The multiple scattering in LEED creates complications: there are additional peaks that cannot be related to a kinematic reflection and also can overlap with other peaks; in addition, the peak locations, peak heights and peak widths are affected by atomic phase shifts and multiple scattering, as well as by inelastic and thermal effects and by structural defects; furthermore, some of the latter effects are not accurately reproduced in theory.

In the R-factor R_2, the peaks are weighted in proportion to the square of their heights. This is often not realistic. While a small peak is less reliably measured, its mere existence and its position on the energy scale imply equally valuable geometrical information as that of a large peak. This is especially true of the higher-energy parts of I(V) curves, where thermal and atomic scattering effects often make all intensities relatively much smaller than at the lower energies, while the energies at which these small peaks occur remain highly important. For these reasons, J. B. Pendry has proposed an R-factor that attempts to treat all peaks (and minima) with equal weights [6.45]. This is based on the logarithmic derivative L of the I(V) curves:

$$L(E) = \frac{1}{I(E)} \cdot \frac{dI(E)}{dE}. \tag{6.11}$$

Pendry has suggested a corresponding R-factor, which avoids singularities when $I(E) = 0$ by defining the function

$$Y(E) = \frac{L^{-1}}{\left(L^{-2} + V_{0i}^2\right)}, \tag{6.12}$$

where $2V_{0i}$ can be taken to be the average peak width of single peaks (as opposed to overlapping peaks). V_{0i} is normally chosen to be the imaginary part of the inner potential, which acts as the damping parameter; usually values between 4 and 5 eV are assumed at medium energies around 100–150 eV. Pendry's R-factor is then defined as:

$$R_P = \frac{\displaystyle\sum_{hk}\sum_i \left(Y_{hk,i}^{th} - Y_{hk,i}^{ex}\right)^2}{\displaystyle\sum_{hk}\sum_i \left(Y_{hk,i}^{th} + Y_{hk,i}^{ex}\right)^2}, \tag{6.13}$$

where the index i labels the energy points of the beam (hk), usually in 1 eV steps. R_P has become the most commonly used R-factor and has become the standard estimate of the quality of a LEED analysis. The definition of this R-factor leads to values of $R_P = 0$ for perfect agreement, $R_P = 1$ for random agreement and $R_P = 2$ for total anti-correlation of I(V) curves.

Values around $R_P = 0.1$ can be considered to indicate a very good agreement. With values between 0.2 and 0.3 the structure model may be qualitatively correct, but details are certainly not correct, giving large error bars. It should be noted that very good agreement can be reached with measurements at low temperature (LN$_2$ temperature and below) and with a large energy range, avoiding the low energy range below 50–60 eV. At room temperature usually only a limited energy range is measurable because of the weak intensities and high background at higher energies. The agreement at room temperature is worse than at lower temperatures because of inadequate treatment of thermal motion. When judging the reliability of an analysis by the minimum R-factor obtained for the best fit result, the temperature of the experimental data should therefore be considered.

Examples for some LEED analyses with low, medium and high level of agreement are shown in Figures 6.2–6.4, respectively.

One disadvantage of R_P is that it requires the calculation of derivatives, which implies that it is sensitive to noise in the experimental data: it is therefore useful to average equivalent data (e.g., I(V) curves measured multiple times or for symmetry-equivalent beams) to reduce the noise. Another minor difficulty arises from the fact that the Y-function may become 0 in both experimental and theoretical $I(E)$ curves for any structure. If the theoretical intensities vanish in a certain energy range and there is background with noise in the experimental data, then R_P becomes 1 in that energy range and may increase the R-factor substantially without indicating a disagreement. The corresponding data points should be omitted. Such a case occurs for example for stepped surfaces: the I(V) curves then exhibit a wide energy range between the Bragg points with theoretically negligible intensity where only background is measured experimentally. The calculated I(V) curve should confirm that in

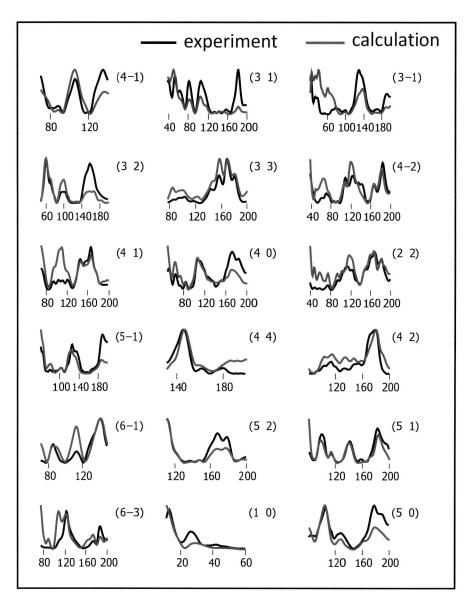

Figure 6.2 Comparison of experimental (black lines) and calculated I(V) curves (red lines) of Cu(111)+(3√3×3√3)R30°–TMB. Intensities of experimental and calculated curves are approximately normalised for each beam separately (with arbitrary units), while energies are given in eV. The average Pendry R-factor of all beams is $R_P = 0.32$. This agreement is not quite satisfactory but is sufficient to identify the structure model and to determine the main parameters; the error bars are large and some structural details are certainly incorrect. TMB stands for 1,3,5-tris(4-mercaptophenyl)-benzene (with chemical formula $C_{24}H_{15}S_3$). Adapted from [6.46] Th. Sirtl, J. Jelic, J. Meyer, K. Das, W. M. Heckl, W. Moritz, J. Rundgren, M. Schmittel, K. Reuter and M. Lackinger, *Phys. Chem. Chem. Phys.*, vol. 15, pp. 11054–11060, 2013, Electronic Supplementary Information (ESI), https://doi.org/10.1039/c3cp50752a, with permission from the PCCP Owner Societies.

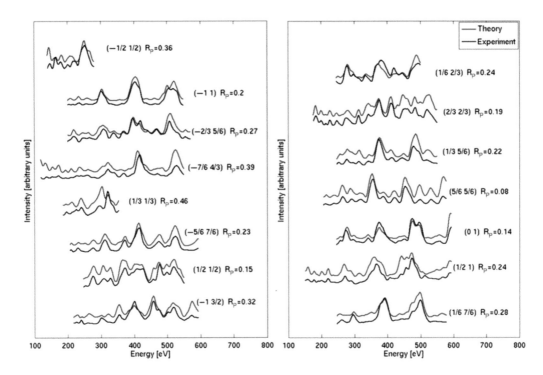

Figure 6.3 Comparison of experimental (black lines) and calculated I(V) curves (red lines) of Ag(111)+(2√3×2√3)R30°–C$_{60}$. The average Pendry R-factor of all beams is $R_P = 0.24$ and indicates good agreement. It is sufficient to identify the structure model and details of the surface structure. A better agreement cannot be expected in view of the complexity of the system with mixing of two orientations of C$_{60}$ molecules. Reprinted with permission from [6.47] K. Pussi, H. I. Li, H. Shin, L. N. Serkovic Loli, A. K. Shukla, J. Ledieu, V. Fournée, L. L. Wang, S. Y. Su, K. E. Marino, M. V. Snyder and R. D. Diehl, *Phys. Rev. B*, vol. 86, p. 205406, 2012; Supplementary Material. https://doi.org/10.1103/PhysRevB.86.205406. © (2012) by the American Physical Society.

this energy range no peak occurs, but in the R-factor analysis this energy range should be omitted.

Pendry's R-factor is not quite metric, meaning that its convergence does not guarantee that the theoretical and experimental curves are necessarily converging together. This non-metric character is due to V_{0i} in the definition of the Y-function, Eq. (6.12), which implies that the R-factor depends on the length of the energy range of the I(V) curves. Test calculations revealed that in all cases the influence on structural results can be neglected.

While many other R-factors were proposed for LEED, only one of these has been used extensively, namely the 'reduced' Zanazzi–Jona R-factor R_{ZJ} [6.49]. This uses the first and second derivatives of $I(E)$ with respect to energy and is defined as:

$$R_{ZJ} = A_{ZJ} \int \left[|I''_{ex} - cI''_{th}| |I'_{ex} - cI'_{th}| / (|I'_{ex}| + \max |I'_{ex}|) \right] dE \qquad (6.14)$$

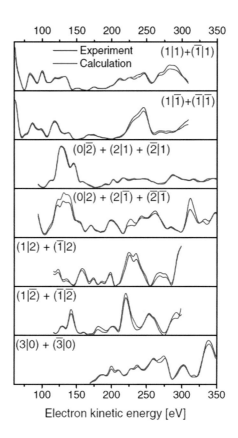

Figure 6.4 Example of an excellent agreement in the LEED I(V) analysis of
$MoO_3/Au(111)$–c(4×2). The final R-factor is $R_P = 0.044$. Reprinted from [6.48] E. Primorac,
H. Kuhlenbeck and H.-J. Freund, *Surf. Sci.*, vol. 649, pp. 90–100, 2016, with permission from
Elsevier.

with

$$A_{ZJ} = 1 \bigg/ \left(0.027 \int I_{ex} dE \right). \tag{6.15}$$

The reduction factor 0.027 is a typical average value that R_{ZJ} would have in the
absence of this factor for unrelated LEED curves. In addition, this factor makes R_{ZJ}
dimensionless. The Zanazzi–Jona R-factor suffers from slightly higher computational
costs and has been found to behave somewhat less predictably than other R-factors in
conventional structural determination, perhaps because of the inclusion of second
derivatives of I(V) curves which enhance sensitivity to experimental noise. This has
been used in earlier structure determinations but was not used in recent LEED I(V)
analyses. We also mention that a series of metric distances have been defined by
J. Philip and J. Rundgren [6.50] which are less sensitive to noise in the data, but they
also require a relatively higher computational effort and are rarely used.

6.1.3   Error Estimates

The LEED I(V) curves can be viewed as a superposition of Lorentzian curves. The peaks are narrow at low energies and become broader at high energies when the curves are drawn as a function of energy, due to the gradually increasing imaginary part of the inner potential V_{0i}, which dominates the peak width. However, when viewed as a function of the wavenumber k, the single peaks have approximately constant width, cf. Figure 6.5.

It is obvious from Figure 6.5 that well separated peaks rarely occur since peaks frequently overlap. The relevant information in the I(V) curves is therefore given by the number of peaks, the peak positions and their heights. Thus, the I(V) curves can be well represented by specifying only the positions, heights and widths of individual Lorentzian peaks. We therefore should count the number of these parameters as the number of independent data points, independent of the number of energy points at which the experimental I(V) curves were measured. Nevertheless, the experimental energy step plays a significant role: a smaller step provides additional data points and thus reduces the experimental noise by allowing better smoothing of the I(V) curves.

The number of independent experimental data points and the number of parameters affect the error bars and the reliability of the analysis. The error bars for the

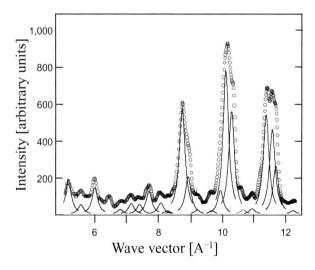

Figure 6.5 Example showing that I(V) curves (with experimental data points shown as circles) are well reproduced by a superposition of Lorentzian peaks (continuous lines) when presented in k-space. The Lorentzian peaks have approximately constant width in k-space. Shown is the $(-1,1)$ beam of $C_{60}/Ag(111)$. Figure drawn with permission from [6.47] K. Pussi, H. I. Li, H. Shin, L. N. Serkovic Loli, A. K. Shukla, J. Ledieu, V. Fournée, L. L. Wang, S. Y. Su, K. E. Marino, M. V. Snyder and R. D. Diehl, *Phys. Rev. B*, vol. 86, p. 205406, 2012; Supplementary Material. https://doi.org/10.1103/PhysRevB.86.205406. © (2012) by the American Physical Society.

model parameters are usually estimated from the expectation values for the R-factors assuming random errors in the data points or from a statistical analysis considering experimental errors and independent parameters. Both ways are used in LEED I(V) analyses.

An estimate of the expectation value of the R-factor for given numbers of data points and parameters has been proposed by J. B. Pendry [6.45]. He assumes that the I(V) curve consists of a superposition of Lorentzian curves of width $w = 2V_{0i}$, where V_{0i} is the optical potential describing the damping of the electron wave in the crystal and w is the HWHM (see Section 6.1.6.5). The separable peaks therefore are spaced on average by an energy interval of $4V_{0i}$ and thus the number of peaks can be estimated to be:

$$N = \frac{\Delta E}{4V_{0i}}, \tag{6.16}$$

where ΔE is the total energy range (the number N includes an estimate of the number of peaks caused by multiple scattering by assuming the maximum possible density of distinct peaks). The statistical error is then

$$RR = \frac{\text{var}R}{\overline{R}} \sim \sqrt{\frac{8V_{0i}}{\Delta E}}, \tag{6.17}$$

where $\text{var}R$ is the variance due to random errors in the experiment and $\overline{R}$ is an average R-factor for uncorrelated data. The meaning of RR, a 'double-reliability factor', is that, for uncorrelated data, random errors will cause the R-factor value to be within the range $\overline{R}(1 \pm RR) = \overline{R} \pm \text{var}R$ with 68% probability. An R-factor value falling outside the range $\overline{R}(1 \pm 1.96RR) = \overline{R} \pm 1.96\,\text{var}R$ has a probability of only 10% of being caused by random fluctuations, for uncorrelated data. A significance factor S can now be defined as

$$S = 1 - \frac{1}{\sqrt{2\pi}} \int_{-T}^{\infty} e^{-\frac{t^2}{2}} dt, \quad \text{where } T = \frac{\overline{R} - R}{RR}. \tag{6.18}$$

The significance factor S is the probability of a random fluctuation of an R-factor to have a value less than R and, by implication, $1 - S$ gives the probability that a structural choice producing the value R is correct [6.45].

It should be noted that V_{0i} is actually energy dependent and becomes relevant in LEED intensity calculations at lower energies, as was first pointed out by J. Rundgren [6.51]. The width of the peaks in the I(V) curves is approximately proportional to the damping parameter V_{0i} while the width of the Lorentzian peaks is approximately constant in k-space. V_{0i} is energy-dependent, smaller at low energies and larger at high energies, as shown in Figure 6.6 for the example of copper. The number of peaks calculated by Eq. (6.16) therefore depends on the energy range. I(V) curves in the low energy range exhibit a higher density of peaks.

The error estimate of Pendry is based on the expectation value for R and neglects the number of parameters and the correlation between parameters. It leads to relatively

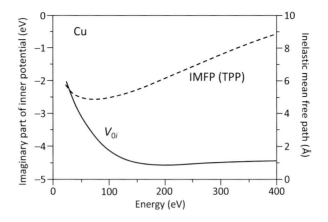

Figure 6.6 Inelastic mean free path (IMFP, dashed line), based on the TPP-2M equation of S. Tanuma, C. J. Powell and D. R. Penn [6.52] and imaginary part of the inner potential (V_{0i}, solid line) for copper. Reprinted with permission from [6.51] J. Rundgren, *Phys. Rev. B*, vol. 59, pp. 5106–5114, 1999. © (1999) by the American Physical Society.

large error bars. On the other hand, the more careful statistical error estimates based on counting statistics as used in X-ray analysis lead to very small error bars because systematic errors and the influence of approximations used in the theory are neglected. This is also the case in LEED, and Pendry's error estimate may compensate for this deficiency so that the relatively large error bars may be quite realistic; they are used in most LEED analyses.

An alternative approach for the error estimate based purely on experimental errors has been proposed by D. L. Adams [6.53]. This follows closely the usual practice in X-ray structure analysis, except that the experimental error cannot be determined from the counting statistics; instead it is derived from the standard deviation determined from repeated measurements and from symmetrically equivalent beams. The standard deviation σ_{hk} is defined for each beam as an average over all energy points:

$$\sigma_{hk} = \frac{1}{n_{hk}} \sum_{i=1}^{n_{hk}} \left[\frac{1}{n_s} \sum_{h'k'} \left(I_{hk,i}^{ex} - I_{h'k',i}^{ex} \right)^2 \right]^{1/2}. \tag{6.19}$$

Here, n_{hk} is the number of energy points in the beam (hk) and n_s is the number of beams $(h'k')$ that are symmetrically equivalent to beam (hk) or the number of independent measurements of the same beam. Adams takes the weighted R-factor comparing experimental and calculated intensities as used in X-ray analysis:

$$R = \frac{\displaystyle\sum_{hk,i} \left(\frac{I_{kh,i}^{ex} - c I_{hk,i}^{th}}{\sigma_{hk}} \right)^2}{\displaystyle\sum_{hk,i} \left(\frac{I_{hk,i}^{ex}}{\sigma_{hk}} \right)^2}. \tag{6.20}$$

The normalisation factor for the calculated intensities is defined as

$$c = \frac{\sum\limits_{hk,i} \left(\frac{I_{hk,i}^{ex} \cdot I_{hk,i}^{th}}{\sigma_{hk}^2} \right)}{\sum\limits_{hk,i} \left(\frac{I_{hk,i}^{th}}{\sigma_{hk}} \right)^2}. \tag{6.21}$$

To obtain the error estimate of a structure parameter, the number of independent data points is needed, which is estimated as the number of Bragg points in the energy range of each beam:

$$N = \sum_{h,k} \frac{\Delta k_{z,hk}(E_2) - \Delta k_{z,hk}(E_1)}{2\pi/d}, \tag{6.22}$$

where d is the interlayer distance in the substrate, while E_1 and E_2 span the energy range of the given beam. The standard deviation of a parameter p_i is then given by:

$$\sigma_i = \left(\frac{R_{min} \, \varepsilon_{ii}}{N - m} \right), \tag{6.23}$$

where m is the number of parameters and ε_{ij} are the matrix elements of the inverse of the error matrix $\boldsymbol{\alpha}$. The error matrix is derived from the curvature of the R-factor near its minimum:

$$\varepsilon_{ij} = \left(\boldsymbol{\alpha}^{-1} \right)_{ij}, \tag{6.24}$$

with

$$\alpha_{ij} = \frac{1}{2} \left(\frac{\partial^2 R}{\partial p_i \partial p_j} \right) \cong \sum_{hk,n} \left(\frac{\partial I_{hk,n}^{th}}{\partial p_i} \right) \cdot \left(\frac{\partial I_{hk,n}^{th}}{\partial p_j} \right). \tag{6.25}$$

The error matrix can be used to check the correlation between parameters. If large off-diagonal terms occur, the error estimates determined from the diagonal terms may be unsafe. The pair correlation coefficients between two parameters i and j can be obtained from the error matrix by

$$P_{ij} = \frac{\varepsilon_{ij}}{\sqrt{\varepsilon_{ii}} \sqrt{\varepsilon_{jj}}}. \tag{6.26}$$

The error estimate proposed by Adams also neglects correlation between parameters. It gives the precision determined from the statistical analysis assuming independent parameters. The true accuracy may be much lower, due to insufficiencies of the multiple scattering theory and further experimental errors.

Another proposal for the error estimates in LEED I(V) analysis was made by C. B. Duke et al. [6.54] based on the χ^2 analysis commonly used in X-ray structure determination. They proposed to use each measured point on the I(V) curve as an independent measurement: this seems inappropriate due to the natural peak width and leads to error bars that are much too small, especially when a small step width in energy has been chosen so that too many points in the I(V) curve are taken as independent whereas they are in reality not independent.

Since Pendry's R-factor R_P is established as standard in LEED I(V) analyses, its value is given in nearly all publications. While the error estimate and the RR-value as defined by Pendry do not consider experimental errors, these could be easily included, as follows. The standard deviation can be defined for each beam as proposed by Adams, Eq. (6.19), and a weighted R-factor based on the Y-function defined by Pendry could be used:

$$
R_w = \frac{\sum\limits_{hk,\,i} \left(\frac{Y^{ex}_{kh,i} - Y^{th}_{hk,i}}{\sigma_{hk}} \right)^2}{\sum\limits_{hk,\,i} \left(\frac{Y^{ex}_{hk,i}}{\sigma_{hk}} \right)^2}. \tag{6.27}
$$

The weights σ_{hk} for each beam, Eq. (6.19), are the standard deviations determined from repeated measurements and from symmetrically equivalent beams. Using this R-factor, the standard deviation for a parameter can be calculated from the error matrix (Eqs. (6.24) and (6.25)). The number of independent points in the I(V) curves is estimated as in Pendry's RR factor from V_{0i} (Eq. (6.16)). The precision determined in a more elaborate statistical analysis frequently underestimates the accuracy of the result but is comparable to the error given in X-ray analyses.

6.1.4   Optimum Energy Range

Clearly, a sufficiently large experimental data set is needed to determine the unknown parameters of a structure. The LEED I(V) measurement is usually made only at normal incidence, because this orientation is easily controlled and all beams having sufficient intensity are measured. Large 2D unit cells imply more parameters to be determined; they also produce a higher density of diffracted beams and therefore a larger amount of measurable intensities that approximately balances the number of parameters. At low temperatures the measurement may reach an upper energy of about 500–600 eV, beyond which intensities become too weak due to thermal effects and defects. At room temperature only a much smaller energy range can be measured.

The data set is not limited to normal incidence and can be extended to measurements at oblique incidence; however, this is rarely done. Off-normal incidence presents a significant disadvantage in the case of symmetrically equivalent domains producing coincident beams, which occur in most structures: LEED calculations must then be performed separately for each equivalent structural domain orientation and then averaged together (possibly with relative weights to be determined), since the incidence directions on different domains will no longer be equivalent. This complicates an automated search procedure considerably.

Redundancy of data, that is, measurement of more data than strictly necessary, is required to perform a reliable structure determination. We may compare the situation in X-ray structure analysis, where in single crystal structure analyses the overdetermination is defined as the number of reflections divided by the number of

Table 6.1 Examples of overdetermination. Each peak is assumed to provide two data points, namely its energy and its height

System	Data set	Overdetermination n
water on $Fe_3O_4(001)$ [6.55]	$\approx$ 110 peaks, 220 data points, 30 free parameters Fit by additional use of restraints (bond lengths)	≈ 7
Cl on Ru(0001) [6.56]	$\approx$ 60 peaks, 120 data points, 10 free parameters	≈ 12
C_{60} on Ag(111) [6.57]	$\approx$ 310 peaks, 620 data points, 15 free parameters	≈ 40
SO_4 on Ag(111) [6.58]	$\approx$ 180 peaks, 360 data points, 107 relevant parameters Substrate relaxation was checked in addition, leading to a total of 163 parameters	≈ 3.5

parameters. An overdetermination of 8 to 10 is recommended for publication in *Acta Crystallographica E*. For powder diffraction the number of peaks follows from the size of the unit cell and the angular range which has been measured. The angular range determines the resolution.

For LEED, the required energy range is not quite so clearly determined, since the number of free parameters depends strongly on the depth down to which substrate relaxations are considered. Depending on the size of the superstructure cell, the number of parameters is much enlarged if relaxations in several substrate layers are included in the analysis. Due to the strong damping of the electron wave the influence of the structural parameters in deeper layers is small and substrate relaxation is better studied by X-ray diffraction than by LEED. With LEED it is useless to refine parameters in layers deeper than about 1 nm, about twice the penetration depth. The data would be overfitted by unrealistic relaxations in the deeper substrate layers which are known to be bulklike. Nevertheless, one should check whether substrate relaxations down to that depth improve the agreement or not without unphysically large relaxations in deeper layers. Detectable substrate relaxations are frequently limited to one or two layers for the case of adsorbates on metals, whereas on oxides and other compounds the situation may be different.

We give some examples of overdetermination for LEED I(V) analyses in Table 6.1, showing that in most cases the number of independent data points is sufficient. The last example, a $(7\sqrt{3}\times\sqrt{3})R30°$ structure of SO_4 on Ag(111), is marginally acceptable due to a large unit cell without symmetry.

6.1.5   Structure Analysis Methods

Nearly all quantitative structure analyses performed by LEED are obtained by the 'trial and error' method. One or more structural models are typically first derived from the possible geometrical arrangements in the surface unit cell, the experimentally known coverage of adsorbate atoms, chemical knowledge of favourable bonding geometries, STM or AFM images, molecular vibration frequencies, prior model calculations and any other information available from other kinds of experiment or theory. These proposed trial models are then evaluated by I(V) calculations and refinement of the parameters, usually producing R-factors that differentiate between the trial models by quantifying their goodness of fit.

Various refinement procedures and search methods are in use in the different LEED computer programs. They can be divided into methods requiring the calculation of a derivative of the intensity with respect to the model parameter and derivative-free methods in connection with R-factors, described in Section 6.1.2. Both have advantages and disadvantages.

All optimisation methods have in common that they can become trapped in local minima and that they are then bound to a model. In that case, the starting model must be sufficiently close to the correct structure. How close must it be? A very rough estimate is that the atoms should be within a few hundredths of a nanometre, that is, a few tenths of an Ångström, from the correct positions. If all the atoms are poorly placed, it is often not possible to start the full simultaneous refinement of all parameters. One then tries to first refine single parameters or blocks of parameters, while keeping other parameters fixed, and to proceed stepwise in this way until in the final step all parameters can be freed and optimised together. For large structures in particular, a block refinement can be used, which however converges more slowly, or constraints can be used to move rigid groups of atoms or molecules in the structure. In any case, one must check that the same minimum is reached when starting with several different trial structures. The radius of convergence, that is, how far from the correct positions one may start an iterative search, is normally larger for simple structures (e.g., determination of the position of a single atom in the unit cell) if the major part of the starting model is correct. In general, many cycles of iteration steps are necessary to locate the true minimum of the R-factor.

We present in Sections 6.1.5.1 and 6.1.5.2 different methods which have been used, starting with local optimisation methods, also called refinement methods, and then global search methods.

6.1.5.1   Local Optimisation Methods

Least-Squares Refinement

In X-ray diffraction the main refinement procedure is the 'least squares' method, which requires the calculation of derivatives of the structure factors with respect to the structural parameters. With X-ray diffraction the calculation of derivatives can be done analytically. In the case of LEED this is not possible or not practical and the calculation must be done numerically [6.59]. Therefore, only minor advantages over derivative-free optimisation methods would remain. It would be helpful if approximations could be used for the calculation of derivatives.

A very efficient optimisation scheme is the so-called expansion method, which is routinely applied in many fields and is also the standard method used in X-ray structure refinement. It is based on the minimisation of a fitting function which is usually the mean square deviation between experimental and theoretical data points. It requires the knowledge of the derivatives of the intensities with respect to the variable parameters. In its original form it works well near the minimum where the gradient is small, because it is based on a linear expansion of the intensities as a function of the variable parameters. Farther away from the minimum, the gradient methods usually work better. Therefore, a combination of both methods has been developed by D. W. Marquardt [6.60] and his approach has in fact become the standard method for optimisation in many fields. Detailed descriptions of the procedure can be found in most textbooks in X-ray crystallography (see for example [6.4]). With LEED this approach has been successfully applied in a number of structure analyses, for example: Ag(111)+(4×4)–O [6.61], Cu(111)+(3√3×3√3)R30°–TMB [6.46], α-Fe$_2$O$_3$(0001), α-Cr$_2$O$_3$(0001) [6.62], Fe$_3$O$_4$(001)+H$_2$O [6.55] and Ru(0001)+ (√3×√3)R30°–Cl [6.56].

We briefly outline the expansion method in the following. The published programs for the expansion method allow the inclusion of a standard deviation for each data point. In the case of LEED, that would be the standard deviation for each beam, σ_{hk}. This quantity is mostly not available or not measured, so that $\sigma_{hk} = 1$ is used corresponding to an unweighted R-factor, R_u. We nevertheless include σ_{hk} in Eq. (6.28) to produce a weighted version, R_w, of the Pendry R-factor, R_P. Minimised is the mean square deviation between Pendry's Y-functions defined in Eq. (6.12), summed over beams and energies, with different beam weights given by σ_{hk}:

$$R_w = \frac{\sum\limits_{hk,i} \left(\frac{Y_{hk,i}^{ex} - Y_{hk,i}^{th}}{\sigma_{hk}} \right)^2}{\sum\limits_{hk,i} \left(\frac{Y_{hk,i}^{ex}}{\sigma_{hk}} \right)^2}. \tag{6.28}$$

The weights σ_{hk} are defined as the standard deviation for each beam,

$$\sigma_{hk} = \frac{1}{n_{hk}} \sum_{i=1}^{n_{hk}} \left[\frac{1}{n_s} \sum_{h'k'} \left(Y_{hk,i}^{ex} - Y_{h'k',i}^{ex} \right)^2 \right]^{1/2}, \tag{6.29}$$

and could be easily determined from the differences between n_s symmetrically equivalent beams which had been measured anyway. In practice this is rarely used and the unweighted R-factor is selected with $\sigma_{hk} = 1$.

A series expansion of the Y-function in terms of a change in parameters $\mathbf{p}_j$ is used, $j = 1, N_p$, where N_p is the number of parameters to be optimised, $\mathbf{p}$ is a vector in the parameter space and $\mathbf{p}_0$ denotes the set of start parameters. We can write:

$$Y_{hk,i}(\mathbf{p} + \Delta\mathbf{p}) = Y_{hk,i}(\mathbf{p}_0) + \sum_{j=1}^{N_p} \frac{\partial Y_{hk,i}(p_j)}{\partial p_j} \Delta p_j + \dots \tag{6.30}$$

The first two terms correspond to the linear expansion and are inserted into the minimum condition

$$\frac{\partial R}{\partial p_j} = 0, \tag{6.31}$$

where R stands for either R_w or $R_u = R_P$. This leads to a set of linear equations that defines the change Δp_j:

$$\sum_{hk,i} \left(\partial Y_{hk,i}^{ex} - \partial Y_{hk,i}^{th} - \sum_{j=1}^{N_p} \frac{\partial Y_{hk,i}^{th}}{\partial p_j} \Delta p_j \right) \frac{\partial Y_{hk,i}^{th}}{\partial p_m} = 0. \tag{6.32}$$

Eq. (6.32) can be written in short form as:

$$\beta_m = \sum_{j=1}^{N_p} \Delta p_j \alpha_{jm}, \tag{6.33}$$

with

$$\alpha_{jm} = \frac{1}{2} \left(\frac{\partial^2 R}{\partial p_j \partial p_m} \right) \simeq \sum_{hk,i} \left(\frac{\partial Y_{hk,i}^{th}}{\partial p_j} \right) \left(\frac{\partial Y_{hk,i}^{th}}{\partial p_m} \right). \tag{6.34}$$

The new parameters $p_j + \Delta p_j$ are obtained by matrix inversion of Eq. (6.33) and the procedure is repeated until convergence is reached. The standard deviation of the i-th parameter is given by

$$\sigma_i = \left(\frac{R_{\min} \varepsilon_{ii}}{N - m} \right), \text{ with } \quad \varepsilon_{ij} = \left(\alpha^{-1} \right)_{ij}, \tag{6.35}$$

where m is the number of parameters and N the number of independent data points. The partial derivatives must be calculated numerically, or else approximately by, for example, tensor LEED. Replacing the coefficients α_{jm} by $\alpha_{jm}^0 = \alpha_{jm}(1 + \delta_{jm}\lambda)$ allows a continuous transition from a gradient-method-like behaviour (at large values of λ) to the expansion method (at small values of λ). The parameter λ is dynamically adjusted. Using the Y-function for comparison of experimental and calculated data as shown in Eq. (6.12), the expansion method is closely related to the linear tensor LEED method [6.63].

The numerical calculation of derivatives of the intensities with respect to the parameters to be optimised is time-consuming and requires approximations since an analytical solution is not practical. Tensor LEED is very efficient but is also an approximation. It shifts the position of an atom but leaves the wave field backscattered from the surrounding atoms unaltered. The optimisation with the expansion method therefore may not find the true minimum; it frequently stops in local minima and may have numerical problems when very small intensity differences occur.

A number of optimisation methods that do not require derivatives have been proposed. We briefly mention in the following some methods which were tried in earlier LEED studies; details of these approaches can be found in the Numerical Recipes by W. H. Press et al. [6.64].

Simplex Method

The simplex method [6.64] is among the simplest optimisation algorithms. It does not require derivatives at all but uses a 'simplex' (set of vertices) of $N + 1$ points in parameter space, if there are N parameters to fit. At each iteration step, the vertex with the highest R-factor is replaced by a new vertex guessed to provide a lower R-factor. The simplex method is very robust, but often slow: in test cases, it tends to be considerably slowed down by long, shallow, twisting valleys in the R-factor hypersurface.

Direction-Set Method

A more effective algorithm is the direction-set method [6.65], which minimises the R-factor along a set of independent directions which are updated as the search proceeds. The minimisation along each direction is done independently, by parabolic interpolation if the function is tested to be parabolic in the region of interest or by simple bracketing if such a test fails. In order to optimise the efficiency of this step, the algorithm updates the directions by trying to converge on a set of so-called conjugate directions, which are such that minimisation along one does not spoil the subsequent minimisation along another. As an example, if we consider minimising a positive, quadratic function, a possible set of conjugate directions (but not the only one) corresponds to the eigenvectors of the matrix A defining the quadratic function, such as the long and short axes of elliptical contours. If these, say N, eigenvectors are identified, the problem is reduced to N one-dimensional independent minimisations, that is, one along each direction.

Rosenbrook Method

The Rosenbrook algorithm [6.66], as well as its modification described in W. H. Press et al. [6.64], is best used when the search has already reached the vicinity of a local minimum. It identifies a set of conjugate directions by repeatedly computing the Hessian matrix of the R-factor (the Hessian is the matrix of partial second derivatives) and by updating the set of conjugate directions to be the set of principal directions of the Hessian. This option returns the position of the minimum together with the principal directions and the corresponding curvatures of the R-factor at the minimum: these are also useful information for evaluating the uncertainties of the structural determination.

Hooke–Jeeves Method

The Hooke–Jeeves approach [6.65; 6.67] is a form of steepest-descent method [6.62]. It explores the local shape of the R-factor hypersurface in the immediate vicinity of a given structure and deduces from it the best direction in which to move in order to reduce the R-factor value. The search proceeds in that direction until the R-factor no longer diminishes. The scheme is then repeated from the new point reached in this manner. This scheme was employed in a LEED structural determination which used full-dynamical calculations for all R-factor evaluations needed in the process [6.68]. It could be considerably accelerated by applying tensor LEED or other methods.

Bounded Optimisation by Quadratic Approximation (BOBYQA)

The optimisation algorithm 'BOBYQA' was developed by M. J. D. Powell [6.69]. It is a derivative-free method to iteratively minimise a cost function (the R-factor in LEED) when upper and lower bounds for the parameters can be defined. The algorithm solves the problem using a trust region method that forms quadratic function models by interpolation. One new point is computed on each iteration, usually by solving a trust region sub-problem subject to the bound constraints or, alternatively, by choosing a point to replace an interpolation point so as to promote good linear independence in the interpolation conditions. The method has been found to locate safely the minimum R-factor region for a large number of parameters if the start model is within ≈ 0.03 nm for all geometrical parameters; an application has been described in R. Wyrwich et al. [6.58].

6.1.5.2   Global Search Methods

Global search methods are optimisation schemes in which the global minimum of the cost function is sought, as opposed to a local minimum. In the case of LEED this cost function is the R-factor hypersurface, which exhibits a high density of local minima and noise [6.70]. Optimisation algorithms like the least squares scheme or simplex method frequently get trapped in a local minimum and usually require a number of trials with different start parameters to reach a minimum which can be considered as global. However, even this is not guaranteed as the parameter space used in local optimisation algorithms is always limited. Global optimisation schemes therefore start with the full parameter space (or in practice at least a reasonably large parameter space) and should find the global minimum of the R-factor hypersurface (within reasonable limits). A global search may go beyond optimising *continuous* variables like atomic coordinates, thermal parameters and site occupancies: a global search may also include optimising so-called *discrete* or *categorical* variables, such as determining the chemical element or even the presence versus absence of atoms (e.g., when their surface coverage is uncertain or periodic vacancies are possible).

Various formulations of global optimisation have been developed and applied with diffraction methods and also in DFT calculations. They require an enormous number of models and therefore have not been applied in LEED to solve unknown structures; however, their applicability has been tested with known structures with a relatively small number of free parameters. With further speeding up of computing times, parallel computation and development of approximations in the multiple scattering calculation, these methods certainly have the potential of more general application. We shall describe three basically different approaches: simulated annealing (SA), genetic algorithms (GAs, also called evolutionary algorithms, EAs) and generalised pattern search (GPS). These methods had been applied in various fields, including some in X-ray diffraction where the large number of structure factor calculations is less of a burden than in LEED.

Simulated Annealing

P. J. Rous [6.70] was the first to apply simulated annealing (SA) to the search problem in LEED. He compared the computational cost with the number of free parameters of the model and found a scaling relation of N^6. The scaling behaviour, that is, the rate of increase of the computational effort with the number of parameters N, is an important factor since the method is hoped to be applicable to systems with a large number of unknown parameters N. This large number of model calculations prohibited the application to structures with more than a few free parameters. Simulated annealing starts with a model with randomly chosen parameters and allows a random increment for each parameter at each step. In the beginning the full parameter space is allowed for each parameter. Using the Metropolis criterion, if the cost function (i.e., the R-factor) has decreased after applying the increments, the move is accepted; an increase of the R-factor is also accepted with a probability determined by a Boltzmann factor controlled by an artificial temperature parameter. The temperature is gradually decreased during the process, analogous to the real annealing of crystals.

In the original SA algorithm used in LEED, a Gaussian function was selected for the probability to accept the step. The convergence of this procedure is very slow. The speed is increased if a Cauchy–Lorentz function is used instead of a Gaussian, which allows occasionally larger steps and a faster temperature decrease than the SA approach: the resulting scaling is an excellent N^1, that is, simple proportionality of computational effort to number of parameters. This method is called fast simulated annealing (FSA) [6.71] and has become the most applied version of the algorithm. The broad wings of the Cauchy–Lorentz function allow a larger step toward a global minimum while the Gaussian function remains in a local minimum, as illustrated in Figure 6.7. Several variants with different statistical distributions are in use [6.72] but will not be discussed here. The number of model calculations can be substantially reduced when fast simulated annealing (FSA) is used, as was shown later by V. B. Nascimento et al. [6.73].

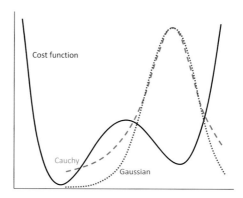

Figure 6.7 Comparison of a Cauchy probability density and a Gaussian probability density over an example cost function. The longer jumps enabled by the Cauchy probability density allow faster escape from local minima. Redrawn from [6.71] H. Szu and R. Hartley, *Phys. Lett. A*, vol. 122, pp. 157–162, 1987, with permission from Elsevier.

FSA and variants with further improvements are the most commonly used approaches in various fields. The effectiveness of the method in LEED, testing different statistics and generalised distribution functions, has been published by E. dos R. Correia et al. [6.74] and has been called 'generalised simulated annealing' (GSA). They found in a theory-to-theory comparison for 2–10 parameters a linear scaling behaviour of N^1 for the example of CdTe(110), as with FSA. The number of required LEED calculations is still enormous in simulated annealing algorithms so that this method has not been applied to solve unknown surface structures.

An approach that is related to simulated annealing was proposed by M. Kottcke and K. Heinz [6.75]. They start from random models and allow random moves of the atoms, but instead of using the Metropolis criterion of SA, they only accept new models with improved R-factors. To avoid getting trapped in local minima, they allow more distant moves with an R-factor-dependent Gaussian step distribution that is wider when the most recent R-factor is higher. Based on tensor LEED, the method can explore structures deviating from a certain preselected reference by atomic shifts, chemical occupation of lattice sites and both isotropic and anisotropic atomic vibrations. This automated optimisation scales as $N^{2.5}$, providing a compromise between global and local search methods. The method is more successful at locating the global minimum than other refinement algorithms but still needs several start structures and many iterations.

Genetic or Evolutionary Algorithms

A sophisticated class of algorithms to locate the global minimum of a cost function has been developed in other fields and is known as genetic or evolutionary algorithms (GAs or EAs). These are inspired by the evolution of biological systems which can be understood as an optimisation process. GAs have been widely applied in very different fields and technological problems.

GAs are based on computing the costs (i.e., R-factors in LEED) of a set of trial structures that form a 'population', in the terminology of this method. Each trial structure is characterised by its 'gene' or 'chromosome', that is, a list of its structural and non-structural parameters that are to be optimised; they are lined up as a string of digits, analogous to biological DNA. A GA proceeds by evolving the population of trial structures, which retains the same size as the starting population, from one 'generation' to the next. This evolution is biased toward increasing the quality of the members of the population and proceeds until the population contains a trial structure of sufficient quality (low enough R-factor). The initial generation of trial structures must be specified by the user, for example, it could be a set of structures deemed to be realistic, but it can also be a set of random structures: the global character of the search should find the correct structure in either case.

Creating a new generation of the population is performed concurrently by various schemes, primarily: mutation, that is, random change of one or more structural or non-structural parameters (or of digits within those parameters); 'crossover', that is,

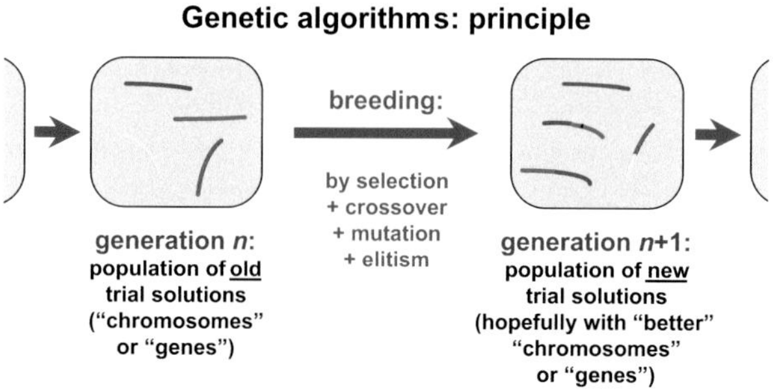

Genetic algorithms: principle

generation *n*:
population of <u>old</u>
trial solutions
("chromosomes"
or "genes")

breeding:

by selection
+ crossover
+ mutation
+ elitism

generation *n*+1:
population of <u>new</u>
trial solutions
(hopefully with "better"
"chromosomes"
or "genes")

Figure 6.8 Schematic representation of the principle of typical genetic or evolutionary algorithms. A population is composed here of four trial solutions shown in different colours, where each coloured string contains all data defining a trial solution, like biological DNA. The population evolves by breeding from generation *n* to generation *n* + 1. Breeding occurs by first selecting some of the best trial solutions from generation *n*, then pairing these to generate new trial solutions for the next generation *n* + 1, by crossover (e.g., the new red/blue and blue/red mixed solutions created from the old red and blue solutions, as exemplified in Figure 6.9) or by mutation (e.g., by randomly changing a gene at the black spot in the new red/blue solution). Elitism can also be applied to allow one or more of the best old trial solutions to survive intact into the new generation (e.g., the red solution here).

intermixing structural features from two parent structures; and elitism, that is, survival of the best structures into the next generation. The breeding process, including crossover, is schematically illustrated in Figure 6.8. Crossover combines pairs of the 'best' trial structures in the population (i.e., structures having relatively better R-factors in LEED) to generate new trial structures: namely, the genes of a pair of so-called parents are mixed (essentially by cut and paste) to give new structures called 'offspring' or 'children'. Figure 6.9 displays a typical flowchart of a GA in LEED, as well as one example of how crossover can be implemented.

A number of variants of GAs exist which differ in the strategy of selecting the offspring by mutation, crossover, etc. Evolutionary processes usually work faster than simulated annealing, as random steps are not used but the experience from previous calculations is included more constructively: for example, while simulated annealing mainly rejects 'bad' solutions, GAs incorporate details of 'better' solutions into new trial structures. The number of LEED calculations required until a minimum of the R-factor is reached remains a limiting factor with GAs, but the method is ideal for parallel computation and, for this reason, it is also more promising than simulated annealing. Genetic algorithms are robust, not very sensitive to noise; they also work well with highly correlated parameters which may trap other optimisation schemes in local minima.

Genetic algorithms were first applied to LEED by R. Döll and M. A. Van Hove [6.77]. Later the method was also applied to photoelectron diffraction [6.78]. Its scaling behaviour has been tested by M. L. Viana et al. [6.79] for the model systems

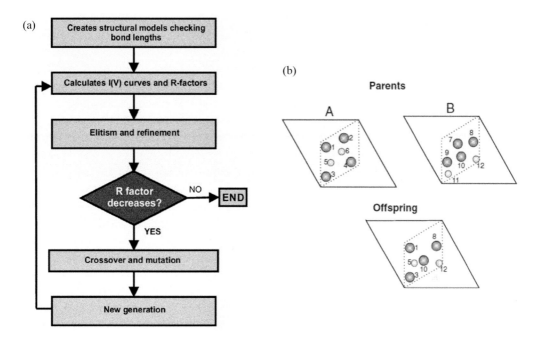

Figure 6.9 (a) Schematic flowchart of a GA algorithm for LEED. (b) Schematic representation of one type of crossover operation within a 2-D surface unit cell from parent structures A and B to a new Offspring structure (exemplifying the crossover shown in Figure 6.8): atoms to the left (at positions 1, 3, 5) inherit their coordinates from parent A, while atoms to the right (at positions 8, 10, 12) inherit their coordinates from parent B. © IOP Publishing. All rights reserved. Reproduced with permission from [6.76] M. L. Viana, D. D. dos Reis, E. A. Soares, M. A. Van Hove, W. Moritz and V. E. de Carvalho, *J. Phys.: Condens. Matter*, vol. 26, p. 225005, 2014; permission conveyed through Copyright Clearance Center, Inc.

Ni(111)+($\sqrt{3}\times\sqrt{3}$)R30°–Sn, InSb(110) and CdTe(110) using 5 to 16 structural and non-structural parameters. For test cases with a relatively small number of parameters a scaling proportional to $N^{1.6}$ was found in R. Döll and M. A. Van Hove [6.77] and $N^{1.3}$ to $N^{1.7}$ in M. L. Viana et al. [6.79], depending on the difficulty to locate a minimum of the R-factor. Further reduction of the computational effort can be achieved by including symmetry and bond length restrictions in the procedure to generate offspring. Models are then immediately rejected before being subjected to a costly LEED calculation if they do not fulfil minimum bond length requirements or given symmetry conditions. The method is called 'novel genetic algorithm for LEED' (NGA-LEED) [6.76] and also differs from the previous application of genetic algorithms by a local refinement step applied to certain intermediate structures to accelerate the search. The atomic positions are located on grid points because the location of a global minimum can then be performed faster with a smaller number of possible structure models and the refinement of parameters can be performed in a final step within the near vicinity of the determined minimum. As a result, the computational effort could be drastically reduced. The method was tested for the previously solved structure of Ag(100)+(4×4)–O and Au(110)–(1×2) using experimental data.

Other variants of evolutionary processes are also very promising. These include differential evolution (DE) [6.80] and covariance matrix adaptation evolutionary strategy (CMA-ES) [6.81]. Differential evolution [6.80] is a rather simple but effective population-based search scheme. It starts with a population of P individuals. Each individual is characterised by a vector consisting of its N structural parameters. The value of P is recommended to be about 5 to 10 times the number of free parameters. The start population is randomly chosen within the full parameter space; no start model needs to be defined. The next generations are produced by mutation, crossover, etc., as in GAs, but the algorithm is quite different to that used in GAs. In the mutation step a weighted difference between two arbitrarily chosen individuals of the population is added to a third arbitrarily chosen individual, thus defining a new individual. In the next step, called crossover, a new population is formed by mixing the components of the new individual with another vector, the target vector. The target vector is initially randomly chosen. The LEED calculation is now performed for the new population and the selection process is the following: if the R-factor for an individual has decreased, this individual then becomes the new target vector, otherwise the old target vector remains. The process is iterated until no further decrease of the R-factor is obtained. Two factors influence this process: one is the factor occurring in the mutation step which defines the weighted difference between two arbitrarily chosen individuals; the other factor occurs in the crossover step and determines the mixing with the target vector. The performance of the optimisation process is determined by the choice of these two parameters. The best choice must be checked by a LEED calculation and depends on the number of parameters and the surface structure. The performance of DE has been investigated by V. B. Nascimento and E. W. Plummer [6.82] for the case of $BaTiO_3(100)$, where they found a scaling behaviour of $N^{1.47}$ for up to 12 free structural parameters. These results show that DE is one of the most promising methods for global optimisation in LEED.

The other variant, the covariance matrix adaptation evolutionary strategy (CMA-ES), is also a population-based method. However, in contrast to GAs, the crossover step is omitted and new individuals are generated solely by mutation. The R-factor hypersurface in the search space is determined in each generation from the R-factors of all individuals in the population. New individuals for the next generation are derived from the multi-variate normal distribution which is given by the covariance matrix. The normal distribution is adapted to the R-factor hypersurface. This means that the choice of new parameters for the individuals of the new population is determined by the shape of the R-factor hypersurface. The speed with which a minimum is reached can be expected to be higher than with GAs, but the process may be trapped in a local minimum of the R-factor hypersurface. The parameters of the CMA-ES algorithm are the numbers of parents l and offspring k, with $k > l$, as well as the initial standard deviations of the algorithm's normal distributions, which also define the distribution of individuals in the initial population.

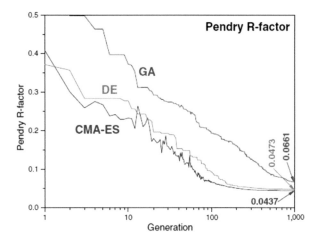

Figure 6.10 Pendry R-factor of the fittest individual as a function of the number of generations for covariance matrix adaptation evolutionary strategy (CMA-ES), differential evolution (DE) and a genetic algorithm (GA). Reprinted from [6.48] E. Primorac, H. Kuhlenbeck and H.-J. Freund, *Surf. Sci.*, vol. 649, pp. 90–100, 2016, with permission from Elsevier.

Different global search methods have been tested and compared by E. Primorac et al. [6.48] for MoO_3 monolayers on Au(111). The results were also compared to a tensor LEED analysis. Three different population-based search methods were employed: CMA-ES, DE and a real-valued GA. The GA is called real-valued because the genome consists of real numbers in the present case. The I(V) curve computations for the individuals of the population were performed in parallel on a multi-processor system which significantly reduced the computation time (but not the overall computational effort). A reduction of the computational effort was achieved by requiring fulfilment of symmetry and of minimum bond length constraints for the models. A total of 28 parameters were optimised, with 24 structural parameters which were refined by the evolutionary processes. The results are shown in Figure 6.10.

The parameters determining the performance of the different algorithms, the convergence radius and the computational effort have been carefully tested. The CMA-ES and DE algorithms turned out to be faster than the GA method and led to a lower R-factor after a certain number of generations. The evolutionary processes DE and CMA-ES seem to be better suited in the last stage of the refinement where very small shifts of the parameters occur. In the CMA-ES method this is probably due to taking into account the local geometry of the R-factor hypersurface. This is not the case in GA, while in DE it is probably indirectly included by the choice of the new target vector in each generation. In the final stage of the refinement, the GA method therefore works more slowly. For CMA-ES several runs with different start models were performed; one run led to the lowest R-factor, showing that the global minimum was reached with an excellent agreement. For comparison, the tensor LEED method leads much faster to a local minimum but, being an approximation, typically yields a larger R-factor. It is therefore useful to combine a relatively coarse global search to identify the probable global minimum region (using, e.g., genetic algorithms) with a fast refinement step to optimise the details of the global minimum itself (using, e.g., tensor LEED); in some cases, the coarse search may suggest several candidates for the global minimum: these can then be distinguished by refinement.

Discrete Variables and Generalised Pattern Search

In optimisation problems like LEED, it may be necessary to optimise not only continuous variables, but also discrete variables: these are often called 'categorical' variables, as they may take only two or a few distinct values, for example, the chemical identities of atoms in a compound system or their presence versus absence. For instance, an adsorbate atom may or may not substitute for a substrate atom which is expelled, so both options may need to be tested. Also, the coverage of an adsorbate may be uncertain, which may be viewed as presence versus absence of such atoms at specific locations, while vacancies may also occur in a substrate. As a further example, a metallic surface alloy may distribute its component atoms differently from the bulk arrangement. Discrete variables preclude differentiation and thus optimisation by steepest descent or similar methods.

Many of the global search methods described in the preceding paragraphs can be generalised to include optimisation of discrete variables, if they do not rely on derivatives of the cost function (R-factor). We here discuss a method that is particularly suited to handle discrete variables at the same time as continuous variables (such as atomic coordinates): the generalised pattern search (GPS) method in its 'mixed variables' version [6.83; 6.84], which is also well adapted for imposing restrictions such as symmetry, bond length and coordination constraints; it is a member of a wider class of generating set search (GSS) methods [6.85]. The basic GPS strategy for continuous variables is simple to explain using a 2-D example, with four compass directions (north, east, south and west): starting from one assumed trial structure, for each variable, the procedure steps by a pre-set amount Δ_0 in the four compass directions until the cost function is reduced and moves the trial structure to that better position; otherwise the step length Δ_0 is halved; these steps are repeated until convergence. For discrete variables, some of their properties must be provided by the user: these properties can vary significantly from one type of application to the next, so the user must select an appropriate method within the code [6.86].

The application of GPS to LEED was tested for a complex ordered surface alloy structure, Ni(100)+(5×5)–Li [6.83]. This structure results from adsorption of Li on Ni (100). The surface structure is illustrated in Figure 6.11, where panels (a–d) show the best fit among the 45 proposed models shown at left, based on prior separate local optimisations by tensor LEED [6.87]. The 45 proposed models differ in the occupation of bulk-like Ni(100) lattice sites with either Ni atoms or Li atoms or neither (i.e., vacancies). These models illustrate the situation where there is uncertainty about the adsorption coverage (density of adsorbate atoms), the penetration of adsorbates into the substrate, the relocation of substrate atoms (substrate reconstruction), the presence of vacancies and even the number of layers affected. Taking as an example only the

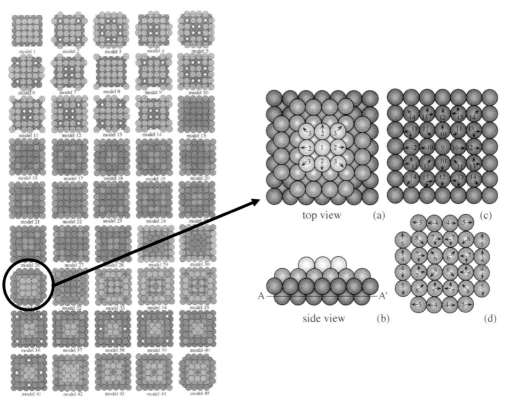

Ni(001)-Li-(5x5) structure models

Figure 6.11 The Ni(100)+(5×5)–Li surface formed by adding Li atoms on a Ni surface. The 45 small panels on the left are structure models built from physical intuition. In these models, the Ni atoms (green) and Li atoms (orange and yellow) are distributed in various arrangements in 1, 2 or 3 layers, with varying numbers of Ni and Li in the (5×5) unit cell. Panels (a) and (b) detail structure model 31, while panels (c) and (d) represent its second and third outermost layers (the first outermost layer is shown in yellow in (a) and (b)): this model has the lowest R-factor among these 45 models, based on prior separate tensor LEED optimisations. Adapted from [6.87] H. Jiang, S. Mizuno and H. Tochihara, *Surf. Sci.*, vol. 380, pp. L506–L512, 1997, with permission from Elsevier.

previously best-fit structure (model 31 highlighted in Figure 6.11(a–d)) and assuming the full p4m point group symmetry of the Ni(100) substrate, there are 14 inequivalent atoms in the cell in the outermost 3 layers (numbered in the figure). In this case, we thus need to optimise 28 continuous structural variables ($3 \times 14 = 42$, but symmetry reduces this number) and 14 discrete variables (each of these atoms can be either Ni or Li). This count ignores thermal parameters as well as the inevitable adjustable inner potential, V_0.

The performance of the GPS method was first compared with that of GAs starting with the model 31 presented in the last paragraph, without optimising the chemical elements (both methods utilise tensor LEED to determine R-factors). A feature of GAs is that their breeding procedures normally create numerous trial structures that are physically totally unrealistic unless suitable constraints are imposed, such as limits on acceptable bond lengths. Therefore, only small atomic relaxations (± 0.04 nm, i.e., ± 0.4 Å) from the best known solution were allowed in the search with GA. GPS produces far fewer such invalid structures and found an improved solution ($R_P = 0.12$) at around 600 function evaluations, while the GA found the model 31 ($R_P = 0.24$) only at around 2,400 function evaluations. Next, atomic positions were fixed, while chemical elements could change. Using the notation 11111222222222 (1 = Li, 2 = Ni) for the 14 chemical identities of the 14 numbered atoms in model 31 (see Figure 6.11), a search was started with the choice 11111122211122 (i.e., some Ni atoms were substituted by Li atoms). GA now needed 280 function evaluations to obtain model 31, while GPS needed only 49 function evaluations. (GPS actually found an improved solution at 135 function evaluations, again with $R_P \sim 0.12$: these solutions with $R_P \sim 0.12$ are actually invalid, as they were found to have an unphysical inner potential determined by the tensor LEED code and not by the GPS procedure; it would be easy to constrain the inner potential within GPS.) Using 20 random initial chemical guesses, GPS reached the same minimum $R_P \sim 0.12$, with an average of 152 function evaluations. Optimising simultaneously the continuous and discrete variables led GPS to a slightly improved $R_P = 0.23$ compared with the prior best fit of model 31 at $R_P = 0.24$, with the largest difference of 0.013 nm = 0.13 Å from the previous best atomic positions [6.87].

6.1.6 Influence of Non-structural Parameters on the Structure Determined by LEED

Several non-structural parameters enter the LEED I(V) analysis and are discussed in this section:

- The scattering potential model. The spherically symmetric muffin-tin (MT) model is most commonly used together with phase shifts.
- The inner potential $V_0(E)$ and its energy dependence. The parameter V_0, often called muffin-tin zero (V_{MTZ}) or muffin-tin constant, is normally taken to be isotropic (independent of the direction of propagation) and often independent of energy; it may vary from one atomic layer to another, especially for overlayers.
- The inelastic mean free path (IMFP) or damping constant is normally given by the imaginary part of the inner potential $V_{0i}(E)$, also called optical potential, and its energy dependence. The damping of the electron wave inside the crystal is usually approximated by the single isotropic parameter V_{0i}, which is sometimes taken as energy dependent. It leads inside the crystal to a complex energy $E + V_0 + iV_{0i}$ and complex wave vectors $\mathbf{k}$ (see also Section 5.1).

The main other non-structural parameters in LEED calculations are the atomic displacement parameters (ADP) which describe displacements from the average atomic positions due to thermal motion and static defects. Their influence on the LEED intensities cannot be neglected and is routinely included in most LEED I(V) analyses by simple isotropic displacements in a Debye model. The theory is described in Section 5.5 for more general cases, in particular for anisotropic displacements.

We investigate here the influence of the potential model and the inner potential. These parameters are usually considered to be less important for the structure analysis, so that frequently average values are used. This is certainly justified when only the structure model must be identified, where an average accuracy of 0.005–0.01 nm (0.05–0.1 Å) is sufficient.

In many cases, however, a higher structural accuracy would be desirable. For example, an important part of quantitative LEED investigations is undertaken for studying catalytic reactions on surfaces. The change of molecular configurations and bond lengths in the adsorbed state compared to the free molecule is of great importance for understanding the reaction. An accuracy of bond lengths in the order of 0.0001 to 0.001 nm (0.001–0.01 Å) would then be required. Such accuracy can be obtained in 3-D X-ray analysis but is not standard in surface studies, including in surface studies by X-ray diffraction.

There is room for improvements in the calculation of LEED intensities in the areas of phase shift calculation and, as mentioned in Sections 5.5 and 6.1.5.2, in anisotropic atomic displacements. Improved calculation of phase shifts is discussed in this section. Their influence is particularly important in the low energy range. Although the muffin-tin model is usually considered to be sufficient in LEED, it may actually be the largest source of theoretical inaccuracy. In DFT calculations, more sophisticated full potential models are required in order to achieve better approximations for the Coulomb potential and the exchange-correlation terms than is possible with a simple muffin-tin approximation, since DFT deals with electrons at lower energies than LEED; see for example a review by K. Schwarz [6.88]. In LEED, higher energies in the range of about 20–500 eV are used, where the LEED electrons are scattered mainly by the Coulomb potential of the atomic cores. The core potential is screened by the bound electrons and therefore depends also on the surrounding atoms. The different approximations used to describe the crystal potential therefore influence the LEED I(V) curves and the accuracy of the structure analysis. It is often observed that excellent agreement between experimental and calculated LEED intensities and small error bars can be obtained for metals with close packed structures. However, for covalent structures and compounds like oxides the R-factors and the error bars are usually larger. This indicates that the muffin-tin model and the phase shift calculation are less appropriate for these structures and leave room for improvement. We first briefly review the development of the construction of the crystal potential and the phase shift calculation.

A note on terminology is needed here. In the relevant literature, the term inner potential is often used interchangeably with muffin-tin zero and muffin-tin constant; these terms are also used in Section 5.2, where the muffin-tin model is illustrated in Figure 5.3. These terms then indicate a spatially constant potential between atomic muffin-tin potentials, although different constant values may apply in different layers, as shown in Figure 5.3. In many publications, however, inner potential refers to the average potential of an electron in a crystal, including the average over the atomic cores (this inner potential can be measured, for example, as an energy shift of Bragg peaks in LEED, although this only gives a very approximate and energy-independent value). In this section, the inner potential $V_0(E)$ will instead designate the spatially constant interstitial potential, which is a theoretical construct that cannot be directly measured; it is referenced to the vacuum level that exists infinitely far from the surface. In Sections 6.1.6.1 and 6.1.6.2, the inner potential V_0 and the muffin-tin zero V_{MTZ} are equivalent, that is, $V_0 = V_{MTZ}$; they may depend on energy, but most LEED calculations have kept them independent of energy. However, in Section 6.1.6.3, while the muffin-tin zero V_{MTZ} will be independent of energy and identical to the vacuum level, the inner potential $V_0(E) = V_{XC}(E)$ will consist of the energy dependent exchange-correlation energy; $V_{XC}(E)$ vanishes at very large energies, where therefore $V_0(\infty) = V_{MTZ}$. In addition, in all cases in LEED, the electron energies are shifted rigidly (by the same constant amount for all potentials or I(V) curves) to best fit the experiment, because the potential models do not properly include other surface-specific effects such as the work function change (due to surface dipoles resulting from spill-over of bound electrons into the vacuum or charge transfer between adsorbate and substrate), which gives rise to the image potential shown in Figure 5.3.

6.1.6.1   The Conventional Muffin-Tin Model

In the multiple scattering theory of LEED, the muffin-tin (MT) model is commonly used to describe the scattering properties of the atoms. The model uses a spherically symmetric potential for each atom and the scattering is described in spherical waves. The phase shifts of the spherical waves are calculated and then enter the multiple scattering theory as parameters, similar to atomic form factors in X-ray diffraction. A short description is given in Section 5.2. The elastic electron–atom scattering theory and scattering phase shift codes used for LEED emanate from the survey 'Augmented Plane Wave Method' by T. L. Loucks [6.89]. From the electron density of the atomic orbitals, the radial spherically symmetric atom potential $V(r)$ of the free atom is calculated for $0 \leq r \leq \infty$, where $V(\infty)$ is taken as zero. To obtain a spherically symmetric potential for one atom after superposition of the potentials from the surrounding atoms, the procedure proposed by L. F. Mattheiss [6.90] is used. In this method the potentials of the neighbouring atoms are expanded in spherical harmonics with the central atom as origin. Only the $l = 0$ term is taken to obtain spherical symmetry. A complete crystal potential comprises MT potentials where nearest neighbour MT spheres approximately touch one another and MT potentials are approximately continuous with a constant muffin-tin zero between the MT spheres, as illustrated in Figure 5.3. In close-packed elemental solids, compounds and adsorbates the crystal potential may be discontinuous, as steps can occur between the different MT spheres and the constant inner potential in the interstitial region; however, such steps are unrealistic and should be minimised, as discussed further in Sections 6.1.6.2 and 6.1.6.3.

The phase shift calculation via the MT potential usually proceeds in three steps:

- Calculation of the electron charge density of the free atom.
- Calculation of the free-atom potential by radial integration of the Poisson equation and construction of the muffin-tin potential by taking the spherically symmetric part of the superposition from the surrounding free atom potentials.

- Calculation of the phase shifts by integration of the non-relativistic Schrödinger equation or the relativistic Dirac equation with the spherically symmetric potential.

In the first step, the electron density of the free atom is calculated. Either the non-relativistic or the relativistic Hartree–Fock equations are solved for the self-consistent atomic orbitals (relativistic effects become important in heavier elements). Some computer programs require input of the atomic orbital structure; other programs require only the nuclear charge to identify the atom: then a table of the orbital structure of the ground state of all atoms is included in the program.

In the second step, using the electron density of the atomic orbitals, the radial atom potential $V(r)$ of the free atom is calculated by integrating the Poisson equation. The part of the electron charge that is absent from the muffin-tin spheres is distributed in the interstitial region and assumed to be spatially constant. Since the early days of LEED, the inner potential has usually been assumed to be independent of energy and of propagation direction. The value of the inner potential is usually determined by fitting to the experimental curves.

In the third step, the Schrödinger equation or the Dirac equation is radially integrated up to a MT radius. For the earliest LEED calculations, the MUFPOT code developed by J. B. Pendry et al. [6.91] was often used; similar codes were developed by other groups. More recently the phase shift code package provided by A. Barbieri and M. A. Van Hove [6.92] has been frequently used to generate phase shifts.

Although the procedure described here has been applied successfully in many structure analyses by LEED, there remain some problems in the construction of the crystal potential. The principal problem is that for structures with *covalent* bonds the MT potential is less appropriate, since for these the charge distributions within and between atoms deviate relatively more from spherical symmetry or constant interstitial values: properly, this would require a 'full potential' LEED calculation (with in particular non-spherically symmetric atomic potentials and a non-constant inner potential, as described in Section 5.2.1 and references [5.6–5.9]). Such calculations, however, would be computationally complex and onerous, especially during a structural optimisation process that would require updating the potential. Nevertheless, the MT model appears to be capable of producing a positional accuracy of at least 0.01 nm (0.1 Å), particularly if low electron energies are avoided; the most recent step-free MT model described in Section 6.1.6.3 delivers a positional accuracy of about 0.002 nm (0.02 Å).

Furthermore, within the MT model, the proper choice of the MT radii has also been problematic in many cases and can be improved, except for elemental metals with close packed structures where the midpoints between nearest neighbours define the MT radius (ignoring the usually small surface relaxations of atomic positions). For non-metallic compounds like oxides, for compounds with covalent bonds and for molecular structures, the choice of half the nearest-neighbour distance for the definition of the MT radius leads to a poor description of the electron density in the interstitial regions and in the outer parts of the MT spheres. The standard prescription for the construction of the MT potential generates artificial steps at the MT radius between the potential within the spheres and the MT zero. A continuous potential would be more realistic.

A frequent question is how to treat ionic compounds. One aspect is whether published ionic radii are appropriate in the MT model. Such ionic radii are used qualitatively to classify structures of ionic compounds like oxides and halides but were not designed to describe electron–ion scattering and thus cannot be assumed to be appropriate, and indeed are questionable, as reported in the following paragraphs. However, ionic radii are useful to estimate interatomic distances in compounds where the structure is not yet known. Another aspect is the long-range electrostatic electron–ion interaction in an ionic lattice. To estimate this interaction, a Madelung sum over the ionic lattice is used, analogous to the Madelung sum used to obtain the lattice energy in ionic solids.

In X-ray crystallography, atomic form factors are used for neutral atoms and for ions, justified by the fact that the scattering factors for X-rays are derived from the number of electrons assigned to a neutral atom or an ion. However, for electron diffraction the situation is less clear. Electrons are mainly scattered by the core potential which is screened by the valence electron density of the atoms or ions, and ionic radii do not seem appropriate to define MT radii. We here briefly describe the LEED results obtained in the past for ionic structures to illustrate the more complex situation. In an early LEED study of the MgO(100) surface, C. G. Kinniburgh [6.93] investigated several potential models and MT radii, with the result that the MT radius had the largest effect on the LEED intensities. The comparison of I(V) curves showed a best fit with equal radii for both Mg^{2+} and O^{2-} ions, while ionic and neutral potentials gave similar agreement with experiment. A subsequent study of the NiO(100) surface by C. G. Kinniburgh and J. A. Walker [6.94] came to a similar conclusion, while finding that an ionic potential gave slightly better agreement with experiment than a neutral potential. These studies were based on a qualitative comparison of experimental and calculated I(V) curves rather than on R-factors. In later LEED studies of adsorbate layers on metal substrates, where I(V) curves were quantitatively compared by an R-factor, MT radii of the adsorbates were mostly derived from neutral atoms. In subsequent LEED studies of metal oxide surfaces and other ionic compounds, the best approach to the phase shift calculation remained an open question.

In a study of TiO$_2$(110) surfaces [6.95], the charge transfer between metal and oxygen ions was approximated by using the ground state electron density of bulk TiO$_2$ obtained from DFT calculations. The MT radii derived from the electron density were approximately equal and about 0.1 nm (1 Å) for both ions; these corresponded to neither atomic nor ionic radii. The LEED I(V) analysis with phase shifts calculated using these MT radii led to a significant improvement of the agreement of the I(V) curves with an R-factor $R_P = 0.29$. The best R-factor obtained with phase shifts calculated from neutral atoms and an altered occupation of atomic states to simulate ionic potentials was about $R_P = 0.6$. The structural results, however, were very similar, with the largest difference of about 0.01 nm (0.1 Å) in the coordinates of an oxygen atom. However, a LEED I(V) analysis of another TiO$_2$ modification, brookite,

$TiO_2(011)$–(1×2), did not find such a large improvement with phase shifts derived from DFT calculations [6.96]. Only small differences in R-factors and structural parameters were found between the results obtained from neutral atoms with symmetrisation by the Mattheiss procedure, compared with using phase shifts derived from self-consistently calculated electron densities.

The presence of potential steps between muffin-tin spheres and inner potential is conceptually not satisfactory; nonetheless, they were present in nearly all LEED I(V) calculations performed so far. It is often suggested that such steps would lead to resonances in the electron wave functions, since electrons could backscatter multiple times between a nucleus and such a nearby potential step. However, this is a misconception, since these potential steps are actually not included as scatterers in the usual LEED theory: only the scattering factors via the phase shifts enter the multiple scattering theory. The electron wave functions are not calculated by solution of the Schrödinger equation for the full crystal potential as required in DFT calculations. Nevertheless, the steps perturb the integration of the Dirac or Schrödinger equation in the calculation of the phase shifts, and the phase shifts export perturbations into the multiple scattering calculation.

The potential steps between vacuum and atomic layers, as well as between inequivalent atomic layers, are normally ignored in the theory, except for their refraction effect, so that resonances in the scattering between atomic layers and potential steps are excluded. This neglect is justified by the fact that a smooth step causes much less reflection than an abrupt step. The potential steps in the MT model have some influence on the LEED I(V) curves because the energy of the electrons is different on both sides of a step, causing changes in kinetic energy and thus also changes in wavelength and propagation direction.

6.1.6.2   An Improved Muffin-Tin Model

To overcome the problem of steps in the MT potential model, an important improvement was introduced by J. Rundgren [6.97; 6.98]: he used MT spheres with preassigned core level shifts and optimised MT radii in order to minimise steps to the interstitial inner potential at the MT radii. Mattheiss' superposition potential was combined with the Hedin–Lundqvist local density functional exchange-correlation potential $V_{XC}(E)$ [6.99; 6.100] to form a total potential for the elastic electron–atom scattering. The energy of the scattered electron inside the solid is $E - V_0(E)$. The MT radii are optimised by the differential evolution (DE) method and then become energy dependent through the energy dependence of the exchange-correlation potential. The potential step between the MT spheres and the MT zero is minimised by coordinated shifts of the MT potentials and optimisation of MT radii. Optimisation leads to steps of the order of 0.1 eV between MT spheres and MT zero. Unfortunately, the resulting MT radii and inner potential depend on incident energy, while in principle the MT radii should be independent of the energy of the incident electron.

V. B. Nascimento et al. [6.101] were the first to apply optimised MT radii in a LEED I(V) analysis, for the $Ca_{1.5}Sr_{0.5}RuO_4(001)$ surface. They applied a number of different methods to calculate the potentials and the MT radii, including charge densities obtained from DFT calculations and also charge densities obtained from a relativistic Hartree–Fock calculation for neutral atoms. Interestingly, they observed undulations in the phase shifts as a function of energy due to energy dependent steps in the MT potential when calculated according to the Mattheiss procedure, while smooth phase shift curves were produced with optimised MT radii. They also applied an energy dependent imaginary part of the inner potential, as proposed by J. Rundgren [6.51]. Numerical data for the inelastic mean free path (IMFP) are available from the National Institute of Standards and Technology (NIST) [6.102]. The relation between the IMFP and the optical potential is described in Section 6.1.6.3 and further references can be found there. The results showed a significant improvement of the R-factor compared to previously used phase shifts [6.101]. The structural results agreed in this case within error estimates with previous results and reduced those error estimates significantly. Phase shifts obtained from optimised MT radii were also used in a number of further LEED I(V) analyses, of which some are mentioned here without intending to give a complete list, for example the studies of: epitaxial thin films of $LaNiO_3(001)$ [6.103] and adsorbate structures like $Cu(111)+(3\sqrt{3}\times3\sqrt{3})$ $R30°$–TMB [6.46], $Ag(111)+(4\times4)$–O [6.61] and $Ag(111)+(7\times\sqrt{3})rect$–$SO_4$ [6.58].

In the method described in this section, non-overlapping MT spheres were used to minimise the potential steps between the MT spheres. However, there remains an energy dependence of the MT radii and the interstitial potential. This is still an inadequate description of the potential, as the potential without the exchange-correlation term should in principle be independent of the energy of the incident electron. The MT radii should likewise be independent of energy. These issues are addressed in Section 6.1.6.3.

6.1.6.3 A New Step-Free Overlapping Muffin-Tin Potential

A new approach to obtain a step-free potential and energy independent MT radii was recently developed by J. Rundgren et al. [6.104]. A step-free potential is characterised by the complete absence of steps and therefore a continuous potential at the MT radii. Referring to Figure 5.3, this implies: there will be a single value of the inner potential $V_0(E)$ for all atomic layers, instead of the layer-dependent muffin-tin zero of Figure 5.3; the inner potential $V_0(E)$ will match without step the spherically symmetric potential within each MT sphere. The step-free method has two important features: (a) the inner potential refers to an energy-independent MT zero V_{MTZ}, which is identical to the vacuum level so that $V_0(E) = V_{MTZ} + V_{XC}(E)$ and $V_0(\infty) = V_{MTZ}$; and (b) the method allows overlapping atomic potentials, such that MT radii can add up to more than the near-neighbour interatomic distances and thus create intersecting spheres. The concept of overlapping MT spheres is not new, but also not common in LEED. These aspects will be further discussed in the following. A schematic illustration of the MT potential with overlapping spheres is given in Figure 6.12.

At high energies, say 200 keV, as used in transmission electron microscopy, the exchange–correlation interaction $V_{XC}(E)$ between an electron and a solid vanishes because of finite-time electron gas relaxation in solids. Then the inner potential $V_0(E)$ becomes energy invariant, and $V_0(E) = V_{MTZ}$; this is determined together with the

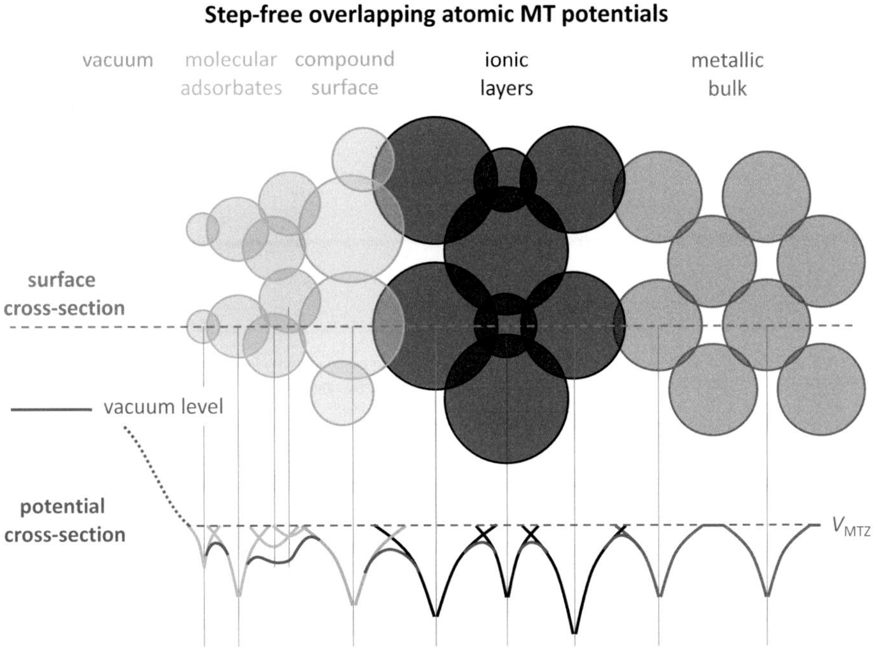

Step-free overlapping atomic MT potentials

Figure 6.12 Sketch of step-free overlapping MT potentials for a hypothetical surface consisting of adsorbates (blue) on a compound surface (green) supported by an ionic film (black) on a metallic bulk (brown), shown in cross-section. The MT radius for each atom is chosen such that a common MT zero V_{MTZ} (red horizontal dashed line) can be defined, with no step between the MT zero and the potentials within the spheres. The MT radii and the MT zero are determined by differential evolution (DE). A schematic cross-section of the potential along the dashed line is shown below (the red curve sums up the other coloured atomic potentials); the dotted red curve represents the transition to the vacuum level.

choice of the MT radii; see for example also the discussion of the 'mean' inner potential by D. K. Saldin and J. C. H. Spence [6.105]. The MT zero V_{MTZ} is shown in Figure 6.13.

Given the low energies used in LEED, the effective scattering potential inside the crystal is set up in the new MT model as the sum of the energy invariant MT zero V_{MTZ} and the local-density-functional exchange-correlation (XC) potential of G. D. Mahan and B. E. Sernelius [6.100]. The spatial average of the XC contribution from the interstitial region gives an energy dependent potential $V_{XC}(E)$ with respect to the MT zero. The MT zero V_{MTZ} and the XC potential $V_{XC}(E)$ add up to the inner potential $V_0(E) = V_{MTZ} + V_{XC}(E)$, referred to the vacuum level, as illustrated in Figures 6.12 and 6.13. The MT zero V_{MTZ} is now completely independent of the energy of the incident electron and applies to the whole surface slab, including both the adsorbate layer and the substrate.

The energy dependent XC potential $V_{XC}(E)$ is calculated from the local density functional theory by G. D. Mahan and B. E. Sernelius [6.100]. For LEED intensity

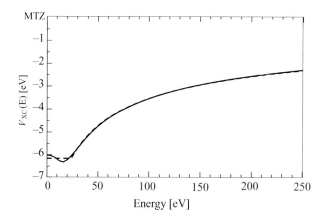

Figure 6.13 Inner potential for the molecular adsorbate system Cu(111)+$(3\sqrt{3}\times3\sqrt{3})$R30°– TMB, calculated with overlapping and optimised MT radii. The MT zero V_{MTZ} is independent of energy. The exchange-correlation potential V_{XC} is calculated according to G. D. Mahan and B. E. Sernelius [6.100] (black curve). Its approximation by four parameters (see Eq. (6.36)) is shown by the dashed curve which is nearly identical with the theoretical curve above a lower energy limit (here 20 eV); below the lower limit, V_{XC} is approximated by a constant. The energy scale is relative to the vacuum level. Reprinted from [6.104] J. Rundgren, W. Moritz and B. E. Sernelius, *J. Phys. Commun.*, vol. 5, p. 105012, 2021 (Open Access).

calculations $V_{\text{XC}}(E)$ is modelled with four adjustable parameters p_1 to p_4, as described in Section 6.1.6.4. It is approximated by a constant lower limit below a minimum energy, as shown in Figure 6.13 by the dashed curve. In the LEED calculation the internal energy of the electron refers to the MT zero which is identical to the vacuum level. However, the experimental external energy (which is measured in the experiment) includes surface specific effects and experimental errors. To relate that internal energy to the external energy, we need the MT zero to be related to the external energy. The parameters used in the LEED calculations therefore include the MT zero. The modulus of the **k**-vector inside the crystal is given by $\sqrt{2m_e(E-V_0(E))}/\hbar$, and the phase shifts extracted from the MT potential are related to $E-V_0(E)$ to allow the easy optimisation of the parameters p_1 to p_4 that model $V_0(E)$. An optimisation of these parameters is still necessary because the calculated $V_0(E)=V_{\text{MTZ}}+V_{\text{XC}}(E)$ does not include work function changes due to surface dipoles and other contributions.

The MT radii of the different atom types are determined by differential evolution (DE) under the condition of achieving a single common V_{MTZ} and different overlap parameters for each atom type. Here, an atom type is defined by the element and similar bonds to the neighbouring atoms; depending on the structure model, atoms of the same element may belong to different types (note that $V_{\text{XC}}(E)$ is common to all atoms).

The overlap of atomic potentials permits great freedom in the optimisation of MT radii combined with energy shift. The MT radii and the common level of the MT zero are found by DE optimisation from the superposition of atomic potentials, as follows.

First, the inner potential V_0 is defined as the spatial average of all atomic potentials in the unit cell outside the MT spheres. The unit cell includes the surface slab and two undistorted substrate layers. V_0 therefore depends on the choice of MT radii R_{MT}. These are next found in an iterative process until all atomic potentials fulfil the condition that $V_i(R_{MT}) = V_0$ for all atom types i; this step truncates the atomic potentials at their individual MT radii R_{MT}, such that each atomic potential merges without step into the constant V_0. The resulting total potential is continuous everywhere. V_{MTZ} is then set to V_0. The details of the procedure are described in J. Rundgren et al. [6.104]. The potentials within the MT spheres are now referenced to the MT zero and the exchange-correlation potential is related to the MT zero. In this scheme, the Mattheiss procedure of superposing atomic potentials is not applied because it assumes non-overlapping MT spheres and fixes the potentials in the MT spheres relative to the vacuum zero; steps in the potential at the MT radius to the interstitial potential cannot be avoided in the Mattheiss procedure.

DE optimisation of overlapping MT sphere radii and a flat MT zero produces potential steps of the order of only 10^{-14} eV (with a computer precision of 10^{-16}). As mentioned in Section 5.2.1, the condition of non-overlapping MT spheres is not essential for the MT model and an overlap is therefore possible [6.106; 6.107]. It has been shown that the use of overlapping spheres leads to the same secular equations as non-overlapping spheres and the same formalism can be used with overlapping spheres, provided that the muffin-tin radii do not exceed the centres of the neighbouring potentials and the interstitial area is treated correctly [6.107]. Increasing the radii of the MT spheres has the advantage of reducing the interstitial volume and allows better representation of bonds between atoms by more accurately including the electron density between the atoms. Overlapping potential spheres have been used, for example, within the exact muffin-tin orbital (EMTO) theory to calculate properties of elemental metal structures and alloys and have led to the same trends as full potential calculations [6.108].

Overlap Parameter

The combination of the energy invariant MT zero with the XC potential and the optimisation of the MT radii implies overlap parameters S_{ovl} which may be different for each element and different bonds. The overlap parameter S_{ovl} is defined as an enhancement of an initially estimated MT radius which is chosen as half of the nearest-neighbour (NN) distance $d(NN)$ of the corresponding atom pair. The condition for the sum of the two MT radii of an atom pair is: $r_{MT} + r_{MT}(NN) \leq d(NN) \times (1 + S_{ovl})$. Here r_{MT} is the radius of the atom of interest. Both r_{MT} and S_{ovl} must be chosen such that this condition is fulfilled for all neighbouring atoms; S_{ovl} may be different for each independent atom in the unit cell, see for example the atoms in the middle of Figure 6.12 where the overlap for atoms with manifestly different radii is shown.

The values for the fractional increase S_{ovl} are determined by the DE procedure to obtain the MT potential and can be optimised by fitting to experimental data in the final step of the analysis. A fit of the overlap parameters S_{ovl} to the experimental data is possible because different MT radii produce different phase shifts and also slight differences in the I(V) curves; V_{MTZ} and $V_{XC}(E)$ also depend slightly on the MT radii,

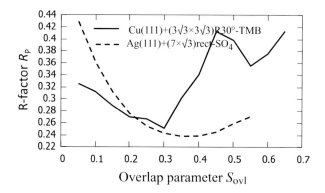

Figure 6.14 R-factor R_P as a function of the overlap parameter S_{ovl} for Cu(111)+$(3\sqrt{3}\times3\sqrt{3})$ R30°–TMB (full line) and Ag(111)+$(7\times\sqrt{3})$rect–SO$_4$ (dashed line). (TMB is defined in the caption of Figure 6.2). The optimum structure is kept constant and the phase shifts are calculated with MT radii fulfilling the condition $r_{MT} + r_{MT}(NN) \leq d(NN) \times (1 + S_{ovl})$. In this example, S_{ovl} is taken to be the same for all atoms. For the bulk atoms the MT radius is fixed to be half the nearest-neighbour distance.

and there is enough freedom in the choice of the overlap parameter S_{ovl} to obtain a flat MT zero. The inner potential $V_0(E) = V_{MTZ} + V_{XC}(E)$ can therefore be fit to the experimental data within limits for the overlap of the MT radii. By design, the MT radii keep the same values throughout the whole energy range. The overlap parameters apply to all atoms; for the bulk atoms in close packed metals the MT radius can be fixed to be half the NN distance.

In actual optimisations [6.104], the DE optimisation leads to an average value of S_{ovl} in the range 0.3–0.4 for all atoms. Only small differences from the average occur when the parameters for individual atom types are fit independently. The R-factor as a function of an average S_{ovl} is shown in Figure 6.14 for two optimisations; there is a clear minimum for S_{ovl} around 0.3. Figure 6.14 shows the results for two relatively complex systems, Cu(111)+$(3\sqrt{3}\times3\sqrt{3})$R30°–TMB and Ag(111)+$(7\times\sqrt{3})$rect–SO$_4$. For the first system the minimum R-factor is at $S_{ovl} = 0.3$, while for the second system it is at $S_{ovl} = 0.35$. Similar results have been found for the simpler systems Ag (111)+(4×4)–O and Ru(0001)+$(\sqrt{3}\times\sqrt{3})$R30°–Cl [6.104]. The parameter S_{ovl} can be refined in the final step of the analysis for the individual atoms to find the minimum R-factor.

Energy Dependence of the Exchange-Correlation Potential

A further parameter f_{XC} is introduced here to take into account possible modifications of the local-density functional XC potential $V_{XC}(E)$ when excitations are present. A factor f_{XC} related to the ground-state exchange-correlation energy $\mu_{XC}(E)$ had been suggested by L. Hedin and B. I. Lundqvist [6.99] and discussed in detail by R. E. Watson et al. [6.109]. This factor should ensure that $V_{XC}(E)$ approaches $\mu_{XC}(E)$ when E approaches zero. The ground-state exchange-correlation energy cannot be determined experimentally. Therefore, a variable factor f_{XC} is used here which can be fitted

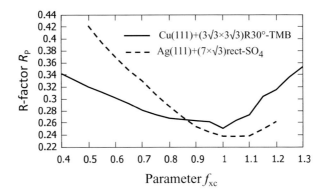

Figure 6.15 R-factor R_P as a function of the correction parameter f_{XC} for the exchange-correlation potential for two systems, Cu(111)+$(3\sqrt{3}\times3\sqrt{3})$R30°–TMB (full line) and Ag(111)+$(7\times\sqrt{3})$ rect–SO$_4$ (dashed line) (TMB is defined in the caption of Figure 6.2).

to the experiment. The factor is certainly not unity, as a comparison has shown between two different calculations of $V_{XC}(E)$ [6.97]. The curves calculated by G. D. Mahan and B. E. Sernelius [6.100] exhibited some differences to the calculation after L. Hedin and B. I. Lundqvist.

A similar parameter is considered in the new potential calculation used here [6.104]. The factor f_{XC} multiplying $V_{XC}(E)$ is used here as a qualitative parameter without considering individual excitation energies. It applies to all atoms. The optimum values for f_{XC} are finally found by fitting to the experimental LEED I(V) curves. In the cases which were tested, this produced a flat minimum in the R-factor with values of f_{XC} close to 1.0, indicating that $V_{XC}(E)$ is quantitatively correct [6.104]. Because $V_{XC}(E)$ describes the energy dependence of the inner potential, the fit of the factor f_{XC} to experimental data includes experimental uncertainties in the measurement of the energy of the primary beam. The influence of f_{XC} on the R-factor is shown in Figure 6.15, where the structural parameters have been kept fixed. In principle, a new refinement of atomic coordinates would be required for each new set of phase shifts to obtain the true minimum of the R-factor, but the resulting differences are small.

The fact that the MT parameters S_{ovl} and f_{XC} can be fit to experimental data does not increase the number of structural parameters in the structure search or the refinement. Since their influence on the I(V) curves is relatively small, the optimisation of S_{ovl} and f_{XC} is not necessary in the beginning of a LEED I(V) analysis. The analysis can therefore start with standard fixed data for the MT parameters. The optimisation of the new MT parameters S_{ovl} and f_{XC} can be done in the final step of the analysis after the structure model is already essentially determined with standard MT parameters: this near-final structure then enters the calculation of new final MT parameters. It leads to a final improved structure with lower R-factor and lower error bars, which result from a new calculation of I(V) curves with a new set of phase shifts using the final MT parameters. The deviation in structural parameters is small, mostly below 0.002 nm (0.02 Å). Although such small deviations are hardly visible in the structure model, they can have a noticeable influence on the R-factor. Indeed, changes of about 50 or more structural parameters in this range can together cause a significant difference in the R-factor, for example for SO_4/Ag(111) from 0.235 to 0.21, or for TMB/Cu(111) from 0.27 to 0.25. Such sensitivity agrees, incidentally, with the observation made by E. Primorac et al. [6.48] in the comparison with different optimisation procedures that the final reduction of the R-factor requires a large number of iterations with small parameter variations.

The optimisation of the MT parameters slightly improves the final result; the influence on the structural parameters is far below the error limits. This shows that the phase shifts calculation can be done with standard parameters for the overlap and V_{XC} correction.

The new method to obtain phase shifts from a step-free potential by optimised MT radii and energy-shifted atomic potentials leads to an improved agreement in the fit of I(V) curves, as is shown in Table 6.2. We summarise the results of the phase shift calculation with the improved MT model as follows: the agreement between calculated and experimental I(V) curves can be significantly improved compared to the calculation using phase shifts from the conventional method depending on the energy range used in the analysis. The structure model and the main structural data are not changed, but the R-factor is lower and the error bars are smaller. A comparison for four different surface structures is shown in Table 6.2. The average difference in the final structural data obtained with and without overlap in the MT potentials lies in the range of 0.003–0.004 nm (0.03–0.04 Å), with a maximum shift of a single atom by about 0.01 nm (0.1 Å). Such larger shifts were found for a few oxygen positions in two structures listed in Table 6.2 and are still below the error estimate.

We should note that the influence of optimised MT radii shows up mainly in the low energy range of the I(V) curves. At energies above about 80 eV, differences in calculating phase shifts have only small effects.

Table 6.2 Comparison of best fit R-factors obtained with phase shifts from MT potential models with and without overlap

Surface structure	Energy range [eV]	Min. R-factor, using optimised MT radii without overlap	Min. R-factor, using optimised MT radii with overlap
Ag(111)+(7×$\sqrt{3}$)rect–SO_4	10–150	0.235 [6.58]	0.211 [6.104]
Cu(111)+(3$\sqrt{3}$×3$\sqrt{3}$)R30°–TMB	11–200	0.32 [6.46]	0.263 [6.104]
Ru(0001)+($\sqrt{3}$×$\sqrt{3}$)R30°–Cl	20–300	0.19 [6.56], 0.137 [6.110]	0.107 [6.104]
Ag(111)+(4×4)–O	25–250	0.34 [6.61]	0.268 [6.104]

The energy range of the experimental data is also listed. For Cl/Ru(0001) the best fit R-factor in the original paper was 0.19, which later could be reduced to 0.137 with new phase shifts for non-overlapping MT spheres and by including more parameters [6.110]. The comparison shows that in all cases the R-factor is noticeably improved by using phase shifts from the step-free MT potential.

6.1.6.4 Energy Dependence of the Inner Potential

In LEED multiple scattering calculations, the energy of the incident electron inside the solid is increased by the inner potential. The inner potential $V_0(E)$ is necessarily negative and the kinetic energy of the electron inside the solid is therefore higher than outside. Most LEED analyses to date have assumed an energy-independent value for the inner potential V_0, which is also often called the MT zero V_{MTZ}. However, the inner potential does depend on the energy through the energy dependence of the electron exchange-correlation potential V_{XC} in the crystal, as discussed in Section 6.1.6.3: thus, $V_0(E) = V_{MTZ} + V_{XC}(E)$.

The calculation of V_{XC} is usually based on the result obtained by L. Hedin and B. I. Lundqvist [6.99] for the free electron gas. They generated diagrams of V_{XC} versus $k = \sqrt{E}$ for a family of atoms with different electron densities. J. Rundgren [6.97] calculated the inner potential $V_{XC}(E)$ for a selection of crystals and observed that $V_{XC}(E)$ curves have a universal shape given by three parameters; here a fourth parameter is used to limit V_0 at low energies. The energy dependence is thus approximated by the four parameters p_1 to p_4 as follows:

$$V_0(E) = \max\left(p_1 + \frac{p_2}{\sqrt{E + p_3}}, p_4\right). \tag{6.36}$$

Here p_1 is the constant shift of the potential, p_2 determines how steeply the potential increases at low energies (p_2 is usually negative), p_3 slightly adjusts that energy dependence, while p_4 sets a lower limit to avoid unphysically low values at very low energies. The energy dependence of the XC potential is well approximated by these parameters. An example is shown in Figure 6.13, where the lower limit set by p_4 causes a constant value below 20 eV of kinetic energy. Depending on the parameters, this range may be narrower or wider and should not cause problems in the I(V) analysis.

The parameters p_1 to p_4 should be fit in the LEED I(V) analysis, especially when the low energy range is used. The fit is done for each iteration step in the optimisation of structural parameters using the theoretically calculated parameters p_1 to p_4 as start parameters. The parameters for the energy dependence of the inner potential are therefore determined in an independent fit. At higher energies, in many cases a constant value of the inner potential is used, and a fit of this value alone is certainly sufficient for the determination of the structure model. As an inherent property of diffraction, the geometrical parameters are measured in terms of the local wavelength of the electrons inside the crystal, so that the accuracy of the wavelength determines the accuracy of the structural result. Therefore, the energy dependence becomes important for precise structure determinations. This has been clearly demonstrated by S. Walter et al. [6.111], who showed that a small deviation of the substrate lattice constant from the bulk value was caused by using an inner potential that did not depend on energy. The influence of the inner potential in cases where the lattice constant is not exactly known, for example, in epitaxial layers which may be distorted, has been investigated by J. Vuorinen et al. [6.112]. They estimate that the precision is limited to about 0.001 nm (0.01 Å) if the inner potential is taken to be independent of energy and fit in the I(V) analysis.

The value of the inner potential V_0 determined in the LEED I(V) analysis is not directly comparable to values determined by other experiments because it depends on the choice of the MT radii, although this is probably not the dominant factor. More important in surface studies are further contributions to the inner potential arising from the structure and orientation of the surface due to the surface dipole (resulting in particular from charge transfers between substrate and adsorbate and from the electron spill-over into vacuum, which is also responsible for the surface image potential). In most LEED I(V) analyses the constant part of the inner potential and frequently also the energy dependence are fit to the experimental data, so that parts of the inner potential not considered in the exchange-correlation potential V_{XC} are included in the experimentally determined V_0.

The different contributions to the 'mean' inner potential of a solid have been reviewed by D. K. Saldin and J. C. H. Spence [6.105]. (Here, 'mean' refers to an average over the full unit cell, including both the atomic cores and the interstitial region.) The mean inner potential at high energies is determined in high-voltage transmission electron microscopy (HTEM); another contribution arises from the surface dipole layer, and a third part from the energy dependent exchange-correlation potential which is important in the low energy range. The value for the mean inner potential determined in that way differs from the inner potential used in the MT model. The MT sphere potential extends for each atom i to a radius $r_{i,MT}$, where it should meet the flat inner potential without step.

6.1.6.5   Inelastic Mean Free Path (IMFP)

The strong damping of the intensity of the electron beam by inelastic processes inside the solid is the main cause of the surface sensitivity of LEED and all other techniques that use electrons to study properties of surfaces, such as XPS, AES and PED. The knowledge of the inelastic mean free path (IMFP) of electrons in a solid is therefore essential for the quantitative analysis of surfaces. The IMFP is in general energy dependent because the inelastic processes depend on the kinetic energy of the diffracted electrons. The energy dependence is roughly described by a universal curve with a minimum around 100 eV, see Figures 2.27 and 6.6. The energy dependence of the IMFP is not related to the energy dependence of the inner potential V_0.

The first database of IMFP values for a series of elemental solids as well as for inorganic and organic compounds was published by M. P. Seah and W. A. Dench [6.113], showing a universal curve with a minimum around 50–100 eV of kinetic energy of the incident electron. The IMFP is different from the effective attenuation length (EAL): the EAL includes the removal of the elastically scattered electrons from the incident beam due to diffraction, in addition to the inelastic scattering effect. The techniques which determine the EAL have been reviewed by A. Jablonski and C. J. Powell [6.114]. Depending on the experimental method, either the EAL or the IMFP is measured, while the EAL may or may not be corrected for diffraction effects. For LEED the IMFP determines the imaginary part of the inner potential or optical potential, called V_{0i}.

The IMFP can be experimentally determined by a variety of methods. A major approach is the 'cover method' where a substrate is covered by a layer of the material of which the IMFP will be determined. The attenuation of the AES or XPS signal from the covered material as a function of the thickness of the layer and the energy of the signal then gives the IMFP. A comparison of different models to calculate the IMFP with experimental data has been given by C. J. Powell and A. Jablonski [6.115]; a comparison of theoretical and experimental data down to low energies has been published by P. de Vera and R. Garcia-Molina [6.116]. Data for elemental solids have been calculated by S. Tanuma et al. [6.117], while experimental data down to energies below 100 eV have been measured by O. Yu. Ridzel et al. [6.118]. Tables with numerical data for 41 elemental solids in the energy range from 10 eV to 30 keV are available from NIST [6.102]. The numerical data can be used in the LEED programs to derive the imaginary part V_{0i} of the inner potential.

Inelastically scattered electrons are in principle filtered out in LEED, with the exceptions that the incident beam itself has a finite energy spread and that the energy resolution in most cases is not sufficient to filter out phonon losses. In the theoretically calculated intensity, strictly only elastically scattered electrons are included while the damping due to phonon losses is described by a Debye–Waller factor. The comparison to experimental intensities which include the phonon losses is justified by the fact that the thermal diffuse scattering arising from phonon losses follows mainly the elastic scattered intensity [6.119]. The other inelastic processes due to interband transitions, excitations, secondary electron emission, etc., with higher energy losses are filtered out in the experiment, while they are included in the theory through a damping parameter: it enters the theory as the imaginary part of the inner potential V_{0i}.

The damping parameter represents all non-phonon inelastic processes. Its description as the imaginary part of the inner potential V_{0i} leads to a complex wave vector **k** inside the solid. The inner potential is usually given as a negative value so that the energy inside the solid is $E - V_0$, where E is the external energy related to the vacuum level (we now write the inner potential V_0 with a real part V_{0r} and an imaginary part V_{0i}):

$$V_0 = V_{0r} + iV_{0i} \tag{6.37}$$

$$\begin{aligned} E - V_0 &= E - V_{0r} - iV_{0i} \\ &= \frac{\hbar^2}{2m_e}k^2 = \frac{\hbar^2}{2m_e}(k_r + ik_i)^2, \end{aligned} \tag{6.38}$$

which gives

$$V_{0i} = -\frac{\hbar^2}{m_e}k_r k_i \tag{6.39}$$

and

$$E - V_{0r} = \frac{\hbar^2}{2m_e}\left(k_r^2 - k_i^2\right). \tag{6.40}$$

It follows that

$$\frac{\hbar^2}{m_e} k_r^2 = E - V_{0r} + \sqrt{(E - V_{0r})^2 + V_{0i}^2} \tag{6.41}$$

and

$$\frac{\hbar^2}{2m_e} k_r^2 \simeq E - V_{0r} \quad \text{when} \quad V_{0i}^2 \ll (E - V_{0r})^2 \quad \text{or} \quad k_i^2 \ll k_r^2. \tag{6.42}$$

The optical potential V_{0i} is related to the inelastic mean free path as follows. The phase factor of the plane wave inside the solid is

$$\exp(i\mathbf{kr}) = \exp(i\mathbf{k}_r\mathbf{r}) \exp(-\mathbf{k}_i\mathbf{r}) = \exp(i\mathbf{k}_r\mathbf{r}) \exp(-k_{i,z}r_z). \tag{6.43}$$

The component of the wave vector normal to the surface has an imaginary part such that the amplitude of the wave is dampened with increasing depth into the crystal. The inelastic mean free path $\lambda_e = |1/k_i|$ thus represents the penetration depth. This leads to the relation between the mean free path and the optical potential, using Eqs. (6.39) and (6.40):

$$V_{0i} = -\frac{\hbar^2}{m_e \lambda_e} \left\{ \frac{2m_e}{\hbar^2}(E - V_{0r}) + \frac{1}{\lambda_e^2} \right\}^{1/2} \tag{6.44}$$

and

$$\lambda_e = |1/k_i| \simeq \sqrt{\frac{\hbar^2}{m_e}(E - V_{0r})} \frac{1}{V_{0i}}. \tag{6.45}$$

When $|V_{0i}|$ is small compared to $E - V_{0r}$, these relations simplify to

$$\frac{\hbar^2}{2m_e} k_r^2 = E - V_{0r} \quad \text{and} \quad k_i = -\frac{V_{0i}}{\sqrt{\frac{\hbar^2}{m_e}(E - V_{0r})}}. \tag{6.46}$$

A small k_i means that the wavelength and the conditions for constructive and destructive interference are hardly affected by the optical potential, except that the peaks in the I(V) curves are broadened. The penetration depth can now be related to the peak widths, as follows. We consider lattice planes parallel to the surface with a mutual spacing d normal to the surface. Assuming for simplicity a mono-atomic lattice, we obtain in the kinematic calculation the diffracted intensity as

$$I(\mathbf{q}) = \left| \sum_{n=0}^{\infty} f_n(\mathbf{q}) \exp(i\mathbf{qr}) \exp(-q_{i,z}nd) \right|^2$$

$$= \left| F(\mathbf{q}) \sum_{n=0}^{\infty} \exp(-q_{i,z}na\lambda_e) \right|^2, \tag{6.47}$$

where $q_{i,z}$ is the imaginary part of the normal component of the $\mathbf{q}$ vector, $\mathbf{q} = \mathbf{k} - \mathbf{k}'$, and the factor α is the layer distance in units of the attenuation length, $\alpha = d/\lambda_e$. The sum in Eq. (6.47) runs over all layers of the semi-infinite crystal. Equation (6.47) describes a broadened Bragg peak of Lorentzian shape. The width is $w \simeq \Delta q_z \simeq 2/\lambda_e$ in momentum space, which means that

$$w \simeq \frac{2V_{0i}}{\sqrt{\frac{\hbar^2}{m_e}(E - V_{0r})}} \tag{6.48}$$

The energy width is then $\Delta E \simeq 2\Delta \mathbf{q}\mathbf{k}_r \simeq 2V_{0i}$. This relationship allows estimating the optical potential from the width of the peaks (but one must beware of overlapping peaks giving a larger apparent width). The IMFP becomes large at low energies, which implies that V_{0i} is small and the peaks become narrow, as observed in the measured I(V) curves. This is illustrated in Figure 6.16. However, the width in reciprocal space remains approximately constant, as is shown in Figure 6.5.

In many LEED I(V) analyses a constant value for the optical potential is chosen. It is assumed that its influence on the structural result is small and is averaged by the energy range of the measurements. A constant value of V_{0i} is in most cases justified at energies above about 40–50 eV. In Figure 6.16 the comparison is shown for two beams, (10) and (−13), of SO_4/Ag(111) in the energy range 10–150 eV; the indices

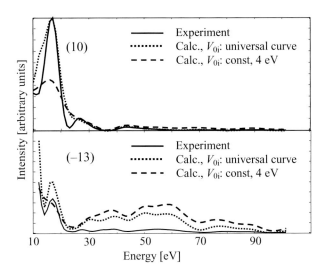

Figure 6.16 I(V) curves for two beams of Ag(111)+(7×√3)rect–SO_4. The pairs of theoretical curves are calculated with V_{0i} energy dependent according to the universal curve from M. P. Seah and W. A. Dench [6.113] (dotted curves) and with an energy independent $V_{0i} = 4$ eV (dashed curves) and identical structural data. The experimental curves (black) are taken from [6.58] R. Wyrwich, T. E. Jones, S. Günther, W. Moritz, M. Ehrensperger, S. Böcklein, P. Zeller, A. Lünser, A. Locatelli, T. O. Menteş, M. Á. Niño, A. Knop-Gericke, R. Schlögl, S. Piccinin and J. Wintterlin, *J. Phys. Chem. C*, vol. 122, pp. 26998–27004, 2018, with permission from the American Chemical Society.

refer to the superstructure. Although the agreement with the experimental data is not very good – the peak heights of the (–13) beam do not match – it is obvious that the theoretical peaks in the range 10–40 eV are too broad when calculated with constant V_{0i}. Above 50 eV the curves calculated with different models for V_{0i} are practically identical.

6.2   Quasicrystals

Quasicrystals are characterised by a state of crystalline order which is not allowed in the classical definition of a crystal. They do not show translation symmetry and exhibit unusual 5-fold, 8-fold, 10-fold, or 12-fold rotational axes which are not compatible with translation symmetry. They possess long range rotational order and produce sharp diffraction reflections which justify the classification as crystals, but they are called quasicrystals because the translational symmetry is missing; see for example the definition of crystals by the IUCr [6.120] or the discussion of three types of aperiodic crystals by W. Steurer and T. Haibach [6.121].

Quasicrystals were first observed in 1982 by D. Shechtman et al. and published in 1984 [6.122]. This new class of materials was intensively investigated in subsequent years. There are excellent reviews and descriptions of properties of quasicrystals (including [6.123–6.128]). Among the general properties of quasicrystals are low electrical conductivity, high mechanical hardness, low surface energy (leading to non-stick behaviour), low friction and high resistance to oxidation, relative to other alloys.

Quasicrystalline phases have been found in a large number of Al-rich transition metal alloys, mostly ternary alloys; a few binary alloys were also found to have quasicrystalline phases. A list is given in a review by W. Steurer [6.124]. Quasicrystals were discovered rather late because they only exist in very narrow ranges of composition, typically within a few percent variation of each component, for example, in Al-Cu-Fe, between 63 and 68% of Al, between 21 and 26% of Cu and between 12 and 13% of Fe.

Two structurally different types of quasicrystals can be distinguished. One type exhibits an icosahedral symmetry without translation symmetry in any direction: this type is often denoted by an 'i' in front of the chemical composition, for example, i-Al-Cu-Fe. A second type exhibits translation symmetry in one direction and consists of a sequence of layers which are quasicrystalline in two directions orthogonal to the first; these are called axial quasicrystals and are sometimes denoted by a prefix p- (pentagonal), o- (octagonal), d- (decagonal) or dd- (dodecagonal) in front of the chemical composition, for example, d-Al-Cu-Co.

The initial quasicrystalline alloys could be produced by very rapid cooling using the melt spinning technique [6.129], thus freezing the solid in the quasicrystalline state. Only micrometre-size quasicrystalline grains were obtained in this way, making surface studies difficult; their thermodynamic stability has not been sufficiently investigated. This situation changed after the discovery of alloys whose quasicrystal-line phase is stable up to the melting point; this allowed using conventional crystal growing techniques like the Bridgman method or a related technique in which a peritectic solidification reaction takes place between different high-temperature crys-talline phases [6.130]. Only a few systems have been proven to exhibit thermo-dynamic stability at high temperatures: these are, among others, the decagonal phases d-Al-Fe-Ni, d-Al-Ni-Co and d-Al-Mn-Pd, and the icosahedral phases i-Al-Cu-Li, i-Al-Cu-Fe and i-Al-Mn-Pd [6.131; 6.132]. In these systems, quasicrystals can be grown with sufficient size for surface studies by LEED, at least a few millimetres across.

We discuss in this section only briefly the quasicrystalline order, the use of *n*-dimensional hyperspace to describe the diffraction by quasicrystals, the indexing of the LEED reflections, approximate multiple scattering calculations that have been performed and results of structural studies.

6.2.1　Quasicrystalline Order

The most prominent geometric example of the quasicrystalline order is the Penrose tiling in two dimensions [6.133]. A space-filling tiling of a plane with a *single* shape of tiles leads to the five symmetrically inequivalent 2-D translationally *periodic* lattices (see Section 2.1.4 for crystalline lattices). While these do not allow 5-fold axes, R. Penrose [6.133] noted that a space-filling tiling and a 5-fold axis are possible by combining *two* tiles of different shapes: an example is shown in Figure 6.17. The two tiles are 'fat' and 'slim' rhombi with acute angles of $360°/5 = 72°$ and $360°/10 = 36°$, respectively, and equal side lengths.

No translation symmetry exists in this pattern, but it exhibits orientational long range order. The pentagons formed by the central points of the fat rhombi are marked in Figure 6.17 with thick black lines, and all exhibit the same orientation or are rotated by $180°$. Also, the tile edges have only five possible orientations: one such orientation is highlighted in blue in Figure 6.17; these orientations differ by multiples of $360°/5 = 72°$. The tile edges line up exactly across the entire infinite plane (no tile edges exist between the blue lines shown in Figure 6.17). The spacings between tile edges form the aperiodic Fibonacci sequence of short (S) and long (L) spacings; this will be further discussed in Section 6.2.2.

The atomic structure of the quasicrystal depends on the atomic contents of the tiles. An example is shown in Figure 6.18 for a 5-fold surface of an $Al_{63}Cu_{24}Fe_{13}$ quasi-crystal, including a multilayer step in the side view, Figure 6.18(b): it exhibits a lack of periodicity in all three dimensions. Its surface structure has been analysed with LEED and STM [6.135]. Five-fold rings can be easily recognised, corresponding to the pentagons of Figure 6.17.

Prominently visible are relatively flat Al-dominated surfaces and interfaces, shown in Figure 6.18(b) as the two upward facing surfaces and two downward

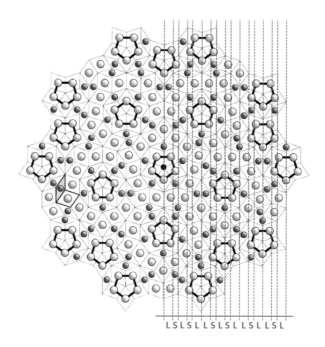

L S L S L L S L S L L S L L S L

Figure 6.17 Aperiodic (quasicrystalline) arrangement of the two Penrose tiles illustrating 5-fold symmetry (emphasised by an approximately circular truncation). The slim and fat tiles are distinguished by red and grey balls, respectively, at the centre of all tiles (one pair of such tiles is highlighted at left). In real quasicrystals the tiles are occupied not by single atoms but by clusters of atoms, see for example Figures 6.16 and 6.18. The black pentagons indicate local 5-fold symmetry, which is normally only valid in the immediate neighbourhood of the 5-fold axis, while the central pentagon (with a dot at its centre) has a 5-fold axial symmetry that is global, extending to infinity in all directions. There are also five global mirror planes through the central pentagon (one of them is shown as the vertical blue dashed line through this pentagon), while the other pentagons have at most one global mirror plane. The blue lines on the right mark one orientation of tile sides and show how these sides line up across the plane (to infinity), with long (L) and short (S) spacings that form an aperiodic Fibonacci sequence. Adapted with permission from [6.134] K. Hermann, *Crystallography and Surface Structure*, 2nd edition, Wiley, 2016. © 2017 Wiley-VCH Verlag GmbH & Co. KGaA, Boschstr. 12, 69469 Weinheim, Germany.

facing surfaces of two multilayer slabs, which have typical icosahedral quasicrystalline structures consisting of high- to low-density layers with variable compositions; Al (red) is seen to strongly dominate in the denser layers, including the slab surfaces: these relatively widely spaced slab surfaces are the preferred terminations at clean surfaces. Step heights in icosahedral quasicrystals exhibit a Fibonacci series of high and low steps, similar to the line spacings in 5-fold planes exhibited in Figure 6.17, including the golden mean ratio τ discussed in Sections 6.2.2 and 6.2.3.

The combination of the two Penrose tiles is not arbitrary, but governed by matching rules, as illustrated in Figure 6.19. There are many ways to construct an

(a)

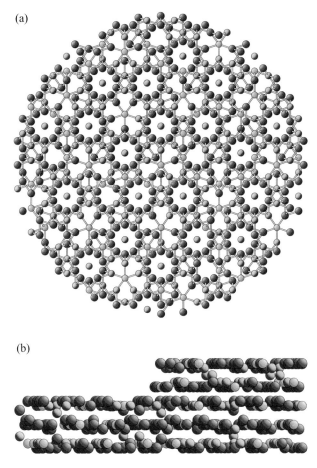

(b)

Figure 6.18 Atomic structure of an Al-Cu-Fe(000001) surface with ideal icosahedral bulk-like atomic positions: about 63% of the atoms are Al (red), about 24% Cu (light blue) and about 13% Fe (green). (a) Top-down view of a thin disk-like slice with a few atomic layers. Prominent are rings of 10 red Al atoms: each ring is not 10-fold but 5-fold symmetrical, as it is composed of two rings of five coplanar atoms in two different planes, forming a 5-fold rotational crown: these rings correspond to the 5-fold centres with pentagons in the Penrose tiling of Figure 6.17. No global 5-fold symmetry axis is present in this surface segment. (b) Side view along a step edge, with the surface at the top. The step connects a lower terrace (at left) to a higher terrace (at right). The view is slightly rotated to show rows of atoms parallel to the surface, receding toward the left. Figures by K. Hermann using his Balsac software, private communication.

aperiodic pattern, but only local (short-range) patterns are repeated across the plane (although not periodically), for example, the pentagons drawn in Figure 6.17. The matching rules must be fulfilled in a quasicrystalline pattern of infinite size, but they are not sufficient to obtain a quasicrystalline lattice by adding tiles step by step: the matching rules do not guarantee the tiling to grow without defects. This raises the fundamental question of how quasicrystals grow,

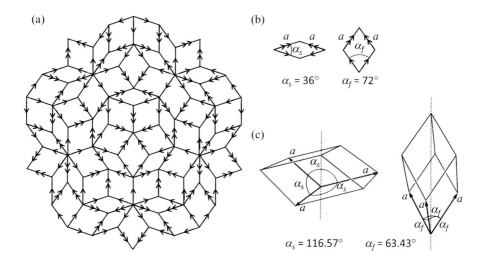

Figure 6.19 (a) Matching rules in the 2-D Penrose tiling (central part of Figure 6.17). The rhombi are combined such that the arrows at the edges match up. (b) Corresponding slim (s) and fat (f) rhombi. (c) Perspective views looking down on the two rhombohedra forming the Penrose tiling in 3-D; the vertical dotted lines indicate 3-fold rotational symmetry axes for these trigonal cells; all side lengths are equal to a to enable matching; further matching rules also exist for a 3-D Penrose tiling. Panel (a) is redrawn with permission from C. Janot, *Quasicrystals: A Primer*, 2nd edition, Clarendon Press, Oxford, 1994. © C. Janot, 1992, 1994. Panel (c) is adapted from K. Hermann, private communication.

since it is generally accepted that long-range direct interactions can be excluded. (The same problem occurs with the large unit cells in Hume–Rothery phases of alloys. It is assumed that these phases are stabilised by electronic effects and the same is probably true for quasicrystals [6.123].) Such questions will not be discussed here, but we note that most quasicrystals exhibit defects, for example, so-called phasons, which limit the resolution of X-ray and neutron diffraction analyses [6.124]. The type of defects occurring in lattices of quasi-crystals is not relevant for LEED where the experimental limit of the resolution is relatively low, while approximations used in the calculation limit the reso-lution even more.

The Penrose tiling can be extended to 3-D: rhombohedra with $\alpha_s = 116.57°$ and $\alpha_f = 63.43°$ are the two building blocks in 3-D [6.123], cf. Figure 6.19. The space is filled such that an icosahedral symmetry is obtained. The structural units in many icosahedral quasicrystals are called Mackay clusters [6.136]. An example of a 'pseudo' Mackay cluster occurring in the system Al-Pd-Mn is shown in Figure 6.20.

Figure 6.21 gives an example of experimentally determined electron densities showing decagonal quasicrystal symmetry.

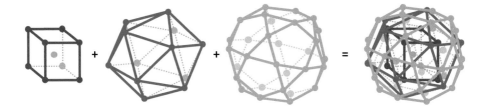

Figure 6.20 'Pseudo' Mackay icosahedron (right) which is considered to be the basic building block of the icosahedral $Al_{70}Pd_{21}Mn_9$ quasicrystal. It consists of an inner shell (partially occupied, hence 'pseudo', small body-centred cube, left), an icosahedron (2nd from left) and an icosidodecahedron (3rd from left), combined at right. The pseudo Mackay cluster contains 51 atoms and is approximately 1 nm in diameter. Adapted with permission from [6.123] C. Janot, *Quasicrystals: A Primer*, 2nd edition, Clarendon Press, Oxford, 1994. © C. Janot, 1992, 1994.

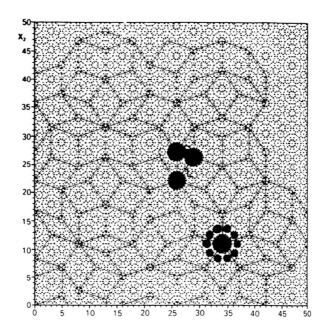

Figure 6.21 Projection of the electron density of the decagonal $Al_{70}Ni_{15}Co_{15}$ phase onto the 10-fold plane, resulting from an X-ray analysis [6.137]. The structure exhibits column-like clusters, some of which are marked as black regions. The Penrose tiling with edge length 1.979 nm is indicated as thin lines in part of the figure. Reproduced with permission of the International Union of Crystallography from [6.137] W. Steurer, T. Haibach, B. Zhang, S. Kek and R. Lück, *Acta Cryst. B*, vol. 49, pp. 661–675, 1993. https://doi.org/10.1107/S0108768193003143.

6.2.2   Structural Principles of Quasicrystals

Here we shall relate the structure of quasicrystals to the mathematical Fibonacci sequence and address a construction method based on projecting a simple periodic lattice in higher dimensions to the aperiodic structure seen in physical space. It is convenient to start in 1-D with a Fibonacci sequence and to describe how it arises by projection of a 2-D periodic lattice onto a 1-D line, forming a 1-D quasicrystalline lattice. The same approach can then be extended to produce a 3-D aperiodic quasicrystalline lattice by a similar projection from a 5-D or 6-D periodic lattice; we will not present this last step, as it is difficult to picture.

In Figure 6.17, dashed vertical lines connect tile sides in a 2-D Penrose tiling: only two spacings occur between these lines, a long (L) and a short (S) spacing, which are determined by the two tile shapes. These spacings follow the Fibonacci sequence, which is aperiodic.

The infinitely long Fibonacci sequence (called Fibonacci word) can be generated as follows: strings f_1, f_2, f_3, ..., composed of the elements L and S, are formed by the recursive rule that the n-th member of the sequence is assembled from the two preceding members by stringing them together as $f_n = f_{n-1}f_{n-2}$; we can build a Fibonacci sequence by starting with $f_1 = L$ and $f_2 = LS$ and then forming $f_3 = f_2f_1 = LSL$, $f_4 = f_3f_2 = LSLLS$, etc. Equivalently, one may start with L alone, then at each recursive step substitute $L \rightarrow LS$ and $S \rightarrow L$. Although the resulting string looks random at first sight (see Figures 6.15 and 6.20, for example), it has no randomness: it is unique by construction, even though it is aperiodic.

The long (L) and short (S) spacings in a quasicrystal have a length ratio that equals the golden mean: $L/S = \tau = (1+\sqrt{5})/2 = 1.618034... = 2\cos 36°$; this results from the shapes of the Penrose tiles, which reflect and induce the 5-fold rotational local symmetry.

Figure 6.22 suggests that the 1-D Fibonacci sequence can also be obtained by the 'cut and project' procedure: it projects onto a 1-D line grid points of a simple square periodic 2-D lattice inclined by a suitable angle α with respect to that 1-D line, namely by a projection of the dots within the shaded band in Figure 6.22 onto the 1-D line. The Fibonacci sequence of long and short spacings is obtained if $L/S = \tau = \cos\alpha/\sin\alpha$, so that $\alpha = 31.72°$. The width Δ of the band is $a(\cos\alpha + \sin\alpha)$, where a is the lattice constant of the square lattice, so that the band spans one 2-D unit cell. The 1-D line is called the real or parallel space ($x_{||}$, or x_{par} in Figure 6.22), while the direction perpendicular to it ($x_\perp$, or x_{perp} in Figure 6.22) is called the perpendicular space. The electron density in 1-D real space $\rho(x_{par})$ is the density parallel to the 1-D real space line, for example, composed of one atom at each dot on the 1-D real space line in Figure 6.22.

The above mentioned 'cut and project' construction reduces the task of describing or producing a 1-D aperiodic quasicrystalline structure (i.e., a Fibonacci sequence) to a simple projection from a periodic 2-D square lattice, where symmetry groups and diffraction conditions can be defined, onto a 1-D line inclined at angle α. This procedure can be generalised to higher dimensions: a 2-D quasicrystal (like the 2-D Penrose tiling) can be obtained by projection from a simple periodic 4-D lattice, while a 3-D quasicrystal can be obtained by projection from a simple 6-D lattice (e.g., a 6-D bcc lattice) in the case of icosahedral quasicrystals; for axial quasicrystals (such as decagonal quasicrystals), one of the 3-D directions is not quasicrystalline but already periodic, so that a 5-D lattice suffices for generating their real 3-D structure.

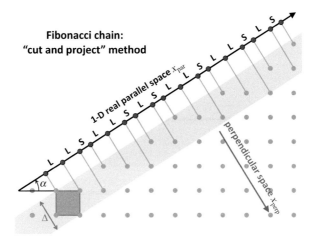

Figure 6.22 A Fibonacci sequence is shown as red dots in 1-D real parallel space. In the 'cut and project' procedure, the red dots result from projection onto the 1-D line called 'real space' or 'parallel space' from the grey dots within the grey strip of width Δ in a 2-D square lattice that extends into the 'perpendicular space' and is rotated by angle α; the strip's width Δ and the angle α are discussed in the text. Long and short segments between dots in the 1-D real space are labelled L and S, respectively.

However, the lattice points of the higher-dimensional space are in general not simple atoms, but 'atomic hypersurfaces' or 'hyperatoms': these could be, for example, simple rods of length Δ, as used in Figure 6.23, or spherical hyperspheres or tricontahedral hypersurfaces in 6-D [6.128], etc. The variety of options for the atomic hypersurfaces allows fitting the resulting interatomic distances and the elemental composition to experiment (typically with bulk X-ray or neutron diffraction). An example is given in Figure 6.23: here the simple rod is split into three parts (shown with three colours for Al, Cu and Fe) according to the elemental composition; in a variation of the cut and project method called 'section method', the colour that intersects the 1-D real space line then determines the element (Al, Cu or Fe) belonging at that position.

Since 3-D quasicrystals are conveniently described in 6-D or 5-D space, it is also practical to express directions, such as planes and beams, in higher dimensions with six or five indices, especially to highlight symmetries [6.121; 6.123; 6.131]; a loose analogy is the use of four indices for periodic hcp lattices. For example, (001) in periodic cubic lattices (sc, fcc, bcc) denotes a plane and surface with 4-fold rotational symmetry: similarly, (0001) is a plane and surface of an hcp surface with 3-fold rotational symmetry, while (000001) in quasicrystalline icosahedral lattices denotes a plane and surface with 5-fold rotational symmetry. In LEED experiments the specular reflection from a periodic crystal surface is often labelled (00), omitting the third index (perpendicular to the surface): likewise, the specular reflection from an icosahedral surface is often labelled (00000) with five instead of six indices, the last one being

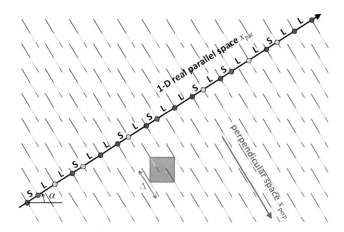

Figure 6.23 Construction of a Fibonacci chain, with 'atomic hypersurfaces' replacing the grey dots of Figure 6.22. Here the hypersurfaces are simple rods of length Δ with three segments of lengths proportional to the chemical composition (e.g., red = Al, green = Cu, blue = Fe). The rod colour that intersects the 1-D 'real' or 'parallel space' determines the atomic element at the intersection point, giving the correct overall composition: red dots become Al atoms, etc.

omitted. Non-specular reflected beams from a cubic surface are labelled (10), (01), $(\bar{1},0)$, $(0,\bar{1})$, (11), etc.; similarly, non-specular beams from a 5-fold symmetrical icosahedral surface are labelled (10000), (01000), (00100), (00010), (00001), $(\bar{1}0000)$, $(00\bar{1}00)$, (10001), $(\bar{1}000\bar{1})$, etc., again omitting the sixth index perpendicular to the surface [6.138; 6.139].

6.2.3   Diffraction from Quasicrystalline Surfaces

One of the most striking features of quasicrystals is that they can produce sharp diffraction spots, despite their lack of periodicity. That this is possible can easily be seen from the one-dimensional example shown in Figure 6.22 by taking the Fourier transform of the grey strip. We can represent the 1-D diffraction pattern $A(q)$ of the Fibonacci sequence shown in Figure 6.22 as the Fourier transform of the convolution of the 2-D square lattice with a 1-D box function $w(x_\perp)$ in the perpendicular space (i.e., a square pulse with $w(x_\perp) = 1$ for $0 \leq x_\perp \leq \Delta$ and 0 elsewhere) and by then selecting only the 1-D cut $A(q_{par})$. The Fourier transform $A(q_{par})$ then results from the product of the Fourier transforms of these two functions: $A(q_{par})$ comes from the product of a simple periodic square 2-D lattice of δ-functions and the 1-D transform $F(w(q_{perp}))$ of the box function centred at each lattice point, both shown in Figure 6.24. This is followed by projecting the result to the 1-D line where $q_{perp} = 0$:

$$A\left(q_{par}\right) = \left| \sum_{h,k} \delta(\mathbf{q} - h\mathbf{a}' - k\mathbf{b}') F\left[w\left(q_{perp}\right)\right] \right|_{q_{perp}=0}. \tag{6.49}$$

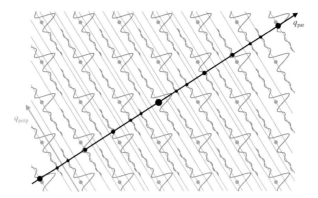

Figure 6.24 Construction of the diffraction pattern of a Fibonacci sequence according to Eq. (6.49). The 2-D square reciprocal lattice of the 2-D lattice in Figure 6.22 is shown as grey dots. The 1-D Fourier transform of the 1-D box function $F[w(q_{perp})]$ of Eq. (6.50) is schematically drawn as a grey wavelet at each 2-D reciprocal lattice point: it decays to zero along the grey lines shown. The cut along q_{par} (shown in black and oriented at angle α according to the irrational golden mean ratio as in Figure 6.22) gives the 1-D Fourier transform of the Fibonacci sequence and hence its diffraction pattern, shown here as black spots with variable intensities. Redrawn with permission from [6.123] C. Janot, *Quasicrystals: A Primer*, 2nd edition, Clarendon Press, Oxford, 1994. © C. Janot, 1992, 1994.

$F[w(q_{perp})]$ rapidly decreases with increasing distance q_{perp} from the lattice points, as schematically illustrated in Figure 6.24:

$$F[w(q_{perp})] = \Delta \frac{\sin(q_{perp}\Delta/2)}{(q_{perp}\Delta/2)}. \tag{6.50}$$

The cut along q_{par} of the Fourier transform shown in Figure 6.24 gives the positions of diffraction maxima of the 1-D Fibonacci sequence. Due to the irrational orientation α of the parallel space (q_{par}), there is an infinitely dense set of intersections (one intersection for each 2-D lattice point) and thus an infinitely dense set of reflections, but only a few of them have appreciable amplitude: amplitudes are shown qualitatively in Figure 6.24 by the size of the black spots and more quantitatively in Figure 6.25(a). Higher-intensity intersections are due to reciprocal lattice points that are closest to the parallel space. The resulting reflections are δ-functions in both the parallel space direction q_{par} and the perpendicular space direction q_{perp} for an infinite sequence.

The corresponding LEED diffraction pattern exhibits 5-fold symmetry and the reflections are arranged in circles around the specular beam at normal incidence, cf. Figure 6.25(b). The diffraction vectors of the strong reflections have the magnitudes $\tau^n q_0$ with powers $n = 0, 1, 2, \ldots$, cf. Figure 6.25(a); here an appropriate q_0 must be selected, since the choice of q_0 is not unique due to the self-similarity of the quasicrystalline lattice (namely, due to its repeated scaling with the factor τ apparent in Figures 6.23(a) and (b)). One of the nearest reflections to the central specular beam

(a)   (b)

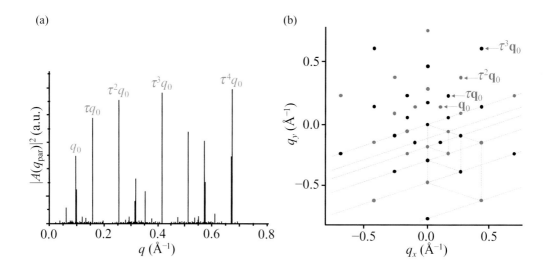

Figure 6.25 (a) 1-D Fourier transform of the Fibonacci sequence, with prominent peaks labelled to show self-similar scaling according to the golden mean τ. Here $q_0 = 2\pi\tau^2/S$, where S is the length of the shorter segment S. (b) A schematic 2-D 5-fold symmetrical LEED pattern, showing prominent spots, which lie on circles with radii $\tau^n q_0$. The dashed lines show relevant geometric relationships between spot positions, as well as the self-similarity with scaling according to the golden mean τ. The grey versus black spots highlight the 5-fold symmetry, while the spot positions exhibit 10-fold rotational symmetry (if we ignore intensities). Additional weaker reflections exist (but are not shown) between the drawn spots, both along radial lines (as in panel a) and between radial lines. Figure redrawn with permission from [6.140] M. Gierer, "Struktur und Fehlordnung an periodischen und aperiodischen Kristalloberflächen", *Habilitation*, Faculty of Geosciences, University of Munich, 2000.

should be labelled the (10000) beam. The diffraction vector $\mathbf{q}_{10000}$ defines the short distance in the sequence: $q_0 = 2\pi\tau^2/S$. Therefore, S is the short distance in the Fibonacci sequence.

An experimental SPA-LEED pattern of a decagonal $Al_{72.1}Ni_{11.5}Co_{16.4}$ quasicrystal is shown in Figure 6.26(a); a section through a line of reflections is shown in Figure 6.26(b), which also indicates the indices of the reflections [6.140]. The indices are chosen such that they match the L and S distances determined from an X-ray analysis. The diffraction pattern exhibits a 10-fold symmetry due to the superposition of two rotated domains: the point group of the decagonal quasicrystal is $\overline{10}\,m_2$ and, due to the ABAB... sequence of the 2-D quasicrystalline layers, two terminations exist that are rotated by $360°/10 = 36°$.

6.2.4 Quantitative Analysis of Quasicrystalline Surfaces

Only a few quantitative LEED I(V) analyses of quasicrystal surfaces have been performed, mainly due to their relatively complex structure, cf. Figure 6.18. A first

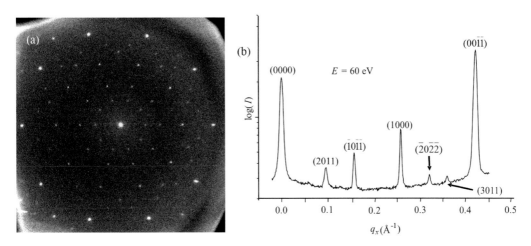

Figure 6.26 (a) SPA-LEED pattern of the 5-fold symmetrical surface of decagonal $Al_{72.1}Ni_{11.5}Co_{16.4}$ at 60 eV. (b) Line scan along $q_x = 0$ through the main reflections of the pattern shown in (a), using a logarithmic scale to emphasise the weaker intensities. Note the 4-index beam labels due to the decagonal bulk, which is periodic perpendicular to the surface. Reprinted from [6.141] M. Gierer, A. Mikkelsen, M. Gräber, P. Gille and W. Moritz, *Surf. Sci. Lett.*, vol. 463, pp. L654–L660, 2000, with permission from Elsevier.

challenge is that the experiment requires an atomically flat surface with an area of at least several square millimetres. Second, the quasicrystallinity severely complicates the multiple scattering theory of LEED, since no periodicity is present. Third, the aperiodicity implies also that there are many distinct local surface structures, as seen in Figure 6.18(a); therefore, many structural parameters should be fit to experiment, including also the alloy composition near the surface and the elemental identity of each atom. Fourth, all bulk terminations (defined as different cuts parallel to the surface) are in principle structurally distinct from each other for icosahedral crystals, since there is no periodicity perpendicular to their surface: many such terminations should therefore be explored. Fifth, steps on the surface can simultaneously expose different bulk-lattice terminations, the diffraction from which should be averaged over. Sixth, two surface orientations are normally possible, related to each other by a rotation of 180°. Seventh, the experimental I(V) database size is similar to that for simple surfaces with small unit cells, since the number of measurable strong beams is not much larger; consequently, there is a relative shortage of data to which the many structural parameters can be fit. For all these reasons, it is necessary to make approximations that in turn degrade the achievable structural detail and accuracy.

The quantitative structure analysis by LEED has the property that the effective scattering factor of a single atom depends on its neighbourhood through multiple scattering, in contrast to X-ray diffraction where the form factor of an atom is independent of its neighbourhood. That would in principle allow determining the chemical composition and atomic arrangement within the local clusters, but the resolution is lower than for normal crystal surfaces due to the lack of periodicity and the need for approximations in the calculation. With X-ray diffraction (or neutron diffraction), on the other hand, the structure determination is also difficult, mostly due to an insufficient number of experimental reflection intensities. In many cases, it is not possible to measure enough weak reflections between the few strong reflections (see Figure 6.25(a)), probably due to disorder [6.128]: the resolution achievable with X-ray and neutron data is then also limited.

To perform LEED calculations, efficient approximations have been developed to overcome the above-mentioned challenges. A suitable approximation for quasicrystalline surfaces is the 'average neighbourhood approximation'. The theory has been described by M. Gierer et al. [6.138; 6.139], the main aspects of which are briefly mentioned here. The structure around each atom is divided into a near cluster, where an exact multiple scattering calculation is performed, and an outer region, where an averaged scattering is assumed to be adequate. The time required for the calculation of one cluster strongly increases with the number of atoms in the cluster. Therefore, the multiple scattering calculation is done for a cluster of very limited size (typically 10 atoms) and only a limited number of different clusters is considered: among many possible clusters, the more likely cluster geometries are selected, based on knowledge of the bulk structure, including, for example, the 5-fold double ring that is prominent in Figure 6.18(a). If the cluster size and the number of distinct clusters are chosen sufficiently small, this approximation leads to acceptable computation times. A structural analysis of the 5-fold surface of i-$Al_{70}Pd_{21}Mn_9$ demonstrated that quasicrystalline structures are indeed accessible to a LEED I(V) analysis [6.138; 6.139]. Further progress would be possible if the techniques developed for nanocrystals, see Section 5.2.4, were applied to quasicrystals. This has not yet been done and NanoLEED had not been developed at the time when quasicrystals were analysed.

We next give a short description of the approximations used in the LEED analyses and show one example of the LEED study of an icosahedral Al-Pd-Mn quasicrystal [6.139].

The average neighbourhood approximation is illustrated in Figure 6.27. The atom positions in the far neighbourhood of each atom were averaged over the azimuthal angle φ. It was assumed that this average is sufficient for the 5-fold surface of the quasicrystal. The initial atomic positions were taken from the result of a bulk X-ray analysis. First, a large number of different terminations of the bulk lattice were considered. For each termination the z-positions (depths below the surface) of the sublayers were optimised while the lateral positions were kept fixed. The Pendry R-factors of the best-fit termination models reached values between 0.45 and 0.8: models with larger R-factors were then omitted. The best-fit R-factors favoured terminations with a large concentration of Al in the outermost two layers and a reduced outermost interlayer distance of 0.38 Å, compared to the bulk value of 0.48 Å, cf. Figure 6.27. In the next step the 10 best-fit terminations were mixed pairwise and the four topmost interlayer distances were optimised. The best-fit R-factor then reached 0.31 for a particular pair of terminations, which pair is concluded to dominate on the experimental surface. The structural result is displayed schematically in Figure 6.28 and a

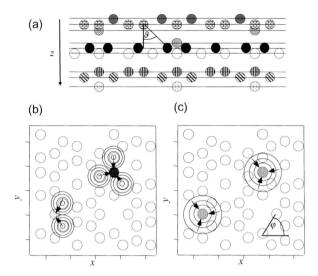

Figure 6.27 Illustration of the average neighbourhood approximation. (a) Schematic drawing of a section through the surface of a quasicrystal. Equal scattering properties of all atoms within a sublayer are assumed, different sublayers being indicated by different shading. (b) The multiple scattering within a cluster of nearest-neighbour or next-nearest-neighbour atoms is calculated exactly. (c) The surroundings further away are approximated by an average: the atom positions depend on the distance and on the polar angle ϑ, as shown in (a), but not on the azimuth φ shown in (c). Figure reprinted with permission from [6.140] M. Gierer, "Struktur und Fehlordnung an periodischen und aperiodischen Kristalloberflächen", *Habilitation*, Faculty of Geosciences, University of Munich, 2000.

section through the pseudo Mackay cluster is presented for comparison in Figure 6.29. The best-fit LEED I(V) curves are shown in Figure 6.30.

Important questions in studies of the structure of quasicrystal surfaces include: Are the surfaces quasicrystalline or reconstructed (due to bond breaking or formation) or do the surface layers exhibit an approximant structure, that is, a periodic structure with a short-range structure similar to that of the corresponding quasicrystal? Are they metallic or do they have low conductivity like the bulk structure? Is the surface composition the same as in the volume or is there surface segregation?

The results mentioned for i-$Al_{70}Pd_{21}Mn_9$ as well as those for some other quasi-crystal surfaces show that these surfaces exhibit quasicrystalline order and their surface composition resembles that of the bulk, while the surface termination favours exposing layers that are inherently rich in Al. These conclusions are not only based on LEED I(V) analyses, but also on LEED patterns, STM images [6.142–6.144] and X-ray photoelectron diffraction (XPD) analysis [6.145]; He atom diffraction and ion scattering spectroscopy (ISS) were also used to establish that the 5-fold surface of Al-Pd-Mn consists mainly of Al atoms, and that its structure is essentially a truncation of the bulk structure [6.139]. Slight deviations in the composition occur due to surface preparation by sputtering and annealing. Preferential sputtering changes the

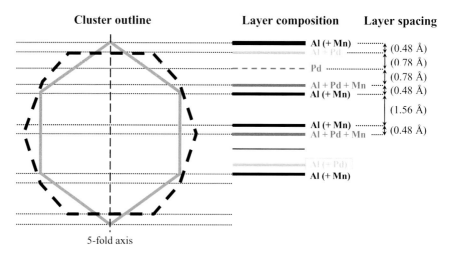

i-Al$_{70}$Pd$_{21}$Mn$_9$(000001) LEED surface structure

Layer composition	Layer depth LEED (bulk)	Layer spacing LEED (bulk)
Al$_{93}$Mn$_7$	$z_1 = 0.06 \pm 0.04$ Å	0.38 Å (0.48 Å)
Al$_{49}$Pd$_{42}$Mn$_9$	$z_2 = 0.44 \pm 0.12$ Å	0.78 Å (0.78 Å)
Pd	$z_3 = 1.22 \pm 0.15$ Å	0.82 Å (0.78 Å)
Al + Pd + Mn	$z_4 = 2.04 \pm 0.15$ Å	0.48 Å (0.48 Å)
Al (+ Mn)	$z_5 = 2.52 \pm 0.05$ Å	(1.56 Å)
Al (+ Mn)	$z_6 = (4.08$ Å$)$	(0.48 Å)
Al + Pd + Mn	$z_7 = (4.56$ Å$)$	

Figure 6.28 Best-fit structure of the 5-fold surface i-Al$_{70}$Pd$_{21}$Mn$_9$(000001) determined by LEED. The bars at the left represent individual atomic layers, with the surface at the top: the bar thickness represents the atomic density in each layer, while the grey level indicates the approximate chemical composition of each layer, spelled out to the right, where parentheses indicate small amounts. Only the composition of the two outermost layers was fit to experiment, with unknown error bars. The layer depth z_i is measured from the outermost layer of the unrelaxed bulk-like structure; values with error bars were fit to experiment by the LEED analysis. Corresponding interlayer spacings are shown on the right. Values in parentheses are for the unrelaxed bulk. Figure redrawn with permission from [6.139] M. Gierer, M. A. Van Hove, A. I. Goldman, Z. Shen, S.-L. Chang, P. J. Pinhero, C. J. Jenks, J. W. Anderegg, C.-M. Zhang and P. A. Thiel, *Phys. Rev. B*, vol. 57, pp. 7628–7641, 1998. https://doi.org/10.1103/PhysRevB.57.7628. © (1998) by the American Physical Society.

AlPdMn pseudo Mackay cluster

Cluster outline	Layer composition	Layer spacing
	Al (+ Mn)	(0.48 Å)
	Al + Pd	(0 78 Å)
	Pd	(0 78 Å)
	Al + Pd + Mn	(0.48 Å)
	Al (+ Mn)	(1.56 Å)
	Al (+ Mn)	(0.48 Å)
	Al + Pd + Mn	
	Al (+ Pd)	
	Al (+ Mn)	

5-fold axis

Figure 6.29 The 'bulk-like' geometry of the pseudo Mackay icosahedron for comparison with Figure 6.28, where the surface is also at the top and the same grey scale is used: the layer sequences near the surface are very similar in terms of spacings and compositions. Figure redrawn with permission from [6.139] M. Gierer, M. A. Van Hove, A. I. Goldman, Z. Shen, S.-L. Chang, P. J. Pinhero, C. J. Jenks, J. W. Anderegg, C.-M. Zhang and P. A. Thiel, *Phys. Rev. B*, vol. 57, pp. 7628–7641, 1998. https://doi.org/10.1103/Phys. RevB57.7628. © (1998) by the American Physical Society.

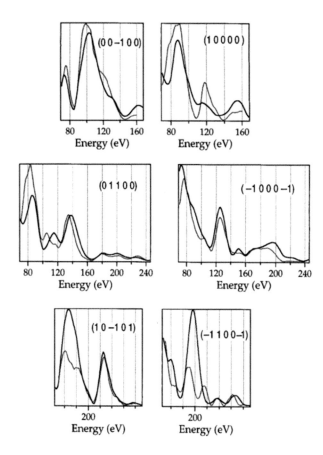

Figure 6.30 Best-fit LEED I(V) curves (experiment: thin lines; theory: thick lines) for six beams of the 5-fold surface of icosahedral $Al_{70}Pd_{21}Mn_9$. Figure reprinted with permission from [6.140] M. Gierer, "Struktur und Fehlordnung an periodischen und aperiodischen Kristalloberflächen." *Habilitation*, Faculty of Geosciences, University of Munich, 2000.

composition, which can be reversed by annealing. The quasicrystalline order is then also restored. The surface is not reconstructed (in the sense of bond breaking and formation) but some relaxation of interlayer distances occurs, which is similar to the relaxations on 'rough' metal surfaces such as fcc(110). These results are consistent with studies on other quasicrystal surfaces [6.146].

We next briefly mention some other LEED and STM results on quasicrystal surfaces for the interested reader. The first structural study of a quasicrystal surface was of decagonal $Al_{65}Co_{20}Cu_{15}$ [6.147]: it was investigated by STM, showing pentagonal clusters that clearly identified a quasicrystalline order of the surface. The high-resolution images showed a quasiperiodic structure similar to a 5-fold Penrose tiling and gave no indication of a reconstruction of the surface structure. Quantitative LEED I(V) analyses exist for 5-fold surfaces of icosahedral phases in the systems $Al_{70}Pd_{21}Mn_9$ [6.139] and $Al_{63.4}Cu_{24.0}Fe_{12.6}$ [6.135]. The structure of the Al-Cu-Fe phase was found to be similar to that in the Al-Pd-Mn phase, with the exception that screw dislocations were identified in the Al-Cu-Fe sample, unlike in the Al-Pd-Mn sample. The LEED structural result for Al-Pd-Mn was qualitatively confirmed by an X-ray photoelectron diffraction analysis [6.145] with a sample of slightly different composition. The 5-fold symmetry of this surface has also been observed by secondary electron emission imaging [6.148]. However, the surfaces of icosahedral quasicrystals are stepped, forming terraces that are structurally inequivalent. A complete atomistic structural characterisation of such a surface is a very complex task indeed: a further discussion is given by P. A. Thiel [6.149].

The characterisation of 10-fold decagonal surfaces is simpler than that of icosahedral surfaces because decagonal structures are periodic in the direction perpendicular to the surface, and therefore only a few distinct bulk terminations are possible. The structure of the 5-fold decagonal $Al_{73}Ni_{10}Co_{17}$ has been determined by a combination of a quantitative LEED I(V) analysis and high resolution STM [6.150], as well as by an analysis of an approximant surface [6.151]. The LEED pattern of the decagonal quasicrystals exhibits 10-fold symmetry due to the $\overline{10}\,m_2$ point symmetry and two terminations by the ABAB... stacking sequence, cf. the SPA-LEED pattern in Figure 6.26 for the same system.

Oxygen adsorption has been studied on the 5-fold surface of icosahedral quasicrystals of $Al_{70}Pd_{21}Mn_9$. Oxygen forms a thin Al-oxide layer (<1 nm) which destroys the quasicrystalline order of the surface. A thin and stable oxide layer passivates the surface; no indication of ordered oxide structures was found [6.152]. The 2-fold and 3-fold surfaces of the same system decompose into facets more readily, indicating qualitatively that they are less stable than the 5-fold surface [6.146].

Many other studies of various aspects of quasicrystal surfaces have been performed with STM and spectroscopic methods which will not be discussed here; reviews of the results can be found elsewhere [6.131; 6.132; 6.153; 6.154]. The present state of knowledge of quasicrystals is also summarised by W. Steurer [6.155]. The surface studies by LEED I(V) analysis remain a challenging subject where further improvements seem possible by the application of calculation methods developed for nanocrystals.

6.3   Modulated Surfaces

We consider here surfaces in which different layers have different lateral 2-D periodicities: the lattice mismatch then produces local variations in atomic environments. When crystalline layers having distinct misfit 2-D periodicities are stacked against each other, the mutual interaction between the layers frequently induces deviations in their atomic positions, vibrations, charge densities, etc. These modulations tend to have large-scale periodic repetitions that form moiré-like patterns. Modulations of atomic positions can occur both parallel and perpendicular to the layers. In particular, a surface composed of an overlayer on a substrate with a different 2-D lattice can

Figure 6.31 Example of a 2-D modulated surface: graphene on Ru(0001). The misfit of the lattice constants leads to an out-of-plane corrugation of the graphene layer together with a slight compression of the C-C bond lengths in the lower flat areas (yellow) and an expansion in the upper corrugation maxima (red). The corrugation is enhanced in the figure for better visibility. A slight induced modulation in the more rigid top substrate layers (blue) is not shown.

exhibit modulations in both the overlayer and the substrate. An example is shown in Figure 6.31: here a graphene layer covers a Ru(0001) substrate, causing perpendicular displacements (but also smaller parallel displacements) in the graphene, as well as minor modulations in the stiffer substrate. Surface modulations occur also in clean reconstructed metal surfaces. (A similar effect is obtained by simply rotating a layer relative to other identical layers, but such rotation is unlikely to be stable since layers with identical lattices are normally aligned as in the bulk crystal.) Aperiodic, in particular incommensurate, modulations may also exist in quasicrystals or composite crystals, for example, but are not discussed here.

The modulations in atomic positions give rise to characteristic satellite reflections in LEED and X-ray diffraction, which can be used to infer the presence, orientation and magnitude of the modulations. Specifically, the LEED patterns of modulated surfaces with overlayers are characterised by satellite reflections in the vicinity of the main reflections from substrate and adsorbate layers. In X-ray diffraction the satellite intensities are usually weak due to small modulation amplitudes, while in LEED multiple scattering effects enhance the satellite intensities. A typical LEED pattern and a short description of the origin of the satellite reflections are given in Section 4.7: see in particular Figure 4.18. In the following we will provide more detailed descriptions and analyses of such diffraction patterns.

A characteristic of modulated structures is the possibility to describe the displacements of all atoms in a large unit cell by a simple function with far fewer parameters than the many individual atomic coordinates (similar to a Fourier transform with only a few terms). The modulation function describes the deviation from a mean value. We will assume here that modulations occur only in the overlayer and top layers of a crystal while the bulk of the substrate is not modulated.

Two different types of modulations can be distinguished: 'displacement waves' and 'density waves'. The displacement waves are similar to position deviations in phonons, but frozen in time. The density waves describe periodic deviations from mean values of occupation factors or thermal vibrations, as well as charge density waves. A combination of both types of modulations occurs frequently, for example, a charge density wave may be combined with a displacive modulation. The two types of waves can be distinguished in 3-D X-ray and neutron diffraction, while analyses with LEED have not been reported until now. These waves are also less distinguishable in LEED than in X-ray diffraction, as will be discussed in Section 6.3.8. We consider only displacive modulations for LEED I(V) calculations in this section.

In X-ray crystallography, modulated structures are usually described by higher dimensional space groups, as mentioned in Section 6.2 for quasicrystals. These methods are not addressed here because the multiple scattering theory has not been worked out in higher dimensional space groups and seems not to be necessary for the cases considered in this section. We instead use a large superstructure cell or coincidence cell of finite size to describe the structure. This means that for LEED intensity calculations a commensurate superstructure is assumed. Real examples of surface modulations frequently appear to be incommensurate (at least within the coherence size of the LEED beam, i.e., on the scale of 10 nm), but for LEED I(V) calculations the approximation by a sufficiently large coincidence cell seems to be sufficient: this approximation forces the two lattices to coincide in 2-D over finite distances, thus forming a periodic moiré pattern with a finite coincidence cell.

6.3.1　Principles of Modulated Structures

The layered stacking of two (or more) different materials (e.g., a substrate and an adsorbate layer) having distinct lattices can mutually induce deviations of atomic positions in each other, leading to lattice modulations. The period of the modulation function is usually much larger than the translation vectors of the undistorted basic lattices of the component materials and may be commensurate or incommensurate with those basic lattices. The terms 'coincidence site lattice' or 'high-order commensurate (HOC) lattice' are frequently used synonymously for commensurate modulated lattices, expressing that a long but finite period exists. In STM investigations the term 'moiré structure' is preferred ('moiré' comes from a French word describing cloth with a rippled appearance). We here select the expression 'modulated lattice' applied to both commensurate and incommensurate structures and use the term 'basic lattice' for the undistorted layers, which may be an adsorbate layer or a substrate layer. In most cases, modulations occur in an adsorbate layer and/or in the top substrate layers, induced by the interaction between the outermost layer and the substrate. The modulation may be negligibly small in some cases.

Examples of 1-D displacive lattice modulations are shown in Figure 6.32. A displacive modulation wave may consist of two transverse and one longitudinal displacement, as in phonons. One transverse displacement is 'in-plane' and one is 'out-of-plane', normal to the surface. The most prominent type of modulation observed with STM or AFM is the height modulation or 'corrugation', which is the

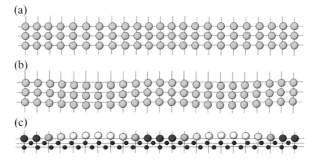

Figure 6.32 Models for three possible displacive 1-D modulations in an adsorbate layer (large spheres) on a substrate (small spheres); two periods of a sinusoidal modulation function are shown from left to right. (a) Longitudinal modulation parallel to the surface, looking down onto the surface; here the horizontal spacings between adsorbate rows varies between small at left, centre and right, and large in between those positions, as can be seen by comparison with the drawn unmodulated grid; (b) transverse modulation parallel to the surface, looking down onto the surface, shown against an unmodulated grid; (c) transverse modulation normal to the surface, looking parallel to the surface. The substrate atoms here have a different average horizontal periodicity, to show a possible cause of the height modulation.

out-of-plane transverse component and may be combined with in-plane transverse and longitudinal components.

Modulations may be 1-D or 2-D. We here use the notation that a 1-D modulation function has one single wave vector, independent of the number of Fourier components or function parameters. Modulations with two modulation waves in two lattice directions are denoted as 2-D modulations, even if the two waves are related by symmetry and in fact only one independent set of parameters exists, such as, for example, in a planar hexagonal lattice with two translation vectors having the same length. The direction of the modulation wave vector is not necessarily the direction of a translation vector, as shown in the examples in Figure 6.33, and may take any oblique direction. In the latter case the symmetry of the basic lattice is not preserved. If the wave vector does not coincide with a lattice vector, incommensurate structures occur, except when the wave vector has a rational relation to a lattice vector.

The modulation wave describes the atomic positions in real space; the direction of the wave vector is conventionally given in reciprocal space because the terminology has been developed for X-ray diffraction where the modulation is observed in the diffraction pattern. At surfaces, modulations can be observed with STM or AFM in real space. To avoid confusion, we always give the indices of the wave vectors in reciprocal space. The calculation of structure factors requires the atom positions to be described by a modulation function in real space. The Fourier transform of the modulation function gives the satellite spots in reciprocal space. Therefore, for the interpretation of the diffraction pattern the indices of the satellite spots are used to describe the modulation wave. This means that the wave vectors of the modulation wave are described in reciprocal space. Modulation functions are described in Section 6.3.5. For clarity the directions in both real and reciprocal space are given in Figure 6.33.

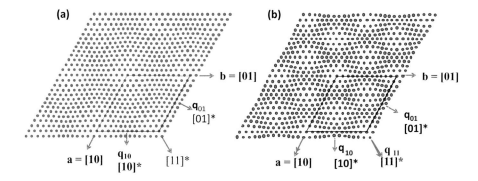

Figure 6.33 Two models of 2-D planar modulations in a hexagonal lattice. The unit cells marked in both panels at the bottom right are those for the simplified structure model for graphene on Ru(0001), with 13 basic graphene unit cells on 12 Ru unit cells in each dimension. The symmetry p3m1 is assumed, which would apply for graphene; for monoatomic structures the models also apply for the symmetry p6mm. (a) Longitudinal waves along the $\mathbf{a} = [10]$ and $\mathbf{b} = [01]$ axes in real space. (b) Longitudinal wave in the [11] and rotationally equivalent directions in real space. The directions in real and reciprocal space are indicated in the figure; for the wave vectors $\mathbf{q}$ the direction in reciprocal space is always used. The * indicates reciprocal space vectors (see Chapter 2 for the crystallographic conventions). Reprinted from [6.156] W. Moritz, S. Günther and K. Pussi, "Quantitative LEED Studies on Graphene", in K. Wandelt, ed., *Encyclopedia of Interfacial Chemistry: Surface Science and Electrochemistry*, vol. 4, pp. 370–377, 2018, with permission from Elsevier.

There exist symmetry restrictions on the possible modulation functions if the symmetry of the unmodulated structure is to be conserved. In the space group p3m1, only longitudinal modulations are allowed in the directions [10] and [11]. In-plane transverse and longitudinal modulations are coupled due to the 3-fold axis and the mirror plane, as can be seen in Figure 6.33. A transverse wave is not compatible with the mirror plane in the p3m1 symmetry. Symmetry restrictions will be discussed in Section 6.3.6.

The modulation period can only be determined within the resolution limit of the LEED instrument or the diffractometer (normally about 10 nm, but more for SPA-LEED, for example). In STM or AFM investigations the resolution limit is frequently given by the domain size or terrace size of the sample. The question whether the modulation is truly incommensurate or high-order commensurate remains, therefore, unresolved in most cases. Often, the modulation period is locally commensurate and varies on a larger scale around average such that it appears to be incommensurate in the diffraction pattern. This may be observable by a characteristic broadening of satellite reflections if the variation of the modulation period is broad enough. In this chapter, we will only treat idealised well-ordered cases.

For modulated surfaces, two different lattice constants must be considered and the two lattices may be commensurate or essentially incommensurate; both cases occur, but here we will only discuss the commensurate case. For quantitative LEED intensity calculations, the essentially incommensurate case is approximated by a coincidence lattice of sufficient size, so that the deviations from a truly incommensurate case are small and can be neglected.

With commensurate modulations, the structure may be described as a superstructure of the substrate lattice. However, the term 'modulated surface' is preferred when the atomic parameters within the superstructure cell are more easily described with a modulation function that has only a few adjustable parameters than with a long list of all individual coordinates. A structure refinement of modulated structures with all atomic coordinates as free parameters will most likely fail because the parameters are highly correlated and too numerous. A reduction of the number of free parameters by use of modulation functions is therefore necessary. In most cases, a few Fourier components are sufficient to describe the modulation function; if this is not the case, other functions can be used, for example, Gaussian functions, step functions or sawtooth functions for periodic antiphase domains and stepped surfaces.

Modulated lattices exhibit long range order and produce sharp reflections in the diffraction pattern. Sharp reflections also occur when the modulations are incommensurate with the substrate lattice, in which case no translation symmetry exists. Both the modulation function and the basic lattice are periodic. The Fourier transform of their combination produces discrete Fourier coefficients, that is, sharp reflections. The diffraction pattern of modulated structures is characterised by the occurrence of satellite reflections around each reflection of the basic lattice. Since two different lattices are combined with the same modulation period, two sets of satellites occur; for commensurate modulations both sets coincide.

6.3.2   Examples of Modulated Surfaces

Modulated surfaces are found in a series of adsorbate layers and some reconstructed clean metal surfaces. The first detected examples of surface modulations were the clean, reconstructed (100) surfaces of Pt [6.157] and Au [6.158]. The reconstruction of Pt(100) was initially identified as a (5×25) superstructure consisting of a quasi-hexagonal top Pt layer on the square Pt substrate lattice [6.159; 6.160]. Two phases could be identified depending on preparation conditions, one aligned to the lattice of the bulk – the so-called unrotated phase – and a slightly rotated phase. The orientation of the quasi-hexagonal layer and its temperature dependence were also investigated by X-ray diffraction, showing rotational transformations at 1,580 K and above 1,685 K [6.161]. A later study by high resolution He diffraction showed that the surface lattice is indeed incommensurate to the substrate [6.162]. The possibility that the second layer is also reconstructed was examined by a detailed STM and theoretical investigation, which found it to be unlikely [6.163]. A high resolution STM image showing the *unrotated* phase of Pt(100) is shown in Figure 6.34(a), while in Figure 6.34(b) the diffraction pattern of the *rotated* phase obtained with SPA-LEED is shown. The LEED pattern shows the superposition of four domains due to the substrate symmetry [6.164] (this study confirmed the existence of several discrete rotation angles depending on annealing temperature, ranging from ~0.75° to 0.94°). The size of the quasi-hexagonal unit cell has been determined from the LEED pattern;

(a)

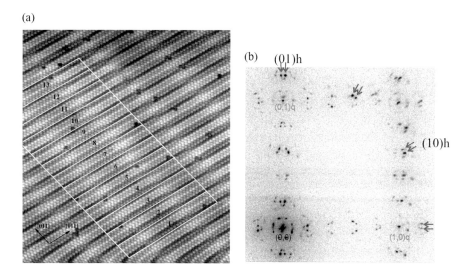

Figure 6.34 (a) Large atom-resolved STM scan of the unrotated hexagonally reconstructed Pt(100) surface. The image size is 20 nm × 20 nm. Thirteen (29×5) unit cells, each containing 30 × 6 top-layer atoms, are outlined and marked. The reconstruction unit cell is slightly smaller than six surface atom distances in the [011] direction, so the cell is shifted in the [011] direction with respect to the cubic substrate, which yields a modulation of the corrugation pattern inside the (29×5) unit cell. From this modulation the periodicity of the superstructure can be estimated. (b) SPA-LEED pattern (in reverse contrast to enhance visibility) of a hexagonally reconstructed Pt(100) surface rotated by 0.94°. The image shows a diffraction area that includes both the (00) and (11) spots (the latter unmarked at upper right). The (10)q and (01) q spots of the square (quadratic) bulk layers are marked in grey in the figure. The grey arrows marked by (10)h and (01)h are the first order reflections of the hexagonal top layer. The characteristic splitting of the spots is due to the rotation by 0.94° and two symmetrically equivalent domains, indicated by the pairs of grey arrows. The other split spots marked by grey arrows belong to two domains rotated by 90°. Panel (a) reproduced with permission from [6.163] G. Ritz, M. Schmid, P. Varga, A. Borg and M. Rønning, *Phys. Rev. B*, vol. 56, pp. 10518–10526, 1997. https://doi.org/10.1103/PhysRevB.56.10518. © (1997) by the American Physical Society. Panel (b) reproduced with permission from [6.164] R. Hammer, K. Meinel, O. Krahn and W. Widdra, *Phys. Rev. B*, vol. 94, p. 195406, 2016. https://doi.org/10.1103/PhysRevB.94.195406. © (2016) by the American Physical Society.

all satellite reflections could be identified but a quantitative analysis of the complicated structure has not been possible until now.

The clean Au(100) surface also exhibits a reconstruction similar to that of Pt(100) with a quasi-hexagonal top layer. From the LEED pattern a c(26×68) reconstruction was determined [6.165]; the structure was later investigated by X-ray diffraction and shown to be incommensurate [6.166–6.169].

A surface modulation resulting from a closer packed top layer was also found on the clean (111) surface of Au; this was unexpected, since the close packed (111) surface is the most stable surface of fcc metals. The reconstruction of Au(111) was first observed by LEED [6.170] and was assigned to a 1-D modulation with a $(\sqrt{3}\times22)$rect unit cell

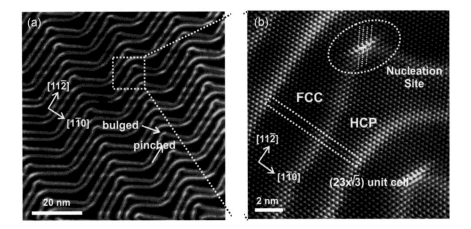

Figure 6.35 (a) STM image of the clean Au(111) surface (80 nm × 80 nm), showing its herringbone or chevron reconstruction. (b) Atom-resolved STM image showing details of the atomic structure of the reconstruction on Au(111) (14 nm × 14 nm), blown up from panel (a). Reprinted from [6.172] F. Besenbacher, J. V. Lauritsen, T. R. Linderoth, E. Lægsgaard, R. T. Vang and S. Wendt, *Surf. Sci.*, vol. 603, pp. 1315–1327, 2009, with permission from Elsevier.

by M. A. Van Hove et al. [6.165]. The resolution with LEED only allowed observing the 1-D modulation with a superposition of three domains; the structure was later named $(23 \times \sqrt{3})$ and called a striped domain phase [6.166]. STM images revealed that indeed a 2-D modulation exists and that it has a herringbone pattern. The herringbone structure was investigated with the higher resolution of X-ray diffraction and phase transitions could be observed at higher temperatures [6.171]. In these papers the structure was called 'chevron structure'. The X-ray results were confirmed by STM images shown in Figure 6.35 [6.172]. The indexing of satellite reflections and possible modulation functions in this surface are discussed in more detail in Section 6.3.5. The basic reconstruction consists of a 1-D compression of the top hexagonal layer in the $[1\bar{1}0]$ direction and a second compression in a symmetrically equivalent direction, either $[\bar{1}01]$ or $[01\bar{1}]$, leading to a herringbone structure, as can be seen in Figure 6.35. Lateral compression of the outermost metal layer is the common feature of the reconstruction of fcc(100) and fcc(111) metal surfaces observed to date.

Modulated structures have also been observed for metal adsorption on other metal surfaces, for example, Cu/Ru(0001) [6.173] and Ag/Ru(0001) [6.174], in physisorbed layers of noble gases on metals [6.175] and for alkali metals on transition metal substrates, for example, K/Ni(100) [6.176]. The cause of the modulation in the latter two cases is the balance between adsorption energy and repulsion between adsorbate atoms; the modulation period therefore depends on coverage or vapour pressure.

An example for lattice modulations occurring in adsorbed molecular layers is given by C_{60} on Pb(111) [6.177]; STM images of two phases at different coverages are shown in Figure 6.36. The moiré pattern indicates a height modulation of the C_{60} molecules, which is probably caused by a reconstruction or modulation of the Pb

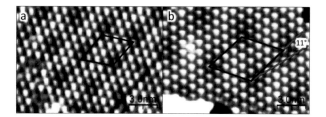

Figure 6.36 Constant current STM images of two different phases of C_{60} on Pb(111): (a) the 3.49 nm moiré structure and (b) the 4.56 nm moiré structure. The fullerenes are in a close-packed structure, each molecule forming a bright spot. The moiré superstructures appear as faint hexagonal height modulations in the STM images. The unit cells of the moiré structures are indicated by rhombi. The 3.4 nm moiré structure is aligned with the C_{60} lattice, whereas the 4.6 nm moiré structure is rotated $11°$ with respect to the C_{60} lattice. Figure reproduced with permission from [6.177] H. I. Li, K. J. Franke, J. I. Pascual, L. W. Bruch and R. D. Diehl, *Phys. Rev. B*, vol. 80, p. 085415, 2009. https://doi.org/10.1103/PhysRevB.80.085415. © (2009) by the American Physical Society.

substrate. An interpretation of the STM images is discussed in Section 6.3.3, see Figure 6.39.

Examples which have attracted much interest are the modulated structures of graphene layers on metal surfaces. The origin of the modulation in these cases is a lattice mismatch between the graphene layer and the substrate. Lattice modulations are easily observed by LEED where satellite reflections occur around the main reflections. As an example, an STM image [6.178] and a LEED pattern of graphene on the Ru (0001) surface are shown in Figure 6.37.

Modulated structures have mostly been observed by STM or AFM, where the modulation is seen directly, and by LEED, where characteristic satellites occur around the main reflections. The diffraction pattern is often rather complicated and, in most cases, it has not been possible to analyse the structures quantitatively. A detailed structure determination of modulated surfaces is rare. Apart from the above-mentioned X-ray analyses of Pt(100), Au(111) and metal on metal systems, only two graphene adsorption systems have been investigated quantitatively. On these surfaces, quantitative LEED I(V), X-ray and DFT analyses have been performed so that the results from different methods can be compared. For graphene on Ru(0001) [6.179; 6.180], qualitative agreement between both experimental methods and the DFT calculation was found, but quantitative differences remained in the amplitude of the corrugation, while the other parameters agreed within the error limits. In the LEED study a height variation up to 0.15 nm was found, while the X-ray results showed 0.08 nm. The discrepancy could not be resolved yet but is probably caused by a small database in the X-ray study and an overestimate of the corrugation in the LEED study due to an insufficient fit and a large R-factor. A further LEED analysis has been made of graphene on Ir(111) [6.181], where the result matches well the structural parameters found in DFT calculations, but some differences remain in the corrugation height found by X-ray diffraction on the same system [6.182].

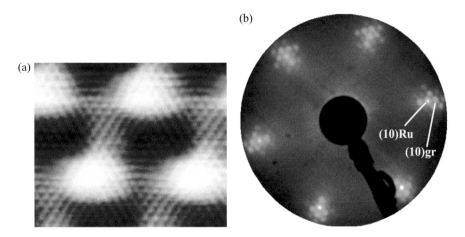

Figure 6.37 (a) STM image of a monolayer of graphene on Ru(0001), 5 nm × 4 nm, $I_t = 1$ nA, $V_{sample} = -0.05$ V. (b) LEED pattern of graphene on Ru(0001), at energy 60 eV. The (10) beams of the substrate and the graphene lattice are marked. Panel (a) reproduced with permission from [6.178] S. Marchini, S. Günther and J. Wintterlin, *Phys. Rev. B*, vol. 76, p. 075429, 2007. https://doi.org/10.1103/PhysRevB.76.075429. © (2007) by the American Physical Society. Panel (b) courtesy of S. Günther, private communication.

It is often extremely useful to combine diffraction methods with STM or AFM investigations. The existence of different domains and the symmetry of a single domain can be directly observed by STM and AFM, as well as fluctuations of the modulation period, which cannot always be simply derived from the diffraction pattern.

While the modulation period and symmetry may be identified from the diffraction pattern, the quantitative analysis is complicated, as will be discussed in the following sections. The most appropriate methods to analyse surfaces with large modulation wavelengths are SXRD and LEED. Standard multiple scattering programs for LEED are limited in the size of the unit cell, but the application of NanoLEED (see Section 5.2.4) could help, while low energy electron microscopy (LEEM) provides sufficient resolution to measure satellite intensities in the case of very large modulation vectors. LEED therefore is appropriate for analysing modulated surfaces. In this respect, both LEED and X-ray diffraction have advantages and disadvantages.

In the case of X-ray diffraction, the intensity of satellite reflections arising from a single atomic layer is extremely low and difficult to measure, so that the experimental database is naturally small. Surface X-ray diffraction requires well-ordered surfaces and a well-collimated primary beam of high intensity. The quantitative analysis is easier than with LEED, since the kinematic theory can be used. A further advantage is that modulations in the substrate can be detected in deeper layers due to the large penetration depth of X-rays. The high instrumental resolution parallel to the surface allows a more precise determination of the modulation period than with a standard LEED system.

LEED, on the other hand, has the advantage of higher surface sensitivity, that is, less deep penetration. The diffraction geometries for both methods are different: for

SXRD mostly grazing incidence angles are used, to enhance the ratio of surface signal to background, while LEED is operated in backscattering geometry nearer normal incidence. The resolution normal to the surface is in principle larger for LEED, while for SXRD the resolution parallel to the surface is larger (see Figures 2.30–2.32, which illustrate the different diffraction geometries for LEED and X-ray diffraction).

6.3.3   Identification of Modulated Lattices from STM Images

We here describe modulated lattices simply as commensurate superstructures. This is the easiest approach for LEED I(V) calculations, as the standard programs can be used, except that the parameters of the modulation functions must be introduced. For truly incommensurate structures, in most cases, an approximation with a commensurate model will be appropriate and sufficient. The more sophisticated theory of higher-dimensional space applied in X-ray diffraction for (3+d)-dimensional space groups [6.183] is not described here and would not be directly applicable in the multiple scattering theory necessary for LEED.

In contrast to 3-D crystals, where in many cases just one basic lattice is modulated, there exist two different modulated 2-D lattices at surfaces: the lattice of the substrate and the lattice of a surface layer, which is usually an adsorbate or a reconstructed layer (for brevity, the latter will not be mentioned explicitly in the following, but assumed implicitly). In 3-D crystals, equivalent systems exist, for example, in intercalation compounds where two different lattices are combined. On modulated surfaces there exist in general three different periodicities, see Figure 6.38:

(1) the bulk-like substrate layer with unmodulated 2-D lattice constants ($\mathbf{a}_{sub}$, $\mathbf{b}_{sub}$);
(2) the adsorbate layer with 2-D lattice constants ($\mathbf{a}_{ad}$, $\mathbf{b}_{ad}$); here we must use the lattice constants of the undistorted adsorbate layer;
(3) the modulation period ($\mathbf{a}_{mod}$, $\mathbf{b}_{mod}$).

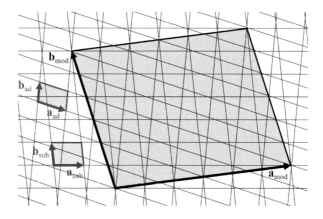

Figure 6.38 Schematic illustration of the basic unit cells of substrate (blue) and adsorbate (red). The coincidence cell (black) defines the modulation vectors.

The modulation period is common to the substrate and adsorbate lattice: if the adsorbate layer is modulated, the underlying substrate layer will be influenced by the adsorbate and will then be modulated with the same period.

Also, in contrast to X-ray studies of modulated 3-D crystals, the modulation lattice can be directly observed by STM or AFM. The moiré pattern occurring by the superposition of two rigid lattices with differing misfit lattice constants or orientations has been described in detail by K. Hermann in Section 6.5 of [6.134], where several examples illustrate how the moiré pattern changes with rotation of one of the lattices or with changes of the lattice constants. The knowledge of these relations is required for the correct interpretation of the STM images, since the orientation of the two basic lattices cannot be directly derived from the moiré pattern alone when they are not resolved in the STM images. In this section we describe how the orientation of the two superposed lattices can be derived when the size and orientation of the modulation vectors have been measured in the STM images and no diffraction pattern is available.

We shall assume that the modulation vectors are known from the moiré pattern observed in the STM images, as well as the lattice constants of the substrate but not the substrate orientation. The lattice constants of the adsorbate are only approximately known because of the possibility of compression or expansion of the adsorbate lattice.

With a procedure similar to that described in the following, the superstructures of C_{60} related to the Pb(111) substrate lattice have been determined by H. I. Li et al. [6.177]. The two different observed hexagonal moiré patterns (visible in Figure 6.36) were found to have lattice constants of 3.49 and 4.56 nm, respectively (the initial estimates were 3.4 and 4.6 nm, respectively). The C_{60} molecules are arranged in a hexagonal lattice with a C_{60}–C_{60} distance of about 1.0 nm, as is common for C_{60} monolayers on other (111) surfaces of metals. The commensurate superstructures related to the Pb(111) lattice are given by $\begin{pmatrix} 2 & -12 \\ 12 & 14 \end{pmatrix}$ for the 3.49 nm moiré pattern shown in Figure 6.36(a) and $\begin{pmatrix} 9 & -14 \\ 14 & 23 \end{pmatrix}$ for the 4.56 nm moiré pattern in Figure 6.36(b). The resulting structures are shown in Figure 6.39(a) and (b).

In the following, we describe a general procedure to derive the superstructures, applicable to all lattices, without assuming hexagonal symmetry. First, we give some relations between the lattices in real and reciprocal space and assume that the orientation of the lattices is known. The relations between the different lattice constants are described by matrices (see Section 2.1.11 for the matrix notation):

$$\begin{pmatrix} \mathbf{a}_{mod} \\ \mathbf{b}_{mod} \end{pmatrix} = \begin{pmatrix} M_{11} M_{12} \\ M_{21} M_{22} \end{pmatrix} \begin{pmatrix} \mathbf{a}_{sub} \\ \mathbf{b}_{sub} \end{pmatrix} = \begin{pmatrix} R_{11} R_{12} \\ R_{21} R_{22} \end{pmatrix} \begin{pmatrix} \mathbf{a}_{ad} \\ \mathbf{b}_{ad} \end{pmatrix},$$

$$= \mathbf{M} \begin{pmatrix} \mathbf{a}_{sub} \\ \mathbf{b}_{sub} \end{pmatrix} = \mathbf{R} \begin{pmatrix} \mathbf{a}_{ad} \\ \mathbf{b}_{ad} \end{pmatrix}, \tag{6.51}$$

where M_{ij} and R_{ij} are integers for commensurate lattices, and at least one matrix element is an irrational number for truly incommensurate lattices.

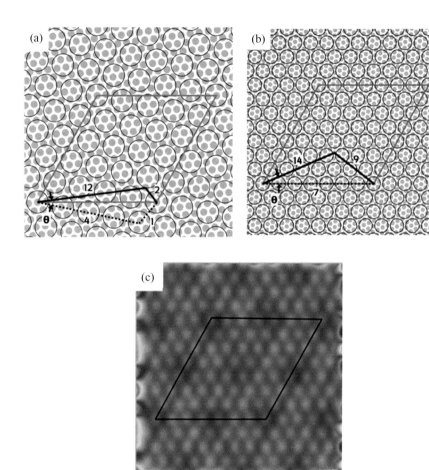

Figure 6.39 Superstructures related to the Pb substrate derived from the moiré patterns shown in Figure 6.36 for C_{60} on Pb(111). The grey dots mark the positions of the Pb atoms; the circles represent the C_{60} molecules. (a) The 4.56 nm moiré structure from Figure 6.36(b); (b) the 3.49 nm moiré structure from Figure 6.36(a); (c) contrast enhanced STM image of the moiré pattern of Figure 6.36(a), showing that the modulation vectors are doubled compared to the initial estimate of 3.4 nm. Figure reproduced with permission from [6.177] H. I. Li, K. J. Franke, J. I. Pascual, L. W. Bruch and R. D. Diehl, *Phys. Rev. B*, vol. 80, p. 085415, 2009. https://doi.org/10.1103/PhysRevB.80.085415. © (2009) by the American Physical Society.

Conventionally, the superstructure is related to the substrate lattice and the matrix **M** characterises the superstructure unit cell, or, if the lattices are incommensurate, the direction and length of the modulation vectors in real space. To calculate atomic positions in the adsorbate layer the matrix **R** is required as well, characterising the relation between the mean adsorbate lattice and the modulation lattice. The matrix **R** can be calculated from **M** if the lattice constants and the orientation of the mean adsorbate lattice with respect to the substrate lattice are known. We use Cartesian coordinates to describe the orientation of the lattice constants by using the following matrix notation:

$$\begin{pmatrix} \mathbf{a}_{sub} \\ \mathbf{b}_{sub} \end{pmatrix} = \begin{pmatrix} a_{x,sub} & a_{y,sub} \\ b_{x,sub} & b_{y,sub} \end{pmatrix} \quad \text{and} \quad \begin{pmatrix} \mathbf{a}_{ad} \\ \mathbf{b}_{ad} \end{pmatrix} = \begin{pmatrix} a_{x,ad} & a_{y,ad} \\ b_{x,ad} & b_{y,ad} \end{pmatrix}. \tag{6.52}$$

The matrix $\mathbf{R}$ is then simply calculated by:

$$\begin{pmatrix} R_{11} R_{12} \\ R_{21} R_{22} \end{pmatrix} = \begin{pmatrix} M_{11} M_{12} \\ M_{21} M_{22} \end{pmatrix} \begin{pmatrix} a_{x,sub} & a_{y,sub} \\ b_{x,sub} & b_{y,sub} \end{pmatrix} \begin{pmatrix} a_{x,ad} & a_{y,ad} \\ b_{x,ad} & b_{y,ad} \end{pmatrix}^{-1}. \tag{6.53}$$

For the lattice vectors in reciprocal and real space the following relations hold (vectors in reciprocal space are indicated by *),

$$\begin{pmatrix} \mathbf{a}_{sub} \\ \mathbf{b}_{sub} \end{pmatrix} = \mathbf{G}_{sub} \begin{pmatrix} \mathbf{a}^*_{sub} \\ \mathbf{b}^*_{sub} \end{pmatrix}, \tag{6.54}$$

where $\mathbf{G}$ is the metric matrix of the corresponding lattice indicated in the subscript (see Appendix C). We need only the 2-D lattice constants here, namely the lengths of the translation vectors $\mathbf{a}$, $\mathbf{b}$ and the angle γ between them:

$$\mathbf{G} = \begin{pmatrix} G_{11} & G_{12} \\ G_{21} & G_{22} \end{pmatrix} = \begin{pmatrix} a^2 & ab\cos\gamma \\ ab\cos\gamma & b^2 \end{pmatrix} \tag{6.55}$$

Finally, the wave vectors of the modulation must be calculated from the observed lattice constants $\mathbf{a}_{mod}$ and $\mathbf{b}_{mod}$. Modulation functions are waves: in this section we generally define the wave vectors in reciprocal space to avoid confusion; the modulation describes the displacements in real space and the period is also measured in real space by STM or AFM. The wave vectors $\mathbf{q}_{mod,1}$ and $\mathbf{q}_{mod,2}$ are identical to the reciprocal lattice vectors of the modulation lattice $\mathbf{a}_{mod}{}^*$ and $\mathbf{b}_{mod}{}^*$. A combination of both may also occur; we assume here

$$\mathbf{q}_{mod,1} = \mathbf{a}_{mod}{}^* \quad \text{and} \quad \mathbf{q}_{mod,2} = \mathbf{b}_{mod}{}^*. \tag{6.56}$$

The wave vectors of the modulation $\mathbf{q}_{mod,1}$ and $\mathbf{q}_{mod,2}$ are usually related to the reciprocal lattice vectors of the substrate $\mathbf{a}_{sub}{}^*$ and $\mathbf{b}_{sub}{}^*$. We can define a matrix $\boldsymbol{\sigma}$ describing the wave vectors $\mathbf{q}$ in terms of the reciprocal lattice of the substrate:

$$\mathbf{q}_{mod,1} = \sigma_{11}\mathbf{a}_{sub}{}^* + \sigma_{12}\mathbf{b}_{sub}{}^*,$$
$$\mathbf{q}_{mod,2} = \sigma_{21}\mathbf{a}_{sub}{}^* + \sigma_{22}\mathbf{b}_{sub}{}^*. \tag{6.57}$$

In the case of a 1-D modulation, only one vector $\mathbf{q}_{mod}$ exists and there is no modulation in another direction. We assume here always 2-D modulations with wave vectors parallel to the surface and the vector $\mathbf{c}$ normal to it. With a matrix $\mathbf{G}_{mod}$ defined for the modulated lattice, analogous to $\mathbf{G}_{sub}$ defined for the substrate lattice in Eq. (6.54), namely

$$\begin{pmatrix} \mathbf{a}_{mod}{}^* \\ \mathbf{b}_{mod}{}^* \end{pmatrix} = \mathbf{G}_{mod}{}^{-1} \begin{pmatrix} \mathbf{a}_{mod} \\ \mathbf{b}_{mod} \end{pmatrix}, \tag{6.58}$$

we obtain

$$\boldsymbol{\sigma} = \mathbf{G}_{\mathrm{mod}}^{-1}. \tag{6.59}$$

The matrix $\boldsymbol{\sigma}$ gives the positions of the first order satellite spots in the reciprocal lattice of the substrate. Equation (6.59) describes the transformation of the matrix $\mathbf{M}$ in real space to the matrix $\boldsymbol{\sigma}$ defining the superstructure in reciprocal space. These general relations hold for commensurate as well as for incommensurate modulated lattices.

Now, if we want to determine the possible orientations of the lattices of substrate and adsorbate which give the measured modulation periods, that is, the moiré structure, we must find the matrices $\mathbf{M}$ and $\mathbf{R}$ (Eq. (6.51)). Frequently, only the modulation lattice, that is, the moiré pattern, is observed. We consider the case where the lattice constants of the substrate are known and those of the adsorbate are approximately known, but neither the orientation of the substrate lattice nor that of the basic adsorbate lattice is resolved in the images. The lattice constants of the adsorbate are only known within certain limits because its lattice probably is distorted in the adsorbed state. It is also not always clear whether the structure is commensurate with the substrate lattice or not. Especially on polycrystalline samples, differently oriented moiré structures often occur with different lattice constants. If a limited set of moiré structures is found with fixed angles relative to the substrate, then it may be assumed that these are commensurate lattices, otherwise arbitrary orientations of the moiré lattice should occur. It can generally be assumed that commensurate lattices are energetically preferred.

If the lattice constants are known for the freestanding overlayer, then limits must be set to allow for elastic distortions in the adsorbed state. There exists a finite set of rotation angles between the two basic lattices which produce a coincidence structure within these limits for the lattice constants of the adsorbate structure. For the superposition of two rigid lattices without elastic distortion, a coincidence lattice does not necessarily exist. A formalism to find the possible relative orientations for coincidence structures has been worked out for hexagonal layers by P. Zeller and coworkers [6.184–6.186] and for general cases by K. Hermann [6.187]. Hermann investigated in detail the geometrical relations of the superposition of two lattices and how the moiré pattern rotates when the angle between the two lattices is changed. The method used by P. Zeller et al. is to calculate the modulated lattice from the product of the two lattice functions, as also used by Hermann. This is done in reciprocal space by the convolution of the Fourier transforms of the two lattices, that is, the two reciprocal lattices. The angle φ between $\mathbf{a}_{\mathrm{sub}}$ and $\mathbf{a}_{\mathrm{ad}}$ is taken as a variable parameter. The smallest vectors occurring in the convolution are the modulation vectors $\mathbf{q}_1 = \mathbf{a}_{\mathrm{mod}}{}^{*}$ and $\mathbf{q}_2 = \mathbf{b}_{\mathrm{mod}}{}^{*}$. The vectors in real space, $\mathbf{a}_{\mathrm{mod}}$ and $\mathbf{b}_{\mathrm{mod}}$, can be derived from these via Eq. (6.58).

If the two layers are hexagonal, then the modulation lattice is also hexagonal; this occurs frequently in graphene layers on (111) surfaces of fcc metals. For two hexagonal lattices with $|\mathbf{a}_{sub}| = |\mathbf{b}_{sub}|$, $|\mathbf{a}_{ad}| = |\mathbf{b}_{ad}|$ and $\gamma_{\mathrm{sub}} = \gamma_{\mathrm{ad}} = 120°$, Eq. (6.53) is reduced to:

$$a_{\text{sub}}\sqrt{M_{11}^2 + M_{12}^2 - M_{11}M_{12}} = a_{\text{ad}}\sqrt{R_{11}^2 + R_{12}^2 - R_{11}R_{12}}. \tag{6.60}$$

The reduction to two parameters allows an analytical solution to find all possible commensurate structures of graphene on hexagonal substrates where only the ratio $a_{\text{sub}}/a_{\text{ad}}$ and the angle φ between the vectors $\mathbf{a}_{\text{sub}}$ and $\mathbf{a}_{\text{ad}}$ appear as parameters. This formalism allows a quick and complete overview of possible commensurate superstructures for hexagonal lattices. It can be extended to square lattices as well. The more general description derived by K. Hermann [6.187] leads to the same results.

We do not describe the above-mentioned methods in detail here, but use a simpler, less sophisticated search method to find the possible orientations applicable for all 2-D space groups. In the following, we assume commensurate structures with known (or approximately known) lattice constants and unknown orientation. The interpretation of the observed moiré pattern then requires the determination of the angles between adsorbate and substrate lattice for which integer values can be expected in the matrices $\mathbf{M}$ and $\mathbf{R}$. We can search the integer values of R_{11} and R_{12} by checking possible combinations of M_{11} and M_{12} in Eq. (6.53). A distortion of the adsorbate layer is included by allowing a small deviation from integer values for R_{11} and R_{12}. Only two of the four elements in matrix $\mathbf{M}$ need to be independently varied. This gives a list of possible values for M_{11} and M_{12}; there is a very small set of possible values, M_{21} and M_{22}, which can be combined with M_{11} and M_{12}, mostly defining symmetrically equivalent unit cells. If the lattices are aligned (having parallel basis vectors), the test of all possible combinations is straightforward. If an arbitrary mutual orientation of the lattices is allowed, a systematic search on a sufficiently fine grid of rotations of the adsorbate lattice becomes necessary. The method is outlined as follows, assuming that the lattice parameters of the substrate $\mathbf{a}_{\text{sub}}$, $\mathbf{b}_{\text{sub}}$ and γ_{sub} are known; we also assume narrow limits for $\mathbf{a}_{\text{ad}}$, $\mathbf{b}_{\text{ad}}$ and γ_{ad} for the adsorbate layer:

(1) The vectors $\mathbf{a}_{\text{ad}}$, $\mathbf{b}_{\text{ad}}$ are calculated assuming an angle φ between $\mathbf{a}_{\text{sub}}$ and $\mathbf{a}_{\text{ad}}$. The vector components in Cartesian coordinates are used. We need only the 2-D matrices.

(2) Equation (6.53) can then be written as:

$$\begin{pmatrix} R_{11} & R_{12} \\ R_{21} & R_{22} \end{pmatrix} = \begin{pmatrix} M_{11} & M_{12} \\ M_{21} & M_{22} \end{pmatrix} \begin{pmatrix} a_{x,\text{sub}} & a_{y,\text{sub}} \\ b_{x,\text{sub}} & b_{y,\text{sub}} \end{pmatrix} \frac{1}{D} \begin{pmatrix} b_{y,\text{ad}} & -a_{y,\text{ad}} \\ -b_{x,\text{ad}} & a_{x,\text{ad}} \end{pmatrix} \tag{6.61}$$

with

$$D = \left(a_{x,\text{ad}}b_{y,\text{ad}} - a_{y,\text{ad}}b_{x,\text{ad}}\right). \tag{6.62}$$

We define

$$\begin{pmatrix} c_{11} & c_{12} \\ c_{21} & c_{22} \end{pmatrix} = \begin{pmatrix} a_{x,\text{mod}} & a_{y,\text{mod}} \\ b_{x,\text{mod}} & b_{y,\text{mod}} \end{pmatrix} \frac{1}{D} \begin{pmatrix} b_{y,\text{sub}} & -a_{y,\text{sub}} \\ -b_{x,\text{sub}} & a_{x,\text{sub}} \end{pmatrix}, \tag{6.63}$$

and obtain four equations for the four elements of matrix $\mathbf{R}$; only two of these need to be checked: $M_{11}c_{11} + M_{12}c_{21} = R_{11}$ and the analogous equation for R_{12}. The values for R_{21} and R_{22} are not independent and are derived from these.

(3) We can assume integer values M_{ij} and look for integer R_{ij}. The possible combinations of integer values M_{11} and M_{12} are easily checked while the maximum values are limited by a reasonable assumption for the size of the unit cell. A tolerance factor can be set for the allowed deviation of R_{11} and R_{12} from integer values. The calculation must be repeated for all combinations of M_{11} and M_{12}.

(4) Steps 1–3 must be repeated with a sufficiently fine grid of φ and, if necessary, for a grid with the lengths a_{ad} and b_{ad} for the translation vectors or with the ratio a_{ad}/b_{ad}, allowing a deformation of the adsorbate lattice within reasonable limits. Finally, the deviation from integer values of $\det(\mathbf{R}) = R_{11}R_{22} - R_{12}R_{21}$ can be plotted as a function of φ to visualise angular regions where coincidence structures can be expected. The quantities $\det(\mathbf{M})$ and $\det(\mathbf{R})$ give the size of the modulation unit cell in terms of the substrate unit cell and the adsorbate unit cell, respectively. These values and $\mathbf{a}_{mod}$, $\mathbf{b}_{mod}$ and γ_{mod} can be checked against the observed values in the moiré pattern.

It should be noted that the acceptable matrices $\mathbf{M}$ with integer components include all possible coincidence structures. The superstructure can be related to the primitive vectors $\mathbf{a}_{sub}$ and $\mathbf{b}_{sub}$ or to a superstructure of the substrate lattice. For example, the matrix $\mathbf{M}$ may have integer components when related to the single hexagonal unit cell but not when related to a $(\sqrt{3} \times \sqrt{3})R30°$ superstructure. Of course, integer components of $\mathbf{M}$ may occur in both cases.

As an example, we show in Figure 6.40 the results for a possible quasi-hexagonal superstructure due to top-layer reconstruction of the square lattice of clean Pt(100). The data are taken from R. Hammer et al. [6.164]. On this surface two reconstructions have been observed, as mentioned in Section 6.3.2: a non-rotated and a rotated quasi-hexagonal structure. The purpose here is to show that we can easily find the lattice constants of the rigid, distorted hexagonal layer and the orientation with respect to the substrate, if the modulation vectors are known; these are seen in the STM images. For the non-rotated case, the authors of the study find a rectangular superstructure cell with approximate dimensions 7.2×32.7 nm aligned to the square lattice of the (100) surface.

We look for possible quasi-hexagonal lattices, and their orientations, which are commensurate with this superstructure. The solution can of course be found directly by a simple calculation, but we want to illustrate the search method and use this as a test case. As input values we take the square lattice of Pt(100) with lattice constants $a_{sub} = 0.2772$ nm, the rectangular modulation lattice with $a_{mod} = 7.2$ nm, $b_{mod} = 32.7$ nm and a hexagonal lattice with $120°$ angle between the lattice vectors; we allow a contraction of the lattice in the range between 1 and 6%, that is, $a_{ad} = b_{ad}$ ranging

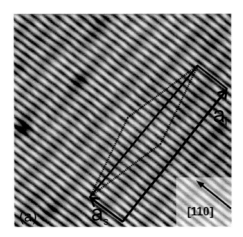

Figure 6.40 STM image of the non-rotated quasi-hexagonal reconstruction of Pt(100), 44 nm × 42 nm. Reproduced with permission from [6.164] R. Hammer, K. Meinel, O. Krahn and W. Widdra, *Phys. Rev. B*, vol. 94, p. 195406, 2016. https://doi.org/10.1103/PhysRevB.94.195406. © (2016) by the American Physical Society.

from 0.2744 to 0.2606 nm, in steps of 0.002 nm. A hexagonal lattice is assumed in the search procedure; a quasi-hexagonal layer or distorted hexagonal layer is obtained by allowing a deviation from integer values in the elements of the matrix **R**.

The search procedure then finds all possible combinations of M_{11} and M_{12} which give R_{11} in the vicinity of an integer value within an error bar of 0.1. The integer values are taken to calculate the vector components $a_{x,\text{ad}}$, $a_{y,\text{ad}}$, $b_{x,\text{ad}}$ and $b_{y,\text{ad}}$ of the distorted hexagonal lattice and the angle between the translation vectors. The corresponding vectors $\mathbf{a}_{\text{mod}}$, $\mathbf{b}_{\text{mod}}$ are also calculated using the calculated vectors $\mathbf{a}_{\text{ad}}$, $\mathbf{b}_{\text{ad}}$ and the unchanged substrate vectors $\mathbf{a}_{\text{sub}}$, $\mathbf{b}_{\text{sub}}$.

Solutions occur only for a hexagonal lattice with 4% contraction: no solutions exist for other values in the grid of $\mathbf{a}_{\text{ad}}$. The distortion of the hexagonal lattice is calculated with the integer values of the matrix elements of **R**. Several possible structures are found: only two structures occur in the vicinity of 90° between $\mathbf{a}_{\text{mod}}$ and $\mathbf{b}_{\text{mod}}$; both are aligned with the **a**-axis of the adsorbate layer parallel to the **a**-axis of the substrate. Six further solutions can be excluded as their modulation lattice deviates too much from the observed rectangular lattice. The details of the two solutions are the following, with lengths given in nanometres and matrices written by rows as (m_{11}, m_{12} / m_{21}, m_{22}):

Solution 1:

$\mathbf{M} = (26, 0 / 1, 118), \mathbf{R} = (27, 0 / 72, 142)$
$\mathbf{a}_{\text{ad}} = (0.26693, 0.00000), \mathbf{b}_{\text{ad}} = (-0.13339, 0.230350), |\mathbf{a}_{\text{ad}}| = 0.26693, |\mathbf{b}_{\text{ad}}| = 0.266190$
$\mathbf{a}_{\text{mod}} = (7.20720, 0.00000), \mathbf{b}_{\text{mod}} = (0.27720, 32.70960), \gamma = 89.5144°.$

Solution 2:

$$\mathbf{M} = (26,0/0,118), \mathbf{R} = (27,0/71,142)$$
$$\mathbf{a}_{ad} = (0.26693, 0.00000), \mathbf{b}_{ad} = (-0.13347, 0.23035), |\mathbf{a}_{ad}| = 0.26693, |\mathbf{b}_{ad}| = 0.26622$$
$$\mathbf{a}_{mod} = (7.20720, 0.00000), \mathbf{b}_{mod} = (0.00000, 32.70960), \gamma = 90°.$$

The second solution is exactly that given in R. Hammer et al. [6.164], the only difference being one of notation: there a centred lattice for the reconstructed layer is used while here a distorted hexagonal lattice has been assumed. In the first solution the vector $\mathbf{b}_{mod}$ is shifted by one lattice constant in the $\mathbf{a}$-direction, which leads to an oblique lattice that is not consistent with the observation but remains within the tolerance limits. The number of possible solutions depends strongly on the tolerance allowed in filtering integer values of the elements in $\mathbf{R}$. A tolerance of 0.1 for the large values of the modulation vectors occurring in this superstructure is fairly small and therefore requires small steps in the variation of the adsorbate lattice constants.

6.3.4   Reflection Indexing

In this section, we address the question of how to label, that is, how to index, the relatively complex sets of reflections that occur with modulated structures. Modulation waves are periodic functions and produce sharp reflections even for incommensurate modulations where no translation symmetry exists. For commensurate modulations a superstructure unit cell exists, so that the reflections could be indexed either according to that superstructure cell by integer indices or with respect to the substrate cell using fractional indices, as is the usual practice in LEED I(V) analyses. The use of superstructure indices is, however, not very convenient in the case of large modulation vectors; for incommensurate modulated structures it is not practical anyway.

When all possible reflections within the reciprocal unit cell are visible, as in the (7×7) reconstruction of Si(111), a description as a modulated structure does not make sense, since the reduction of the number of parameters is not possible by use of modulation functions. This reconstruction is therefore not considered to be a modulated structure.

Modulated surface structures are characterised by satellite reflections visible in the vicinity of main reflections; we define the main reflections as being due to the undistorted substrate lattice, while a secondary set of reflections is due to the undistorted adsorbate lattice. For the diffraction pattern of modulated surfaces, we use an indexing scheme following the usual practice in X-ray diffraction, where additional indices are used to describe the positions of satellite reflections relative to the associated main reflections. For 2-D modulated surfaces, two additional indices are used. In surface X-ray diffraction conventionally three indices are used for the main reflections, while for LEED usually only two indices are used for the reflections (since the component of the scattering vector normal to the surface can be derived from the electron energy and is therefore usually omitted). To avoid misunderstandings, we propose here to separate the three indices of the main X-ray reflection or the two indices of the LEED beam from the indices of the satellites by a semicolon; commas can be used to separate the indices when ambiguities still occur. As an example, assume that the 3-D diffraction vector of the reflection $I(\mathbf{k}' - \mathbf{k}_0)$ is:

$$\mathbf{k}' - \mathbf{k}_0 = h\mathbf{a}^*_{sub} + k\mathbf{b}^*_{sub} + l\mathbf{c}^*_{sub} + m_1\mathbf{q}_1 + m_2\mathbf{q}_2. \tag{6.64}$$

The three indices h, k and l define the main reflections, while the satellite reflection indices are m_1 and m_2: then the index is written as $(hkl; m_1m_2)$ or $(h, k, l; m_1, m_2)$ in X-ray diffraction. As the index l is usually omitted in LEED analyses, the index notations $(hk; m_1m_2)$ or $(h, k; m_1, m_2)$ can be used, as illustrated in Figure 6.41. For a 1-D modulation only one modulation wave exists, with one index m, yielding the index notations $(h, k; m)$ or $(hk; m)$ in LEED. This notation is used for commensurately as well as for incommensurately modulated lattices. The alternative for commensurate lattices when the modulation period is not too large is, of course, to use superstructure indices. In order to ensure a unique notation of the experimental data, the substrate reciprocal lattice is always used as a reference.

As an example with satellite reflections in large modulated cells, we consider the reconstruction of the Au(111) surface. At first, we consider a simplified model of the so-called $(\sqrt{3}\times22)$rect reconstruction of Au(111) [6.165]. The notation $(\sqrt{3}\times22)$rect defines a rectangular unit cell of dimensions $\sqrt{3}\times22$ in units of the nearest-neighbour

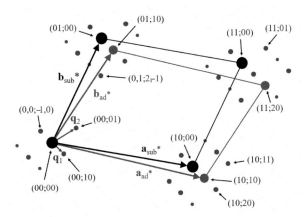

Figure 6.41 Example of indexing satellite reflections (red and blue spots) in LEED for a commensurate adsorbed overlayer on a substrate. The indices (defined in the text) refer to the main substrate reflections (black spots) for uniqueness. The red spots are reflections of the unmodulated adsorbate layer. The reciprocal lattice vectors of the substrate ($\mathbf{a}_{sub}^*$ and $\mathbf{b}_{sub}^*$) are coloured black, while those of the overlayer ($\mathbf{a}_{ad}^*$ and $\mathbf{b}_{ad}^*$) are red and those of the modulation ($\mathbf{q}_1$ and $\mathbf{q}_2$) are blue. The reciprocal lattice of the modulation is shown as a grey grid: all its lattice points are potential diffraction spots, but spot intensities tend to decay away from the main reflections, most being too weak to detect, which is symbolised here as smaller spots. Black indices refer to the reciprocal lattice of the substrate, while blue indices refer to the reciprocal lattice of the modulation.

distance on the Au(111) surface. However, this label does not give the full picture of the reconstruction, since it was chosen based on the initial observations by LEED, which had limited resolution. The model proposed then consisted of a simple 1-D compression of the top layer and neglected the complex 'herringbone' or 'chevron' structure which was later found to correspond to a 2-D modulation. STM images of the actual structure are shown in Figure 6.35.

The model for the earlier 1-D modulation is shown in Figure 6.42(a): there are 24 unit cells in the top layer covering 23 unit cells in the second layer (the numbers 24 and 23 are based on later more accurate determinations). This corresponds to a 1-D compression by about 4.2%. The compression is in the $[1\bar{1}0]$ direction, referring to the cubic lattice of Au. The $[1\bar{1}0]$ direction is taken as the x direction. The stripes occur in the direction normal to it, that is, in the $[11\bar{2}]$ direction. (We use here the notation $(23 \times \sqrt{3})$rect for comparison with the STM image shown in Figure 6.35).

The observed corrugation with two stripes in the $(23 \times \sqrt{3})$rect unit cell can be easily modelled by two parameters. The compressed layer is modulated in three directions: there occur lateral shifts in the x and y directions and vertical shifts in the z direction, the latter being the corrugation observed in the STM images. In Figure 6.42(a) the corrugation is not calculated by a global modulation function but by assuming a local rigid ball model so that top Au atoms in bridge sites are higher than in fcc and hcp hollow sites. The corrugation is therefore determined indirectly by the lateral modulations. The lattice constants of the compressed top layer are $a' = \frac{23}{24}a_0\sqrt{2}$ and $b' = \sqrt{3}a_0$, where a_0 is the cubic lattice constant. A transverse modulation by a cosine function with an amplitude $(1/12)b'$ shifts the atoms from an fcc site to an hcp site. As is obvious from the STM images in Figure 6.35, the region with hcp sites is smaller than that with fcc sites (because fcc sites are energetically more favourable than hcp sites). This can be simulated by an additional longitudinal modulation with a sine function: in Figure 6.42(a) and (b), an amplitude of 0.14 nm has been used. The diffraction pattern from a single domain is shown in Figure 6.42(c) and the superposition due to three domain orientations in Figure 6.42(d). The indices refer to the substrate unit cell; the satellite reflection can be indexed by a single index counting the distance from the main reflections. There are no cross-satellites visible – such as (10;21) – indicating that we have a superposition pattern from three separate 1-D modulated domains.

Figure 6.42(a) shows a simplified model with a 1-D modulation: the actual structure model exhibits a 2-D modulation, with a second modulation vector in the $[10\bar{1}]$ or $[01\bar{1}]$ direction of the cubic lattice. The satellite reflections corresponding to the 2-D modulation have been measured by A. R. Sandy et al. [6.171] with X-ray diffraction. A schematic structure model with two symmetrically equivalent 1-D modulated domains and the corresponding pattern with satellite reflections around the main reflections are shown in Figure 6.43. The structure model is taken from [6.171] and the directions indicated in the figure refer to the surface unit cell.

The modulation vectors of the Au(111) reconstruction have been determined in the detailed X-ray study by A. R. Sandy et al. [6.171]. The modulation periods are $L_K = 112.4a$ (32.43 nm) and $L_D = 22.5a$–$22.9a$ (6.49–6.61 nm), where a is the NN-distance between Au atoms in the bulk, $a = a_0/\sqrt{2}$ and a_0 is the cubic lattice constant of Au.

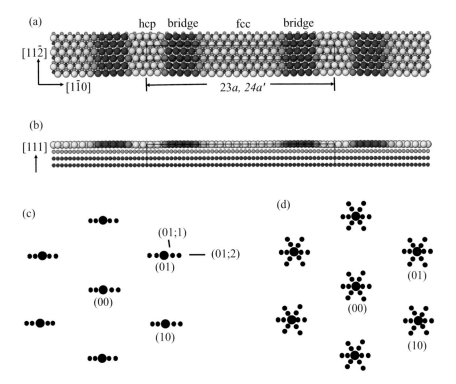

Figure 6.42 (a) Model of Au(111)-(23×√3)rect with 1-D compression and transverse and longitudinal sinusoidal modulations, in top view. The modulation parameters are explained in the text, where the lengths a and a' are also defined. (b) Side view of the corrugation derived from a rigid ball model, which is small and almost invisible in the drawing but indicated by the colours in the outermost layer: the Au atoms at bridge sites (red) are 0.023 nm (0.23 Å) higher that at fcc and hcp hollow sites (yellow). (c) Diffraction pattern of a single domain (using spot indexing for a single modulation direction). (d) Superposition of diffraction patterns for three domain orientations.

To obtain a unique identification, the satellite reflections are related to the main reflections of the substrate and a single domain. Some indices are given in Figures 6.39 and 6.40. The use of superstructure indices would not be useful; furthermore, the use of scattering vectors gives the position in reciprocal space but offers no hint by which modulation vectors the satellite intensities are generated. The use of two additional indices m_1 and m_2 provides the easiest way to identify satellite reflections.

6.3.5   Modulation Functions

The modulation function of a lattice is frequently represented by a Fourier series as the easiest way to describe a modulation wave, even though this is not the only

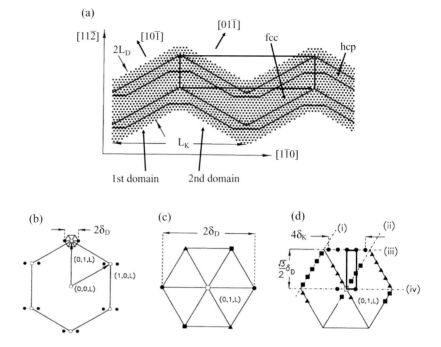

Figure 6.43 (a) Real space model of the 2-D reconstruction of Au(111) consisting of a regular sequence of two domains of the model shown in Figure 6.42 connected with rounded corners that give a 'herringbone' or 'chevron' appearance. The modulated superstructure cell is shown by the blue rectangle, with side lengths L_D and L_K. There exist three rotational domains of this model. (b–d) Reciprocal space representations near certain integer-order reflections: (b) First-order satellite reflections from a single domain of the striped phase, as in Figure 6.42(c). (c) Blow-up from panel (b) showing first-order satellite reflections around the $(0, 1, L)$ reflection from three domains of the striped phase. The distance to the main reflection is $\delta_D = 2\pi/L_D$. The modulation length of the striped phase, L_D, is shown in panel (a). (d) Satellites due to the 'herringbone' or 'chevron' structure around the (01) reflection measured by scans in different directions, in the same region shown in panel (c). The reciprocal cell of the modulation is indicated by the black rectangle. The wave vectors of the modulation are $q_1 = \delta_K = 2\pi/L_K$ and $q_2 = 2\pi/(\sqrt{3}/2L_D)$. Panels (b–d) use X-ray diffraction notation for the main reflections. The same patterns would be observable in LEED. Figure adapted with permission from [6.171] A. R. Sandy, S. G. J. Mochrie, D. M. Zehner, K. G. Huang and D. Gibbs, *Phys. Rev. B*, vol. 43, pp. 4667–4687, 1991. https://doi.org/10.1103/PhysRevB.43.4667. © (1991) by the American Physical Society.

possibility: other representations of the shift of the atoms may be preferred to avoid high-order Fourier coefficients in the Fourier series. Modulation functions are waves and are described by wave vectors **q** in reciprocal space. For 2-D modulations two vectors are required and for each modulation vector one longitudinal and two transverse displacement directions are possible, as illustrated in Figure 6.32 for a 1-D modulation. On surfaces one of the two transverse waves describes the corrugation (or z modulation) of the layer. The second transverse modulation describes the in-plane lateral displacement, while the longitudinal wave describes the in-plane compression or expansion. For simplicity, we consider here only the modulation within an adsorbate layer; the modulation of the substrate layers can be described analogously.

There are N atoms in the unit cell of the modulated lattice, with positions

$$\mathbf{r}_n = x_n \mathbf{a}_{\text{mod}} + y_n \mathbf{b}_{\text{mod}} + \Delta \mathbf{r}_n. \tag{6.65}$$

Here x_n and y_n are the atomic positions in units of the translation vector of the modulated lattice, ranging from 0 to 1; the perpendicular coordinate z_n is set to 0, as it is constant in the layer. The N displacements $\Delta \mathbf{r}_n$ of the N atoms are described by a Fourier series in which we use coefficients A and B for the cosine and sine terms, respectively; alternatively, a sine term and a phase could be used. The displacement of the n-th atom is in general given by:

$$\Delta \mathbf{r}_n = \{\Delta_n^x, \Delta_n^y, \Delta_n^z\}, \text{with}$$
$$\Delta_n^j(x_n, y_n) = A_{00}^j + \sum_{s,t} \{A_{st}^j \cos\left(2\pi(sx_n + ty_n)\right) + B_{st}^j \sin\left(2\pi(sx_n + ty_n)\right)\}. \tag{6.66}$$

The integer-valued indices s and t of the Fourier coefficients A_{st}^j and B_{st}^j refer to the wave vector $\mathbf{q}_{st}$, where s and t both range in the interval $\{0, \ldots, M-1\}$ and M is the highest useful frequency, given by the number of basic cells in the superstructure. Usually low-order Fourier coefficients suffice. The index j, with $j = 1, 2, 3$ (or x, y, z) indicates the type of displacement, that is, one longitudinal and two transverse displacements in two possible directions. It should be noted that for transverse modulations the shift vector is normal to the wave vector. An equivalent formulation of the Fourier series by sine terms and phases is frequently used in the literature on modulated structures:

$$\Delta_n^j(x_n, y_n) = A_{00}^j + \sum_{s,t} \left\{A_{st}'^j \sin\left(2\pi\left(sx_n + ty_n + \Phi_{st}^j\right)\right)\right\}, \tag{6.67}$$

with

$$\tan\left(\Phi_{st}^j\right) = \frac{A_{st}^j}{B_{st}^j} \quad \text{and} \quad A_{st}'^j = \sqrt{\left(A_{st}^j\right)^2 + \left(B_{st}^j\right)^2}. \tag{6.68}$$

In the following, we shall use the description by coefficients A and B for the cosine and sine terms, respectively.

In structures with more than one atom per primitive unit cell of the adsorbate layer, each atom may have its own amplitude and phase of the modulation. For instance, the graphene layer has two atoms per unit cell, so that the number of atoms in the superstructure cell is doubled. The coordinates of the k-th atom in the n-th unit cell are:

$$\mathbf{r}_{k,n} = \{x_{k,n}, y_{k,n}, z_{k,n}\} = \{(m_1 + x_k)\mathbf{a}_{ad}, (m_2 + y_k)\mathbf{b}_{ad}, \Delta z(x_n, y_n)\}, \tag{6.69}$$

with $k = \{1, 2\}$, while m_1 and $m_2 = \{0, \ldots, M-1\}$ and $n = \{1, \ldots, 2M^2\}$ label the unit cells; a $(M \times M)$ superstructure is assumed. The coordinates x_k and y_k refer here to the mean lattice of the adsorbate layer, $\mathbf{a}_{ad}$ and $\mathbf{b}_{ad}$. The shift of the z position of the k-th atom in the n-th unit cell is given by:

$$\Delta_{k,n}^z(x_{k,n}, y_{k,n}) = A_{00} + \sum_{s,t}\left\{A_{s,t}^z \cos\left(2\pi\left(sx_{k,n} + ty_{k,n}\right)\right) + B_{s,t}^z \sin\left(2\pi\left(sx_{k,n} + ty_{k,n}\right)\right)\right\}.$$

$$(6.70)$$

In Eq. (6.70) we have used the same Fourier coefficients for all atoms of the layer, which is probably appropriate for graphene. It might be different for a modulated compound layer. For example, in a layer of hexagonal boron nitride on Rh(111) the modulation of the boron and nitrogen sublattices may have different Fourier components as the interaction with the substrate is expected to be different.

For less smooth modulations, such as a step function or sawtooth function, high-order Fourier coefficients would be required, but in these cases simpler functions would be more appropriate. Even though Fourier coefficients are linearly independent, their influence on the structure and on the diffracted intensity is highly correlated, especially for LEED where multiple scattering effects play a dominant role. Therefore, it may be appropriate to use Gaussian functions for a corrugation instead of Fourier coefficients where the height and width of a maximum is more easily controlled by two parameters than with many Fourier coefficients. Whatever function is used to describe the modulation, the Fourier series is in general valid for the cases occurring in structure analyses and we use it here to illustrate the principal characteristics of the diffraction pattern: for our qualitative discussion, the kinematic theory is adequate and will thus be used.

To illustrate modulation functions in a 2-D lattice with symmetry p3m1, two examples for the possible modulation vectors $\mathbf{q}_{10}$ and $\mathbf{q}_{11}$ with sinusoidal functions are shown in Figure 6.33. It should be noted that the wave functions describe the displacement of atoms in real space (Eq. (6.66)) and the waves are also observed in real space in the STM images, while the wave vectors are conventionally described in reciprocal space because they are observed in the diffraction pattern. The directions in real and reciprocal space for the hexagonal lattice are indicated in Figures 6.31(a) and (b).

6.3.6   Symmetry Restrictions of the Modulation

Symmetry restricts the Fourier coefficients A and B for the cosine and sine terms of Eqs. (6.66) and (6.70) or for the phases of these waves, depending on the space group of the modulated surface. For the symmetry p3m1 shown in Figure 6.33, the indices of the basis wave vectors are (10) and (01): the 3-fold symmetry makes them mutually equivalent. Besides multiples like (20), (30), etc., representing higher-order Fourier coefficients, combinations like (12), etc., may also occur in the symmetry group p3 but not in p3m1. We can distinguish the symmetries of the (10) and the (11) waves. There are three symmetrically equivalent directions for the (10) wave and six for the (11) wave. The (11) wave produces a 6-fold modulation pattern. For monoatomic lattices as shown in Figure 6.33 the (10) wave also produces a 6-fold pattern. The 3-fold symmetry observed in the diffraction pattern results from the bulk crystal structure, that is, the underlying layers. Therefore, both directions are distinguished in the tables of the symmetry groups of modulated lattices (see for example in the International Tables [6.188]). This is of special interest for incommensurate structures. The symmetry of the commensurate superstructures considered here is not changed, of course: it remains p3m1. In this space group, only the z modulation and the longitudinal displacement are allowed. A longitudinal shift in the [10] direction implies a transverse shift in the [01] direction and vice versa, see for example Figure 6.33. An independent in-plane transverse wave in the [10] direction therefore does not occur; the same holds for the square lattices. Further symmetry restrictions exist for the cosine and sine terms of the Fourier coefficients or for the phases of the waves, respectively. The symmetry properties of the cosine and sine functions require that some coefficients vanish:

$$
\begin{aligned}
(10) \quad & x \text{ displacement} & A_{10} = 0, B_{10} \neq 0 \\
& z \text{ displacement} & A_{10} \neq 0, B_{10} \neq 0, & \qquad (6.71) \\
(11) \quad & x \text{ displacement} & A_{11} = 0, B_{11} \neq 0 \\
& z \text{ displacement} & A_{11} \neq 0, B_{11} = 0.
\end{aligned}
$$

Similar restrictions exist for the other 2-D space groups. A list can be found in the International Tables [6.188]: see, for example, Table 9.8.3.4(b), entitled '(2+2)-Dimensional superspace groups'.

6.3.7 Diffraction from Modulated Surfaces

We consider first the characteristic features of the diffraction pattern in the kinematic theory, as these can be analytically calculated. A brief description has been given in Section 4.7 in connection with the interpretation of the diffraction pattern. The characteristic feature of the diffraction pattern from a modulated surface is a set of satellite reflections surrounding main reflections. We derive this feature in the kinematic theory, but the main characteristics of the diffraction pattern remain valid for LEED and in the multiple scattering theory.

The theory is only very briefly outlined here to explain the occurrence of satellite intensities and the characteristics of their intensity. The reader is referred to a textbook for a full description of the theory [6.183]. In the kinematic theory the amplitude of the diffracted wave is given by

$$
A(\mathbf{k}' - \mathbf{k}_0) = \sum_n f_n(\mathbf{k}' - \mathbf{k}_0) e^{i(\mathbf{k}' - \mathbf{k}_0)\mathbf{r}_n}. \qquad (6.72)
$$

We consider displacive modulations. The atomic position vectors are then given by the position in the undistorted lattice and a shift which is described by the Fourier series in Eq. (6.66):

$$
\mathbf{r}_n = \mathbf{r}_{0,n} + \Delta\mathbf{r}_n. \qquad (6.73)
$$

The diffraction amplitude can thus be written as

$$A(\mathbf{k}' - \mathbf{k}_0) = \sum_n f_n(\mathbf{k}' - \mathbf{k}_0) e^{i(\mathbf{k}'-\mathbf{k}_0)\mathbf{r}_{0,n}} e^{i(\mathbf{k}'-\mathbf{k}_0)\Delta\mathbf{r}_n}, \tag{6.74}$$

or, using the indices h, k, l for the main reflections,

$$A(hkl) = \sum_n f_n(hkl) e^{2\pi i\left(hx_{0,n}+ky_{0,n}+lz_{0,n}\right)} e^{2\pi i\left(h\Delta_n^x+k\Delta_n^y+l\Delta_n^z\right)}. \tag{6.75}$$

The first phase factor in the sum of Eq. (6.75) refers to the undistorted lattice and gives the structure factor of the main reflections. The second phase factor contains the shift vectors and in a Fourier series expansion the sine and cosine terms occur in the exponential. These can be expanded into a sum over Bessel functions as

$$e^{iy\sin\vartheta} = \sum_{p=-\infty}^{\infty} J_p(y) e^{ip\vartheta}, \tag{6.76}$$

where J_p is the Bessel function of order p. After some rearrangement of the different factors, we obtain the reflection intensity for the simple example of a mono-atomic lattice and a 1-D sinusoidal modulation with the period M:

$$I(hkl) = |F(hkl)|^2 \cdot \left| \sum_{h,k,l} \sum_p \delta\left(h + \frac{p}{M}\right) \delta\left(k + \frac{p}{M}\right) \delta\left(l + \frac{p}{M}\right) \cdot J_p(D_x) J_p(D_y) J_p(D_z) \right|^2, \tag{6.77}$$

with $D_x = 2\pi h A^x$, $D_y = 2\pi k A^y$ and $D_z = 2\pi l A^z$. Here A^x is the amplitude of the sine wave in the x direction and analogously for A^y and A^z.

The formulas for the general case and for higher-order Fourier coefficients look more complicated, but the characteristic features of the diffraction pattern remain and can be concluded from Eq. (6.77) and the properties of the Bessel functions shown in Figure 6.44, namely:

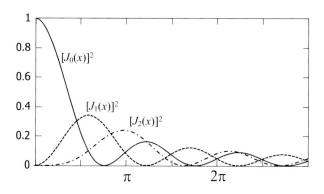

Figure 6.44 Square of the first three Bessel functions $\left[J_p(x)\right]^2$

- Each reflection is surrounded by an infinite set of satellite reflections in the case of incommensurate lattices. For commensurate superstructures the satellites will coincide with higher-order main reflections and their satellites, resulting in a finite density of reflections.
- The intensity of each satellite decreases (non-monotonously) with increasing distance from the nearest main reflection.
- The intensity of each satellite follows the intensity of the main reflection.
- The intensities of the main reflections decay as J_0^2 with increasing scattering vector while those of the satellites increase. This behaviour is illustrated in Figure 6.45.

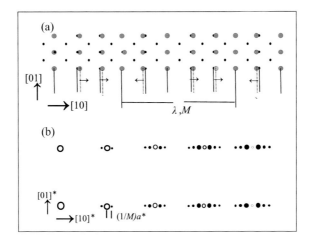

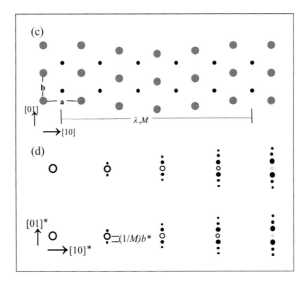

Figure 6.45 Characteristic diffraction pattern of two 1-D sinusoidal modulations: (a) longitudinal modulation and (b) corresponding diffraction pattern; (c) transverse modulation and (d) corresponding diffraction pattern. In (a) and (c) the black dots serve as a periodic grid to highlight the modulations (grey positions). In (b) and (d) the open circles are main reflections, while black dots are satellite reflections; dot size suggests average spot intensity.

The Bessel function of order 0 is a cosine-like function with decaying amplitude and characterises the intensity decay of the main reflections with increasing scattering vector. The higher-order Bessel functions are decaying sine-like functions where the maximum is shifted to higher values of the argument with increasing order. The characteristic diffraction patterns of two 1-D displacive modulations are shown in Figure 6.45.

Satellite reflections occur in the reciprocal lattice in the direction of the modulation vector. There are no satellites around the $(0k0)$ plane in reciprocal space for a longitudinal modulation in the [100] direction and no satellites around the $(h00)$ plane for a transverse modulation with wave vector in [100] direction and shift in the [010] direction.

For modulations of the scattering amplitudes (as opposed to modulations of atomic positions) the diffraction pattern looks different. This occurs, for example, with a modulation of the occupation factor in X-ray diffraction. In LEED the effective scattering amplitude of an atom depends on its surroundings and is therefore always coupled to a displacive modulation. A periodic deviation from the mean atomic scattering factor may occur by a modulation of the occupation factor in compounds, by vacancies, by charge density waves or by other factors that do not change the atomic position. For a 1-D modulation with a period M the atomic scattering factor of the n-th atom can be written as:

$$f_n(hkl) = f_{o,n}(hkl) + \sum_s \Delta f_s e^{2\pi i\, sn/M}, \tag{6.78}$$

where Δf_s are the Fourier coefficients; the complex-number formulation of the Fourier series is used here. Equation (6.78) can be inserted into the structure factor equation (Eq. (6.75)) and it is directly seen that satellites occur around each reflection. Each Fourier component produces one satellite. The diffraction pattern is schematically shown in Figure 6.46.

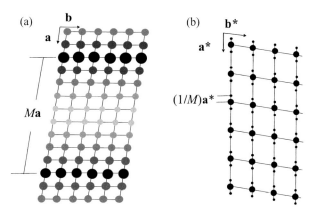

Figure 6.46 (a) Example of a modulation of the scattering factor with a periodic occupation of lattice points with different atoms. (b) Schematic drawing of the diffraction pattern for a sinusoidal modulation, showing only the nearest satellites.

6.3.8   Multiple Scattering Effects in LEED from Modulated Structures

Multiple scattering effects in LEED from modulated structures introduce several differences compared to X-ray diffraction. Because of multiple scattering, the effective scattering amplitude of an atom depends on its surroundings. This is illustrated in Figure 6.47. The effect is not small, as the amplitude and the phase of the effective scattering amplitude vary substantially when the interatomic distances and angles to the surrounding atoms are changed. The result is that any displacive modulation is accompanied by a modulation of the effective scattering factor. We may consider the superposition of two rigid lattices with different lattice constants. In X-ray diffraction the superposition of two reciprocal lattices is illustrated in Figure 6.48(a) and (b). Satellite reflections occur only when a modulation exists. This is not the case for multiple scattering: the superposition of two rigid lattices can produce non-zero intensity at all combinations of the two reciprocal nets. This can be explained by double and multiple diffraction from the two lattices, as illustrated in Figure 6.48(c), or by the variation of the effective scattering amplitude, as in Figure 6.47(b). Both interpretations are completely equivalent.

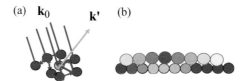

Figure 6.47 Illustration of multiple scattering in LEED. (a) The effective scattering amplitude relates the amplitude of the incoming wave $\mathbf{k}_0$ to that of the scattered wave $\mathbf{k}'$ leaving the surface without further scattering (drawn in red are various multiple scattering paths that end at the same green atom). The effective scattering amplitude therefore depends on the surroundings of an atom. (b) Varying surroundings are symbolised by different colours of atoms in inequivalent positions.

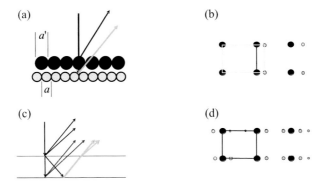

Figure 6.48 (a) Combination of two unmodulated lattices with lattice constants a and a'. (b) Diffraction pattern in X-ray diffraction showing the superposition of reflections at the points of the two corresponding reciprocal lattices. (c) Various multiple scattering paths in LEED (for normal incidence). (d) Diffraction pattern observed in LEED.

The consequence for LEED investigations of modulated surfaces is that it is not possible to distinguish whether reflections result from multiple diffraction or from surface modulation. By contrast, in X-ray diffraction it is possible to distinguish displacive modulations from density modulations by the observation of satellite reflections in the diffraction pattern.

In the kinematic scattering theory, the intensity of the satellite intensities follows the intensity of the main reflections and for displacive modulations the intensity increases with increasing distance from the main reflection, as illustrated in Figure 6.45. In X-ray diffraction the intensity of satellite reflections as a function of the scattering vector can be used to estimate modulation amplitudes; Patterson methods can be applied for that purpose [6.189]. However, this is not possible for LEED, although it remains in general true that the intensity of satellites increases with the modulation amplitude.

Due to multiple scattering in LEED, a modulation of the scattering amplitude occurs in addition to a displacive modulation. Therefore, more satellites are visible in LEED than in X-ray diffraction and occur also around the (00) spot; for a purely displacive modulation this would not be observed in X-ray diffraction for lateral modulations, see Figure 6.45. In the LEED pattern of modulated surfaces, the main factor determining the intensity of the satellites is a height modulation, that is, a transverse modulation normal to the plane of the surface. Since in LEED the measurement is usually made near normal incidence and at a high momentum transfer normal to the surface, the region in reciprocal space with low momentum transfer normal to the surface is rarely observed. Satellites near the (00) beam always occur at high momentum transfer normal to the surface and do not indicate a density modulation.

7 Surface X-ray Diffraction

X-ray diffraction is the main tool used to obtain the atomic structure of 3-D crystals. The relatively weak interaction of the X-ray beam with matter and the resulting large penetration depth makes it insensitive to structural details in a small surface area: the surface is therefore usually neglected in structure determinations of 3-D crystals. This has changed with the development of synchrotron radiation as an X-ray source which provided new applications in X-ray crystallography. The very high intensity and angular resolution of the synchrotron beam allow the study of numerous effects which had been considered too weak to detect with laboratory X-ray sources. It has been shown, however, that with intensive X-ray sources that are available now, the structure analysis at surfaces is also possible in the laboratory.

The study of the structure of surfaces with X-rays started in the early 1980s, when the UHV technique and the reliable preparation and control of well-defined surfaces both became routinely available and could be combined with X-ray diffraction. Many methods were subsequently developed to study surface properties with X-rays. We consider in this chapter only structure analysis techniques, and within this field only the development and application of direct methods, which provide a structural model for subsequent refinement. The surface structure determination with X-ray diffraction has been described in several review articles and chapters in books on surface properties and need not be repeated here. The structure refinement mainly follows the methods well established for analysis of 3-D crystal structures. This is, however, not the case for the structure solution by direct methods. The specific properties of diffraction at grazing incidence require and allow different methods not applicable for 3-D crystals. We therefore focus on the development of direct methods and give only a short overview of specific properties of surface X-ray diffraction (SXRD) and the related experimental methods. A full description of X-ray diffraction from surfaces and possible applications is beyond the scope of this book.

To facilitate the understanding of the specific aspects of X-ray diffraction at grazing incidence, Section 7.1 gives a brief description of some general properties of X-ray scattering and of surface diffraction. In Section 7.2, some representative experimental setups and experimental stations at synchrotron beam lines are presented. Section 7.3 discusses in more detail the use of direct methods in the analysis of diffraction data from surfaces. The specific diffraction conditions at surfaces allow the application of a set of direct methods that are different from the direct methods developed for X-ray diffraction from 3-D crystals.

7.1   X-ray Diffraction Methods

X-ray diffraction at surfaces offers a wide range of applications which exceed the capabilities of surface structure determination by LEED. In particular, the application of the kinematic theory of diffraction makes the analysis of diffraction data much easier than with LEED; surfaces of insulators can be more easily studied; magnetic scattering is observable; and resonant soft X-ray scattering allows the study of charge ordering and electron states. The experimental effort, on the other hand, is much larger than with LEED and the systems investigated so far by surface X-ray diffraction and by LEED overlap only partially. Since the beam time at synchrotrons is expensive, surface structures which could be solved with LEED are rarely studied with X-ray diffraction. The large penetration depth of X-rays also allows studying buried interfaces. The study of liquid–gas, liquid–solid and liquid–liquid interfaces is possible as well. Another major difference is that X-ray diffraction is not bound to UHV: this allows the study of reactions at surfaces under realistic environmental conditions (such as high pressures). We will next briefly present different techniques that produce structural information of free surfaces and interfaces. Since the focus of this book is on diffraction and surface structure determination, spectroscopic techniques are not discussed.

Two equivalent names are used for the study of surface structure with X-rays: surface X-ray diffraction (SXRD) and grazing incidence X-ray diffraction (GIXD or GIXRD). Overviews are available in review articles by I. K. Robinson and co-workers [7.1; 7.2], R. Feidenhans'l [7.3], and A. Stierle and E. Vlieg [7.4]. An overview of applications can be found in a review article by M. Sauvage-Simkin [7.5]. The strong dependency of the penetration depth on the incident and exit angles in the vicinity of the angle of total external reflection can be used to study depth dependent phenomena at surfaces; a description has been given by H. Dosch [7.6]. The study of buried interfaces from crystal truncation rod intensities was recently reviewed by A. S. Disa et al. [7.7].

Further X-ray techniques to study the surface structure are X-ray standing waves (XSW) and grazing incidence small angle X-ray scattering (GISAXS). An excellent review of the XSW technique has been given by J. Zegenhagen and A. Kazimirov [7.8]. GISAXS is mainly used when the focus is on the surface morphology and the shape of nanoparticles on surfaces, that is, on a larger scale than atoms; the measurement and the analysis methods have been reviewed by G. Renaud et al. [7.9] and, for nanoparticles, recently by T. Li et al. [7.10]. If hard X-rays are used with an energy of up to 80 keV (wavelength $\lambda \approx 0.015$ nm), this is emphasised by the name high energy surface X-ray diffraction (HESXRD). It has recently been shown that the measurement of a complete data set could be substantially sped up with hard X-rays in combination with a wide-angle area detector: see Section 7.2 for the description of different experimental techniques and references.

Widely used are also X-ray reflectivity measurements to study layered structures: these are reviewed by E. Chason and T. M. Mayer [7.11] and M. Tolan and W. Press [7.12].

At grazing incidence in the vicinity of the critical angle, the diffraction of X-rays cannot be treated strictly by the kinematic theory, because refraction then becomes important but is not included in the kinematic theory. The measurement and analysis methods then differ from the structure analysis method applied in SXRD and GIXD. The density profile normal to the surface, needed to properly treat refraction, is obtained by measuring reflectivity; the fundamentals have been described by J. Als-Nielsen et al. [7.13].

We mention for completeness that a number of other X-ray based techniques have been developed with the purpose of determining partial or complete structural data such as bond lengths, bond orientations and 3-D coordinates. For example, surface extended X-ray absorption fine structure (SEXAFS), which is a variant of EXAFS, has long been applied to study details in 3-D crystal structures. Reviews of these techniques applied to surfaces can be found in [7.14–7.17]. A related technique is near edge X-ray absorption fine structure (NEXAFS), also called X-ray absorption near-edge spectroscopy (XANES); reviews of NEXAFS/XANES have been published by G. Hähner [7.18] and by H. Ade and A. P. Hitchcock [7.19]. Techniques using X-ray induced photoelectrons include several versions of X-ray photoelectron spectroscopy (XPS), also called X-ray photoelectron diffraction (XPD), reviewed by C. S. Fadley et al. [7.20].

An important field is the study of magnetic ordering and charge ordering at surfaces. Magnetic effects are not treated in this book, but we mention for the interested reader some review articles about magnetic X-ray scattering and the technique of X-ray magnetic circular dichroism (XMCD), namely: J. Bansmann et al. [7.21], D. Sander [7.22] and W. G. Stirling and M. J. Cooper [7.23]. Closely related is the technique of resonant soft X-ray scattering (RSXS), by which charge, spin and orbital ordering can be detected; reviews have been published by R. Comin and A. Damascelli [7.24] and by J. Fink et al. [7.25].

7.1.1   General Properties of X-ray Scattering

A very short description of X-ray diffraction has been given in Chapter 2. Before we go into details of some specific aspects of X-ray diffraction at surfaces, we must first introduce several general features of X-ray scattering.

X-rays are transverse electromagnetic waves. In homogeneous, isotropic media, the electric and magnetic field vectors are normal to the wave propagation vector. The electric field of the incident beam, often also called primary beam, forces the electrons of an atom to oscillate and the moving electron creates scattered intensity without energy loss. This is called Thomson scattering. There is a fixed phase difference of π between the scattered wave and the incident beam. This phase relation simplifies the interference of the elastically scattered amplitudes from different atoms and contributes to the relatively easy determination of the atom positions, compared with LEED.

A second process occurs in which the collision of a photon with an electron results in an energy loss. This process is called Compton scattering: this inelastic process does not

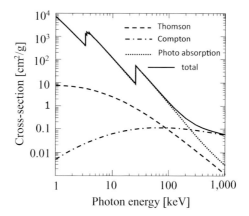

Figure 7.1 Cross-sections of various processes occurring in the interaction of X-rays with matter as a function of photon energy, for Ag as an example. The dominant processes are Thomson scattering and photo absorption at lower energies and Compton scattering at higher energies. The data are available from NIST [7.27] (Public Domain).

have a coherent phase relation to the incident beam. Consequently, no interference is observable in Compton scattering and the inelastic scattered photons contribute to the diffuse background. In addition to Thomson scattering and Compton scattering, other inelastic scattering processes occur which can be measured in various spectroscopic techniques: phonon scattering, interband transitions and electronic excitations.

Inelastic processes together with the elastic scattered intensity into different beams cause an attenuation of the incident beam and a finite penetration depth into the crystal. The penetration depth is an average quantity; the actual value depends on the intensity and direction of the exiting beams. In specific cases the true penetration length may be much larger than the average value. Information about X-ray scattering for all elements, absorption factors and their energy dependence, etc., can be found in the X-ray data booklet [7.26] and in the tables published by NIST [7.27]. Figure 7.1 shows an example of relative intensities of Thomson scattering, Compton scattering and the total cross-section as a function of photon energy, for a particular element, Ru. The dominant processes are Thomson scattering and photo absorption at the lower energies, which are normally used for diffraction. At higher energies Compton scattering becomes the dominant process.

Measurements of surface X-ray diffraction are mainly performed at high intensity sources, namely synchrotrons using X-rays with wavelengths in the range of ~0.04 nm to 0.3 nm. Usually, the energy of the photons is given; the energy can be converted to wavelength using the relation:

$$12.3984 \, [\text{keV}] \leftrightarrow 0.1 \, [\text{nm}] = 1 \, [\text{Å}].$$

For the purpose of surface structure determination, X-ray energies in the range of 12–18 keV are used at most beam lines, corresponding to a wavelength in the range of 0.1–0.07 nm. Shorter wavelengths occur with hard X-rays.

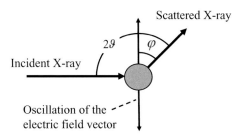

Figure 7.2 Illustration of the scattering geometry of X-rays by an electron. The oscillation of the electric field induces oscillation of electrons within atoms, which then emit radiation in all directions except along the E-field, according to Eq. (7.1).

We next consider elastic X-ray scattering from atoms. The intensity of the Thomson scattering from a single electron at a point with distance r from the electron is:

$$I_{el} = I_0 \frac{1}{r^2} \left(\frac{e^2}{m_e c^2} \right)^2 \sin^2 \varphi, \tag{7.1}$$

where φ is the polar angle between the direction of the E-field of the incident beam and the direction of the scattered X-ray, as shown in Figure 7.2, m_e is the electron mass, c is the speed of light and r is the distance between the scattering centre and the detector. No intensity is observed in the polarisation direction of the incident beam, which is defined by $\varphi = 0$. We note that Eq. (7.1) applies to all particles with charge e, for example, also to protons, where the appropriate mass has to be taken. It should also be noted that the scattered intensity does not depend on the wavelength of the incident beam and is only proportional to its intensity.

The Thomson scattered intensity in the plane normal to the oscillation of the E-field ($\varphi = 90°$) at distance r from the scattering centre is:

$$I_{el} = I_0 \frac{1}{r^2} \left(\frac{e^2}{m_e c^2} \right)^2. \tag{7.2}$$

The scattering angle between the incident beam and the diffracted beam is 2ϑ (ϑ is the Bragg angle, not to be confused with the incident angle in LEED). For an unpolarised incident beam, that is, a beam having random azimuthal orientation of the E-field around the incident direction ($\vartheta = 0$), the scattered intensity is:

$$I_{el} = I_0 \frac{1}{r^2} \left(\frac{e^2}{m_e c^2} \right)^2 \frac{(1 + \cos^2 2\vartheta)}{2}. \tag{7.3}$$

The factor $P = 0.5(1 + \cos^2 2\vartheta)$ is called the polarisation factor for unpolarised incident beams. Synchrotron radiation is polarised in the plane of the synchrotron ring, but the beam arriving at the sample is only partially polarised due to the diffraction in the monochromator and reflection at mirrors in the beam line. A polarisation factor specific for the beam line and the diffraction geometry must be taken into account when analysing the data. For unpolarised beams the intensity has a maximum in the direction of the incident beam and a minimum in the plane perpendicular to the incident beam. The total scattered intensity from one electron is given by the integral over all directions:

$$I_{tot} = I_0 \frac{1}{r^2} \left(\frac{e^2}{m_e c^2}\right)^2 \int \frac{(1 + \cos^2 2\vartheta)}{2} 2\pi r^2 \sin(2\vartheta)d\vartheta = I_0 \frac{8\pi}{3} \left(\frac{e^2}{m_e c^2}\right)^2. \qquad (7.4)$$

The total scattering cross-section I_{tot}/I_0 amounts to 6.7×10^{-25} cm^2 per electron.

As can be seen in Eq. (7.4), the scattered X-ray intensity is inversely proportional to m_e^2, so that the scattering from protons is 1837^2 times smaller than for scattering from electrons and can be neglected. Thus, X-ray diffraction from crystals determines the electron density, in contrast to LEED where the scattering is dominated by the potential of the nucleus. Magnetic effects are usually neglected in X-ray diffraction; they are only considered in cases where magnetic properties are studied with wavelengths at the absorption edge of the magnetic materials.

The question arises whether there is enough intensity to measure diffraction intensities from surfaces. Reflection from a monolayer is about 10^5 to 10^6 times less intense than reflection from the substrate. To get an impression of how much the intensities measured in a surface diffraction experiment differ from intensities obtained from a typical 3-D crystal, we may compare the diffracting volumes in both cases. For SXRD, let us assume 10 mm^2 of active surface area and 1 nm of penetration depth into the surface. This gives an active diffracting volume of 10^4 μm^3. A 3-D crystal about 100 μm in size has a volume of 10^6 μm^3, corresponding to a small organic crystal. The diffracting volume in the 3-D crystal is thus about 100-times larger than that in the surface diffraction measurement. Therefore, enough intensity for the surface diffraction is available thanks to the high beam intensity at synchrotrons. Even diffraction from much smaller surface areas would be measurable.

We should mention that synchrotron sources are not required in all cases. With intense laboratory sources and a focused incident beam, X-ray structure determination of surfaces is also possible in the laboratory. Examples of diffractometers used in the laboratory are shown in Section 7.2.

7.1.2   Reflection and Transmission of X-rays

For the study of surface structures, usually small incidence angles are chosen in order to minimise the background intensity arising from Compton scattering, defect scattering and inelastic scattering from excitation processes. The background intensity arises from the substrate: with small incidence angles the illuminated volume in the substrate is minimised.

At grazing incidence angles, the refraction index for X-rays becomes important. For structure analysis of 3-D crystals the refraction index is usually neglected because the deviation in the diffraction angles resulting from refraction is too small at high incidence angles and reflection from surfaces is even smaller. In the kinematic theory which is applied in 3-D X-ray structure analysis, refraction does not occur. The refraction index is a result of dynamical effects, that is, multiple scattering effects.

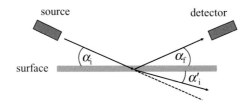

Figure 7.3 Schematic drawing of diffraction at a surface, showing the incidence angle α_i, the exit angle of the diffracted beam α_f and the internal incidence angle α_i'.

In surface X-ray diffraction the refraction index cannot be neglected. It is nevertheless not necessary to apply the dynamical theory for the surface structure analysis methods by X-rays considered in this book; in many cases it is sufficient to combine refraction with the kinematic theory. That means taking into account the critical angle, the angles inside and outside the crystal, and the penetration depth, as well as using the structure factors calculated by the kinematic theory. That is possible when the incident angle is larger than the critical angle. When the incident angle is smaller or equal to the critical angle, it must be checked whether the surface reflectivity needs to be included, as will be discussed in Section 7.1.3.

In the energy range used for X-ray diffraction, the refraction index for photons is smaller than unity, in contrast to optical photons where the refraction index is larger than 1. This leads to refraction of the incident beam away from the surface normal, as shown in Figure 7.3. Throughout this chapter, the incidence angle α_i is defined as the angle to the surface, not the polar angle, thus differing from the custom in LEED. The refraction index n depends on the electron density in the material and on the wavelength. It is only slightly smaller than unity: the deviation is usually on the order of 10^{-6}–10^{-5}.

The refraction index is related to the electron density and the scattering factor for X-rays. For crystalline materials with N atoms per unit cell the refraction index n is given by (see for example [7.28]):

$$n = \sqrt{1 - \sum_{i=1}^{N} \lambda^2 \frac{f_i(0)e^2}{4\pi^2 \varepsilon_0 m_e c^2 V_c}} \approx 1 - \sum_{i=1}^{N} \lambda^2 \frac{f_i(0)e^2}{8\pi^2 \varepsilon_0 m_e c^2 V_c}, \tag{7.5}$$

where ε_0 is the dielectric constant, λ is the wavelength of the photon, V_c is the volume of the unit cell, and m_e is the electron mass; $f_i(0)$ is the scattering factor of atom i in the forward direction. With the classical electron radius

$$r_e = \frac{e^2}{4\pi \varepsilon_0 m_e c^2} \tag{7.6}$$

and the number n_j of atoms of element j per unit cell, the refraction index is given approximately by

$$n \approx 1 - r_e \frac{\lambda^2}{2\pi} \sum_{j=1}^{Z} n_j f_j(0), \tag{7.7}$$

where the sum runs over all elements in the unit cell. The atomic scattering factor of an atom with Z electrons, also frequently named atomic form factor, is defined as:

$$f(\mathbf{q}) = 4\pi \sum_{j=1}^{Z} \int \rho_j(\mathbf{r}) e^{i\mathbf{q}\mathbf{r}} d\mathbf{r}, \tag{7.8}$$

where $\rho_j(\mathbf{r})$ is the electron density of an atom of element j. The calculation using the electron density in the different shells of an atom shows that $f(0)$ is not quite the sum over the scattering factors of free electrons. The deviation is expressed by the dispersion correction with a real part f' and an imaginary part f''; it depends on the wavelength and the element. The atomic form factor in forward direction is given by:

$$f(0) = f_0(0) + f'(0) + if''(0), \tag{7.9}$$

where $f_0(0) = Z$. The real part is related to the total elastic cross-section of the atom with Z electrons. The imaginary part is derived from the linear absorption coefficient μ_l measured from the attenuation of the X-ray beam in a sample of thickness d:

$$I = I_0 e^{-\mu_l d}. \tag{7.10}$$

The imaginary part f'' describes the phase shift of the scattered beam when the elastically scattered photon is reemitted from the atom and is related to the linear absorption coefficient. The refractive index, Eq. (7.5), becomes, using Eq. (7.9):

$$n = 1 - \delta - i\beta, \tag{7.11}$$

$$\delta = r_e \frac{\lambda^2}{2\pi} \sum_{j=1}^{Z} n_j \big(f_j(0) + f_j'(0) \big), \tag{7.12}$$

$$\beta = r_e \frac{\lambda^2}{2\pi} \sum_{j=1}^{Z} n_j f_j''(0), \tag{7.13}$$

or, with $\mu_l = 2 r_e \lambda f''(0)$:

$$\beta = \frac{\lambda}{4\pi} \mu_l. \tag{7.14}$$

Here δ is on the order of 10^{-5}, while β is usually even smaller if the energy of the X-rays is far from an absorption edge; at the absorption edge, β can become large (see Figure 7.1). Data for f' and f'' can be obtained from NIST [7.27] or, for selected wavelengths, from the International Tables of Crystallography [7.29]. The linear absorption coefficient μ_l depends on the wavelength; tables of μ_l can also be found in the X-ray data booklet [7.26].

The refraction index for X-rays implies a small critical angle α_c for total external reflection: α_c is usually some tenths of a degree for most materials. According to Snell's law, the incidence angle inside the crystal is

$$\cos \alpha_i' = \frac{\cos \alpha_i}{n}, \tag{7.15}$$

where the refraction index is assumed to be 1 for the medium outside the crystal. For the reflected wave we have $\cos \alpha_f = \cos \alpha_i$. Total external reflection occurs when the incidence angle becomes smaller than the critical angle α_c and $\alpha_i' = 0$. Then,

$$\cos \alpha_c = n \cos \alpha_i' = n \quad \text{and} \quad \alpha_c = \sqrt{2\delta} = \lambda \sqrt{\frac{r_e}{\pi} \sum_{j=1}^{Z} n_j \big(f_j(0) + f_j'(0)\big)}, \quad (7.16)$$

if we neglect the very small term β away from an absorption edge. We should note that β cannot be neglected with incidence angles at or below the critical angle. Equation (7.16) shows that the critical angle increases with λ and the electron density: $\alpha_c \propto \lambda\sqrt{Z}$. The transmitted wave for $\alpha_i \leq \alpha_c$ is exponentially damped; this can be seen from the relations

$$n \cos \alpha_i' = \cos \alpha_i,$$
$$n = \cos \alpha_c, \quad (7.17)$$
$$\cos \alpha_c \cdot \cos \alpha_i' = \cos \alpha_i,$$

which lead, at small angles α_i and α_c, to:

$$\alpha_i'^2 \approx \alpha_i^2 - \alpha_c^2,$$
$$\alpha_i' \approx \sqrt{\alpha_i^2 - \alpha_c^2} = \sqrt{\alpha_i^2 - 2\delta - 2i\beta}. \quad (7.18)$$

We thus find that α_i' becomes imaginary for $\alpha_i \leq \alpha_c$. The wave vector in the crystal is

$$|\mathbf{k}'| = n|\mathbf{k}|, \quad (7.19)$$

where vacuum has been assumed for the external side, with $n_{\text{vac}} = 1$. The wave vector outside the crystal is real, while it becomes complex with complex n inside the crystal. The parallel component of the wave vector remains unchanged by transmission of the wave through the surface, which means (for small angles):

$$|\mathbf{k}|\alpha_i = |\mathbf{k}'|\alpha_i'. \quad (7.20)$$

The normal component of the wave vector inside the crystal becomes:

$$k_z' = |\mathbf{k}'| \sin \alpha_i' \approx |\mathbf{k}'| \, \alpha_i'. \quad (7.21)$$

With Eqs. (7.18)–(7.21) we get for the imaginary part:

$$\begin{aligned}
\text{Im}(\mathbf{k}') &= \sqrt{|\mathbf{k}'^2| - \mathbf{k}^2} \\
&= \frac{2\pi}{\lambda\sqrt{2}} \left(\sqrt{\left(\alpha_i^2 - \alpha_i'^2\right)^2 + 4\beta^2} - \alpha_i'^2 + \alpha_i^2 \right)^{1/2}.
\end{aligned} \quad (7.22)$$

The transmitted wave is thus exponentially damped inside the crystal. The imaginary part of $\mathbf{k}'$ occurs in the z-component, $\mathbf{k}' = \{k_x', k_y', k_{z,1}' + ik_{z,2}'\}$, so that:

$$e^{i\mathbf{k}'\mathbf{r}} = e^{i\left(k_x' r_x + k_y' r_y\right)} e^{ik_{z,1}' r_z - k_{z,2}' r_z}. \quad (7.23)$$

The penetration depth Λ is defined by:

$$k'_{z,2}\Lambda = \frac{1}{2} \quad \text{or} \quad \Lambda^{-1} = \frac{4\pi}{\lambda\sqrt{2}} \left(\sqrt{(\alpha_i^2 - \alpha_i'^2)^2 + 4\beta^2} - \alpha_i'^2 + \alpha_i^2 \right)^{1/2}. \quad (7.24)$$

For $\alpha_i \ll \alpha_c$, we have $\alpha_i' = i\alpha_c$ and therefore:

$$\Lambda_{min}^{-1} = \frac{4\pi}{\lambda}\alpha_c. \quad (7.25)$$

The minimum penetration depth Λ_{min} at $\alpha_i \ll \alpha_c$ depends on the material: for example, Si has $\Lambda_{min} = 3.2$ nm, while Au has $\Lambda_{min} = 1.2$ nm. It can be seen in Figure 7.4 that the penetration depth does not become zero at $\alpha_i = 0$.

The transmitted wave is scattered within the crystal, which means there is structural information from a thin surface slab at grazing incidence. The penetration depth can be controlled by varying α_i in the vicinity of α_c.

The depth probed by a diffraction experiment, defined here as effective scattering depth Λ_{eff}, depends on both the incidence angle and the exit angle, as illustrated in Figure 7.5.

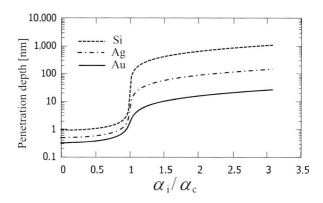

Figure 7.4 Penetration depth as a function of α_i/α_c, for three elements.

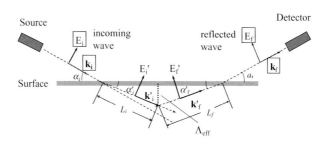

Figure 7.5 The effective scattering depth Λ_{eff} is defined by the incidence angle α_i and the exit angle α_f. The penetration depth of the incident wave at a scattering point is $L_i\cos(\alpha_i')$ and the scattered beam needs the depth $L_f\cos(\alpha_f')$ to leave the crystal.

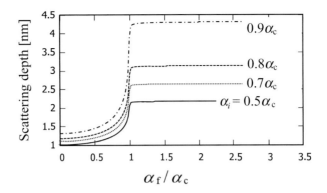

Figure 7.6 Effective scattering depth as a function of α_f/α_c for different incidence angles α_i below the critical angle α_c. The incidence angle limits the effective scattering depth.

The effective scattering depth Λ_{eff} is given by:

$$\Lambda_{eff}^{-1} = -2|\mathbf{k}|(B_i + B_f), \tag{7.26}$$

$$B_{i,f} = \frac{1}{\sqrt{2}} \left[\sqrt{\left\{ (\alpha_{i,f}^2 - \alpha_c^2)^2 + 4\beta^2 \right\}} - \alpha_{i,f}^2 + \alpha_c^2 \right]^{1/2}. \tag{7.27}$$

Equations (7.26) and (7.27) show that complete equivalence exists between the incidence and exit angles. In Figure 7.6 the effective scattering depth as a function of the exit angle is shown for some incidence angles below the critical angle. If the exit angle exceeds α_c, the effective scattering depth is limited by α_i. The analogous limit holds for the case where the exit angle is kept at or below the critical angle and the incidence angle is large: this diffraction geometry can also be chosen in some surface diffractometers.

The effective scattering depth can be controlled by α_i and α_f. This fact can be used to study depth-dependent phenomena at surfaces, for example surface melting [7.6], or ordering in epitaxial layers [7.11].

7.1.3 Reflection and Transmission Coefficients near the Critical Angle

The Fresnel theory of reflection and transmission of electromagnetic waves [7.30] at interfaces is based on the fact that the components of the electric field parallel to the surface are continuous at the interface, cf. Figure 7.7:

$$E_i \sin \alpha_i + E_r \sin \alpha_r = E_i' \sin \alpha_i'. \tag{7.28}$$

The same holds for the components of the magnetic field. At small angles we get for the amplitudes of the transmitted (refracted) wave E_i' and the reflected wave E_r, with transmission and reflection coefficients $T(\alpha_i)$ and $R(\alpha_i)$, respectively:

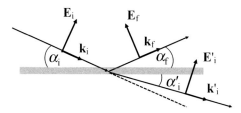

Figure 7.7 Illustration of the electric field vectors at the interface

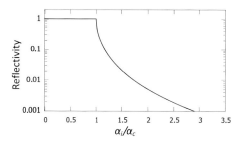

Figure 7.8 Fresnel reflectivity of a perfectly flat surface. Assumed is $\beta = 0$. With non-zero absorption, the abrupt drop of the reflectivity at $\alpha_i/\alpha_c = 1$ is rounded.

$$E_i' = \frac{2\alpha_i}{\alpha_i + \alpha_i'} E_i = T(\alpha_i) E_i \quad \text{and} \quad E_r = \frac{\alpha_i - \alpha_i'}{\alpha_i + \alpha_i'} E_i = R(\alpha_i) E_i. \tag{7.29}$$

The Fresnel theory assumes a perfectly flat interface and homogeneous media on both sides of the interface; α_i' becomes 0 when $\alpha_i = \alpha_c$ and $\beta = 0$. At the critical angle the amplitude of the transmitted wave becomes $2E_i$: the intensity is thus in principle enhanced by a factor of 4. The reflection coefficient for a perfectly flat surface is given by:

$$R(\alpha_i) = \frac{\alpha_i - \alpha_i'}{\alpha_i + \alpha_i'}. \tag{7.30}$$

Its square describes the reflected intensity from the surface and holds strictly for $\alpha_f = \alpha_i$. It follows with $\alpha_i'^2 \approx \alpha_i^2 - \alpha_c^2$ (Eq. (7.18)) that the directly reflected intensity falls off rapidly as

$$(2\alpha_i/\alpha_c)^{-4} \text{ for } \alpha_i >> \alpha_c. \tag{7.31}$$

Without absorption and for $\alpha_i' = 0$ at $\alpha_i \leq \alpha_c$ we have $R = 1$. The reflectivity is shown in Figure 7.8. If the surface exhibits some roughness the decrease of the reflected intensity will be even stronger.

The idealised picture of a perfectly flat surface and homogeneous media is not realised in crystals with an atomic structure and some roughness in the surface.

A homogeneous medium produces only specular reflection, with no other scattered waves and no diffracted beams. The evanescent incident beam at incidence angles below the critical angle does not carry flux into a homogeneous non-absorbing medium. This is not the case at crystal surfaces. We observe diffracted intensity at incidence angles smaller than the critical angle, which means that the incident intensity cannot be quite totally reflected, but the difference is small and usually not observed as the abrupt decrease of the reflectivity is in any case rounded by absorption. The rapid decrease of the directly reflected intensity at $\alpha_f = \alpha_i$ can be observed in reflectivity measurements. Since the directly reflected intensity falls off rapidly with increasing angle of incidence, it is usually not observable in 3-D crystal diffraction with its high incidence angles.

The transmitted wave cannot propagate into the crystal but extends only to a certain depth inside the crystal: it decays exponentially with increasing depth. This is described by a complex α_i' and $\mathbf{k}_i'$, see Eqs. (7.18) and (7.22). The penetration depth is illustrated in Figure 7.4. At incidence angles $\alpha_i > \alpha_c$ the transmitted wave is damped by the absorption coefficient. When $\alpha_i < \alpha_c$ then α_i' becomes imaginary and the penetration depth drops rapidly to a very small value. The fact that the incident wave is not abruptly cut off at the surface leads to diffracted intensity arising from the area within the penetration depth.

According to Eq. (7.29), the transmitted wave is in principle doubled at the critical angle, so the intensity is enhanced by a factor 4. This can be observed, for example, in the fractional-order superstructure reflections from a reconstructed surface, as shown in Figure 7.9. The enhancement factor measured in the experiment is actually less than 4 due to the limited angular resolution of the incident beam and detector, as well as due to absorption.

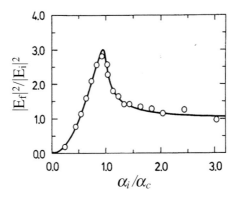

Figure 7.9 Transmission coefficient $|T(\alpha_i)|^2$ as a function of α_i/α_c. The curve is shown for an InSb(111) surface and a wavelength $\lambda = 0.12$ nm; the critical angle is $\alpha_c = 0.25°$. The experimental points are from the (4/3,0) reflection of the InSb(111)–(3×3) surface normalised to the correct scale. The intensity of a superstructure reflection is a measure of the intensity of the evanescent wave and hence of the transmission coefficient. Reprinted from [7.3] R. Feidenhans'l, *Surf. Sci. Rep.*, vol. 10, 105–188, 1989, with permission from Elsevier.

The enhanced intensity at the critical angle is missing at the angles below the critical angle. The measurement at the critical angle requires a very high accuracy of the orientation of the surface and of the setting of the angles in the diffractometer. To avoid errors due to misalignment, the measurement of intensities for surface structure determination is usually not performed at the critical angle, but at $\alpha_i \approx 2\alpha_c$. The theory applied in structure analysis of surfaces is the kinematic theory, where the angles inside the crystal are corrected by the refraction index and a limited penetration depth is included due to incidence angles being in the vicinity of the critical angle.

It should be noted that the kinematic theory is not applicable when $\alpha_i \approx \alpha_c$ or $\alpha_f \approx \alpha_c$ as the refraction index n does not occur in that theory. However, it must be included in reflectivity measurements at or near the critical angle where the density profile normal to the surface is measured [7.11]. It plays a role in reflectivity measurements which are performed with incidence angles at or near the critical angle and are used to determine the electron density profile normal to the surface in layered crystals [7.12], or for example to study capillary waves at liquid surfaces, see J. Als-Nielsen et al. [7.13].

An extension of the kinematic theory is called the distorted-wave Born approximation [7.31]. This combines the dynamical treatment of the reflection and transmission between layers with the kinematic structure factors for the diffraction within layers. It is used in the analysis of reflectivity measurements. We do not discuss this formalism here but include, for the purpose of surface structure determination, only the deviation of the incidence angle inside the crystal and the limited penetration depth, while using the kinematic theory.

7.1.4   X-ray Diffraction at Grazing Incidence

Surface diffraction experiments are usually performed at grazing angles of incidence or exit. The reason is to reduce the background intensity by reducing the illuminated volume. The illuminated area at the surface is defined by the entrance and exit slits (see Figure 7.10 and O. Robach et al. [7.32]) and the illuminated depth depends on the incidence angle. The background intensity arises from thermal diffuse scattering, Compton scattering and defect scattering. Due to the large penetration depth at higher incidence angles, the background from the substrate then exceeds the small surface signal.

It should be mentioned, however, that grazing incidence or exit angles are not necessary if the illuminated volume which is visible in the detector can be sufficiently reduced, for example by setting appropriate entrance and exit slits and using a microfocus beam.

The incidence angle α_i is usually in the range of $\approx 0.1°-1°$, while the exit angle, normally denoted as γ in the description of diffractometer circles, is limited to remain below about $60°$ in most diffractometers. The reflection intensity becomes too small at higher exit angles due to the polarisation factor. For the study of surfaces of solids, the diffraction geometry is usually chosen such that the surface normal is horizontal, which means that for synchrotron radiation the polarisation vector is normal to the surface, thus allowing a full rotation of the sample around the surface normal. For the

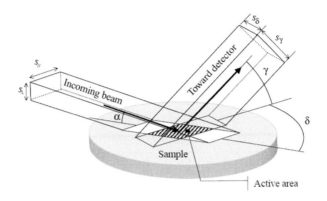

Figure 7.10 Illustration of the active surface area determined by the entrance and exit slits. Reprinted from [7.32] O. Robach et al., *J. Appl. Cryst.*, vol. 33, pp. 1006–1018, 2000, with permission from John Wiley and Sons.

study of liquid surfaces, usually with a horizontal orientation of the surface, the polarisation vector lies in the surface plane; then the range of diffraction angles in the horizontal plane is restricted.

The diffracted intensity is given by:

$$I_{hkl} = I_0 \sigma_e LP |F_{hkl}|^2 |G_{hkl}|^2, \tag{7.32}$$

where $|G_{hkl}|^2$ is the lattice factor and P is the polarisation factor. The Lorentz factor L must be defined for the diffraction geometry at the diffractometer and differs from the usual Lorentz factor used for 3-D crystals at a four-circle diffractometer; it usually includes a geometrical factor for the measurement along the rod in reciprocal space. The Lorentz factor will be discussed in Section 7.2. The factor σ_e takes into account the scattering of a photon by an electron:

$$\sigma_e = \frac{e^4}{m_e^2 c^4}. \tag{7.33}$$

For the semi-infinite crystal we must treat the lattice factor normal to the surface differently from the lattice sum parallel to the surface:

$$G_{hkl} = g(h)g(k)g(l), \quad \text{with} \tag{7.34}$$

$$|g(h)|^2 = \left| \sum_{n=-N/2}^{N/2} e^{2\pi i h n} \right|^2 = \frac{\sin^2(\pi N h)}{\sin^2(\pi h)}. \tag{7.35}$$

An analogous result applies to $|g(k)|^2$. Normal to the surface of the semi-infinite crystal we get, ignoring damping:

$$|g(l)|^2 = \left| \sum_{n=0}^{\infty} e^{2\pi i n l} \right|^2 = \frac{1}{4 \sin^2(\pi l)}. \tag{7.36}$$

Including a limited penetration depth, we can define a damping factor

$$\alpha = \frac{c}{2\Lambda_{eff}}, \tag{7.37}$$

by which the incident beam is damped when propagating one lattice constant c normal to the surface. Therefore (see also Figure 7.11):

$$g(l) = \sum_{n=0}^{\infty} e^{(2\pi i l - \alpha)n},$$

$$|g(l)|^2 = \frac{1}{(1 - e^{-\alpha})^2 + 4e^{-\alpha}\sin^2\pi l}. \tag{7.38}$$

The crystal truncation rods shown in Figure 7.11 are broadened in the l-direction due to the semi-infinite crystal, even without damping. A damping factor leads to lower maxima at the Bragg points, higher intensities at the minima between the Bragg points and additional broadening. Without damping the minimum intensity between the Bragg points is about 10^5- to 10^6-times lower than the intensity at the Bragg points. In Figure 7.11, a relatively strong damping of $\alpha = 0.2$ has been assumed to make the effect visible, as may occur at incidence angles below the critical angle. The damping constant α in Eq. (7.38) depends on the effective scattering depth (Eq. (7.26)) at the given incidence and exit angles and on the absorption factor of the material. The lattice truncation rod (also called crystal truncation rod) with damping can be considered as a superposition of Lorentzian functions.

Surface roughness has a similar effect as damping. A model which is widely used in the analysis of surface structures was developed and first applied to the roughness of W(100) by I. K. Robinson [7.33]. It is assumed that a fully occupied substrate layer exists and each subsequent layer above it has a coverage $\beta < 1$. This leads to a lattice factor

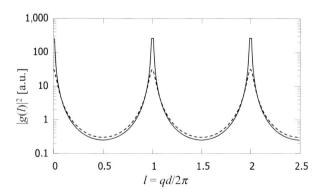

Figure 7.11 Lattice factor $|g(l)|^2$ of the crystal truncation rod; a primitive lattice is assumed; q is the normal component of the diffraction vector and d the layer spacing. Solid line: lattice factor without damping, from Eq. (7.36) (singularities are truncated). Dashed line: lattice factor with damping of $\alpha = 0.2$, from Eq. (7.38). The lattice factor applies to all primitive lattices; with centred lattices some integer l-values do not correspond to a Bragg point.

$$|g(l)|^2 = \frac{(1-\beta)^2}{\left(1 - 2\beta\cos 2\pi l + \beta^2\right)}. \tag{7.39}$$

The model is valid if the effective scattering depth is much larger than the asperity height. This is sketched in Figure 7.12(a), while the lattice factor for the rough surface together with that for the ideal surface are shown in Figure 7.12(b). In contrast to damping of the incident beam, the intensity falls off more rapidly and the intensity between the Bragg points is smaller than for the ideal surface. If no crystal truncation rod intensity can be measured between Bragg points, then the surface is very probably not well prepared and too rough.

At clean surfaces, relaxation of interlayer spacings at the surface can be observed in X-ray diffraction: this produces characteristic changes in the intensity profile, as illustrated in Figure 7.13. The truncation rod intensity is quite sensitive to small relaxations in the surface layers because the phase contrast between the contributions from the top layers and the substrate show up in the truncation rods. For adsorbate layers the intensity of the truncation rod includes contributions from both the adsorbate structure and the substrate. On the other hand, relaxation in the substrate induced by strain from the adsorbate cannot be easily separated from the adsorbate and requires an accurate knowledge of the adsorbate structure and a comparison with the clean surface. The surface slab consists of the adsorbate layers or reconstructed layers as well as distorted substrate layers down to a depth where deviations from the substrate periodicity can still be detected

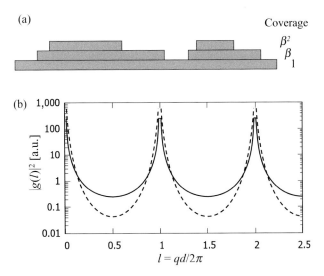

Figure 7.12 (a) Schematic profile of a rough surface: in each successive layer above a fully occupied layer, the coverage is reduced by another factor β. This leads to an exponential decrease of the coverage in the layers above a fully covered semi-infinite crystal. (b) Full line: lattice factor $|g(l)|^2$ of the ideal lattice; dashed line: lattice factor for a roughness parameter $\beta = 0.7$.

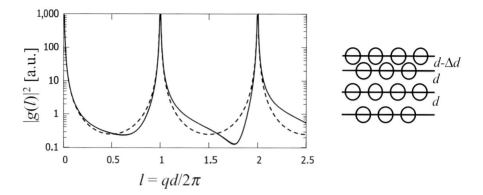

Figure 7.13 Lattice factor $|g(l)|^2$ of the lattice truncation rod. Dashed line: ideal lattice. Solid line: top interlayer spacing d contracted by 5%. The right panel illustrates the top interlayer spacing contraction.

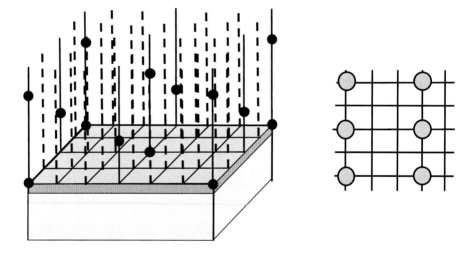

Figure 7.14 Schematic reciprocal space for a (3×2) superstructure on a square surface lattice. The dashed lines are fractional-order superstructure rods; the solid lines are the substrate lattice truncation rods with the Bragg points. In the right panel the 2-D unit cell of the superstructure is shown schematically.

in the diffracted intensities. These may reach fairly deep into the substrate – depending on the substrate and the precision of the measurement. In practice, distortions of the substrate layers are included until the error limit in the intensities is reached.

Two types of reflections occur at surfaces with a superstructure, see Figure 7.14. The intensity along the crystal truncation rods I_{CTR} contains a contribution from the substrate and another from the surface:

$$I_{CTR}(h_b k_b l_b) = |F_{subst}(h_b k_b l_b) + F_{surf}(h_b k_b l_b)|^2. \qquad (7.40)$$

The superstructure rod reflections arise only from the surface:

$$|F_{surf}(h_s k_s l_s)|^2 = \left| \int \rho_{surf}(\mathbf{r}) e^{2\pi i \mathbf{h} \mathbf{r}} d\mathbf{r} \right|^2. \qquad (7.41)$$

The fact that two types of reflections exist has consequences for the interpretation of the Patterson function and for the application of direct methods, as will be discussed in Section 7.3. When relaxation in the substrate layers is small or can be neglected, the analysis of the superstructure reflection gives only the internal structure of the surface slab: no information about the bonding to the substrate is then available.

7.2   Experimental Setup

X-ray diffraction at surfaces requires a larger effort to set up the technical equipment than a LEED experiment. Higher precision is required for the orientation and manipulation of the sample. Since many measurements require UHV to avoid uncontrolled contamination of the surface, a goniometer is required which can carry a heavy load and keep the centre of the circles within a few micrometres. For surface diffraction the incidence angle must be controlled and the full rotation of the sample around the surface normal is necessary. Two circles are needed to align the surface normal parallel to the rotation axis of the sample. Two more circles are needed to move the detector to the reflection position. Therefore, six circles are required, two more than the four circles of the conventional X-ray diffractometer. For surface X-ray diffraction, the diffractometers are usually described as six-circle diffractometers even if more circles exist to control entrance and exit slits and polarisation of the incident beam; also, the lateral movement of the sample in three directions must be possible. Some of the circles, namely the two circles orienting the sample, are in practice replaced by a hexapod or octopod with which the sample can be positioned and oriented with high precision while carrying a heavy load; an example will be shown later in Figure 7.20.

Measurements are mostly performed in the z-axis mode. The diffraction geometry is shown in Figure 7.15 and the principle of the z-axis mode is schematically drawn in Figure 7.16.

For the surfaces of solids, the orientation of the surface plane is usually vertical. The surface normal and the ω-axis of the diffractometer lie in the horizontal plane, because the synchrotron beam is linearly polarised in that plane. This orientation keeps the polarisation factor constant during the full rotation of the sample around the surface normal. This is, of course, not the case for liquid surfaces where a horizontal surface is required. To allow both measurements in the same diffractometer, several surface science beam lines are equipped with diffractometers in which either a horizontal or a vertical ω-axis can be chosen.

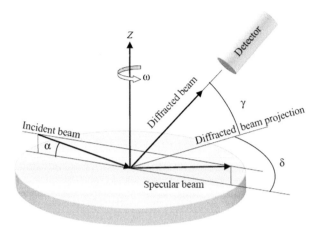

Figure 7.15 Illustration of the diffraction geometry in the z-axis mode. The incidence angle α is small, near the critical angle of total external reflection. The exit angle γ is usually limited to about $60°$ because at higher exit angles the polarisation factor becomes too small when synchrotron radiation is used. Reprinted from [7.32] O. Robach et al., *J. Appl. Cryst.*, vol. 33, pp. 1006–1018, 2000, with permission from John Wiley and Sons.

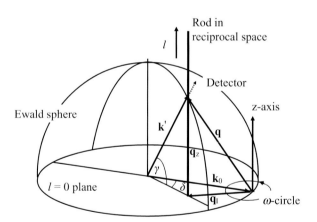

Figure 7.16 Measurement of the intensity along a lattice rod in the z-axis mode. The crystal surface (horizontal in this view, with the surface normal parallel to the z-axis) is rotated around the surface normal, which shifts the point where the lattice rod intersects the Ewald sphere. The detector must be moved in two circles, δ and γ, to follow the reflection when the sample is rotated around the surface normal in the ω-circle. The incidence angle is assumed to be zero for simplicity.

Another diffraction geometry is also possible: the so called χ-circle geometry where the incidence angle may be large and the exit direction is kept at grazing angles. This diffraction geometry with a different definition of the angles was used in the first experiments [7.34; 7.35] but seems to have been used less frequently since then. It has advantages when equipment for surface preparation and control has to be positioned in front of the sample and allows switching from a vertical to a horizontal orientation of the sample [7.36].

In many experimental stations the surfaces are prepared in a separate chamber; they are then moved to a transfer chamber which can be mounted on a diffractometer and is equipped with a Be dome that allows the measurement of the diffracted beams in all directions. The preparation in a separate chamber and the measurement in a transfer chamber is advantageous at synchrotron beam lines where the beam time is limited: the time-consuming preparation step thereby does not block the beam line.

The aforementioned kind of experiment is not always possible when reactions at surfaces or growth processes have to be followed in situ by X-ray diffraction. For such cases experimental setups exist where the surface preparation takes place in the X-ray chamber. Several such stations exist, some of which are described in Section 7.2.1.

We need to emphasise that X-ray diffraction is not bound to UHV conditions. Liquid surfaces and solid–liquid interfaces can be studied. It is also possible to study reactions in environmental conditions or even at high pressures. The term 'high pressures' is used for pressures of several mbar to several bar. The requirement of cleanliness and controlled contamination, as is standard under UHV conditions, must be maintained for experiments done at higher pressures or for solid–liquid interfaces.

7.2.1   Diffractometer Types

Different types of diffractometers have been constructed for surface X-ray diffraction experiments using the z-axis mode shown schematically in Figures 7.15 and 7.16. Figure 7.17 shows the principle of the z-axis mode diffractometer and the notation of its circles [7.37].

The z-axis mode diffractometer has three axes for the sample (φ, χ, ω) and two for the detector (δ, γ). The sixth circle (α or α_i in Section 7.1) is the incidence angle, set by rotating the whole diffractometer. Different names are in use for the different combinations of circles in various constructions, but usually the angles defining the diffraction vector, the resolution function and the correction factors are transformed to the angles and notation of the circles of the z-axis mode. The convention is to use α for the incidence angle and γ for the exit angle (α_f in Section 7.1). The rotation around the surface normal is the ω-circle and the surface normal is adjusted by the two angles φ and χ; the two angles defining the detector position are δ and γ. The diffraction angle 2ϑ does not occur in these circles and has to be derived from the six angles. The diffractometer shown in Figure 7.17 is also called a $(3+2)^*1$-type diffractometer.

The z-axis mode is also shown in Figure 7.18(a). An alternative combination of circles is shown for comparison in Figure 7.18(b) and is called a $(2+2)$-type diffractometer, first proposed by K. W. Evans-Lutterodt and M.-T. Tang [7.38]. Both constructions have a vertical α-circle defining the incidence angle. The difference in the two types lies in the γ-circle defining the exit angle. The γ-circle is decoupled from

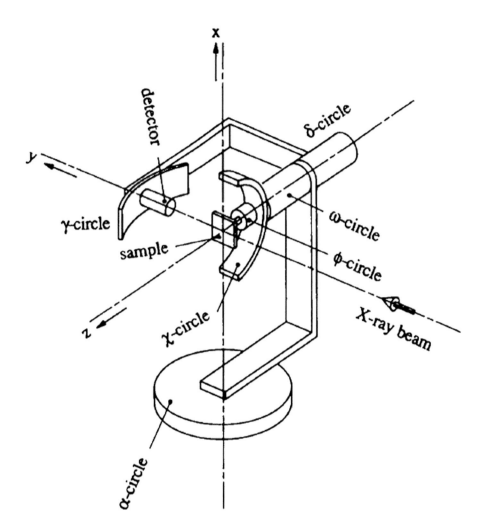

Figure 7.17 Schematic drawing of a z-axis diffractometer with a horizontal ω-axis, and a vertical axis for the α-circle by which the incidence angle is set. Reprinted from [7.37] E. Vlieg, *J. Appl. Cryst.*, vol. 30, pp. 532–543, 1997, with permission from John Wiley and Sons.

the α-circle while in the $(3+2)^*1$ type the axis of the γ-circle is moved with the δ-circle. The angle calculation and the difference between rotations in the two diffractometer types have been discussed by E. Vlieg [7.39]. The independent γ-circle requires an additional circle rotating the detector slits (the ν-circle shown in Figure 7.19) in order to maintain the definition of the active surface area and the correction factors to be the same as in the z-axis mode (see Figure 7.10). This has been called a (4+2)-type diffractometer by M. Takahashi and J. Mizuki [7.40], who used an additional angle for the orientation of the sample, in the so called κ-geometry (not shown in Figure 7.18).

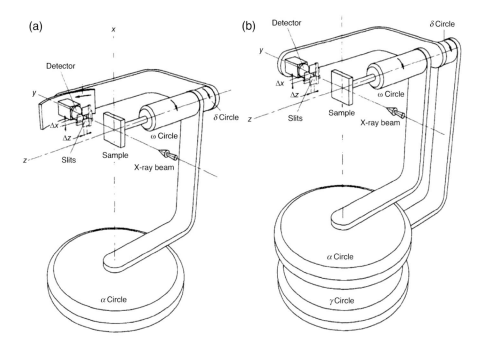

Figure 7.18 Schematic drawing of (a) a z-axis diffractometer and (b) a (2+2)-type diffractometer. The two axes φ and χ orient the surface normal in the horizontal plane. Reprinted from [7.39] E. Vlieg, *J. Appl. Cryst.*, vol. 31, pp. 198–203, 1998, with permission from John Wiley and Sons.

The decoupling of the γ-circle from the δ-circle has the advantage that the horizontal and vertical orientations can be more easily exchanged on the same diffractometer. A further construction combining the two types has been proposed by O. Bunk and M. M. Nielsen [7.41], who called it a z-axis/(2S+2D) hybrid diffractometer and also discussed the advantages and disadvantages of the different diffractometer constructions.

We show as an example the diffractometer of the Materials Science Beamline X04SA at the Swiss Light Source (SLS) in Figure 7.19 [7.42]. This allows measurements with horizontal or vertical orientation of the sample. The orientation of the surface normal by the χ- and φ-circles is realised by a hexapod on which a UHV transfer cell can be mounted. Two examples of UHV transfer chambers for measurements at high pressure or low temperature are shown in Figure 7.20 [7.43]. The preparation of the sample is frequently performed in a separate UHV chamber (other constructions designed to study in situ reactions at surfaces are shown in Section 7.2.3). The diffractometer is equipped with an area detector which allows fast measurement of the intensities; area detectors are used in most modern diffractometers. The orientation of the sample is in most cases performed by a hexapod or an octopod which simulate the φ- and χ-circles in Figure 7.17.

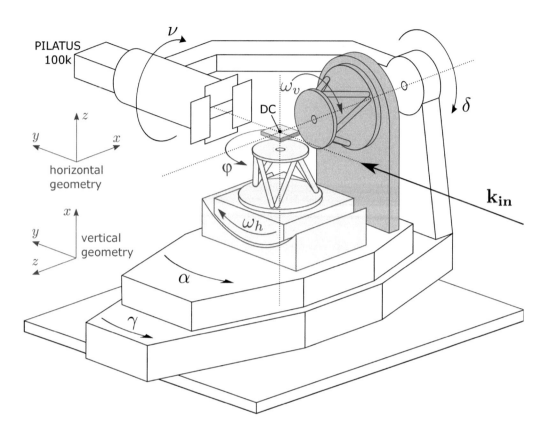

Figure 7.19 Schematic of the Newport $(2+3)$ circle diffractometer with all circles shown at their zero positions. The two sample circles can be configured in one of two alternative modes. (i) In the 'vertical geometry' the sample surface lies in a vertical plane, hence the surface normal is horizontal. Here, ω and α provide the azimuthal and polar degrees of freedom, respectively. (ii) In the 'horizontal geometry' the sample surface is kept approximately horizontal, while the surface normal points upwards. The azimuthal and polar rotations are now provided by the φ- and ω-circles. The detector has three degrees of freedom: γ and δ are used to position the detector in the direction of the diffracted X-ray beam, while the ν-axis provides a rotation of the detector slits around this direction. All rotation axes intersect at the diffractometer centre (DC). Reprinted from [7.42] C. M. Schlepütz et al., *J. Appl. Cryst.*, vol. 44, pp. 73–83, 2011, with permission from John Wiley and Sons.

7.2.2   Measurement

In earlier times, the measurement of the intensities along a lattice rod in reciprocal space was done by a point detector and ω-scans, so-called rocking scans. For each l-value on the lattice rod, a scan of about 50–100 scan points is required. Scans are necessary to obtain the integrated intensity and to determine the background. The measurement with a point detector is time consuming and, since 2-D area detectors became available, the latter are used at most beam lines when possible. Point detectors are used when an area detector cannot be installed. In general, 2-D area detectors have

(a)   (b)

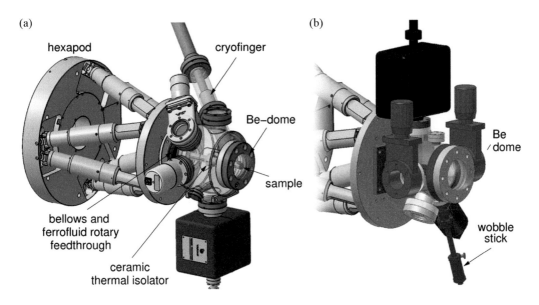

Figure 7.20 (a) Schematic drawing of chambers available at the Materials Science Beamline X04SA at the Swiss Light Source (SLS). (a) The UHV transfer chamber for experiments requiring UHV conditions. It is equipped with an X-ray-transparent beryllium dome with an inner radius of 31.5 mm and a thickness of 0.4 mm. The samples are transferred via a DN40CF flange from another UHV chamber while maintaining vacuum. A small manipulator is used to fix the sample holder within the chamber and to attach a heater contact. A battery supply of 2×12 V provides power to an ion pump for more than 20 hours. (b) A UHV cryostat chamber for low temperature experiments as low as 13 K. It is equipped with a similar Be dome. The sample movements are decoupled from the vacuum vessels using bellows and a ferroid rotary feedthrough. The sample mount is made of copper. A ceramic spacer thermally insulates the sample mount from the rest of the chamber. A cryostat is connected to the sample holder via copper braids. An ion pump ensures vacuum in the mid 10^{-10} mbar range. Reprinted from [7.43] P. R. Willmott et al., *J. Synchrotron Rad.*, vol. 20, pp. 667–682, 2013 (Open Access).

a large aperture, while the resolution is determined by the distance to the sample and the pixel size.

The purpose of the measurement of the integrated intensity within the width of the reflection is to determine the structure factors $|F(hkl)|^2$. The integrated intensity of reflections from 3-D crystals has been discussed by B. E. Warren [7.44], and for surface measurements by E. Vlieg [7.37]. Following the formulation given by Vlieg, the total integrated intensity of a reflection (hkl) can be written as:

$$I_{\text{int}} = I_0 \frac{A}{A_{\text{u}}} \int r_e^2 P |F(hkl)|^2 u(\mathbf{Q}) dt d\delta d\gamma, \tag{7.42}$$

where I_0 is the incident flux, A the active surface area of the sample, A_{u} the area of the unit cell, P the polarisation factor, r_e the classical electron radius, $F(hkl)$ the structure factor of the reflection and $u(\mathbf{Q})$ the shape function of the reflection (the latter two depend on measuring time); $\mathbf{Q}$ is the scattering vector. The integral over the measuring time is replaced by the sum over the number of points in the rocking curve and the measuring time T_ω for each point. For an ω-scan with N_ω discrete angles, where the crystal is continuously rotated around the surface normal, the integrated intensity is given by:

$$I_{\text{int}} = I_0 T_\omega \frac{A}{A_{\text{u}}} \sum_{i=1}^{N_\omega} \int r_e^2 P |F(hkl)|^2 u(\mathbf{Q}) d\delta d\gamma \Delta \omega_i. \tag{7.43}$$

The integral over the line shape function is unity and is included here because a correction factor is needed in cases where only the incomplete shape of the reflection is measured. That may occur due to limitations of the detector aperture or insufficient setting of the post sample slits. It may also be that the reflection profile is too broad. This must be checked in any case, regardless of whether the full profile is measured or not.

The integral over the volume in real space $\Delta\delta\Delta\gamma\Delta\omega$ can be transformed into an integral over a volume in reciprocal space by

$$\int I(hkl) d\delta d\gamma d\omega = \int I(hkl) \left| \frac{\partial(\delta, \gamma, \omega)}{\partial(h, k, l)} \right| dh dk dl, \tag{7.44}$$

where $\left| \frac{\partial(\delta, \gamma, \omega)}{\partial(h, k, l)} \right| = L$ is the Lorentz factor transforming the volume in real space into the volume in reciprocal space; see for example the definition of the Lorentz factor by W. Clegge et al. [7.45] and R. Shayduk [7.46] for the derivation in 3-D crystals. For 3-D single crystals the Lorentz factor becomes:

$$L = \frac{\lambda^3}{v_{\text{u}}} \frac{1}{\sin 2\vartheta}, \tag{7.45}$$

where ϑ is the Bragg angle and v_{u} the volume of the unit cell. The Lorentz factor depends on the diffraction geometry and for surfaces we have:

$$L = \frac{\lambda^2}{A_{\text{u}}} \frac{1}{\cos\alpha \cos\gamma \sin\delta}, \tag{7.46}$$

as derived by E. Vlieg [7.37]. With $\int dl = \Delta l$ and $\Delta l = \frac{2\pi}{\lambda} \Delta\gamma \cos\gamma$ we obtain:

$$I_{\text{int}} = I_0 T_{\text{t}} \frac{A}{A_{\text{u}}} \lambda^2 r_e^2 P |F(hkl)|^2 C_{\text{area}} C_{\text{det}} C_{\text{beam}} \frac{\Delta\gamma \cos\gamma}{\cos\alpha \cos\gamma \sin\delta}. \tag{7.47}$$

Here T_{t} is the total measuring time; C_{area} is the area correction as A may change with φ, γ and δ; A is the illuminated area; A_{u} is the area of the unit cell; C_{det} is a correction factor for the case when the detector does not include the full width of the beam; C_{beam} takes into account the intensity distribution in the incident beam in case this is not uniform; and $\Delta\gamma \cos\gamma$ is the intercept of the rod with the Ewald sphere (see Figure 7.21). The structure factors are obtained from the integrated intensity and by applying correction factors.

As the incident intensity I_0 is usually not measured, only the relative intensities of the structure factors are determined and $I_0 = 1$ is used. For surfaces a continuous

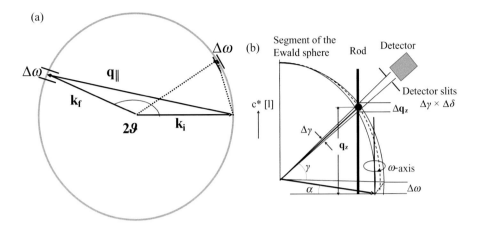

Figure 7.21 Illustration of the ω-scan. (a) Ewald construction of the scan of in-plane reflections. The width $\Delta\omega$ in which a reflection of finite width intersects the Ewald sphere increases with $\mathbf{q}_{\parallel}$. The step width of the ω-rotation should be adapted to the vector $\mathbf{q}_{\parallel}$ and a correction factor has to be applied. (b) Perspective view of an ω-scan at a higher exit angle γ. Rotation of the crystal by $\Delta\omega$ shifts the intersection point of the rod with the Ewald sphere by $\Delta\mathbf{q}_z$. This shift becomes smaller with increasing exit angle γ. In order to measure intensity in constant intervals $\Delta\mathbf{q}_z$ along the rod, the integrated intensity derived from the ω-scan must be multiplied by a correction factor.

intensity occurs along rods in reciprocal space and for the integral over l a limited range of l must be used. This range has to be small enough so that the structure factor can be assumed to be constant within the integration range.

Point Detector
We first discuss the measurement by means of the point detector. The most used scan to obtain the structure factors is the ω-scan where the crystal is rotated around the surface normal. An illustration is given in Figure 7.21(a) for the 'in-plane' measurements and in Figure 7.21(b) for high exit angles. While ω is rotated by $\Delta\omega$ the detector position is kept fixed. This is called a rocking scan. The scan is required to collect the intensity within the width of the reflection. Several other scan types are possible along the h, k or l directions or along an arbitrary direction in reciprocal space. Such scans require the simultaneous movement of the detector position with the rotation of the crystal around a direction in reciprocal space. These scans are used for the measurement of peak profiles or the detection of satellite or superstructure reflections. The measurement of the Bragg peaks must be avoided either by omitting these points, or by damping the incident intensity by a filter, as the high intensity in the Bragg peaks would destroy the scintillation crystal.

The integration of rocking scans requires in any case the application of correction factors, because the raw data measured with the detector depend systematically on the exit angle and $q_{\parallel}$, the length of the in-plane component of the diffraction vector, as illustrated in Figure 7.21. The correction factors for the integration of measurements with a point detector have been discussed in several publications: C. Schamper et al. [7.47], E. Vlieg [7.37], O. Robach et al. [7.32], and N. Jedrecy [7.48].

Figures 7.21–7.24 demonstrate only qualitatively the necessity and importance of the correction factors and refer to the original literature where the correction factors for different cases are derived in detail. The correction factors are specific to the diffractometer, the setting of the collimator or the primary slits and the setting of the detector slits. It must also be considered whether one or two sets of slits are used to collimate the beam. Soller slits may be used to collimate the incident beam and also influence the correction factors.

O. Robach et al. [7.32] have pointed out three main factors influencing the width $\Delta\omega$ of the scan: (i) The rod interception of the detector aperture, Δq_z or Δl, depends on the exit angle γ. A correction factor $1/\cos\gamma$ is required to obtain the intensity for a constant interval Δl, assuming $|F(hkl)|^2$ to be constant within the interval Δl. The increase of the width $\Delta\omega$ with the exit angle γ is shown in Figure 7.22. (ii) The intrinsic width of the rod is affected by a mosaic spread of the crystal, by disordered surfaces and by rough surfaces. The latter causes a varying width along the rod. A mosaic spread causes a width that increases with growing q_z. Depending on the type of disorder, the shape of the reflection may be round or elongated. These cases must be taken into account and the width of the ω-scan has to be adapted accordingly. (iii) The corresponding area factor depends on the shape of the sample and on ω, see Figure 7.23 illustrating this effect. A further area correction factor may occur when the detector at the beginning and end of the ω-scan does not see the full area illuminated by the incident beam, as illustrated in Figure 7.24.

The structure factor is integrated in constant intervals Δq_z along the rod, with

$$\Delta q_z = |\mathbf{k}|\Delta\gamma \cos\gamma, \tag{7.48}$$

where $\Delta\gamma$ is set by the detector slits and kept constant during the scan; Δq_z is also set by the width of the ω-scan. We have

$$\Delta\gamma = \Delta\omega \frac{\sin\alpha_0 \sin\delta_0}{\sin\gamma}. \tag{7.49}$$

To obtain the integrated intensity, correction factors are required to take into account the angular width of the scan which depends on $\mathbf{q}_{\parallel}$, namely the projection of the vector $\mathbf{q}$ onto the plane $l = 0$. The second factor is a geometrical factor which takes care of the fact that the change of γ per ω-step decreases with increasing γ, that is, the rocking scan becomes broader. Both factors are sometimes combined as the Lorentz factor; occasionally the second factor is named separately as L_{d}. The correction factors can become very large, depending on q and γ, and need to be calculated carefully for each setting of the entrance and detector slits. The determination of the integrated intensity from rocking scans requires that the aperture of the detector is wide enough to accept the whole reflection. If this is not the case the standard procedure to obtain integrated intensities cannot be applied. R. Shayduk [7.46] proposed different scans with small detector apertures through the reflections and assumed Gaussian shapes of the reflection. Further correction factors are the area factor and the polarisation factor.

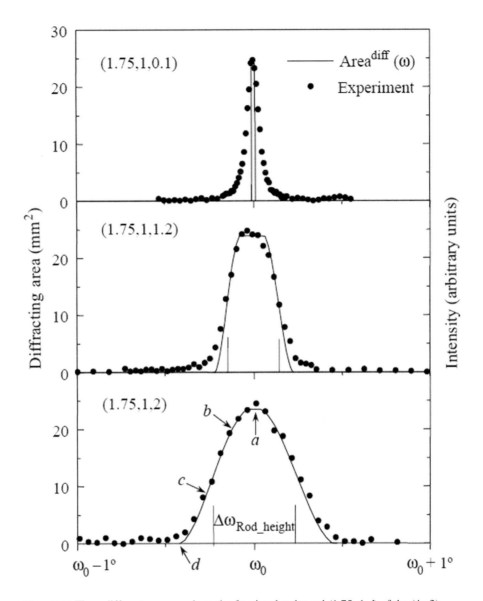

Figure 7.22 Three different ω-scans along the fractional-order rod (1.75, 1, l) of the (4×2) reconstructed InAs(001) surface. The experimental line shapes (filled circles) are compared, after applying a scale factor, to the evolution of the diffracting area labelled $\text{Area}^{\text{diff}}$ (solid line). In the plane ($q \approx 0$), the width ($\Delta Q = 3.5 \times 10^{-2}$ nm^{-1} = 3.5×10^{-3} Å^{-1}) is mainly determined by the finite size of the reconstruction domains. Reprinted from [7.32] O. Robach et al., *J. Appl. Cryst.*, vol. 33, pp. 1006 1018, 2000, with permission from John Wiley and Sons.

During the ω-scan the angle γ changes slightly as the intersection point of the rod and the Ewald sphere move along the rod. If the detector is kept fixed, this has the effect that the projection of the detector slits along the diffracted beam no longer points to the centre of the active area, causing an additional correction to the area

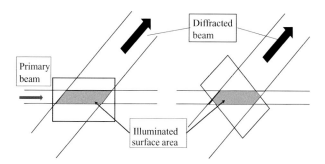

Figure 7.23 Illustration of the active surface area, looking down onto the surface plane. The rotation of the sample (shown as a rectangle) may change the active surface area, depending on the sample size and the settings of the entrance and exit slits (represented by beam widths).

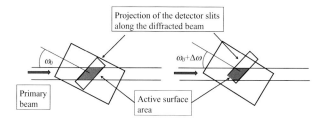

Figure 7.24 Change of the active surface area during an ω-scan at higher exit angles γ. Here ω_0 is the angle for which the detector position is at the intersection point of the reciprocal rod with the Ewald sphere. The rotation of the sample moves this point up (or down) along the rod: the exit angle of the diffracted beam changes, but the detector remains fixed and may not match the centre of the active area anymore.

factor. This point has been discussed by O. Robach et al. [7.32]. This can be avoided if the detector slits are set large enough or Soller slits are applied instead of detector slits.

The measurement with the point detector is very time-consuming, as for each point along the rod a scan with about 50 to 100 measurements is required. To speed up the measurement, X. Torrelles and J. Rius [7.49] have proposed to measure the integrated intensity with an l-scan where only one ω-scan is required for each interval Δl. This scan is obtained if the detector position is not kept fixed, but follows the intersection point of the lattice rod with the Ewald sphere during the rotation of ω. The measurement is fast but can only be applied if the rod is narrow, the detector aperture is large enough to collect the whole intensity of the reflection and the background is sufficiently constant so that one representative rocking scan can be used.

Area Detector
Most surface diffractometers at synchrotron beam lines and in the laboratory are now equipped with an area detector which speeds up the measurement substantially

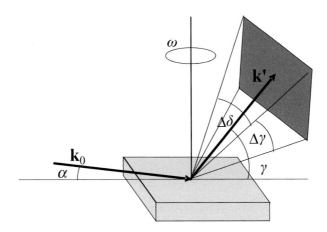

Figure 7.25 Schematic drawing of the diffraction geometry in real space. The solid angle of acceptance $\Delta\delta\times\Delta\gamma$ of the 2-D pixel detector is usually much larger than that of the reflection.

compared to the time required by point detectors. With an area detector the measurement of the integrated intensity is possible with a single image without a rocking scan if the area of the pixel detector is large enough to accept the full size of the reflection and the background next to it. The angles and the diffraction geometry are shown in Figure 7.25.

The measurement with an area detector and the necessary correction factors to obtain the structure factors have been described by C. M. Schlepütz et al. [7.50], S. O. Mariager et al. [7.51], J. Drnec et al. [7.52] and S. J. Leake et al. [7.53]. A data reduction and analysis software for 2-D detectors has been published by S. Roobol et al. [7.54].

The size of the area detector usually exceeds the size of the reflection, but this may not be the case at low exit angles. Then an ω-scan is required while at higher exit angles the detector and the crystal are kept fixed. The situation at small l-values is illustrated in Figure 7.26.

The area detector is divided into a sequence of 'regions of interest' (RoI) which contain the interval Δl of the crystal truncation rod (CTR) or superstructure rod (SSR). At small l-values a rocking scan may be required. The area detector is divided into several regions of interest. The intensity in each region of interest is integrated and $I(l)$ is obtained.

At higher exit angles the full integrated intensity can be obtained from a single image where the sample and the detector are kept fixed. An example is shown in Figure 7.27.

A specific feature of the measurement with an area detector is the measurement of the background. Each pixel of the area detector accepts scattered intensity from an area larger than the active area on the sample. The background intensity may arise from the sample holder if parts of it are hit by the incident beam or from the Be window, as shown in Figure 7.28. The background may become anisotropic and depends on the incident and exit angles. The effect does not occur with a point

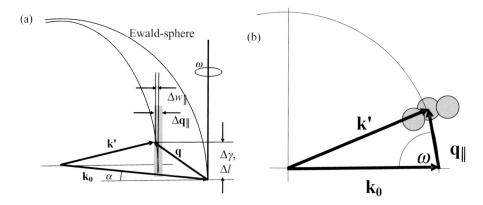

Figure 7.26 Ewald construction of the measurement at small exit angles. $\Delta\gamma$ is the height of the area detector or the region of interest (RoI) in the area detector. This defines the interval Δl used in the integration of the intensity along the rod in reciprocal space. $\Delta w_{\parallel}$ is the projection of the width of the area detector onto the plane $l = 0$. $\Delta q_{\parallel}$ is the width of the rod; if this is larger than $\Delta w_{\parallel}$, then only part of the intensity is collected in the area detector, and an ω-scan is required to obtain the full integrated intensity.

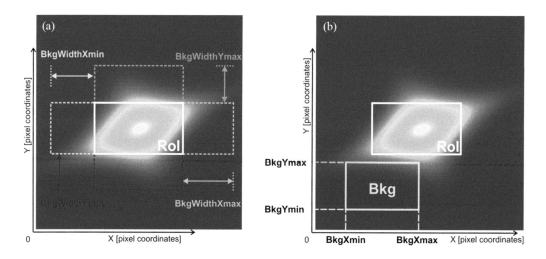

Figure 7.27 Different ways of selecting the scattered X-ray background: (a) rectangles adjacent to each RoI edge; (b) rectangular area outside the RoI. Reprinted from [7.52] J. Drnec et al., *J. Appl. Cryst.*, vol. 47, pp. 365–377, 2014, with permission from John Wiley and Sons.

detector, where the entrance and exit slits limit the active area on the sample. In that case the background can only arise from the active area of the sample, except for background excited in the post sample Be window which can usually be neglected.

For a 2-D area detector another difficulty arises: the so-called divergence problem. The whole illuminated surface area contributes to the background and each pixel at the

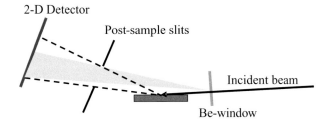

Figure 7.28 The wide opening angle of the slits, which is required to capture a region in reciprocal space with a 2-D area detector, results in a nonuniform background across the detector (the angular range of the background arising from the Be window is indicated in grey in the figure). This background originates from scatterers other than the sample, for example a beryllium window. This means that when taking two images at slightly different detector positions, such that there is some overlap between the two captured regions, the background intensity in the overlapping region is not constant.

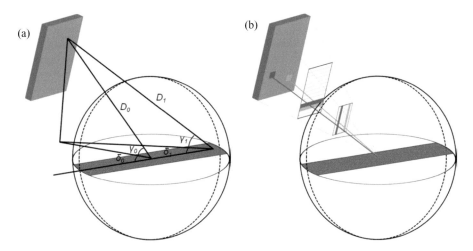

Figure 7.29 (a) Graphic representation of the divergence problem. Each pixel in the 2-D detector receives photons scattered from a different part of the sample at different scattering angles, both in plane and out of plane. (b) The use of two 1-D slits positioned perpendicular to each other solves the divergence problem and the border issue. Reprinted from [7.52] J. Drnec et al., *J. Appl. Cryst.*, vol. 47, pp. 365–377, 2014, with permission from John Wiley and Sons.

detector receives background intensity integrated over a range of scattering angles. This range of scattering angles is not constant in the region of interest, as shown in Figure 7.29(a). The evaluation of the intensity may also become difficult if the sample is not homogeneous and has a border that scatters very differently from the centre of the sample. The use of two linear slits perpendicular to each other as illustrated in Figure 7.29(b) avoids that problem [7.52].

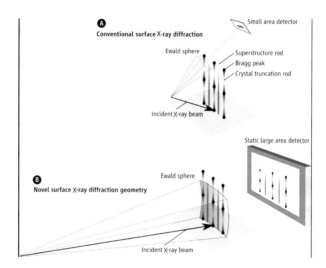

Figure 7.30 High energy X-rays (e.g., 85 keV) and a static large area detector allow the simultaneous observation of some reciprocal lattice rods. Fast processes can be captured by observing the intensity changes along some rods while keeping the sample fixed. The quantitative measurement of the intensities along the rods requires a tilting of the sample around an axis which is in the surface and also normal to the incident beam. (a) Conventional surface X-ray diffraction using lower energy X-rays (typically 10–20 keV) and a small 2-D detector which has to be movable. (b) Novel diffraction geometry using high energy X-rays and a large static area detector. Reprinted from [7.55] C. Nicklin, *Science*, vol 343, pp. 739–740, 2014, with permission from AAAS (online readers may view, browse and/or download material for temporary copying purposes only, provided these uses are for noncommercial personal purposes. Except as provided by law, this material may not be further reproduced, distributed, transmitted, modified, adapted, performed, displayed, published or sold in whole or in part, without prior written permission from the publisher).

Further Techniques
Using high energy X-rays (of order 85 keV) and a wide angle 2-D pixel detector, the measurement of the crystal truncation rods (CTR) and superstructure rods (SSR) can be substantially sped up, as described by C. Nicklin [7.55] and J. Gustafson et al. [7.56]. The diffraction geometry corresponds to that of reflection high energy electron diffraction (RHEED) and is shown in Figure 7.30. The high energy of the incident beam leads to a large area where the Ewald sphere intersects the reciprocal lattice. The intensities along the rod in reciprocal space can be measured by tilting the crystal around an axis parallel to the surface and normal to the incident beam.

7.2.3   Examples of Surface X-ray Diffractometers at Synchrotron Sources

In the following, some surface X-ray diffractometers and UHV chambers are shown and briefly described. The different instruments are designed for studying the growth of layers by molecular beam epitaxy (MBE), reactions at surfaces at high pressures and different environmental conditions, or liquid surfaces and interfaces. Only a few representative instruments are selected to illustrate different constructions and applications. We also include the beam lines which played an important role in the early development of the technique.

APS (Advanced Photon Source, Argonne National Laboratory, USA)
An UHV/molecular beam epitaxy (MBE) chamber combined with a six-circle diffractometer has been installed at an undulator beam line of Sector-33-ID-D,E at APS, as documented by J. H. Lee et al. [7.57] and H. Hong and T.-C. Chiang [7.58]. An image of the system is shown in Figure 7.31. As mentioned by J. H. Lee et al., the recently upgraded system allows in situ studies of the growth of oxide layers and is equipped with several deposition sources and a pure ozone delivery system for oxide MBE growth. The sample position can be controlled by a three-axis translation stage.

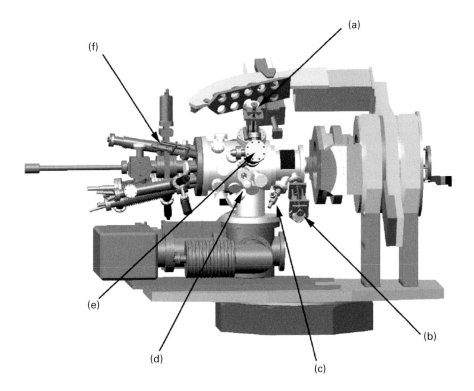

Figure 7.31 3-D rendering of the oxide MBE chamber on the six-circle diffractometer at undulator beam line of Sector-33-ID-D,E at APS, with the various components to monitor and control the oxide layer growth. (a) Film thickness monitor #2 with linear translation. (b) Film thickness monitor #1, also with linear translation. (c) Ozone injector #1. (d) Wobble stick for sample transfer. (e) RHEED gun. (f) Ozone injector #2 with linear translation. (a) and (f) are for a sample at the RHEED position. (b) and (c) are for the X-ray position. Reprinted from [7.57] J. H. Lee et al., *Rev. Sci. Instrum.*, vol. 87, p. 013901, 2016, with the permission of AIP Publishing.

The chamber is equipped with a reflection high-energy electron diffraction (RHEED) system for a second diagnostic control of the oxide layer deposition. Two separate thickness monitors are installed. The sample can be cooled by liquid nitrogen and heated to high temperatures by electron beam bombardment or direct current heating. During film growth and processing, the sample temperature is monitored by thermocouples and can be continuously and rapidly varied between 110 K to above room temperature.

NSLS (National Synchrotron Light Source, Upton, New York, USA)
At beam line X16A the first surface diffraction experiments using synchrotron radiation were performed. The beam line was built by I. K. Robinson. A second beam line X10A was also dedicated to surface diffraction. They are now closed. The synchrotron has been reconstructed. The new machine NSLS-II currently does not provide surface diffraction.

ESRF (European Synchrotron Radiation Facility, Grenoble, France): Surface diffraction beam line ID03
ID03 was a pioneering user facility dedicated to surface diffraction and hosted more of such experiments than all surface diffraction beam lines taken together at other facilities. It is now closed and will become the new EBSL beam line for hard X-ray microscopy.

ESRF (European Synchrotron Radiation Facility, Grenoble, France): Spanish Beam Line SPLINE, BM25
At the first focal point a multi-purpose diffractometer is installed at the Spanish beam line described by J. Rubio-Zuazo et al. [7.59]. The diffractometer is operated in vertical geometry as shown in Figure 7.32. This geometry provides low vertical divergence and high horizontal polarisation of the synchrotron radiation. Several environmental setups with a load of up to 50 kg can be mounted on the diffractometer. The beam line offers its users four different sample environments:

(1) A multipurpose portable UHV chamber specially conceived to carry out grazing incidence X-ray diffraction under different environmental conditions. The chamber can be used from UHV, 10^{-10} mbar, up to 1 bar with a wide sample temperature range from 60 to 1,000 K.
(2) A gas/solid, liquid/solid and electrochemistry cell. The cell is equipped with a thermostatic bath to control the temperature and to modify the liquid composition in situ. The sample can be covered either by a thin glass dome or by a Mylar foil.
(3) A highly reactive gas cell. The cell has been conceived for in situ X-ray diffraction studies of surface modification driven by highly reactive gases in the bar pressure range. The sample can be heated up to 800 K at pressures between mbar and 2 bar. The reactive gas pressure is accurately controlled by a high precision gas valve and a high-pressure gauge.
(4) A portable powder–liquid high-corrosion-resistant reaction cell. The main body of the cell is built of Teflon providing high corrosion resistance and allowing a

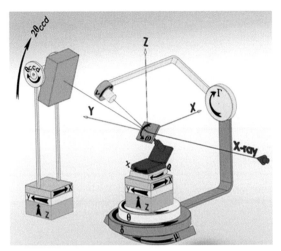

Figure 7.32 Six-circle UHV diffractometer at the Spanish beam line, SPLINE, BM25, ESRF. 2-D area detector. Reprinted from [7.59] J. Rubio-Zuazo et al., *Nuclear Instr. Methods Phys. Res. A*, vol. 716, pp. 23–28, 2013, with permission from Elsevier.

wide range of chemical reactions. The outside dimensions of the body are $125 \times 30 \times 51$ mm height. The cell body is covered with stainless steel to offer physical resistance and heat protection. Crystallisation processes and heterogeneous catalytic processes can be followed.

At the second focal point a diffractometer is installed combining X-ray diffraction (X-ray reflectivity, surface X-ray diffraction, grazing incidence X-ray diffraction and reciprocal space maps) with hard X-ray photoelectron spectroscopy (HAXPES). Both techniques can be operated simultaneously on the same sample and using the same excitation source [7.60]. The set-up includes an ultra-high vacuum chamber equipped with a unique photoelectron spectrometer (few eV < electron kinetic energy < 15 keV), X-ray tube (Mg/Ti), 15 keV electron gun and auxiliary standard surface facilities (molecular beam epitaxy evaporator, ion gun, low-energy electron diffraction, sample heating/cooling system, leak valves, load-lock sample transfer, etc.). This end-station offers the unique possibility of performing simultaneous HAXPES and X-ray diffraction studies. The diffractometer is shown in Figure 7.33.

ESRF (European Synchrotron Radiation Facility, Grenoble, France):
French beam line, BM 32
Two instruments are available to offer a complementary characterisation of surfaces and interfaces in soft and hard condensed matter:

(1) In situ study of atomic layers growth under ultra-high vacuum is allowed in the In situ Nanostructures and Surfaces (INS) hutch.

(2) The Multitechnic Goniometer (GMT) working in air offers a wide range of possibilities to investigate interfaces between different elements or states of matter.

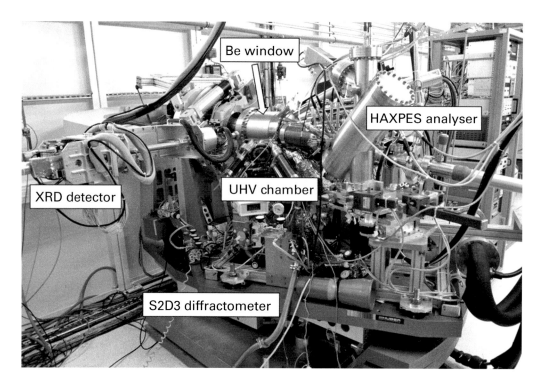

Figure 7.33 Six-circle UHV diffractometer at the Spanish beam line, SPLINE, BM25, ESRF, combining SXRD with HAXPES. Courtesy of J. Rubio-Zuazo.

We show in Figure 7.34 the X-ray reactor chamber designed for both UHV and reactive gas environments, described by M.-C. Saint-Lager et al. [7.61]. This chamber can be mounted on the multi-technique goniometer (GMT). This diffractometer is a versatile instrument dedicated to surface and interface studies. Although it can handle heavy and large experimental setups, the available space is limited. A special feature of the setup is the possibility of easily moving it between the laboratory and synchrotron environments, thus allowing classical studies (sample preparation and reaction conditions) prior to allocated beam time at ESRF.

The aim of this experimental setup is thus to study model catalysts which are prepared under UHV and then analysed by SXRD or GISAXS under catalytic reaction conditions. To meet this purpose a geometry with two horizontally connected chambers was chosen: the X-ray reactor chamber and a UHV chamber. The sample is prepared and characterised in the UHV chamber. It is then transferred under UHV conditions into the X-ray reactor, designed for both UHV and reactive gas environments to perform *in operando* studies. This means that special attention was paid to the choice of the materials inside the reactor. A detailed description of the whole setup is given in [7.61].

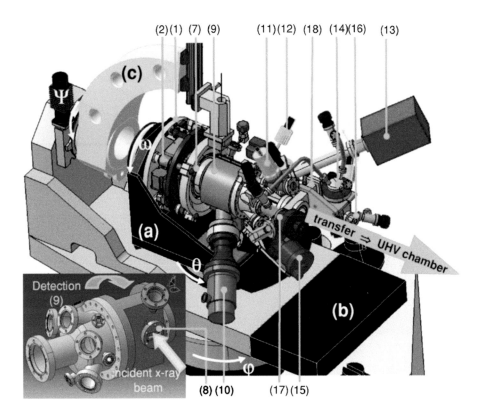

Figure 7.34 Overview of the reactor mounted on the GMT diffractometer of the French beam line, BM 32, at ESRF. The reactor is set on the θ table (a), while (b) is the φ detector table and (c) is the $d\psi$ detector arm. The large arrow indicates the UHV transfer to the preparation chamber. The function of the different components is described in the original publication. Reprinted from [7.61] M.-C. Saint-Lager et al., *Rev. Sci. Instrum.*, vol. 78, p. 083902, 2007, with the permission of AIP Publishing.

PETRA III, DESY (Deutsches Elektronen Synchrotron, Hamburg, Germany)
PETRA III is designed to deliver hard X-ray beams with very high brilliance. A novel X-ray diffractometer for studies of liquid–liquid interfaces has been installed at beam line P08. This beam line is specialised in X-ray scattering and diffraction experiments on solids and liquids where extremely high resolution in reciprocal space is required, cf. O. H. Seeck et al. [7.62]. A high precision six-circle diffractometer for solid samples and a specially designed liquid diffractometer are installed in the experimental hutch, cf. B. M. Murphy et al. [7.63]. We show the liquid diffractometer in Figures 7.35 and 7.36. The study of liquid–liquid interfaces with X-ray scattering methods requires special instrumental considerations. A dedicated liquid surface diffractometer was designed, employing a tilting double-crystal monochromator in Bragg geometry. This diffractometer allows reflectivity and grazing incidence scattering measurements of an immobile and mechanically completely decoupled liquid sample, providing high mechanical stability.

(a)

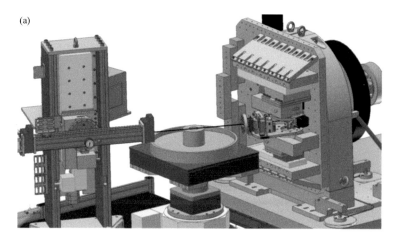

(b)

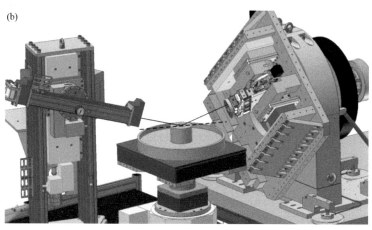

(c)

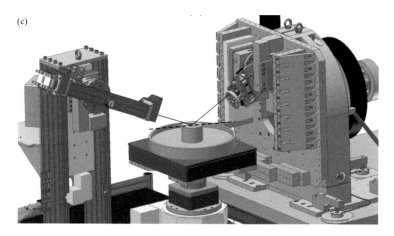

Figure 7.35 Illustration of the beam-tilter rotation during reflectivity measurements at beam line P08, Petra III, DESY. The position of the beam-tilter rotation (called angle mchi) at right is shown for: (a) mchi $= 0°$, $\alpha = \beta = 0°$; (b) mchi $= 45°$, $\alpha = \beta = 4.1°$; and (c) mchi $= 90°$, $\alpha = \beta = 5.8°$ at the maximum position of the angles α and β. The detector height, vertical and horizontal translation and rotations follow the reflected beam. Reprinted from [7.63] B. M. Murphy et al., *J. Synchrotron Rad.*, vol. 21, pp. 45–56, 2014 (Open Access).

(a)

(b)

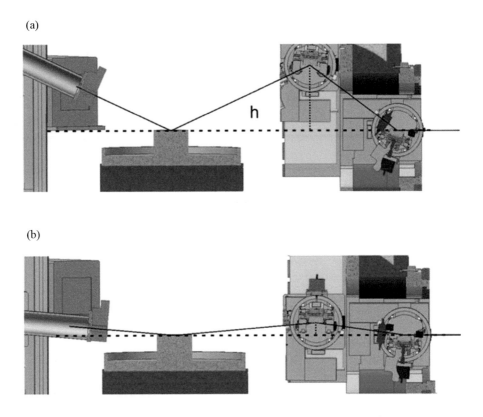

Figure 7.36 Height and angles for detector arm (left) and tilting crystal (right) are shown for (a) 29.4 keV and (b) 6.4 keV. To select an energy, both crystals are rotated to the required Bragg angle and the second crystal is translated to the appropriate height h. The detector angle and height are chosen accordingly. The black line shows the X-ray beam path; the horizontal direction is shown with a dashed line. The Si crystals mounted on the beam-tilter are shown in orange. Reprinted from [7.63] B. M. Murphy et al., *J. Synchrotron Rad.*, vol. 21, pp. 45–56, 2014 (Open Access).

Diamond Light Source (Harwell Science and Innovation Campus, Oxfordshire, UK)

Beam line I07 at the Diamond Light Source is dedicated to the study of the structure of surfaces and interfaces for a wide range of sample types, from soft matter at different ambient conditions to samples requiring ultra-high vacuum. It has two endstations. In the first station, EH1, a (2+3) diffractometer is installed, which acts as a versatile platform for grazing incidence techniques including surface X-ray diffraction, grazing incidence small- (and wide-) angle X-ray scattering, X-ray reflectivity and grazing incidence X-ray diffraction. The second station, EH2, contains a similar diffractometer with a large environmental chamber mounted on it, dedicated to in-situ ultra-high vacuum studies. The UHV chamber is shown in Figure 7.37. It is equipped with scanning tunnelling microscopy, low-energy electron diffraction and X-ray

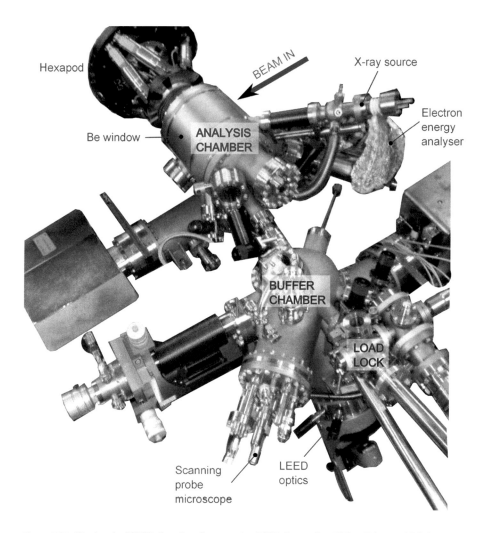

Figure 7.37 The in situ UHV chamber from station EH2, beam line I07 at Diamond Light Source, with key components labelled. Reprinted from [7.64] C. Nicklin et al., *J. Synchrotron Rad.*, vol. 23, pp. 1245–1253, 2016 (Open Access).

photoelectron spectroscopy, ensuring that correlations between the different techniques can be performed on the same sample and in the same chamber. This endstation allows accurate determination of well-ordered structures and measurement of growth behaviour during molecular beam epitaxy; it has also been used to measure coherent X-ray diffraction from nanoparticles during alloying. The beam line and the experimental stations are described in C. Nicklin et al. [7.64].

SOLEIL (Gif-sur-Yvette, France)
A multi-purpose diffractometer for in situ measurements has been installed at SOLEIL's SIXS (surface interface X-ray scattering) beam line dedicated to the study

of grazing incidence X-ray scattering from surfaces and interfaces of hard and soft matter in various environments in the 5–20 keV energy range. The beam line is equipped with two experimental hutches dedicated to:

(1) In situ studies under UHV conditions.
(2) Studies in various environments (catalysis chambers, soft matter, electrochemical cells). A diffractometer coupled with exchangeable chambers allows sample surfaces both in vertical and horizontal geometries.

Both facilities allow performing grazing incidence X-ray diffraction (GIXD), grazing incidence small angle X-ray scattering (GISAXS), anomalous surface X-ray scattering, X-ray reflectivity (XRR), magnetic surface X-ray scattering and coherent scattering experiments.

SPring-8 (Sayo, Japan)
Beam line BL13XU at SPring-8 is dedicated to reveal structures of surface layers on solids and thin films at the atomic scale by grazing angle X-ray diffraction, crystal truncation rod scattering, reflectivity, microbeam diffraction, and reciprocal space mapping in vacuum as well as in air. A photograph of the diffractometer where some main components are marked is shown in Figure 7.38.

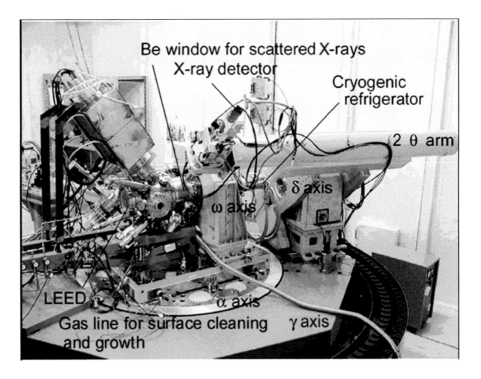

Figure 7.38 Photograph of the UHV diffractometer at beam line BL13XU at SPring-8. Reproduced with permission from SPring-8/JASRI.

A UHV chamber is mounted on a six-circle diffractometer. The chamber is equipped with tools for prior sample preparation and surface analysis. Target materials range widely from hard matter (such as metals and inorganic materials) to soft matter (such as organic semiconductors). Recent measurements include diffraction from nanostructures such as atomic wires, nanodots and ultra-thin films [7.65].

7.2.4   Examples of Surface X-ray Diffractometers for Use as Laboratory Sources

It is not always necessary to use well collimated and highly intense synchrotron beams for SXRD with the purpose of structure determination. Intense laboratory sources like rotating anodes, microfocus sources or Ga-jet sources provide sufficient intensity to measure surface signals. Focussing monochromators can be used, such as graphite monochromators or Göbel mirrors, because the rods in reciprocal space always intersect the Ewald sphere. For 3-D single crystals with no – or very small – mosaic spread the use of a focussed beam would have no advantages. In contrast to powder diffraction, where the focus of the incident beam is on the detector, with surface X-ray diffraction the focus can be put on the sample. This keeps the illuminated area at the sample small and leads to a broader reflection which can be tolerated when only the integrated intensity is measured. A high angular resolution and a high brilliance are not required for many applications.

Laboratory Sources

The first surface diffraction experiment was performed by P. Eisenberger and W. C. Marra [7.66] using a rotating anode. Since then, most experiments have been done with synchrotron radiation. Only a few surface X-ray diffractometers using laboratory sources have been built in the past. Y. Fujii et al. [7.67] describe a four-circle diffractometer combined with a rotating stage and an 18-kW rotating anode. At the Institute of Crystallography, University of Munich, a six-circle diffractometer was built which could be operated in the laboratory with an 18-kW rotating anode as well as at the synchrotron at HASYLAB in Hamburg. It was later operated by H. L. Meyerheim at the Max Planck Institute of Microstructure Physics in Halle, Germany, for about 15 years. M. Albrecht et al. [7.68] described a six-circle diffractometer using an 18-kW rotating anode and a focussing graphite monochromator. This diffractometer placed a semiconductor detector in the UHV chamber with the advantage that the detector has a high counting efficiency and practically no noise.

We show in Figure 7.39 a recent six-circle diffractometer at the MPI-Halle [7.69]. It uses a Ga-jet source [7.70] and an octopod to align the orientation of the sample surface. It allows in situ preparation of adsorbate structures and MBE layers.

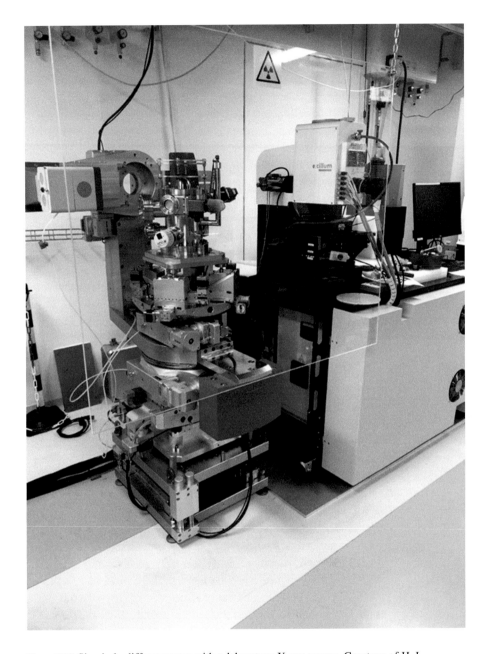

Figure 7.39 Six-circle diffractometer with a laboratory X-ray source. Courtesy of H. L. Meyerheim [7.69].

7.2.5   Synchrotron Radiation Facilities

The following is a selected list of synchrotron radiation facilities with beam lines dedicated to surface X-ray diffraction, as of 2022. A full list of synchrotron radiation facilities can be found online [7.71]:

ANKA (Angströmquelle Karlsruhe), Germany: Beam line MPI-MF.

APS (Advanced Photon Source) at Argonne National Laboratory, Illinois, USA: Beam line XOR/UNI.

DIAMOND (Diamond Light Source) at Harwell Science and Innovation Campus, Oxfordshire, UK: Beam line I07.

ELETTRA (Elettra Sincrotrone Trieste) Trieste, Italy: Beam line ALOISA.

ESRF (European Synchrotron Radiation Facility) Grenoble, France: Beam lines BM32, ID32, BM25.

NSLS (National Synchrotron Light Source) at Brookhaven National Laboratory (BNL), Upton, New York, USA.

PETRA III, at DESY (Deutsches Elektronen Synchrotron), Hamburg, Germany: Beam line P08.

SLS (Swiss Light Source) at Paul Scherrer Institut (PSI), Switzerland: Materials science beam line.

SOLEIL (Source optimisée de lumière d'énergie intermédiaire du LURE, Laboratoire pour l'utilisation du rayonnement électromagnétique) Gif-sur-Yvette, France: Beam line SIXS.

SPring-8 (Super Photon ring-8 GeV), Sayo, Japan: Beam line BL13XU.

7.3   Data Analysis

The structure analysis of surfaces by X-ray diffraction follows mainly the methods developed for 3-D X-ray crystallography with the amendments of refraction and limited penetration depth as described in Section 7.1. Most frequently, the trial and error approach is applied, as in LEED, starting with reasonable structure models and refining the most promising ones. The structure refinement requires a structure model sufficiently close to the solution so that a least squares optimisation converges. There are several methods in use in bulk crystallography to find a reasonable model, but not all can be used with surface diffraction. However, the special conditions provided by surface diffraction also offer different analysis methods. The advances in experimental and theoretical techniques have been recently reviewed by H. Tajiri [7.72]. Substantial progress has been made in the last decades in the solution of the phase problem and the development of direct methods, by which is meant the finding of the structure model (prior to refinement) directly from the diffraction data. This section mainly concentrates on the development of these methods.

In the early years of surface structure analysis, a reasonable model for adsorbate layers or reconstructed surfaces was obtained from the Patterson function, mostly with in-plane reflections which give the Patterson function of the 2-D projection of the structure. The Patterson function, however, is not very useful with large unit cells and a limited data set. There are also restrictions because only superstructure reflections can be used, as will be discussed in Section 7.3.1. Nowadays, the Patterson function is practically no longer used since direct methods have been developed. Finding a structure model is frequently also based on information from other methods like STM or AFM images, or spectroscopic investigations of chemical bonds and atomic environments, as well as first-principles surface structure calculations. Once a reasonable structure model has been found, the analysis proceeds with the well-established methods from bulk crystallography, but these require some modifications in the refinement programs.

For surface structure refinement, the very useful ROD program has been used in most analyses; it is available online together with a manual [7.73]. Included is also a data reduction program, ANAROD, which transforms the data of the rocking scans to structure factors. This program has also been updated to include data from 2-D area detectors: we have presented this approach in Section 7.2, where the measurement techniques are also briefly described. We do not discuss here the conventional structure refinement procedures, as these follow mainly the well-documented methods used in bulk crystallography: see for example the textbooks by U. Shmueli [7.74] or G. Giacovazzo [7.75], while the specific surface diffraction conditions are described in detail in the manual of the ROD program.

We thus restrict this section about structure solution to the description of direct methods: these produce initial models for subsequent refinement. By direct methods we mean that the initial model is derived model-free from the diffracted intensities. Several approaches have been developed for different applications and are discussed in Sections 7.3.1–7.3.8. Since these methods result in phases of the structure factors or in images of the electron density, a subsequent refinement step is necessary to determine the precise atomic coordinates, atomic displacement parameters (including thermal parameters) and occupation factors. This can be done by the ROD program mentioned earlier in this section. Other programs are also in use in different groups. The refinement step may still be tedious and requires experience to produce reliable results, but the strategies are well established. Similar to LEED, the finding of a correct structure model is the first and most important step while, in contrast to LEED, direct methods are available in SXRD.

First, for comparison, we summarise the situation in 3-D (i.e., bulk) structure analysis by X-ray diffraction, where two main approaches are in use. The first approach involves phase estimates derived from the relation between intensities and phases based on the positivity of the electron density and on Sayre's equation resulting from the atomicity of the electron density. A detailed description of the phasing methods has been given by C. Giacovazzo [7.76]. These methods are routinely applied for large molecules and usually result in a number of structure models which can then be used as starting structures for refinement. These methods cannot be directly applied to surface diffraction because they do not allow the use of truncation rod data; furthermore, the Fourier transform of the structure factors from the superstructure alone may exhibit not only positive but also negative values, as is shown in Section 7.3.1, and thus does not correspond to an image of the structure. For surfaces, therefore, a modification of such methods is required.

The second approach uses the so-called pixel methods, in which the electron density is directly determined by an iterative process. These methods are based on the Fienup-Gerchberg-Saxton algorithm [7.77] which was originally developed for the interpretation of radar data. The application in X-ray analysis is known as the 'charge flipping' method [7.78]. Here, the electron densities initially obtained from the Fourier transform of the observed structure factors with zero or random phases are simply made positive by changing the sign of negative values below a certain level. It is successfully and routinely used mostly for small molecules. This method can be used for surface structures with some modification to include the truncation rod data, as will be described in Section 7.3.5.

In principle, the Patterson method can also be viewed as a direct method. However, it does not give an image of the structure, but a superposition of all interatomic distances from all atoms. It requires a search procedure to derive a structure model from the Patterson function. For surface X-ray diffraction the complication occurs that two different kinds of reflections exist: truncation rod data and superstructure data (see Figure 7.14). That leads to specific features of the Patterson function from superstructure reflections.

7.3.1 Patterson Function

To illustrate the consequences of the fact that two different sets of structure factors exist in SXRD, we first briefly discuss the Patterson function. The corresponding specific feature of the Fourier transform of the superstructure reflection is also used by the direct methods discussed in Sections 7.3.2–7.3.8.

The Patterson function is the autocorrelation function of the electron density and is calculated as the Fourier transform of the diffracted intensities. We use $I(hkl) = |F(hkl)|^2$ here and in the following, and omit all pre-factors defining the absolute intensity. $I(hkl)$ is the measured integrated intensity on an arbitrary scale, because the incident intensity is usually not measured. See B. E. Warren [7.44] for a detailed discussion of the meaning of absolute and integrated intensity which is the only measurable quantity. $F(hkl)$ is the structure factor

$$F(hkl) = \sum_v f_v(hkl)e^{2\pi i(hx_v+ky_v+lz_v)}T_D(hkl)$$

$$= |F(hkl)|e^{i\varphi(hkl)}, \tag{7.50}$$

where $f_v(hkl)$ are the atomic scattering factors, the sum over v runs over all atoms in the unit cell and $T_D(hkl)$ is the Debye–Waller factor taking care of the atomic displacement parameters (see Section 5.5 and Appendix I for definitions). The Debye–Waller factor is omitted in the discussion of direct methods, since for finding the model only the atom position vectors $\mathbf{r}_v = x_v\mathbf{a} + y_v\mathbf{b} + z_v\mathbf{c}$ need to be determined; however, it is required for structure refinement as well as an occupation factor which is not considered here. The structure factor is also defined by the Fourier transform of the electron density:

$$F(hkl) = \int_V \rho(x,y,z)e^{2\pi i(hx+ky+lz)}dxdydz. \tag{7.51}$$

If the phases $\varphi(hkl)$ of the structure factors are known, the electron density could be calculated from the inverse Fourier transform. The determination of the phases is the main subject of the direct methods. For 3-D crystals the Patterson function is given by the Fourier transform of the intensities which can also be written as the autocorrelation function of the electron density:

$$
\begin{aligned}
P(uvw) &= \sum_{hkl} I(hkl) e^{-2\pi i(hu+kv+lw)} \\
&= \sum_{hkl} 2|F(hkl)|^2 \cos\left(2\pi(hu+kv+lw)\right) \\
&= \int \rho(x,y,z)\rho(x+u,y+v,z+w)dxdydz,
\end{aligned}
\tag{7.52}
$$

where the sum is over all reflections. The Patterson function must be positive; therefore, since the transmitted beam $I(000)$ is usually not measured and thus omitted from the sum, the Patterson function is made positive by adding a constant term. For SXRD only the superstructure reflections can be used, which has the consequence that both positive and negative values of the Patterson function can occur. This can be easily seen by splitting the sum over all reflections (hkl) into two sums: one over $(h_i k_i l_i)$ with integer indices for the crystal truncation rod data, producing a form factor F_{CTR}; and one over $(h_s k_s l_s)$ with fractional indices for the superstructure rod data, producing a form factor F_{SSR}. To separate the contributions from these two sets of reflections, the electron density is written as the sum of the average electron density of the (1×1) structure and the deviation from the (1×1) structure:

$$
\rho(\mathbf{r}) = \rho_{1\times 1}(\mathbf{r}) + \Delta\rho(\mathbf{r}).
\tag{7.53}
$$

The Patterson function becomes:

$$
\begin{aligned}
P(\mathbf{u}) &= \int \langle\rho_{1\times 1}(\mathbf{r})\rangle\langle\rho_{1\times 1}(\mathbf{r}+\mathbf{u})\rangle d\mathbf{r} + \int \Delta\rho(\mathbf{r})\Delta\rho(\mathbf{r}+\mathbf{u})d\mathbf{r} \\
&= \sum_{h_i k_i l_i} 2|F(h_i k_i l_i)|^2 \cos\left(2\pi(h_i u + k_i v + l_i w)\right) \\
&\quad + \sum_{h_s k_s l_s} 2|F(h_s k_s l_s)|^2 \cos\left(2\pi(h_s u + k_s v + l_s w)\right) \\
&= P_{1\times 1}(\mathbf{u}) + P_s(\mathbf{u}).
\end{aligned}
\tag{7.54}
$$

$P_s(\mathbf{u})$ is the Patterson function of the difference electron density and shows positive and negative maxima; its average is zero. The difference electron density has the same symmetry as the superstructure. Interpretation of the Patterson function shows differences due to shifts of atoms and missing atoms; see, for example, J. Rius et al. [7.79], Y. Takeuchi [7.80] and, for the general properties, M. J. Buerger [7.81].

A schematic drawing for a (2×1) superstructure is shown in Figure 7.40. The difference electron density (Figure 7.40(e)) exhibits positive and negative values. The positive values correspond to atomic positions in the superstructure, while

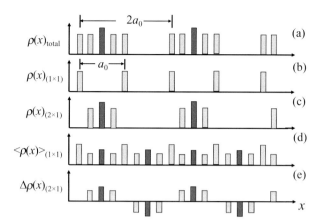

Figure 7.40 Schematic drawing of the electron density of a (2×1) superstructure along the surface in the direction where doubling occurs. (a) Total electron density of the (2×1) structure. (b) Electron density of the (1×1) substrate structure. (c) Electron density of the (2×1) superstructure. (d) Electron density of the average (1×1) structure, obtained as (b) + 1/2(c) into each (1×1) unit cell. (e) Difference electron density, (a) – (d).

the negative values correspond to missing atoms. The interpretation of the Patterson function may be ambiguous because the product of two negative maxima results also in a positive maximum. The Patterson function calculated from the superstructure reflections of a (2×1) superstructure on a square lattice is shown in Figure 7.41. For larger unit cells of the superstructure, the Patterson function calculated from the superstructure reflections is dominated by the interatomic vectors of the superstructure.

7.3.2   Direct Methods for Surface Structure Analysis

SXRD has the advantage that the structure of the substrate is usually known and the interference between the substrate and the surface slab can be used to determine the phases of the structure factors in the superstructure rods (F_{SSR}). Knowing the phases of the superstructure reflections and of the reflections in the crystal truncation rods (F_{CTR}) the electron density of the adsorbate layer can be determined. The interference between substrate and surface is related to the 'heavy atom method' used in 3-D crystallography, see for example [7.74]. The substrate plays the role of the heavy atom. As has long been noticed, the heavy atom method is related to holography. The interference of the diffracted waves from the surface slab with those of the substrate can be interpreted as a generalised hologram, as has been pointed out by T. Takahashi et al. [7.82].

Several analytical methods have been developed independently and approximately at the same time to find the structure solution for the surface slab using the interference between substrate and surface slab. An approach based on conventional phasing

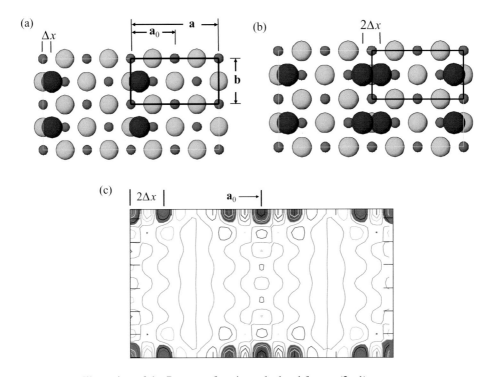

Figure 7.41 Illustration of the Patterson function calculated from a (2×1) superstructure on a square substrate lattice. (a) The structure model shows an adsorbed atom on a (100) surface of the fcc-lattice shifted by Δx off the top site. The (2×1) unit cell is marked; the symmetry is p1m1. (b) The Patterson function adds an inversion centre and the symmetry becomes p2mm. (c) Patterson function calculated from the superstructure reflections of model data. Blue dots represent positive maxima and red ones represent negative maxima: these correspond to the missing atom in the middle of the (2×1) unit cell.

methods in combination with genetic algorithms was proposed by L. D. Marks et al. [7.83; 7.84]. In this method, the phases of some strong superstructure reflections are first estimated; from these, some possible sets of phases for further reflections are derived; the most probable sets of phases, those with the best figure of merit, 'FOM', are found by optimisation with a genetic algorithm. The method has been successfully applied to a number of surface structures [7.85]. We do not further discuss this approach here but focus in the following on methods that use the interference between the bulk and the surface contribution to the structure factor. The methods applied so far are listed here and explained in more detail in Sections 7.3.3–7.3.8:

(1) COBRA (coherent Bragg rod analysis), developed by Y. Yacoby et al. [7.86]. This method uses the factors F_{CTR} only and has been applied to determine the structure of buried interfaces, as well as adsorbate systems.

(2) MSF (modulus sum function), developed by J. Rius and co-workers [7.87; 7.88]. This method derives phase estimates for superstructure reflections from the relation between intensities and phases and uses sums of products of three structure factors instead of two structure factors; two are used in the conventional phasing methods.

(3) PARADIGM (phase and amplitude recovery and diffraction image generation method), developed by D. K. Saldin, W. Moritz and co-workers [7.89; 7.90]. It uses the interference between substrate and superstructure to determine the electron density of the superstructure by an iteration process.

(4) The DCAF (difference map using the constraints of atomicity and film shift) method, introduced by M. Björck et al. [7.91]. The DCAF method uses the truncation rod data to derive the electron density of the surface slab with an iteration scheme of Fourier and inverse Fourier transforms and different constraints, namely atomicity in real space, not derived from structure factors as is the case in Sayre's equation, see for example C. Giacovazzo [7.76].

(5) Analysis of density profiles, proposed by I. K. Robinson et al. [7.92]. The control of the thickness of layers and interdiffusion at boundaries is of high interest in the preparation of layered semiconductor structures. A perturbation method is proposed to determine the density profile normal to the surface directly from the measured structure factors.

It should be mentioned that the phases $\varphi(hkl)$ can be measured in principle. One possibility is to use anomalous X-ray diffraction [7.93]. The main method in bulk crystallography uses dynamical effects in strong reflections and is used in structure determination of organic crystals but is not always applicable. For surfaces it has been shown that dynamical effects occur with well-ordered structures [7.94], but the method has not been applied to structure determination.

7.3.3   Coherent Bragg Rod Analysis (COBRA)

The COBRA algorithm was developed by Y. Yacoby et al. [7.86] and uses the truncation rod data only. There exist two variants of this method. The first uses the phase differences along the crystal truncation rods measured by a double diffraction experiment. It requires the evaporation of a thin gold film from which the incident beam is reflected at grazing incidence and serves as a second incident beam with a different incidence angle. The purpose of this arrangement is that two beams with a definite and known phase relation are diffracted by the surface slab or the interface. The difference in the scattering vectors is small so that the derivative of the phase of the structure factor can be measured. An illustration of the experimental arrangement is shown in Figure 7.42. This variant is generally applicable, as long as the experimental conditions allow the deposition of a gold film. It further requires a well collimated incident beam.

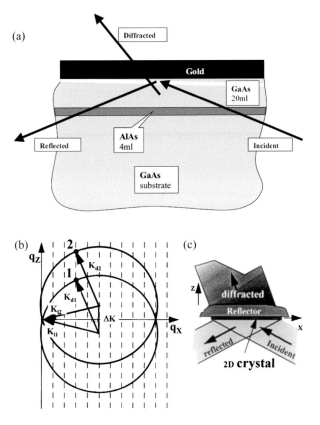

Figure 7.42 (a) A schematic diagram of the sample and scattering geometry for the measurement of the derivative of the phase of the crystal truncation rods of a 2-D crystal of AlAs sandwiched between a substrate and a film of GaAs. (b) Reciprocal space representation of the two-beam diffraction interference experiment. The dashed lines represent the Bragg rods; the circles represent the Ewald spheres. (c) Schematic diagram of the superposition of wave fields in a two-beam diffraction experiment. (a) © IOP Publishing. All rights reserved. Reproduced with permission from [7.86] Y. Yacoby et al., J. *Phys.: Condens. Matter*, vol. 12, pp. 3929–3938, 2000; permission conveyed through Copyright Clearance Center, Inc. (b) and (c) are reprinted from [7.98] H. Baltes et al., *Phys. Rev. Lett.*, vol. 79, pp. 1285–1288, 1997, https://journals.aps .org/prl/abstract/10.1103/PhysRevLett.79.1285. © (1997) by the American Physical Society.

The second variant applies an iterative determination of the phases of the truncation rod structure factor from the interference between the surface and bulk contributions. The method requires an initial model, for example for buried interface layers the bulk structure of the interface material. The initial model is iteratively modified until it matches the measured data. It uses the positivity of the electron density and the known substrate structure. It is further based on the assumption that the phase of the unknown part, the surface slab, should vary more slowly than the phase of the bulk contribution. The method has been applied to a number of surface and interface structures [7.95–7.98]. A superstructure in the surface has to be derived from the average in the (1×1) structure as only the CTR data are used.

We first explain the second variant. The structure factor of the surface slab and its phase are determined iteratively starting with the model structure. Following the terminology of Y. Yacoby et al. [7.86], we use for the structure factors:

$T(h,k,l)$: the measured structure factor of the CTR;
$U(h,k,l)$: the unknown structure factor of the surface (or interface) slab;
$S(h,k,l)$: the known structure factor of the substrate.

The measured intensity is given by:

$$|T(h,k,l)|^2 = |U(h,k,l) + S(h,k,l)|^2. \tag{7.55}$$

The phase of $T(h,k,l)$ is unknown but it is assumed that it varies slowly and continuously between neighbouring points on the truncation rod. $S(h,k,l)$ is calculated from the known bulk structure and $U(h,k,l)$ is iteratively optimised. Close to a Bragg point, the large value of $S(h,k,l)$ determines the phase of the measured structure factor $T(h,k,l)$. The two points next to the Bragg point can be used as starting points in the iteration. In the initial sequence, the structure factor $U(h,k,l)$ and its phase are calculated from the starting model:

$$S(h,k,l) + U(h,k,l) = |T(h,k,l)|e^{i\varphi_1}, \tag{7.56}$$

for the first point and with φ_2 for the next point:

$$S(h,k,l+\Delta l) + U(h,k,l+\Delta l) = |T(h,k,l+\Delta l)|e^{i(\varphi_1 + \Delta\varphi)}, \tag{7.57}$$

where $\Delta\varphi = \varphi_2 - \varphi_1$. Here $U(h,k,l+\Delta l)$ is calculated as

$$U(h,k,l+\Delta l) = |Q|e^{i\varphi_Q}|U(h,k,l)|, \tag{7.58}$$

with the complex ratio Q from the preceding iteration:

$$Q = \frac{U(h,k,l+\Delta l)}{U(h,k,l)}. \tag{7.59}$$

The structure factor $S(h,k,l)$ and its phase are known because the structure of the substrate is known. The modulus of the truncation rod structure factor $|T(h,k,l)|$ is measured and the phase difference $\Delta\varphi = \varphi_2 - \varphi_1$ is initially known because the structure factor of the starting model is known. Using the initial approximation $Q \approx 1$, the two Eqs. (7.56) and (7.57) are solved for $U(h,k,l)$, φ_1 and φ_Q for all values l. There are two solutions for $U(h,k,l)$ at each value l and the smaller variation of $U(h,k,l)$ is chosen. With the calculated $U(h,k,l)$ for all values l, the procedure is repeated until convergence of $U(h,k,l)$ occurs. The procedure has been found to converge after less than six iterations in all cases.

A different approach is the first variant mentioned at the beginning of this section: the experimental determination of the derivative of the phase along the CTR. It requires the evaporation of a thin gold film on the sample. The reflected beam from the gold film provides a second measurement of the same point in reciprocal space with a slightly different value of $\Delta\mathbf{k}$. The scattering geometry in real and reciprocal space is illustrated in Figure 7.42. If the phase difference or $\Delta\mathbf{k}$ is small enough, the derivative of the phase can be determined. The single diffracted intensity $I_S(\mathbf{q})$ transmitted through the gold film and the second diffracted intensity $I_D(\mathbf{q})$ reflected from the gold film and then diffracted by the interface layer are measured. The phase difference $\varphi(q_z) - \varphi(q_z + \Delta\mathbf{k})$ can then be derived from Eq. (7.58). The intensities of the two diffracted beams I_S and I_D are given by H. Baltes et al. [7.98] as:

$$I_S(q_z) = P_D|T(q_z)|^2,$$

$$I_D(q_z) = P_D\left\{\begin{array}{l}|T(q_z)|^2 + |R|^2|T(q_z + \Delta\mathbf{k})|^2 + \\ 2R|T(q_z)||T(q_z + \Delta\mathbf{k})|\cos(\varphi(q_z) - \varphi(q_z + \Delta\mathbf{k}) - \Phi)\end{array}\right\}, \quad (7.60)$$

where P_D is a proportionality constant independent of $\mathbf{q}$. The modulus $|R|$ of the reflection coefficient R and its phase Φ can be measured. $T(q_z)$ is the complex structure factor of the truncation rod, while $\varphi(q_z)$ is its phase. Only the phase difference $\varphi(q_z) - \varphi(q_z + \Delta\mathbf{k})$ can be measured with respect to an arbitrary reference point. If $\Delta\mathbf{k}$ is small enough, it gives directly the derivative of the phase. The phase derivatives are used to solve Eqs. (7.56) and (7.57). The direct integration of the derivatives would in principle also be possible, but the absolute value of the phase is required at least at one point of the truncation rod; however, noise in the data would cause large errors.

The second variant of the COBRA method described in this section does not use an integration of derivatives but applies an iterative optimisation of the phase relation Q in Eqs. (7.58) and (7.59), starting with $Q \approx 1$. Figure 7.43 shows the moduli of the complex scattering factors of the GaAs substrate and four monolayers of AlAs on top of it. The phase of the surface structure factor $U(h, k, l)$ varies much more slowly than the phase of the bulk structure factor, which justifies the use of $Q \approx 1$ as an initial value in the iteration process.

The method is general and does not depend on the symmetry properties of the system; it also provides the 3-D structure of the 2-D system; however, it is much simpler than the experimental measurement of the phase derivatives and its computational complexity scales only linearly with the number of atoms. The method allows investigating interface structures as well as adsorbate layers and has been applied to a number of systems, for example a study of the influence of metal electrode layers on $BaTiO_3$ films.

We show here as an example the analysis of interdiffusion in $La_{2-x}Sr_xCuO_4$ films [7.96]. It was performed in connection with the study of interface superconductivity that occurs in metal-insulator (M–I) bilayers where M $= La_{1.55}Sr_{0.45}CuO_4$, a non-superconducting metal, and I $= La_2CuO_4$, an antiferromagnetic insulator. The study should solve the question of whether the interface superconductivity is related to charge transfer or to interdiffusion. Three samples were prepared by MBE growth of six unit cells of $La_{2-x}Sr_xCuO_4$ films on a $LaSrAlO_4$ substrate. That corresponds to

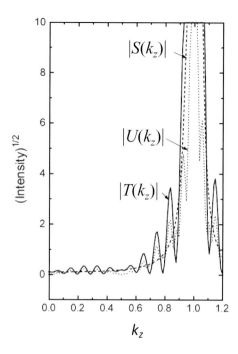

Figure 7.43 An example of the moduli of the complex scattering factors of the semi-infinite GaAs substrate, $|S(h, k, l)|$, of the sample shown in Figure 7.42(a) with four AlAs monolayers, $|T(h, k, l)|$, and of the difference between the two, $|U(h, k, l)|$, over a fraction of a Bragg rod. The large peak at $k_z = 1$ corresponds to a Bragg point. © IOP Publishing. All rights reserved. Reproduced with permission from [7.86] Y. Yacoby et al., *J. Phys.: Condens. Matter*, vol. 12, pp. 3929–3938, 2000; permission conveyed through Copyright Clearance Center, Inc.

12 $La_{2-x}Sr_xCuO_4$ formula units. (The unit cell is doubled by a glide plane and contains two formula units.) The Sr dopant atoms were deposited above, below, or on both sides of each CuO_2 layer. The structure of the samples is shown in Figure 7.44.

The diffraction intensity along the [00L] Bragg rod of the sample in Figure 7.44(b) is shown in Figure 7.45(a). The data were analysed using the COBRA method using a difference map to take care of the weak scattering contrast between La and Sr. The difference map requires the measurement of the intensities along the CTR rods at two different energies. The details of the analysis are described in Y. Yacoby et al. [7.96]. The structure factors obtained by the COBRA method were used in a refinement. The electron density along the [0 0 Z] line, perpendicular to the surface, is shown as an example in Figure 7.45(b). This line goes through La/Sr, O, Al and Cu atoms. Notice that the La/Sr peaks above (to the right of) the Cu atoms are smaller than those below (to the left), suggesting that the Sr concentration in the layers immediately above the CuO_2 planes is larger than in those below. The results of the analysis of all three samples show that the concentration of Sr in La/Sr layers just

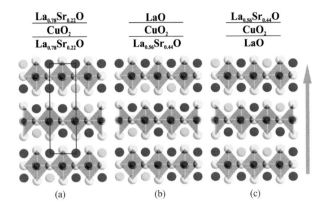

Figure 7.44 Sample growth sequences. Schematics of the structure of three $La_{2-x}Sr_xCuO_4$ samples with nominal distribution of Sr dopant atoms as targeted by the atomic layer deposition sequences. At room temperature, the structure is tetragonal and the space group is I4/mmm. The unit cell is delineated by the rectangular box. La: blue; Sr: green; Cu: red; O: yellow. (a) $La_{2-x}Sr_xCuO_4$ with a random distribution of Sr atoms across the structure; (b) $La_{2-x}Sr_xCuO_4$ with a preferential distribution of Sr atoms only below the CuO_2 plane; (c) $La_{2-x}Sr_xCuO_4$ with a preferential distribution of Sr atoms only above the CuO_2 plane. The film growth direction is indicated by the arrow. Figure reproduced with permission from [7.96] Y. Yacoby et al., *Phys. Rev. B*, vol. 87, p. 014108, 2013. https://journals.aps.org/prb/abstract/10 .1103/PhysRevB.87.014108. © (2013) by the American Physical Society.

above the CuO_2 layers is much larger than in layers just below them, irrespective of the deposition sequence, thus drastically breaking the inversion symmetry.

Both variants of the COBRA method – the experimental measurement of the phase derivatives and the iterative determination of the phases – require the interference in the CTRs and therefore determine the average (1×1) structure. In some cases, the structure of the epitaxial film can be unravelled knowing the bulk structure and possible sites of the epitaxial film, as has been shown for Gd_2O_3 layers on GaAs (100). Gd_2O_3 grows epitaxially on GaAs(100) with three epitaxial unit cells on 16 substrate unit cells, and a $\begin{pmatrix} 4 & 4 \\ -2 & 2 \end{pmatrix}$ superstructure. The structure of the Gd_2O_3 film could be derived from the knowledge of the possible bonds and sites of Gd on the GaAs substrate [7.97].

7.3.4 Modulus Sum Function (MSF)

The conventional phase estimates for 3-D structures use the fact that $\rho(\mathbf{r})$ and $\rho^2(\mathbf{r})$ describe similar structures, as a consequence of the atomicity of the structure. The methods rely also on the fact that the electron density has to be non-negative. However, as mentioned in Section 7.3.1, the Fourier transform of the superstructure reflections leads to a difference electron density $\Delta\rho(\mathbf{r})$ which has both positive and negative values and therefore $\rho^2(\mathbf{r})$ cannot be used. To overcome that problem, J. Rius

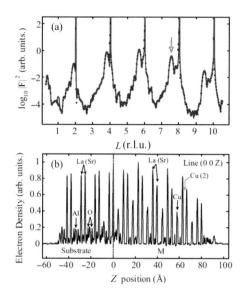

Figure 7.45 Bragg rod measurement, COBRA determined electron density and fit for the sample shown in Figure 7.44(b). (a) A representative example of the dependence of specular diffraction intensity on the momentum transfer along the (00L) Bragg rod measured on a single-layer metallic cuprate sample, corresponding to Figure 7.44(b) (dots) and calculated from the COBRA determined electron density (solid line). The arrow at $L = 7.55$ reciprocal lattice units (r.l.u.) indicates the point at which the diffraction intensity was measured as a function of energy. (b) The electron density along the [0 0 Z] line through La/Sr, O, Al and Cu atoms. The dashed line represents the nominal substrate/metal interface. To the left of the dashed line the electron density of the LaSrAlO$_4$ substrate is shown. The topmost four unit cells of the substrate are included in the structure refinement. To the right of the dashed line the metallic La$_{2-x}$Sr$_x$CuO$_4$ layer is shown. Cu(2) represents the positions of Cu atoms along the [0.5 0.5 Z] line. Figure reproduced with permission from [7.96] Y. Yacoby et al., *Phys. Rev. B*, vol. 87, p. 014108, 2013. https://journals.aps.org/prb/abstract/10.1103/PhysRevB.87.014108. © (2013) by the American Physical Society.

has developed a modification of the phase estimate methods, the 'modulus sum function' (MSF), by using $\Delta\rho^3(\mathbf{r})$ and assuming that this is similar to $\Delta\rho(\mathbf{r})$. The similarity is a consequence of the peaked nature of the electron density, that is, of the atomicity of the structure. We shall briefly describe the basics of this theory, following the notation of J. Rius and co-workers [7.87; 7.88]. Normalised structure factors $E_s(\mathbf{h})$ are used, where for normalisation only the set of superstructure reflections is included. We use in the following derivation $|E_s(\mathbf{h})|e^{i\varphi_S(\mathbf{h})}$ for the structure factors of the difference electron density and $|C(\mathbf{h})|e^{i\alpha(\mathbf{h})}$ for the structure factors of $\Delta\rho^3(\mathbf{r})$. It is assumed that

$$\varphi_S(\mathbf{h}) \cong \alpha(\mathbf{h}). \tag{7.61}$$

The set $E_s(\mathbf{h})$ is divided into two subsets, namely the subset containing the normalised structure factors with largest E_s magnitudes (reflections $\mathbf{k}$) and the second part containing the remaining structure factors (reflections **l**). Let Φ denote the set of phases of the reflections **k**, and let's assume that the resolution of the intensity data is high enough to produce well resolved peaks in $\Delta\rho^3(\mathbf{r})$. $E_s(\mathbf{h})$ and $C_\mathbf{h}(\Phi)$ are the Fourier transforms of $\Delta\rho(\mathbf{r})$ and $\Delta\rho^3(\mathbf{r})$, respectively. It follows that

$$C_\mathbf{h}(\Phi) = \sum_{\mathbf{k}'}\sum_{\mathbf{k}''}E_s(\mathbf{k}')E_s(\mathbf{k}'')E_s(\mathbf{h} - \mathbf{k}' - \mathbf{k}'').\tag{7.62}$$

Use has been made of the well-known fact that products in real space result in a convolution in reciprocal space. The structure factors are discrete functions defined at the nodes of the reciprocal space. It follows that the convolution of structure factors occurring in the Fourier transform of $\Delta\rho^3(\mathbf{r})$ becomes a sum of products in Eq. (7.62). The amplitude of $C_\mathbf{h}(\Phi)$ in terms of Φ can be approximated by:

$$
\begin{aligned}
|C_\mathbf{h}(\Phi)| &= C_\mathbf{h}(\Phi)e^{i\alpha(-\mathbf{h})}\\
&\cong e^{i\varphi_s(-\mathbf{h})}\sum_{\mathbf{k}'}\sum_{\mathbf{k}''}E_s(\mathbf{k}')E_s(\mathbf{k}'')E_s(\mathbf{h} - \mathbf{k}' - \mathbf{k}'').
\end{aligned}\tag{7.63}
$$

In order to determine the phases $\varphi_s(\mathbf{h})$, we define Fourier coefficients $\delta P'(\mathbf{u})$ and $\delta P(\mathbf{u}, \Phi)$:

$$\delta P'(\mathbf{u}) = V^{-1}\sum_\mathbf{h}(E_s(\mathbf{h}) - \langle E_s(\mathbf{h})\rangle)\exp(-2\pi i\mathbf{h}\cdot\mathbf{u}),\tag{7.64}$$

$$\delta P(\mathbf{u}, \Phi) = V^{-1}\sum_\mathbf{h}C(\Phi)\exp(-2\pi i\mathbf{h}\cdot\mathbf{u}).\tag{7.65}$$

The first function, $\delta P'(\mathbf{u})$, is the Fourier transform of $E_s(\mathbf{h})$, where the peak at the origin is removed and the second function, $\delta P(\mathbf{u}, \Phi)$, expresses $\Delta\rho^3(\mathbf{r})$ as a function of Φ.

It has been shown by J. Rius [7.88] that the product function $Z_{\delta P'}(\Phi)$, defined in Eq. (7.66), has a maximum for the correct set of phases Φ. In that case, maxima in $\Delta\rho(\mathbf{r})$ and $\Delta\rho^3(\mathbf{r})$ coincide.

$$
\begin{aligned}
Z_{\delta P'}(\Phi) &= V\int\delta P'(\mathbf{u})\delta P(\mathbf{u}, \Phi)d\mathbf{u}\\
&= \sum_\mathbf{h}(E_s(\mathbf{h}) - \langle E_s\rangle)C_\mathbf{h}(\Phi).
\end{aligned}\tag{7.66}
$$

The maximum of $Z_{\delta P'}(\Phi)$ is determined by

$$\partial Z_{\delta P'}(\Phi)/\partial\varphi_s(\mathbf{k}) = 0\tag{7.67}$$

for all reflections **k**. In the calculation of $Z_{\delta P'}(\Phi)$ the division of the set of superstructure reflections into strong reflections **k** and the remaining weaker reflections **l** has to be considered. We note here only the final result and refer the interested reader to J. Rius and colleagues [7.87; 7.88] where the derivation is given. The set of phases $\varphi_s(\mathbf{k})$ is iteratively determined, the phases $\alpha(l)$ being recalculated in each iteration cycle.

$$\varphi_s(\mathbf{k}) = \text{phase of} \left\{ \sum_{\mathbf{k}'} E_s(\mathbf{k}') \sum_{\mathbf{k}''} X(\mathbf{k}, \mathbf{k}', \mathbf{k}'') E_s(\mathbf{k}'') E_s(\mathbf{k} - \mathbf{k}' - \mathbf{k}'') \right.$$
$$\left. + \sum_{\mathbf{l}} (E_s(\mathbf{l}) - \langle E_s \rangle) e^{i\alpha(l)} \sum_{\mathbf{k}''} E_s(\mathbf{k}'') E_s(\mathbf{k} - \mathbf{k}'' - \mathbf{l}) \right\}, \quad (7.68)$$

where the term $X(\mathbf{k}, \mathbf{k}', \mathbf{k}'')$ is given by:

$$X(\mathbf{k}, \mathbf{k}', \mathbf{k}'') = 1 - \frac{\langle E_s \rangle}{4} \left\{ E_s(-\mathbf{k})^{-1} + E_s(\mathbf{k}')^{-1} + E_s(\mathbf{k}'')^{-1} + E_s(\mathbf{k} - \mathbf{k}' - \mathbf{k}'')^{-1} \right\}.$$
$$(7.69)$$

The sum function $Z_{\delta P'}(\Phi)$ uses only quartets and quintets of structure factors. There exists a large number of four-phase relationships, which limits the applicability of Eq. (7.68). The solution of Eq. (7.68) becomes lengthy for 3-D problems with their large data sets but is easily applied to surfaces when the $(hk0)$ reflections are used for the projection of the electron density. Rius notes that in comparison with $\rho(\mathbf{r})$, the difference structure $\Delta\rho(\mathbf{r})$ contains fewer peaks. Consequently, the phases obtained by maximising $Z_{\delta P'}(\Phi)$ are expected to be effective in spite of the increased complexity arising from the use of $\Delta\rho^3(\mathbf{r})$. Finally, it should be mentioned that the refinement of phases maximizing $Z_{\delta P'}(\Phi)$ cannot distinguish between $\Delta\rho(\mathbf{r})$ and its negative replica $-\Delta\rho(\mathbf{r})$. That means both solutions $\varphi(\mathbf{h})$ and $\varphi(\mathbf{h}) + \pi$ are equally probable.

The modulus sum function has been applied to a number of surface structures where the difference Patterson map obtained from the in-plane data could not be interpreted and led to incorrect results. It is found in general that for large unit cells and complex superstructures the Patterson map is not very useful. Examples where the direct method by MSF led to a solution have been described by X. Torrelles et al. [7.99]. The surface structures investigated are $In_{0.04}Ga_{0.96}As(001)$–p($4\times2$), Ge(001)–c($4\times2$) [7.100], Ge(113)+($2\times2$)–Sb [7.101] and Au(110)+p(6×5)–C_{60} [7.102].

We show in Figures 7.46 and 7.47 the results for Ge(113)+(2×2)–Sb. Figure 7.46(a) displays the difference Patterson function with, for comparison, the $\Delta\rho(\mathbf{r})$ map in Figure 7.46(c) resulting from the MSF analysis. Some interatomic vectors occurring in the Patterson function are marked in the final model in Figure 7.46(b). The model could not be derived from the Patterson function, but the $\Delta\rho(\mathbf{r})$ map allowed conclusions about the Sb positions. The $\Delta\rho(\mathbf{r})$ map contains the average structural information of the c(2×2) superstructure projected into a (1×2) unit cell, since only these reflections could be used in the MSF analysis. The stronger peaks in the $\Delta\rho(\mathbf{r})$ map correspond to adsorbed Sb atoms and to topmost Ge atoms directly bonded to them, showing maximum displacements from their 'ideal' bulk positions. From this interpretation a projected model containing three adsorbed Sb

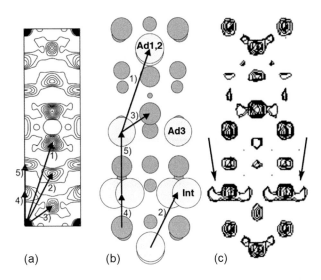

(a)   (b)   (c)

Figure 7.46 (a) 2-D Patterson map obtained from the in-plane data of the Ge(113)+p(1×2)–Sb superstructure. The difference electron density map does not allow deriving the model shown in (b) because positive and negative electron densities cancel in the difference map. (b) Top view projection of the model obtained after interpretation of the $\Delta\rho(\mathbf{r})$ map. (c) Contour map of $\Delta\rho(\mathbf{r})$ obtained from in-plane data. © IOP Publishing. All rights reserved. Reproduced with permission from [7.99] X. Torrelles et al., *J. Phys.: Condens. Matter*, vol. 14, pp. 4075–4086, 2002; permission conveyed through Copyright Clearance Center, Inc.

atoms can be easily obtained, shown in Figure 7.46(b). The Sb atoms, labelled Ad1,2 and Ad3, are 3-fold coordinated to Ge, while the interstitial Sb-Sb dimers, labelled Int, are 2-fold coordinated to Ge.

It is remarkable that the $\Delta\rho(\mathbf{r})$ map clearly shows the split positions of Sb forming a dimer. The apparent split positions in the (1×2) unit cell arise from the fact that only reflections corresponding to the smaller (1×2) unit cell could be used in the analysis. This leads to the projection of the c(2×2) superstructure into a (1×2) cell. In the c(2×2) cell no split positions occur but an Sb dimer exists: the model is shown in Figure 7.47(c).

The modulus sum function has been applied up to now only with in-plane superstructure reflections because of the difficulty to measure a sufficient number of reflections on superstructure rods at higher exit angles.

7.3.5   Phase and Amplitude Recovery and Diffraction Image Generation Method (PARADIGM)

The PARADIGM method uses structure factors from both truncation rods and superstructure rods. By contrast, the two methods discussed in the preceding sections use either the (1×1) reflections (COBRA) or only superstructure reflections (MSF).

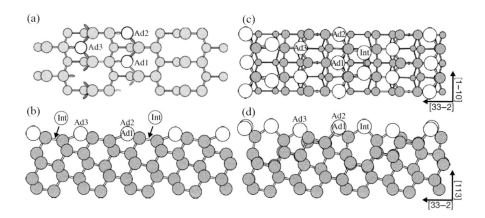

Figure 7.47 (a) Top view of the unreconstructed Ge(113) surface. Three-fold coordinated adsorption sites are marked by Ad1–3. For some top Ge atoms one or two dangling bonds are indicated. (b) Side view of (a): Sb interstitial atoms (labelled Int) break stretched Ge–Ge bonds (marked by arrows). (c) Top view of Ge(113)+c(2×2)–Sb reconstruction. (d) Side view of (c). Reproduced with permission from [7.101] A. Hirnet et al., *Phys. Rev. Lett.*, vol. 88, p. 226102, 2002. https://journals.aps.org/prl/abstract/10.1103/PhysRevLett.88.226102. © (2002) by the American Physical Society.

The projection of the superstructure into a (1×1) substrate cell can be determined from the (1×1) reflections. The unravelling of the 'averaged' (1×1) structure into an image of the superstructure may not be possible in all cases. The MSF method, on the other hand, determines the difference electron density from the superstructure reflections, which, particularly for large superstructures, gives a very useful image from which the superstructure can be derived. For small unit cells, the application of direct methods is mostly not necessary anyway. But the requirement of a sufficiently large number of out of plane diffraction data to obtain a 3-D image of the surface slab limits the applicability of the MSF method.

PARADIGM uses both kinds of reflections, SSR and CTR, and determines the full superstructure [7.89; 7.90; 7.103–7.106]. The electron density in the surface slab is determined iteratively under the constraint that the structure factors should match the experimental data for the phases of the CTR and the SSR structure factors. The structure factors of the substrate are assumed to be known. Neither the composition nor the thickness of the surface region needs to be known exactly. The substrate region starts at a depth where it can be assumed to be not or only negligibly distorted by the adsorbate and/or surface reconstruction. We consider here the case where the substrate layers have the undistorted bulk structure and the surface region exhibits a superstructure. The intensity along the CTRs contains contributions from the bulk and the surface:

$$I_{\mathrm{CTR}}(hkl) = |F_{\mathrm{bulk}}(hkl)g(l) + F_{\mathrm{surf}}(hkl)|^2 = |F_{\mathrm{CTR}}(hkl)|^2. \qquad (7.70)$$

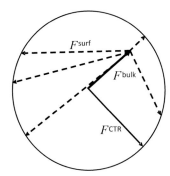

Figure 7.48 Illustration of the bulk and surface contribution to F_{CTR}. F_{bulk} and its phase are known; the modulus of F_{CTR} is measured, but its phase is unknown. There exist limits for the modulus of F_{surf}, while modulus and phase of F_{surf} are coupled, with an ambiguity of two solutions.

The structure factors from the bulk, $F_{bulk}(hkl)$, must be multiplied by the lattice factor for the semi-infinite crystal, see Eq. (7.38):

$$g(l) = \sum_{n=0}^{\infty} e^{(2\pi i l - \alpha)n}, \quad |g(l)|^2 = \frac{1}{(1 - e^{-\alpha})^2 + 4e^{-\alpha} \sin^2 \pi l}, \tag{7.71}$$

where α is the damping factor, that is, the reciprocal of the penetration depth in units of the lattice constant c, see Eq. (7.37). The intensity along the superstructure rod $I_{surf}(hkl)$ arises from the surface region only. The phases of the surface structure factors F_{surf} are required to determine the electron density in the surface slab. Both sets of reflections, superstructure rods and CTRs were required to produce the full image of the electron density. Modulus and phase are not known for $F_{surf}(hkl)$ in the CTR, but $I_{CTR}(hkl)$ is measured. This limits the modulus and phase of $F_{surf}(hkl)$ in the CTR, as illustrated in Figure 7.48.

The method is related to the 'heavy atom method', where the substrate now plays the role of the heavy atom; see for example the description of this method in a textbook [7.75]. The heavy atom method is known to be related to holography; the analysis of the interference between bulk and surface as holography has been discussed by D. K. Saldin and V. L. Shneerson [7.105]. The bulk contribution plays the role of the reference wave, while the surface contribution is the object wave. In holographic images the interference between the object wave and the reference wave is used (see an analogous treatment of holographic LEED in Section 6.1.1.2):

$$I(hkl) = |A_{ref}(hkl) + A_{obj}(hkl)|^2$$
$$= |A_{ref}(hkl)|^2 + 2|A_{ref}(hkl)||A_{obj}(hkl)| \cos(\varphi_{obj} - \varphi_{ref}) + |A_{obj}(hkl)|^2. \tag{7.72}$$

If the object wave is weak, the object term $|A_{obj}(hkl)|^2$ is neglected in holographic reconstructions. This is not possible here as the object wave is in general not weak, so the full object term must be used. The analysis method PARADIGM can therefore be considered as generalised holography. That the CTR data can be interpreted as a hologram has also been shown by T. Takahashi et al. [7.82] with model calculations.

The electron density of the surface slab is determined in PARADIGM by a simple iterative scheme. The electron density of the surface slab is updated in each iteration step until the calculated structure factors closely match the experimental values. Different criteria are possible to adjust the surface electron density. One approach is to minimise the difference electron density: it is calculated from the observed and calculated structure factors by giving the measured structure factors the phase of the calculated ones. Another approach is simply to restrict the electron density to positive values in each iteration step, which is one of the methods proposed by J. R. Fienup [7.107] and applied by D. K. Saldin et al. [7.103]. Also, the charge flipping method can be used [7.78], where negative values of the electron density are simply made positive; a limit for small values must be set so that only the charge of the stronger peaks is changed. Examples for the different approaches are given in the remainder of this section.

We use in the following the notation $\mathbf{h}_B$ for the scattering vectors of CTR reflections and $\mathbf{h}_S$ for the superstructure reflections. $I_{\text{surf}}(\mathbf{h}_S)$ is the observed intensity in the superstructure reflections and $I_{\text{CTR}}(\mathbf{h}_B)$ the observed intensity in the truncation rods. The quantities determined in the iteration scheme are the electron density in the surface slab, $\rho_{\text{surf}}(\mathbf{r})$, and the calculated surface structure factors; their phases are obtained from the Fourier transform of the electron density:

$$F_{\text{surf}}(\mathbf{h}) = \int \rho_{\text{surf}}(\mathbf{r}) \exp(2\pi i \mathbf{h} \mathbf{r}) d\mathbf{r}, \tag{7.73}$$

$$e^{i\varphi(\mathbf{h}_S)} = \frac{F_{\text{surf}}(\mathbf{h}_S)}{|F_{\text{surf}}(\mathbf{h}_S)|}. \tag{7.74}$$

The bulk electron density is known from the bulk structure factors and is obtained from the inverse Fourier transform (IFT):

$$\rho_{\text{bulk}}(\mathbf{r}) = \frac{1}{N_p} \sum_{\mathbf{h}_B} F_{\text{bulk}}(\mathbf{h}_B) \exp(-2\pi i \mathbf{h}_B \mathbf{r}). \tag{7.75}$$

Here, N_p is the number of pixels used, while the sum runs over the available CTR reflections (those that are not measured are set to zero). The intensities in the truncation rods must include the lattice factor (Eq. (7.71)) so that they can be compared to the experimental data:

$$|F_{\text{CTR}}(\mathbf{h}_B)|^2 = |F_{\text{surf}}(\mathbf{h}_B) + F_{\text{bulk}}(\mathbf{h}_B) g(l)|^2, \tag{7.76}$$

where F_{surf} is calculated from the current $\rho(\mathbf{r})$ in the iteration scheme. The calculated surface structure factors have to be scaled to the measured structure factors. The scaling factor is determined by a least squares minimisation of the differences between measured and calculated data, leading to a scaling factor:

$$c = \frac{\sum_{\mathbf{h}_S} |F_{obs}(\mathbf{h}_S)||F_{calc}(\mathbf{h}_S)|}{\left|\sum_{\mathbf{h}_S} F_{calc}(\mathbf{h}_S)\right|^2} \tag{7.77}$$

for the superstructure reflections. Here F_{obs} means the observed surface structure factors and F_{calc} the calculated ones. Different scaling factors may be defined for the superstructure reflections ($\mathbf{h}_S$) and for the truncation rod data, which are (1×1) reflections ($\mathbf{h}_B$). We describe in the following the iteration schemes for the cases of single domains and the frequently occurring case of multiple domains, including possible different strategies. The various strategies arise from different choices to optimise $\rho(r)$, the choice of the starting data and the choice of using the full data set in one step or using CTR data and SSR data sequentially. Which strategy is most appropriate depends probably on the data set, but in any case should lead to the same result.

Iteration Scheme for a Single Domain

In the iteration cycle the electron density is modified until the difference electron density is minimised. In the n-th iteration step we calculate the current surface structure factor $F_{surf}^{(n)}(\mathbf{h}_B)$ from the current surface electron density. The iteration scheme can be performed in two ways, using the (1×1) reflections alone in the first step and then including the superstructure reflections in the second step or using both sets of reflections from the beginning. The latter case starts with

$$F_{surf}^{(n)}(\mathbf{h}_B) = FT\left\{\rho_{surf}^{(n)}(\mathbf{r})\right\}, \tag{7.78}$$

and calculates a truncation rod structure factor with its phase:

$$F_{CTR}^{(n)}(\mathbf{h}_B) = F_{surf}^{(n)}(\mathbf{h}_B) + F_{bulk}(\mathbf{h}_B), \tag{7.79}$$

$$e^{i\varphi^{(n)}(\mathbf{h}_B)} = \frac{F_{CTR}^{(n)}(\mathbf{h}_B)}{\left|F_{CTR}^{(n)}(\mathbf{h}_B)\right|}. \tag{7.80}$$

The measured surface structure factors of the (1×1) reflections are then given by:

$$F_{exp}^{(n)}(\mathbf{h}_B) = \sqrt{I_{CTR}(\mathbf{h}_B)}e^{i\varphi^{(n)}(\mathbf{h}_B)} - cF_{bulk}(\mathbf{h}_B). \tag{7.81}$$

The new electron density $\tau(\mathbf{r})$ becomes

$$\tau(\mathbf{r}) = IFT\left\{F_{exp}^{(n)}(\mathbf{h}_B) + F_{surf}^{(n)}(\mathbf{h}_S)\right\}. \tag{7.82}$$

The electron density derived from both sets of reflections must be positive. In principle it would be possible to use the superstructure reflections alone in an iteration

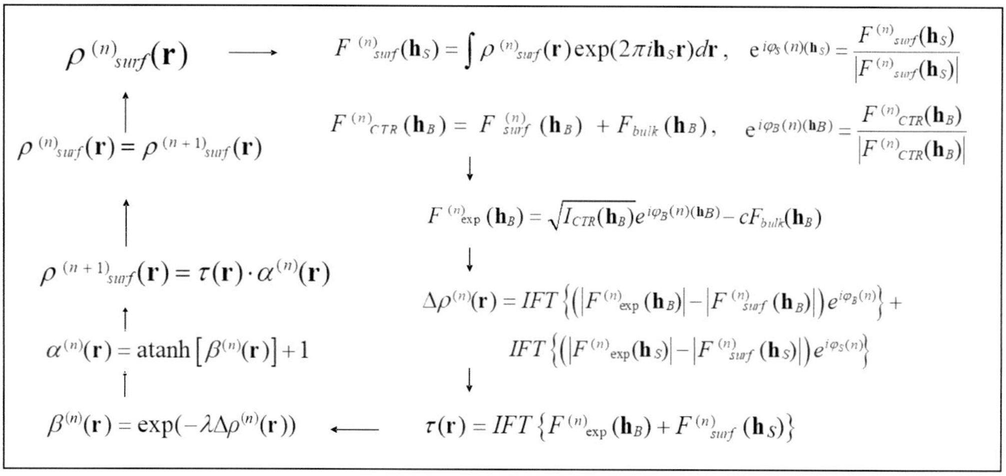

$$\rho^{(n)}_{surf}(\mathbf{r}) \longrightarrow F^{(n)}_{surf}(\mathbf{h}_S) = \int \rho^{(n)}_{surf}(\mathbf{r})\exp(2\pi i\mathbf{h}_S\mathbf{r})d\mathbf{r}, \quad e^{i\varphi_S(n)(\mathbf{h}_S)} = \frac{F^{(n)}_{surf}(\mathbf{h}_S)}{|F^{(n)}_{surf}(\mathbf{h}_S)|}$$

$$\uparrow \qquad\qquad\qquad F^{(n)}_{CTR}(\mathbf{h}_B) = F^{(n)}_{surf}(\mathbf{h}_B) + F_{bulk}(\mathbf{h}_B), \quad e^{i\varphi_B(n)(\mathbf{h}_B)} = \frac{F^{(n)}_{CTR}(\mathbf{h}_B)}{|F^{(n)}_{CTR}(\mathbf{h}_B)|}$$

$$\rho^{(n)}_{surf}(\mathbf{r}) = \rho^{(n+1)}_{surf}(\mathbf{r}) \qquad\qquad\qquad\qquad \downarrow$$

$$F^{(n)}_{exp}(\mathbf{h}_B) = \sqrt{I_{CTR}(\mathbf{h}_B)}\,e^{i\varphi_B(n)(\mathbf{h}_B)} - cF_{bulk}(\mathbf{h}_B)$$

$$\uparrow \qquad\qquad\qquad\qquad\qquad \downarrow$$

$$\rho^{(n+1)}_{surf}(\mathbf{r}) = \tau(\mathbf{r})\cdot\alpha^{(n)}(\mathbf{r}) \qquad \Delta\rho^{(n)}(\mathbf{r}) = IFT\left\{\left(|F^{(n)}_{exp}(\mathbf{h}_B)| - |F^{(n)}_{surf}(\mathbf{h}_B)|\right)e^{i\varphi_B(n)}\right\} +$$

$$\uparrow \qquad\qquad\qquad\qquad IFT\left\{\left(|F^{(n)}_{exp}(\mathbf{h}_S)| - |F^{(n)}_{surf}(\mathbf{h}_S)|\right)e^{i\varphi_S(n)}\right\}$$

$$\alpha^{(n)}(\mathbf{r}) = \text{atanh}\left[\beta^{(n)}(\mathbf{r})\right] + 1 \qquad\qquad \downarrow$$

$$\uparrow$$

$$\beta^{(n)}(\mathbf{r}) = \exp(-\lambda\Delta\rho^{(n)}(\mathbf{r})) \longleftarrow \tau(\mathbf{r}) = IFT\left\{F^{(n)}_{exp}(\mathbf{h}_B) + F^{(n)}_{surf}(\mathbf{h}_S)\right\}$$

Figure 7.49 Flow chart for single domains. The modification of $\tau(\mathbf{r})$ is done through $\beta(\mathbf{r}) = \exp(-\lambda\Delta\rho(\mathbf{r}))$ and $\alpha(\mathbf{r}) = \text{atanh}[\beta(\mathbf{r})] + 1$. The factor λ is an enhancement factor to obtain efficient optimisation, since, depending on the number of pixels, $\rho(\mathbf{r})$ can become a very small number.

scheme, but the positivity of the electron density could then not be used. If the difference electron density is required for updating $\tau(\mathbf{r})$, there are two possibilities. One option is to use the (1×1) reflections alone; then $\Delta\rho(\mathbf{r})$ is:

$$\Delta\rho^{(n)}(\mathbf{r}) = IFT\left\{F^{(n)}_{exp}(\mathbf{h}_B) - F^{(n)}_{surf}(\mathbf{h}_B)\right\}. \tag{7.83}$$

Alternatively, the superstructure reflections are included; then $\Delta\rho(\mathbf{r})$ is:

$$\Delta\rho^{(n)}(\mathbf{r}) - IFT\left\{F^{(n)}_{exp}(\mathbf{h}_B) - cF^{(n)}_{surf}(\mathbf{h}_B) + |F^{(n)}_{exp}(\mathbf{h}_S)|e^{i\varphi^{(n)}_{surf}} - cF^{(n)}_{surf}(\mathbf{h}_S)\right\}. \tag{7.84}$$

The iteration scheme for the case of a single domain is detailed in Figure 7.49.

In Figure 7.49, $\rho(\mathbf{r})$ is found by minimisation of the difference electron density $\Delta\rho(\mathbf{r})$. It is efficient to use the atanh function for the correction factor $\alpha(\mathbf{r})$ to avoid divergence in the iteration.

An alternative method is to skip the calculation of the difference electron density and use the method proposed by J. R. Fienup [7.107]:

$$\rho^{(n+1)}(\mathbf{r}) = \begin{cases} \tau(\mathbf{r}) & \text{if } \tau(r) > 0 \\ 0 & \text{otherwise,} \end{cases} \tag{7.85}$$

which works equally well and has been used by D. K. Saldin and co-workers in several examples [7.103–7.106]. To start the iteration, we need an initial electron density and have the following two different options:

(1) Start with random phases. The measured structure factors $|F^{(0)}_{surf}(\mathbf{h}_S)|$ get random phases and the initial electron density is calculated from the inverse Fourier transform: $\rho^{(0)}_{surf}(\mathbf{r}) = \sum_{\mathbf{h}_S}|F^{exp}_{surf}(\mathbf{h}_S)|e^{i\varphi_\mathbf{r}}e^{-2\pi i\mathbf{h}_S\,\mathbf{r}}$, where $\varphi_\mathbf{r}$ is an initially random

phase. The iteration in Figure 7.49 starts with $\rho_{\text{surf}}^{(0)}(\mathbf{r})$ and the phases are adapted in the next iteration steps.

(2) Start with a constant electron density, $\rho_{\text{surf}}^{(0)}(\mathbf{r}) = N_e/N_p$, where N_e is the total number of electrons in the surface region and N_p the number of pixels. The number N_e is unknown and is therefore initially estimated. It has to be multiplied in each iteration step by a scale factor to match the calculated structure factors.

The second possibility leads to initially calculated surface structure factors $|F_{\text{surf}}^{(0)}(\mathbf{h_S})| = 0$, and in the next iteration $\rho_{\text{surf}}^{(1)}(\mathbf{r})$ becomes the electron density of the substrate. This choice of the starting values also offers the possibility to include some fixed atom positions in the surface area if such are known. The electron density from the fixed atoms is always added to the calculated surface electron density. Further conditions can be introduced to influence the result if the iteration process does not lead to a reasonable solution. The phases of certain structure factors are zero or fixed due to symmetry restrictions and the choice of the origin. This always applies to some (1×1) reflections and their phases $e^{i\varphi(\mathbf{h_B})}$, and, depending on the symmetry, also to some superstructure reflections and their phases $e^{i\varphi(\mathbf{h_S})}$. The choice of the fixed phases is not discussed here, but detailed descriptions can be found in the literature about direct methods, for example in C. Giacovazzo [7.76]. As the origin is in most cases set by the structure data of the substrate, the choice of the origin is not free; also, the setting of the phase of some reflections does not have the same importance as in the direct methods in bulk crystallography.

Iteration Scheme for Multiple Domains
Most surfaces have domains due to terraces or may have twin and antiphase domains in superstructures. In structure analysis, the intensities or moduli of the structure factors are compared with experimental data. Usually, the integral intensity is measured so that the antiphase character of such domains can be neglected, since a phase factor does not change the intensities. Twin domains are taken into account by averaging the corresponding intensities. With averaging, sometimes the question arises in which cases coherent versus incoherent averaging is correct. In general, the incoherent average must be taken. The *integral* intensity is measured in X-ray diffraction as well as in LEED; see for example the discussion about the integral intensity by B. E. Warren [7.44]. Domains lead to a broadening of the reflections and the measurement of the integral intensity thus leads to incoherent averaging; see Appendix J for a detailed discussion. An exception is coherent diffraction, of course, but this requires a different technique and is applied in a different field.

Domains due to steps and terraces occur in every surface: whether they are antiphase domains or twin domains depends on the surface orientation and space group of the substrate. If the space group has a screw axis as the principal axis, then different terraces are twin domains if the surface plane is normal to the principal axis. Other orientations may exhibit different terminations, or not, depending on the orientation and the structure. For example, the (0001) surface of the hcp lattice, with space group $P6_3/mmc$, exhibits twin domains in the terraces. A single domain in the (0001) surface produces a

3-fold diffraction pattern, while a second terrace is rotated by $60°$, so that the combined diffraction pattern of both terminations has 6-fold symmetry.

We consider the case of rotated domains where the superstructure is the same in each domain but rotated with the substrate symmetry. The algorithm can be easily extended to more complicated cases [7.103]. We use the superscript d to identify the domain number among N_D domains and assume that this is the same for the superstructure and the substrate:

$$I_{\text{surf}}(\mathbf{h}_S) = \frac{1}{N_D} \sum_{d=1}^{N_D} \left| F^d(\mathbf{h}_S) \right|^2, \tag{7.86}$$

$$I_{\text{CTR}}(\mathbf{h}_B) = \frac{1}{N_D} \sum_{d=1}^{N_D} \left(\left| F_S^d(\mathbf{h}_B) \right|^2 + \left| F_B^d(\mathbf{h}_B) \right|^2 \right), \tag{7.87}$$

assuming equal areas for each domain. The modulus of the structure factor of the first domain in the truncation rod is then:

$$\sqrt{\left| F_S^1(\mathbf{h}_B) \right|^2 + \left| F_B^1(\mathbf{h}_B) \right|^2} = \left[I_{\text{CTR}}(\mathbf{h}_B) - \frac{1}{N_D} \sum_{d=2}^{N_D} \left(\left| F_S^d(\mathbf{h}_B) \right|^2 + \left| F_B^d(\mathbf{h}_B) \right|^2 \right) \right]^{1/2}. \tag{7.88}$$

In the iteration scheme, detailed in Figure 7.50, we optimise the electron density of domain 1 and replace $F_{\text{surf}}^{(n)}(\mathbf{h}_B)$ by $F_{\text{surf}}^{(1,n)}(\mathbf{h}_B)$, and the other quantities accordingly:

$$F_{\text{surf}}^{(1,n)}(\mathbf{h}_B) = FT\left\{ \rho_{\text{surf}}^{(1,n)}(\mathbf{r}) \right\}, \tag{7.89}$$

$$F_{\text{exp}}^{(1,n)}(\mathbf{h}_B) = \sqrt{I_{\text{CTR}}(\mathbf{h}_B)} e^{i\varphi^{(n)}(\mathbf{h}_B)} - \sum_{d=2}^{N_D} cF_{\text{bulk}}^{(d,n)}(\mathbf{h}_B). \tag{7.90}$$

PARADIGM has been applied to a number of structures, for example Au(110)+c $(2{\times}2)$–Sb [7.108], Au(110)+$(\sqrt{3}{\times}\sqrt{3})$R54.7°–Sb [7.109], SrTiO$_3$(106) [7.110] and some examples with model data; an overview is given in [7.104].

The results for SrTiO$_3$(106) are shown in Figure 7.51. The lattice constant of SrTiO$_3$ is $a_0 = 0.3905$ nm $= 3.905$ Å and an orthorhombic surface unit cell with the vector $\mathbf{c}$ normal to the surface would have an inconveniently large number of atoms. Therefore, a monoclinic unit cell was used with $\mathbf{a} = [6a_0, 0, -a_0]$, $\mathbf{b} = [0, a_0, 0]$ and an oblique vector $\mathbf{c} = [a_0, 0, -a_0]$; it has 62 atoms per unit cell. The model is shown in Figure 7.52. The electron density resulting from the direct method clearly shows the existence of single steps and the two alternating terminations with Ti and Sr atoms. The structure factors are nearly perfectly fit, with the exception of three intense beams showing major differences between the experimental and calculated structure factors. This may be due to experimental errors. In fact, such deviations are not uncommon for strong reflections because the strong reflections are subject to dynamic effects in nearly perfect crystals. The results of the direct method were confirmed with a refinement using the ROD [7.73] program. The refinement determined the atomic relaxations and the occupancies; relaxations from the bulk were found to be negligibly small.

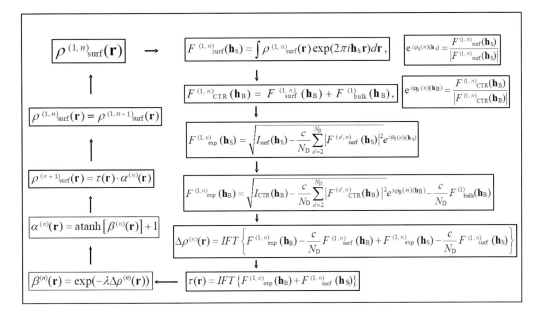

Figure 7.50 Flow chart for multiple twin domains; N_D is the number of domains. It is assumed that the twin symmetry operation applies to substrate and adsorbate and each domain is covered with the same superstructure. The resulting quantity is the electron density of a single domain.

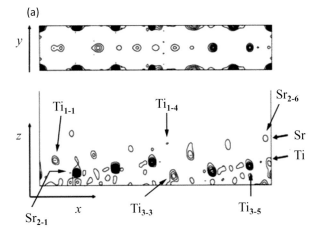

Figure 7.51 (a, upper panel) Top view (x, y plane) and (a, lower panel) side view (x, z plane) of the electron density resulting from PARADIGM for $SrTiO_3(106)$. The termination of the surface with Sr and Ti terraces is obvious. The heavier Sr is prominent in the electron density map, while oxygen is not resolved. (b) Comparison of measured structure factors, $F(obs)$, with the calculated ones, $F(calc)$, from the electron density determined by the direct method. Reprinted from [7.110] X. Torrelles et al., *Surf. Sci.*, vol. 589, pp. 184–191, 2005, with permission from Elsevier.

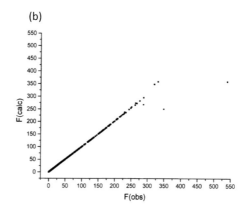

Figure 7.51 (*cont.*)

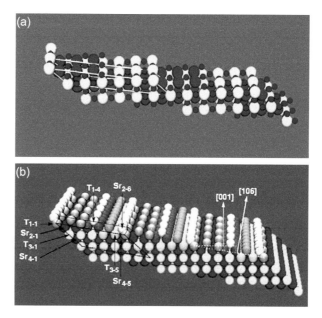

Figure 7.52 (a) The ideal SrO and TiO_2 terminated $SrTiO_3(106)$ surface with Sr (red), Ti (yellow) and O (dark blue). (b) Result of the structural refinement of $SrTiO_3(106)$ showing the occupancies of Ti and Sr positions; oxygen is not shown. Indicated in the model is the monoclinic unit cell (black lines). Fully occupied Ti and Sr sites are represented by yellow and red colours, respectively. Lighter colours indicate lower occupancies. Atomic labels T_{1-i}, Sr_{2-i}, etc. refer to the outermost Ti in a (001) plane, second layer Sr in a (001) plane, etc., within the surface unit cell; here i refers to the atom sequence within the layer. Reprinted from [7.110] X. Torrelles et al., *Surf. Sci.*, vol. 589, pp. 184–191, 2005, with permission from Elsevier.

As an example of the superposition of two domains, we describe here the results for model data of reconstructed Ge(001)–(1×2). This surface has two domains with asymmetric dimers: (1×2) and (2×1), like the Si(100) surface reconstruction. The average structure from the CTR data has a (2×2) periodicity. Figure 7.53(a) shows the

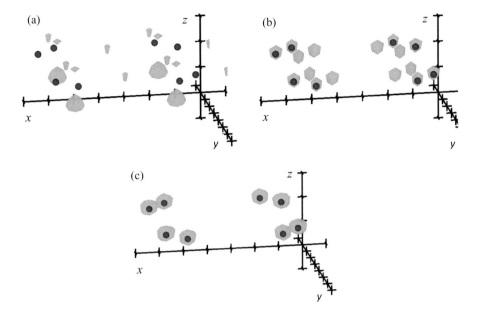

Figure 7.53 (a) A perspective view of iso-surfaces of electron density representing the starting electron distribution $\rho(1)$ (the difference Fourier estimate) after a single pass of the flow chart of Figure 7.49. Note that the translucent green iso-surfaces give little indication of the positions (red and blue dots) of the atoms in a (2×2) unit cell from the two mutually rotated domains. (b) Electron density iso-surfaces in the same unit cell after inclusion of data from just the CTRs and the execution of 175 iterations of the recursion algorithm described in Figure 7.50. Green lobes are now seen to surround the positions of atoms from both domains, but such iso-surfaces are also seen around the positionally averaged sites in the (2×2) unit cell. (c) Electron-density iso-surfaces in the same unit cell after inclusion of data from both the CTRs and the superstructure rods and the execution of a further 500 iterations of the recursion algorithm. Green lobes are now seen to surround only the positions of atoms in the two domains, thus recovering the correct orientationally averaged structure in the (2×2) unit cell. © IOP Publishing. All rights reserved. Reproduced with permission from [7.104] D. K. Saldin et al., *J. Phys.: Condens. Matter*, vol. 14, pp. 4087–4100, 2002; permission conveyed through Copyright Clearance Center, Inc.

electron density after the first iteration. The red and blue circles indicate the position of the dimer atoms. The green areas mark enhanced electron density. Figure 7.53(b) shows the result using the CTR data only, which leads to the average (2×2) structure with a superposition of the two domains. In Figure 7.53(c) the final result from the superstructure rods and the CTR data shows the electron density from a single domain.

7.3.6   Difference Map Using the Constraints of Atomicity and Film Shift (DCAF)

An alternative way to retrieve the electron density of a thin film adsorbed on a substrate has been developed by M. Björck et al. [7.91]. They use the truncation rod data, where the structure factors of the film and the substrate are superimposed, to

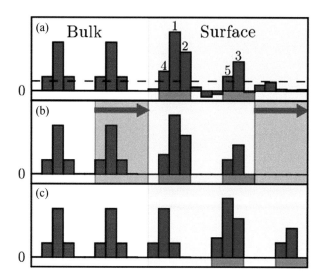

Figure 7.54 A schematic illustration of the real space projection steps of the DCAF method. (a) The starting electron density distribution before a shift of the surface slab is executed. The area beneath the two surface atoms found is highlighted. (b) The result of the atomicity and positivity projection. (c) The electron density after the application of the film shifting projection (marked by red arrows in (b)). © IOP Publishing. All rights reserved. Reproduced with permission from [7.91] M. Björck et al., *J. Phys.: Condens. Matter*, vol. 20, p. 445006, 2008; permission conveyed through Copyright Clearance Center, Inc.

extract the electron density of the film. The method, called 'difference map using the constraints of atomicity and film shift' (DCAF), is an alternative to the COBRA method described in Section 7.3.3. It is based on the difference map iteration scheme proposed by V. Elser [7.111] for use in bulk crystallography. The difference map uses, like the other 'pixel methods', iteration of Fourier and inverse Fourier transforms of the electron density with independent constraints due to measured structure factors, positivity of the electron density and atomicity. For the condition of atomicity an algorithm in real space is applied. A threshold level is used to identify atoms. The dashed line in Figure 7.54(a) marks the threshold level.

The authors introduce, as additional constraint, the 'film shift', which relaxes the film thickness constraint as the film thickness is unknown and necessarily uncertain due to intermixing and roughness. The film shift constraint shifts the electron density of the film normal to the surface by a unit cell of the underlying bulk substrate if the topmost region of the same thickness contains insignificant electron density. In that way the thickness of the surface slab is variable and the boundary to the substrate is not fixed. It may occur that a substrate layer is included in the surface slab: that does not change the results. In the iteration scheme, which fits the Fourier transform of the electron density to the measured structure factors, a threshold is set for the difference between subsequent iterations in order to avoid being trapped in local minima. The starting point is chosen with random phases assigned to the measured structure factors.

The region where the unknown part of the electron density is considered to be non-zero is denoted as the support.

An example for the film shift in real space is schematically shown in Figure 7.54. In Figure 7.54(a), four atom positions are shown: voxels 1 and 3 belong to two non-overlapping atoms. The overlap is checked laterally as well as between layers. Voxels 2 and 4, for example, above the threshold, are not counted as separate atoms because they overlap with voxel 1, while voxel 5 overlaps with voxel 3. In Figure 7.54(b), the electron density below a certain level is set to zero. Figure 7.54(c) shows the result for the electron density of the film where the boundary between the film and the substrate is shifted due to the principle of the maximum filling of the unknown region. The surface slab thickness always remains at three layers. The empty top layer in Figure 7.54(b) is not counted and the top substrate layer is now assigned to the surface slab.

The DCAF method was applied to two systems, a five-monolayer-thick film of LaAlO$_3$ on SrTiO$_3$ and a three-monolayer-thick film of La$_{1-x}$Sr$_x$MnO$_3$ on SrTiO$_3$ [7.91], as well as to YBa$_2$Cu$_3$O$_{7-x}$ (YBCO) films on SrTiO$_3$(001) (STO) [7.112]. The results were in good agreement with subsequent refinement.

Like all other direct methods, the DCAF method provides results for further refinement and the electron density must be interpreted to identify the atoms. In Figure 7.55 the electron density is shown as obtained from the iteration process.

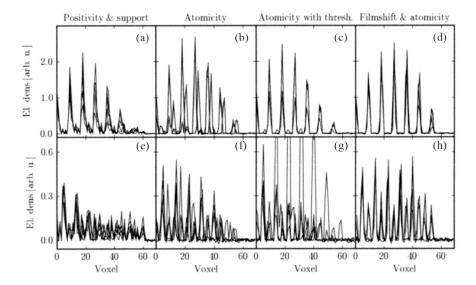

Figure 7.55 (a)–(d) As-retrieved electron density distributions for the Sr/La sites and (e)–(h) for the Ti/Al sites for a five-monolayer film of LaAlO$_3$ on SrTiO$_3$. The surface and vacuum are on the right. (a) and (e) show the use of positivity and support; (b) and (f) include atomicity with a fixed number of atoms; (c) and (g) include atomicity with thresholding; and (d) and (h) include atomicity with thresholding and film shift. Each panel shows five different retrieval runs. © IOP Publishing. All rights reserved. Reproduced with permission from [7.91] M. Björck et al., *J. Phys.: Condens. Matter*, vol. 20, p. 445006, 2008; permission conveyed through Copyright Clearance Center, Inc.

Figures 7.55(a–d) display line scans through the electron density in the out-of-plane direction for the first sample of five monolayers of $LaAlO_3$ on $SrTiO_3$. The line scans pass through the centres of La/Sr atoms. Figures 7.55(e–h) show line scans through the Ti/Al sites. The influence of the different restraints on the resulting electron density is clearly demonstrated. In Figures 7.55(a–d) the oxygen positions are visible as small peaks between the La/Sr positions. Figures 7.55(a) and (e) correspond to PARADIGM, described in Section 7.3.5, where only the positivity of the electron density is considered. Support denotes the region where the electron density is considered to be non-zero.

7.3.7 Perturbation Method of Analysis of Density Profiles from Crystal Truncation Rod Data

An important application of surface X-ray diffraction is the determination of the density profile in layered semiconductor surfaces and the interdiffusion at boundaries. X-ray diffraction is the most appropriate method to study concentration profiles of two or more species as it is non-destructive and has sufficient spatial resolution. The relatively small difference in the atomic form factors allows the treatment of the change in the structure factors as perturbation of an average value. A direct inversion method of the crystal truncation rod (CTR) data is possible and has been proposed by I. K. Robinson et al. [7.92]. The analysis of the density profile of layered structures has been applied to InP/GaInAs/InP heterostructures. Density modulations also cause strain in the surface region. The intensity of the truncation rods is quite sensitive to small layer displacements and the capability of the method to analyse strain connected with density profiles has been demonstrated.

The method to analyse the small density change in a surface region compared to the density in the underlying substrate is based on the fact that phase change can be neglected and the modulus of the structure factor is modulated. The modulation function can be easily evaluated. The analysis of strain uses the analogous approach, by assuming an average constant density and small phase changes in the surface region. The structure factor can also be factored into an ideal truncation rod and a modulation function. The combination of both modulations (density and displacements) is also possible but requires a more difficult evaluation. In Figure 7.56, results are shown for five samples with 45 layers of InP covering a variable number of layers of $Ga_{47}In_{53}$ on an InP substrate [7.92]. This composition of the GaIn layers was chosen to produce the same lattice constant as InP. The absence of strain makes it possible to apply the direct inversion method of the modulation function. In Figure 7.56, the apparent deviation from the nominal top 45 InP layers may be caused by a slightly incorrect calibration of the evaporation source. The occurrence of the second dip in the top layers to the left of the straight orange line is not completely understood; it is probably an artefact of the analysis. That this step shifts with the number of GaInAs monolayers may indicate a breakdown of the perturbation approximations.

The samples chosen to analyse the density profiles were designed to be essentially strain-free to avoid the combination of strain and density modulations. Whether strain

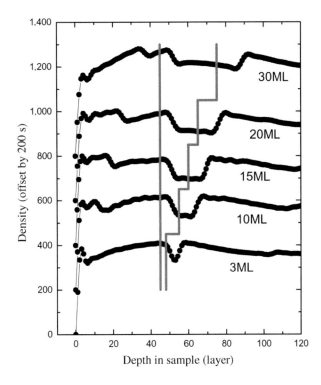

Figure 7.56 Derived density profiles of the five samples investigated for an InP/Ga$_{47}$In$_{53}$/InP film structure with five different Ga$_{47}$In$_{53}$ thicknesses. The nominal thickness of the GaInAs film is specified by the number of monolayers (ML) indicated by the label near each curve, followed by 45 monolayers of InP. The curves are offset by 200 units on the vertical scale. The expected positions of the steps in the profiles are also indicated by grey vertical lines. Reprinted from [7.92] I. K. Robinson et al., *J. Appl. Cryst.*, vol. 38, pp. 299–305, 2005, with permission from John Wiley and Sons.

can be analysed with the same method has been checked with model data as no samples were available. As model heterostructure a Si crystal was chosen on which a stack of layers was deposited with interlayer spacing increased by Δd over 20 layers, followed by a 'cap' of 26 more Si layers with the bulk spacing. Three values, $\Delta d/d =$ 0.25%, 0.5%, and 0.75%, were tested. The intensities of 400 equally spaced data points were calculated over the range $1.6 < L < 2.4$, around the (002) reflection, assuming a primitive structure with two layers per unit cell. These simulated data were divided by an ideal CTR profile to extract the modulation function, filtered and finally Fourier transformed. The limited data set introduces 'cut-off' effects in the Fourier transform which can be smoothed by a filter parameter. The influence of this parameter has also been investigated. The details of the analysis are described in ref. [7.92] and are not discussed here. The results are shown in Figure 7.57 as a function of the depth, given by the layer number. The upper panel shows the influence of the filter parameter. If the data are cut off too abruptly, extra ripples in the strain distribution occur; excessive filtering rounds the profiles too much. The lower panel shows an

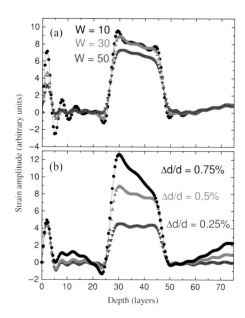

Figure 7.57 Strain profiles derived from test calculations for a buried slab of 20 layers of strained Si material with constant density buried under a cap of 26 layers. Plotted on the vertical axis is the imaginary part of the Fourier transform of the modulation function for a variety of test conditions. (a) Variation of the filter parameter used to smooth the cut-off at the end of the data range, W, measured in grid points of the $N = 2{,}048$ FFT employed. (b) Variation of the response to different amounts of strain, $\Delta d/d$, within the 20-layer slab, showing the onset of distortions. Reprinted from [7.92] I. K. Robinson et al., *J. Appl. Cryst.*, vol. 38, pp. 299–305, 2005, with permission from John Wiley and Sons.

asymmetry in the profile, the origin of which is not yet understood; probably the approximation breaks down for the larger strain.

The sharpness of the steps is a relevant technological problem and the precision with which it can be measured is an important question. The results in Figures 7.56 and 7.57 demonstrate the applicability of the perturbation approximation, the limits up to which it is valid, and the influence of analysis parameters.

7.3.8   Further Methods for Experimental Phase Determination

Besides the phase retrieval methods for the measured structure factors and the measurement of the derivative of phases in the COBRA method there exist further experimental methods to measure the phase, that is, the phase differences between the different structure factors, as has been shown by S. A. Pauli et al. [7.93]. The measurement must be done at two or more wavelengths around the absorption edge of one of the elements in the surface layer and applies the phase shift in the form factor for this atom due to the so-called anomalous scattering. The method has been applied to a thin film of $SrTiO_3$ grown on $NdGaO_3(110)$.

Another approach would be to use dynamical effects, that is, the interference between diffracted beams, from which the phase of some beams can be determined. The method requires well-ordered crystals and facilitates the use of direct methods. It is frequently used in bulk crystallography. For surfaces the indirect excitation of a surface reflection can occur via a bulk reflection. The possibility to observe such effects has been shown by V. M. Kaganer et al. [7.94] for Ge(113) but the method has not yet been applied for structure determination.

Appendices

Appendix A Lists of Books Related to Surface Science

A.1   General Books and Proceedings

- Link to list of *Useful Books on Surface Analysis*: http://www.xpsdata.com/useful_
  books.htm
R. Caudano, J.-M. Gilles, and A. A. Lucas, eds., *Vibrations at Surfaces*, Plenum,
  New York, 1982.
M.-C. Desjonquères and D. Spanjaard, *Concepts in Surface Physics*, Springer,
  Berlin, 1996.
C. B. Duke and E. W. Plummer, eds., *Frontiers in Surface Science and Interface
  Science*, North-Holland, Amsterdam, 2002.
G. Ertl and J. Küppers, *Low Energy Electrons and Surface Chemistry*, 2nd ed.,
  Verlag Chemie, Weinheim, 1985.
T. Fauster, L. Hammer, K. Heinz and M. A. Schneider, *Surface Physics:
  Fundamentals and Methods*, de Gruyter, Berlin, 2020.
K. Heinz, K. Mueller, T. Engel, and K. H. Rieder, *Structural Studies of Surfaces*,
  Springer, Berlin, 1992.
H. Ibach, *Physics of Surfaces and Interfaces*, Springer, Berlin, 2006.
M. Lannoo and P. Friedel, *Atomic and Electronic Structure of Surfaces*, Springer,
  Berlin, 1991.
H. Lüth, *Solid Surfaces, Interfaces and Thin Films*, Springer, Berlin, 2015.
K. Oura, V. G. Lifshits, A. Saranin, A. V. Zotov, and M. Katayama, *Surface Science:
  An Introduction*, Springer, Berlin, 2003.
G. A. Somorjai, *Chemistry in Two Dimensions*, Cornell University Press, Ithaca,
  NY, 1981.
G. A. Somorjai, *Principles of Surface Chemistry*, Prentice Hall, Englewood Cliffs,
  NJ, 1972.
G. A. Somorjai and Y. Li, Introduction to Surface Chemistry and Catalysis, Wiley,
  Weinheim, 2010.
K. Wandelt, ed., *Encyclopedia of Interfacial Chemistry: Surface Science and
  Electrochemistry*, Elsevier, Amsterdam, 2018.
K. Wandelt, ed., *Surface and Interface Science*, vols. 1–10, Wiley, Weinheim, 2012
  and later.
A. Zangwill, *Physics at Surfaces*, Cambridge University Press, Cambridge, 1988.

A.2   Books and Proceedings on Phenomena

M. Grunze and H. J. Kreuzer, eds., *Adhesion and Friction*, Springer, Berlin, 1989.

V. E. Henrich and P. A. Cox, *The Surface Science of Metal Oxides*, Cambridge University Press, Cambridge, 1994.

K. Hermann, *Crystallography and Surface Structure: An Introduction for Surface Scientists and Nanoscientists*, 2nd ed., Wiley, Weinheim, 2016.

D. A. King and D. P. Woodruff, eds., *The Chemical Physics of Solid Surfaces and Heterogeneous Catalysis, Vol. 1: Clean Solid Surfaces*, Elsevier, Amsterdam, 1981.

D. A. King and D. P. Woodruff, eds., *The Chemical Physics of Solid Surfaces and Heterogeneous Catalysis, Vol. 2: Adsorption at Surfaces*, Elsevier, Amsterdam, 1983.

D. A. King and D. P. Woodruff, eds., *The Chemical Physics of Solid Surfaces and Heterogeneous Catalysis, Vol. 3A: Chemisorption Systems, Part A*, Elsevier, Amsterdam, 1991.

D. A. King and D. P. Woodruff, eds., *The Chemical Physics of Solid Surfaces and Heterogeneous Catalysis, Vol. 3B: Chemisorption Systems, Part B*, Elsevier, Amsterdam, 1984.

D. A. King and D. P. Woodruff, eds., *The Chemical Physics of Solid Surfaces and Heterogeneous Catalysis, Vol. 4: Fundamental Studies of Heterogeneous Catalysis*, Elsevier, Amsterdam, 1984.

D. A. King and D. P. Woodruff, eds., *The Chemical Physics of Solid Surfaces and Heterogeneous Catalysis, Vol. 5: Surface Properties of Electronic Materials*, Elsevier, Amsterdam, 1988.

D. A. King and D. P. Woodruff, eds., *The Chemical Physics of Solid Surfaces, Vol. 6: Coadsorption, Promoters and Poisons*, Elsevier, Amsterdam, 1993.

D. A. King and D. P. Woodruff, eds., *The Chemical Physics of Solid Surfaces, Vol. 7: Phase Transitions and Adsorbate Restructuring at Metal Surfaces*, Elsevier, Amsterdam, 1994.

D. A. King and D. P. Woodruff, eds., *The Chemical Physics of Solid Surfaces, Vol. 8: Growth and Properties of Ultrathin Epitaxial Layers*, Elsevier, Amsterdam, 1997.

K.-H. Meiwes-Broer, *Metal Clusters at Surfaces: Structure, Quantum Properties, Physical Chemistry*, Springer, Berlin, 2000.

M. Michailov, ed., *Nanophenomena at Surfaces*, Springer, Berlin, 2011.

W. Mönch, *Electronic Properties of Semiconductor Interfaces*, Springer, Berlin, 2004.

W. Mönch, *Semiconductor Surfaces and Interfaces*, Springer, Berlin, 2001.

E. Ogryzlo, K. A. R. Mitchell, and M. A. Van Hove, eds., *The Structure of Surfaces VI* (Proceedings ICSOS-VI, Vancouver, July 26–30, 1999), *Surf. Rev. Lett.*, vol. 6, nos. 5–6, 1999.

P. Soukiassian, G. Le Lay, and M. A. Van Hove, eds., *The Structure of Surfaces V* (Proceedings ICSOS-V, Aix-en-Provence, July 8–12, 1996), *Surf. Rev. Lett.*, vol. 4, nos. 5–6, 1997; vol. 5, no. 1, 1998.

S. Y. Tong, M. A. Van Hove, K. Takayanagi, and X. D. Xie, eds., *The Structure of Surfaces III* (Proceedings ICSOS-III Milwaukee, July 9–12, 1990), Springer, Berlin, 1991.

J. F. van der Veen and M. A. Van Hove, eds., *The Structure of Surfaces II* (Proceedings ICSOS-II, Amsterdam, June 22–25, 1987), Springer, Berlin, 1988.

M. A. Van Hove and S. Y. Tong, eds., *The Structure of Surfaces* (Proceedings ICSOS-I, Berkeley August 13–16, 1984), Springer, Berlin, 1985.

P. R. Watson, M. A. Van Hove, and K. Hermann, *Atlas of Surface Structures, Based on the NIST Surface Structure Database (SSD)*, *J. Phys. Chem. Ref. Data*, Monograph No. 5, vols. 1A and 1B, pp. 1–907, 1994.

D. P. Woodruff, *The Solid-Liquid Interface*, Cambridge University Press, Cambridge, 1973.

D. P. Woodruff, ed., *The Chemical Physics of Solid Surfaces, Vol. 9: Oxide Surfaces*, Elsevier, Amsterdam, 2001.

D. P. Woodruff, ed., *The Chemical Physics of Solid Surfaces, Vol. 10: Surface Alloys and Alloy Surfaces*, Elsevier, Amsterdam, 2002.

D. P. Woodruff, ed., *The Chemical Physics of Solid Surfaces, Vol. 11: Surface Dynamics*, Elsevier, Amsterdam, 2003.

D. P. Woodruff, ed., *The Chemical Physics of Solid Surfaces, Vol. 12: Atomic Clusters: from Gas Phase to Deposited*, Elsevier, Amsterdam, 2007.

X. D. Xie, S. Y. Tong, and M. A. Van Hove, eds., *The Structure of Surfaces IV* (Proceedings ICSOS-IV Shanghai, August 16–19, 1993), World Scientific, Singapore, 1994.

A.3   Books and Proceedings on Techniques

C. L. Bai, *Scanning Tunneling Microscopy and Its Application*, Springer, Berlin, 2000.

E. Bauer, *Surface Microscopy with Low Energy Electrons*, Springer, Berlin, 2014.

C. Becker, "UHV Surface Preparation Methods", in *Encyclopedia of Interfacial Chemistry: Surface Science*, Elsevier, Amsterdam, 2015.

G. Benedek and J. P. Toennies, *Atomic Scale Dynamics of Surfaces: Theory and Experimental Studies with Helium Atom Scattering*, Springer, Berlin, 2018.

D. A. Bonnell, *Scanning Probe Microscopy and Spectroscopy: Theory, Techniques and Applications*, Wiley, Weinheim, 2001.

D. A. Bonnell, *Scanning Tunneling Microscopy and Spectroscopy: Theory, Techniques and Applications*, Wiley, Chichester, 1993.

C. J. Chen, *Introduction to Scanning Tunneling Microscopy*, Oxford University Press, Oxford, 1993.

L. J. Clarke, *Surface Crystallography, An Introduction to LEED*, Wiley, Chichester, 1985, out of print.

R. Feder, *Polarized Electrons in Surface Physics*, Springer, Berlin, 1986.

H.-J. Güntherodt and R. Wiesendanger, eds., *Scanning Tunneling Microscopy I: General Principles and Applications to Clean and Adsorbate-Covered Surfaces*, Springer, Berlin, 1994.

S. Hofmann, *Auger- and X-Ray Photoelectron Spectroscopy in Materials Science,* Springer, Berlin, 2012.

E. Hulpke, *Helium Atom Scattering from Surfaces*, Springer, Berlin, 1992.

H. Ibach, ed., *Electron Spectroscopy for Surface Analysis*, Springer, Berlin, 1977.

H. Ibach, "Surfaces in Ultrahigh Vacuum", in *Physics of Surfaces and Interfaces*, Chapter 2.2, Springer-Verlag, Berlin, 2006.

S. Kalinin and A. Gruverman, eds., *Scanning Probe Microscopy Electrical and Electromechanical Phenomena at the Nanoscale: Fundamentals*, Springer, Berlin, 2006.

S. N. Magonov and M.-H. Whangbo, *Surface Analysis with STM and AFM: Experimental and Theoretical Aspects of Image Analysis*, VCH-Wiley, Weinheim, 1996.

E. Meyer, H. J. Hug, and R. Bennewitz, *Scanning Probe Microscopy: The Lab on a Tip*, Springer, Berlin, 2004.

S. Morita, R. Wiesendanger, and E. Meyer, eds., *Noncontact Atomic Force Microscopy: Vol. 1*, Springer, Berlin, 2002.

E. W. Mueller and T. T. Tsong, *Field Ion Microscopy*, American Elsevier, New York, 1969.

D. J. O'Connor, B. A. Sexton, and R. S. C. Smart, eds., *Surface Analysis Methods in Materials Science*, Springer, Berlin, 2003.

J. B. Pendry, *Low Energy Electron Diffraction: The Theory and its Application to Determination of Surface Structure*, Academic Press, London, 1974, out of print.

P. Samori, *Scanning Probe Microscopies Beyond Imaging: Manipulation of Molecules and Nanostructures*, Wiley, Weinheim, 2006.

D. Sarid, *Scanning Force Microscopy*, Oxford University Press, Oxford, 1994.

W. Schattke and M. A. Van Hove, eds., *Solid-State Photoemission and Related Methods: Theory and Experiment*, Wiley-VCH, Weinheim, 2003.

J. Stoehr, *NEXAFS Spectroscopy*, Springer, Berlin, 1992.

J. A. Stroscio and W. J. Kaiser, *Scanning Tunneling Microscopy*, Academic Press, London, 1993.

M. A. Van Hove and S. Y. Tong, *Surface Crystallography by Low Energy Electron Diffraction: Theory, Computation and Structural Results*, Springer, Berlin, 1979, out of print.

M. A. Van Hove, W. H. Weinberg, and C.-M. Chan, *Low-Energy Electron Diffraction: Experiment, Theory and Structural Determination*, Springer, Berlin, 1986.

R. Wiesendanger, *Scanning Probe Microscopy and Spectroscopy: Methods and Applications*, Cambridge University Press, Cambridge, 1994.

R. Wiesendanger, ed., *Scanning Probe Microscopy: Analytical Methods*, Springer, Berlin, 1998.

R. F. Willis, ed., *Vibrational Spectroscopy of Adsorbates*, Springer, Berlin, 1980.

D. P. Woodruff and T. A. Delchar, *Modern Techniques of Surface Science*, 2nd ed., Cambridge University Press, Cambridge, 1994, and 2010 online.

Appendix B Lists of Surface Science Websites, LEED Codes and Related Data

Note: Since weblinks change frequently, we cannot guarantee the correctness or existence of the following links.

B.1 General Websites

- Lecture Notes on Surface Science by Philip Hofmann, Aarhus University (PDF file): https://adam.unibas.ch/goto_adam_file_89155_download.html
- Surface Science by Roger M. Nix, Queen Mary, University of London: https://chem.libretexts.org/Bookshelves/Physical_and_Theoretical_Chemistry_Textbook_Maps/Book%3A_Surface_Science_(Nix)
- The Nanotube Site by David Tománek, Michigan State University: https://nanoten.com/NTSite/

B.2 Website Related to Surface Structures

- Gallery of Surface Structures by Klaus Hermann, Fritz-Haber-Institut: www.fhi-berlin.mpg.de/KHsoftware/Balsac/pictures.html

B.3 Websites Related to Techniques

- The I(V) Data Repository (LEED) by Franco Jona and Jim Quinn, SUNY at Stony Brook: http://doll.eng.sunysb.edu/ivdata/

B.4 LEED Codes and Related Data

B.4.1 Structure Determination from LEED Intensities

AQuaLEED (J. Lachnitt: Automated Quantitative Low-Energy Electron Diffraction; uses SATLEED)
Information and download: https://physics.mff.cuni.cz/kfpp/povrchy/software

CLEED (G. Held et al: Automated Surface Structure Determination from LEED-IV Curves)

Obtainable from: georg.held@diamond.ac.uk

DL_LEED (A. Wander)

Information and download: https://doi.org/10.1016/S0010-4655(01)00168-0

Reference:

A. Wander, "A New Modular Low Energy Electron Diffraction Package – DL_LEED", *Comp. Phys. Commun.*, vol. 137, pp. 4–11, 2001.

eeasisss (J. Rundgren, W. Moritz, and E. W. Plummer: Elastic Electron-Atom Scattering in Solids and Solid Surfaces, A Code for Atomic LEED Phase Shifts)

Obtainable from: J. Rundgren, jru@kth.se; and W. Moritz, wolfgang.moritz@lrz.uni-muenchen.de

References:

J. Rundgren, "Optimized Surface-Slab Excited-State Muffin-Tin Potential and Surface Core Level Shifts", *Phys. Rev. B*, vol. 68, p. 125405, 2003; vol. 76, pp. 125405, 2003.

J. Rundgren, "Elastic Electron-Atom Scattering in Amplitude-Phase Representation with Application to Electron Diffraction and Spectroscopy", *Phys. Rev. B*, vol. 76, p. 195441, 2007.

J. Rundgren, B. E. Sernelius, and W. Moritz, "Low-Energy Electron Diffraction with Signal Electron Carrier-Wave Wavenumber Modulated by Signal Exchange–Correlation Interaction", *J. Phys. Commun.*, vol. 5, p. 105012, 2021.

LEEDFIT (W. Moritz et al.)

Obtainable from: W. Moritz, wolfgang.moritz@lrz.uni-muenchen.de

References:

G. Kleinle, W. Moritz, and G. Ertl, "An Efficient Method for LEED Crystallography". *Surf. Sci.*, vol. 238, pp. 119–131, 1990.

W. Moritz, "Effective Calculation of LEED Intensities Using Symmetry-Adapted Functions", *J. Phys. C: Solid State Physics*, vol. 17, pp. 353–362, 1984.

H. Over, U. Ketterl, W. Moritz, and G. Ertl, "Optimization Methods and Their Use in Low-Energy Electron–Diffraction Calculations", *Phys. Rev. B*, vol. 46, pp. 15438–15446, 1992.

SATLEED, etc. (M. A. Van Hove, A. Barbieri, et al: A suite of codes tailored for various situations)

Information and download:

https://www.icts.hkbu.edu.hk/VanHove_files/leed/leedpack.html

TensErLEED (V. Blum and K. Heinz), see also ViPErLEED in Section B.4.2

Information and download: https://elsevier.digitalcommonsdata.com/datasets/js36dx77rz/1

Updates available from: L. Hammer, lutz.hammer@fau.de

Reference:

V. Blum and K. Heinz, "Fast LEED Intensity Calculations for Surface Crystallography Using Tensor LEED", *Comput. Phys. Commun.*, vol. 134, pp. 392–425, 2001.

B.4.2   Experimental I(V) Curve Generation from LEED Patterns

EasyLEED (A. Mayer et al.: Automated extraction of intensity–energy spectra from LEED patterns)
Information and download: https://andim.github.io/easyleed/
Reference:
A. Mayer, H. Salopaasi, K. Pussi, and R. D. Diehl, "A Novel Method for the Extraction of Intensity–Energy Spectra from Low-Energy Electron Diffraction Patterns", *Comput. Phys. Commun.*, vol. 183, pp. 1443–1447, 2012.

PLEASE (M. Grady, Z. Dai, and K. Pohl: The Python Low-energy Electron Analysis SuitE – Enabling Rapid Analysis of LEEM and LEED Data, e.g., to Produce I(V) Curves)
Information and download: http://doi.org/10.5334/jors.191
Reference:
M. Grady, Z. Dai, and K. Pohl, "PLEASE: The Python Low-Energy Electron Analysis SuitE – Enabling Rapid Analysis of LEEM and LEED Data", *J. Open Res. Software*, vol. 6(1), p. 7, 2018.

ViPErLEED (F. Kraushofer, M. Schmid, A. M. Imre, T. Kisslinger, U. Diebold, L. Hammer, M. Riva: Vienna Package for TensErLEED, including software and hardware solution for acquisition of LEED videos, program for automated spot tracking and extraction/processing of I(V) curves, and easy to use front-end for TensErLEED)
Information and download: https://github.com/viperleed
Contacts: M. Riva riva@iap.tuwien.ac.at and L. Hammer lutz.hammer@fau.de

B.4.3   LEED I(V) Data

The I(V) Data Repository (F. Jona and J. Quinn)
Information and download: http://DoL1.eng.sunysb.edu/ivdata/

B.4.4   Surface Structure Data

Database of surface structures (free) (P. R. Watson, M. A. Van Hove and K. Hermann), formerly known as SSD distributed by NIST
Information and download: www.fhi-berlin.mpg.de/KHsoftware/oSSD/index.html

Gallery of Surface Structures (K. Hermann)
Site: www.fhi-berlin.mpg.de/KHsoftware/Balsac/pictures.html

Surface Structure Information (M. A. Van Hove: includes links to Online interactive software, Software and database for download, Surface structure publications and Other useful websites)
Site: www.icts.hkbu.edu.hk/vanhove/

B.4.5   LEED Pattern Simulation

Graphical LEED pattern simulator LEEDpat (K. Hermann and M. A. Van Hove)
Site: www.fhi-berlin.mpg.de/KHsoftware/LEEDpat/index.html

B.4.6 Visualisation Software for Surface Structures

Surface graphics software Balsac (K. Hermann, PC-based software to construct lattice sections with extended graphical display and analysis options)
Site: www.fhi-berlin.mpg.de/KHsoftware/Balsac/index.html

Surface Explorer (F. Rammer and K. Hermann, online interactive software for imaging bulk-terminated crystalline surfaces with any Miller indices)
Site: http://surfexp.fhi-berlin.mpg.de/

Appendix C Vector Calculation in Oblique Lattices

Let a_1, a_2, a_3 and α, β, γ be the lattice constants of an oblique lattice with $a_1 = |\mathbf{a}_1|$, $a_2 = |\mathbf{a}_2|$, $a_3 = |\mathbf{a}_3|$. The angles between the basis vectors are defined by

$$\mathbf{a}_1 \cdot \mathbf{a}_2 = |\mathbf{a}_1||\mathbf{a}_2| \cos \gamma, \quad \mathbf{a}_2 \cdot \mathbf{a}_3 = |\mathbf{a}_2||\mathbf{a}_3| \cos \alpha, \quad \mathbf{a}_1 \cdot \mathbf{a}_3 = |\mathbf{a}_1||\mathbf{a}_3| \cos \beta. \tag{C.1}$$

A vector $\mathbf{u}$ is described in terms of its components u_i by $\mathbf{u} = \sum_{i=1}^{3} u_i \mathbf{a}_i$. The scalar product of two vectors $\mathbf{u}$ and $\mathbf{v}$ can be calculated with the metric matrix:

$$\mathbf{u} \cdot \mathbf{v} = |\mathbf{u}| |\mathbf{v}| \cos (\varphi)$$

$$= \sum_{i=1}^{N} \sum_{j=1}^{N} u_i \mathbf{a}_i \cdot \mathbf{a}_j v_j$$

$$= \sum_{i=1}^{N} \sum_{j=1}^{N} u_i G_{ij} v_j, \tag{C.2}$$

where $G_{ij} = \mathbf{a}_i \cdot \mathbf{a}_j = G_{ji}$ are the components of the metric matrix

$$\mathbf{G} = \begin{pmatrix} G_{11} & G_{12} & G_{13} \\ G_{21} & G_{22} & G_{23} \\ G_{31} & G_{32} & G_{33} \end{pmatrix} = \begin{pmatrix} a_1^2 & a_1 a_2 \cos \gamma & a_1 a_3 \cos \beta \\ a_1 a_2 \cos \gamma & a_2^2 & a_2 a_3 \cos \alpha \\ a_1 a_3 \cos \beta & a_2 a_3 \cos \alpha & a_3^2 \end{pmatrix}. \tag{C.3}$$

An example is the hexagonal system with $a_1 = a_2 = 0.4$ nm, $a_3 = 0.5$ nm, $\alpha = \beta = 90°$ and $\gamma = 120°$:

$$\mathbf{G} = \begin{pmatrix} 0.16e^2 & -0.08e^2 & 0e^2 \\ -0.08e^2 & 0.16e^2 & 0e^2 \\ 0e^2 & 0e^2 & 0.25e^2 \end{pmatrix}, \tag{C.4}$$

where $e = 1$ is the length of the unit vector (e.g., 1 nm or 1 Å or in any other units). For the cubic system we get

$$\mathbf{G} = a_0^2 \begin{pmatrix} 1 & 0 & 0 \\ 0 & 1 & 0 \\ 0 & 0 & 1 \end{pmatrix}, \tag{C.5}$$

where a_0 is the cubic lattice constant.

The length of a vector $|\mathbf{u}| = \sqrt{\mathbf{u} \cdot \mathbf{u}}$ is given by

$$\mathbf{u} \cdot \mathbf{u} = \sum_{i=1}^{3}\sum_{j=1}^{3} u_i G_{ij} u_j = \sum_{i=1}^{3}\sum_{j=1}^{3} u_i \mathbf{a}_i \mathbf{a}_j u_j. \tag{C.6}$$

The angle between two vectors $\mathbf{u}$ and $\mathbf{v}$ is given by:

$$\cos\varphi = \frac{\mathbf{u} \cdot \mathbf{v}}{|\mathbf{u}|\,|\mathbf{v}|} = \frac{\sum_{i=1}^{N}\sum_{j=1}^{N} u_i G_{ij} v_j}{|\mathbf{u}|\,|\mathbf{v}|}. \tag{C.7}$$

The vector product of two vectors $\mathbf{u}$ and $\mathbf{v}$ is a vector normal to the plane defined by $\mathbf{u}$ and $\mathbf{v}$:

$$\mathbf{u} \times \mathbf{v} = -\mathbf{v} \times \mathbf{u}$$
$$|\mathbf{u} \times \mathbf{v}| = |\mathbf{u}||\mathbf{v}| \sin\varphi. \tag{C.8}$$

Here φ is the angle between the two vectors. The vector product $\mathbf{u} \times \mathbf{v}$ is defined in such a way that $\mathbf{u}$, $\mathbf{v}$ and $\mathbf{u} \times \mathbf{v}$ form a right-handed coordinate system, while $\mathbf{u}$, $\mathbf{v}$ and $\mathbf{v} \times \mathbf{u}$ form a left-handed coordinate system. We obtain for $\mathbf{u} = (u_1\mathbf{a}_1 + u_2\mathbf{a}_2 + u_3\mathbf{a}_3)$ and $\mathbf{v} = (v_1\mathbf{a}_1 + v_2\mathbf{a}_2 + v_3\mathbf{a}_3)$:

$$\mathbf{u} \times \mathbf{v} = (u_1\mathbf{a}_1 + u_2\mathbf{a}_2 + u_3\mathbf{a}_3) \times (v_1\mathbf{a}_1 + v_2\mathbf{a}_2 + v_3\mathbf{a}_3), \tag{C.9}$$

$$\mathbf{u} \times \mathbf{v} = (u_1 v_2 - u_2 v_1)(\mathbf{a}_1 \times \mathbf{a}_2)$$
$$+ (u_2 v_3 - u_3 v_2)(\mathbf{a}_2 \times \mathbf{a}_3)$$
$$+ (u_3 v_1 - u_1 v_3)(\mathbf{a}_3 \times \mathbf{a}_1). \tag{C.10}$$

The vectors $\mathbf{a}_1^* = (\mathbf{a}_2 \times \mathbf{a}_3)$, $\mathbf{a}_2^* = (\mathbf{a}_3 \times \mathbf{a}_1)$ and $\mathbf{a}_3^* = (\mathbf{a}_1 \times \mathbf{a}_2)$ are the basis vectors of the reciprocal lattice, and the vector $\mathbf{u} \times \mathbf{v}$ can be expressed as a vector in the reciprocal space:

$$\mathbf{u} \times \mathbf{v} = (u_2 v_3 - u_3 v_2)\mathbf{a}_1^* + (u_3 v_1 - u_1 v_3)\mathbf{a}_2^* + (u_1 v_2 - u_2 v_1)\mathbf{a}_3^*$$
$$= h\mathbf{a}_1^* + k\mathbf{a}_2^* + l\mathbf{a}_3^*. \tag{C.11}$$

Similarly, the vector product of two vectors in reciprocal space can be expressed as a vector in real space. To describe $\mathbf{u} \times \mathbf{v}$ in real space we need the coordinates of the reciprocal basis in real space. It is easy to derive that:

$$\mathbf{a}_i^* = \sum_{j=1}^{3} G_{ij}^* \mathbf{a}_j, \quad \mathbf{a}_i = \sum_{j=1}^{3} G_{ij}\mathbf{a}_j^* \quad \text{and} \quad \mathbf{G}^* = \mathbf{G}^{-1}. \tag{C.12}$$

The volume of the unit cell is $V = |\mathbf{a}_1 \cdot (\mathbf{a}_2 \times \mathbf{a}_3)|$, with an analogous expression for V^*. With $G_{ij} = \mathbf{a}_i \cdot \mathbf{a}_j = |\mathbf{a}_i||\mathbf{a}_j| \cos{(\mathbf{a}_i, \mathbf{a}_j)}$, we can derive that

$$|\mathbf{G}| = (a_1 a_2 a_3)^2 \left(1 - \cos^2\alpha - \cos^2\beta - \cos^2\gamma + 2\cos\alpha\cos\beta\cos\gamma\right) = V^2 \quad \text{(C.13)}$$

and therefore:

$$V = \sqrt{|\mathbf{G}|}, \quad V^* = \sqrt{|\mathbf{G}^*|} \text{ and } V^* = V^{-1}. \quad \text{(C.14)}$$

For further descriptions of vector calculations in oblique lattices, see D. E. Sands [2.13].

Appendix D Distance between Neighbouring Lattice Planes in the Bravais Lattices

Table D.1 gives general relations that allow one to calculate the d-spacings between neighbouring lattice planes in all crystal systems, for any allowed choice of Miller indices.

Table D.1 Relation between the lattice constants and d-spacings d_{hkl} (defined as the distance perpendicular to the planes) for 3-D lattices

Crystal system	$1/d_{hkl}{}^2$
Cubic	$\dfrac{1}{a_0^2}\left(h^2 + k^2 + l^2\right)$
Tetragonal	$\dfrac{\left(h^2 + k^2\right)}{a^2} + \dfrac{l^2}{c^2}$
Orthorhombic	$\dfrac{h^2}{a^2} + \dfrac{k^2}{b^2} + \dfrac{l^2}{c^2}$
Hexagonal and trigonal (P)	$\dfrac{4}{3a^2}\left(h^2 + k^2 + hk\right) + \dfrac{l^2}{c^2}$
Trigonal (R)	$\dfrac{1}{a_0^2}\left\{\dfrac{\left(h^2 + k^2 + l^2\right)\sin^2\alpha + 2(hk + hl + kl)\left(\cos^2\alpha - \cos\alpha\right)}{\left(1 - 2\cos^3(\alpha) + 3\cos^2(\alpha)\right)}\right\}$
Monoclinic	$\dfrac{h^2}{a^2\sin^2\beta} + \dfrac{k^2}{b^2} + \dfrac{l^2}{c^2\sin^2\beta} - \dfrac{2hl\cos\beta}{ac\sin^2\beta}$
Triclinic	$\left(1 - \cos 2\alpha - \cos 2\beta - \cos 2\gamma + 2\cos\alpha\cos\beta\cos\gamma\right)^{-1}$ $\times\left\{\dfrac{h^2}{a^2}\sin^2\alpha + \dfrac{k^2}{b^2}\sin^2\beta + \dfrac{l^2}{c^2}\sin^2\gamma + \dfrac{2kl}{bc}\left(\cos\beta\cos\gamma - \cos\alpha\right)\right.$ $\left. + \dfrac{2lh}{ca}\left(\cos\gamma\cos\alpha - \cos\beta\right) + \dfrac{2hk}{ab}\left(\cos\alpha\cos\beta - \cos\gamma\right)\right\}$

Note: for centred lattices not all combinations of h, k, l are allowed, see Table 2.2

Appendix E Lattice Transformations

A lattice transformation is required if the same structure must be described with different translation vectors. For surface structure analysis it is often required to use two translation vectors parallel to the surface with the third vector normal to the surface; in special cases a translation vector oblique to the surface may also be used. The translation vectors of the surface unit cell therefore differ in most cases from those used for the 3-D unit cell. The atomic coordinates are conventionally described in fractional coordinates relative to the translation vectors and can be found in crystallographic databases. Several surface studies have been performed that require non-trivial lattice transformations, for example for the oxides Al_2O_3, Fe_2O_3, Fe_3O_4, TiO_2 and V_2O_5, some of which contain many atoms in the unit cell; the use of a computer program is recommended in such cases. The surface structure analysis with LEED or surface X-ray diffraction requires the description of the bulk structure related to the translation vectors of the surface unit cell. The symmetry of the surface unit cell is one of the 17 2-D groups and this symmetry is probably reduced compared to that of the 3-D structure. The number of atoms in the surface unit cell is therefore usually different from that in the 3-D unit cell. We describe below a relatively simple approach; a more general method is given by K. Hermann in section 3.2 of [2.6].

First, the translation vectors of the surface unit cell must be found. At low index surfaces they may be determined directly from the 3-D unit cell. For high index and stepped surfaces, a method has been described in Section 2.1.12. In other cases, the translation vectors may be determined from the diffraction pattern. If a commensurate superlattice exists and the translation vectors of the substrate are known, then the translation vectors of the superlattice can be determined. An appropriate choice must be made for the third vector: normal or oblique to the surface. This vector is needed for the semi-infinite substrate.

If the translation vectors and the symmetry of the surface unit cell are chosen, then the transformation of the coordinates from the 3-D substrate unit cell to the surface unit cell can be performed with the equations given in Section 2.1.9; it is easier to use a computer program, for example Balsac [2.14] or CRYSCON [2.15] which can be found on the internet. These programs need as input the data of the 3-D unit cell (e.g., in the format of a CIF file for CRYSCON), as well as the translation vectors of the 2-D unit cell and a vector **c**. The vector **c** of the surface unit cell must be chosen in such a way that all atoms in the 3-D unit cell are included. In Balsac, the transformed unit cell is evaluated internally based on corresponding Miller indices (hkl). In CRYSCON, the transformation matrix from the 3-D unit cell to the surface unit cell with **a**, **b** parallel to the surface and an appropriate vector **c** is required. In cases where no vector **c** normal to the surface is possible, an oblique vector must be chosen.

In the following, we give some frequently occurring examples and some standard transformations, for example, the hexagonal setting of a fcc and bcc metal crystal required for their (111) surfaces.

The general transformation rules are as follows. Let a, b, c and α, β, γ be the lattice constants of the 3-D crystal lattice. The lattice vectors of the transformed lattice $\mathbf{a}'$, $\mathbf{b}'$, $\mathbf{c}'$ are given by a transformation matrix M:

$$\begin{pmatrix} \mathbf{a}' \\ \mathbf{b}' \\ \mathbf{c}' \end{pmatrix} = \begin{pmatrix} M_{11} & M_{21} & M_{31} \\ M_{12} & M_{22} & M_{32} \\ M_{13} & M_{23} & M_{33} \end{pmatrix} \begin{pmatrix} \mathbf{a} \\ \mathbf{b} \\ \mathbf{c} \end{pmatrix}. \tag{E.1}$$

The coordinates are transformed as

$$\mathbf{r}' = \left(\mathbf{M}^{\mathrm{T}}\right)^{-1} \cdot \mathbf{r}, \tag{E.2}$$

where $\mathbf{M}^{\mathrm{T}}$ is the transpose of **M** and the coordinates are given in relative units. If it is required to transform reflection indices (e.g., for X-ray diffraction) the transformed indices are given by

$$\begin{pmatrix} h' \\ k' \\ l' \end{pmatrix} = \begin{pmatrix} M_{11} & M_{21} & M_{31} \\ M_{12} & M_{22} & M_{32} \\ M_{13} & M_{23} & M_{33} \end{pmatrix} \begin{pmatrix} h \\ k \\ l \end{pmatrix}. \tag{E.3}$$

The transformation matrix is in general not unitary. In LEED calculations the surface structure is frequently described as a stacking of layers and the bulk interlayer distance vector is required. This is given in the examples in Sections E.1–E.4.

E.1 The (111) Surface of the fcc Lattice

The hexagonal setting of the fcc unit cell is:

$$\mathbf{M} = \begin{pmatrix} 1/2 & -1/2 & 0 \\ 0 & 1/2 & -1/2 \\ 1 & 1 & 1 \end{pmatrix}. \tag{E.4}$$

Other choices are possible, where the vector $\mathbf{c}'$ points in another direction. A three-layer stacking sequence occurs. The volume of the unit cell is $V_{\mathrm{hex}} = 3/4\ V_{\mathrm{fcc}}$. The layer distance normal to the surface plane is $c'/3 = a_0\ \sqrt{3}/3$. An oblique distance vector in the mirror plane is $d = [1/2, 0, 1/2]$. The 2-D space group of the surface unit cell is p3m1.

E.2   The (111) Surface of the bcc Lattice

The hexagonal setting of the bcc unit cell is:

$$M = \begin{pmatrix} 1 & -1 & 0 \\ 0 & 1 & -1 \\ 1/2 & 1/2 & 1/2 \end{pmatrix}. \tag{E.5}$$

The volume of the unit cell is $V_{hex} = 1.5\ V_{bcc}$. A three-layer stacking sequence occurs. The layer distance normal to the surface is $d = a_0\ \sqrt{3}/6$. The symmetry of the surface unit cell is p3m1.

E.3   Hexagonal to Rhombohedral Transformation

Given are a triple hexagonal cell and the rhombohedral space group. Required is a primitive rhombohedral cell.

We define the rhombohedral lattice with $\mathbf{a'}, \mathbf{b'}, \mathbf{c'}, \alpha'$, and the hexagonal lattice with $\mathbf{a}, \mathbf{b}, \mathbf{c}, \alpha = \beta = 90°, \gamma = 120°$.

The relations between the cell parameters $\mathbf{a}, \mathbf{b}, \mathbf{c}$ of the triple hexagonal cell and the cell parameters $\mathbf{a'}, \mathbf{b'}, \mathbf{c'}$ are:

$$\begin{pmatrix} \mathbf{a'} \\ \mathbf{b'} \\ \mathbf{c'} \end{pmatrix} = \begin{pmatrix} 2/3 & 1/3 & 1/3 \\ -1/3 & 1/3 & 1/3 \\ -1/3 & -2/3 & 1/3 \end{pmatrix} \begin{pmatrix} \mathbf{a} \\ \mathbf{b} \\ \mathbf{c} \end{pmatrix}. \tag{E.6}$$

The matrix $\mathbf{Q} = (\mathbf{M}^T)^{-1}$ transforms the atomic coordinates, $\mathbf{r'} = \mathbf{Qr}$, where

$$\mathbf{Q} = \begin{pmatrix} 1 & 0 & 1 \\ -1 & 1 & 1 \\ 0 & -1 & 1 \end{pmatrix}. \tag{E.7}$$

The lengths of the translation vectors in the rhombohedral cell and the angle α' are given by:

$$a' = \frac{1}{3}\sqrt{3a^2 + c^2}, \tag{E.8}$$

$$\sin\frac{\alpha'}{2} = \frac{3}{2\sqrt{3 + (c^2/a^2)}}, \quad \text{or} \quad \cos\alpha' = \frac{(c^2/a^2) - 3/2}{(c^2/a^2) + 3}. \tag{E.9}$$

The volume of the unit cell is $V_{rhomb} = 1/3\ V_{hex}$. The layer distance normal to the surface is $d = c/3$.

E.4   Rhombohedral to Hexagonal Transformation

Given is a primitive rhombohedral cell. Required is a triple hexagonal cell, with obverse setting.

We define the rhombohedral lattice with $\mathbf{a}$, $\mathbf{b}$, $\mathbf{c}$, α, and a hexagonal lattice with $\mathbf{a}'$, $\mathbf{b}'$, $\mathbf{c}'$, $\alpha' = \beta' = 90°$, $\gamma' = 120°$.

The relations between the cell parameters $\mathbf{a}'$, $\mathbf{b}'$, $\mathbf{c}'$ of the triple hexagonal cell and the cell parameters $\mathbf{a}$, $\mathbf{b}$, $\mathbf{c}$ of the primitive rhombohedral cell are:

$$\begin{pmatrix} \mathbf{a}' \\ \mathbf{b}' \\ \mathbf{c}' \end{pmatrix} = \begin{pmatrix} 1 & -1 & 0 \\ 0 & 1 & -1 \\ 1 & 1 & 1 \end{pmatrix} \begin{pmatrix} \mathbf{a} \\ \mathbf{b} \\ \mathbf{c} \end{pmatrix}. \tag{E.10}$$

The matrix $\mathbf{Q}$ transforms the atomic coordinates, $\mathbf{r}' = \mathbf{Q}\mathbf{r}$, where

$$\mathbf{Q} = \begin{pmatrix} 2/3 & -1/3 & -1/3 \\ 1/3 & 1/3 & -2/3 \\ 1/3 & 1/3 & 1/3 \end{pmatrix}. \tag{E.11}$$

The lengths of the translation vectors are:

$$a' = a\sqrt{2}\sqrt{1 - \cos\alpha} = 2a\sin\frac{\alpha}{2}, \quad c' = a\sqrt{3}\sqrt{1 + 2\cos\alpha}, \tag{E.12}$$

$$\frac{c'}{a'} = \sqrt{\frac{3}{2}}\sqrt{\frac{1 + 2\cos\alpha}{1 - \cos\alpha}} = \sqrt{\frac{9}{4\sin^2\frac{\alpha}{2}} - 3}. \tag{E.13}$$

The volume of the unit cell is $V_{\text{hex}} = 3\,V_{\text{rhomb}}$. The layer distance normal to the surface is $d = c'/2$.

Appendix F Calculation of Translation Vectors for Stepped Surfaces

The formalism for calculating translation vectors at stepped surfaces is quite simple and is described here for the general case of oblique lattices. Another general approach that is more direct uses the additivity theorem for Miller indices, see Section 4.3 of K. Hermann [2.6]. We here decompose the Miller indices of the stepped surface in terms of terrace and step indices:

$$(h, k, l) = n_t (h_t, k_t, l_t) + n_s (h_s, k_s, l_s). \tag{F.1}$$

The direction of the step edge $\mathbf{s}$ is given by the vector product

$$[s_1, s_2, s_3] = (h_t, k_t, l_t) \times (h, k, l). \tag{F.2}$$

Here, $\mathbf{s}$ is a vector in real space, while (h, k, l) and (h_t, k_t, l_t) are vectors in reciprocal space. The components of $\mathbf{s}$ can possibly be reduced by dividing by a common integer factor. The reduced $\mathbf{s}$ can then be taken to be the translation vector $\mathbf{a}_s$ of the (h, k, l) surface. If $\mathbf{b}_s$ is the second translation vector in the surface plane, the vector product $\mathbf{a}_s \times \mathbf{b}_s$ gives the direction of the surface normal:

$$[a_{s1}, a_{s2}, a_{s3}] \times [b_{s1}, b_{s2}, b_{s3}] = (h, k, l). \tag{F.3}$$

The second translation vector $\mathbf{b}_s$ can thus be determined by first calculating the vector of the normal distance $\mathbf{w}'$ between step edges from the cross product:

$$(a_{s1}, a_{s2}, a_{s3}) \times (h, k, l) = \left[w_1', w_2', w_3' \right]. \tag{F.4}$$

Dividing by a common integer factor in the components of $\mathbf{w}'$ gives a vector $\mathbf{w}$. A further factor n can be divided out if the product

$$(\mathbf{a}_s \times \mathbf{w}) \cdot d_{hkl} = nV \tag{F.5}$$

gives a multiple of the volume of the unit cell. The vector $\mathbf{w}/n$ is not necessarily a lattice point and does not necessarily have integer components: $|\mathbf{w}/n|$ is the normal distance between step edges. If the lattice has a mirror plane in the direction of $\mathbf{a}_s$ or normal to it, the second translation vector $\mathbf{b}_s$ must be normal to $\mathbf{a}_s$ and we can take $\mathbf{w}/n$ as the translation vector. In the general case without symmetry, the translation vector $\mathbf{b}_s$ is found by adding a fraction f of $\mathbf{a}_s$ to $\mathbf{w}/n$ such that integer components occur and the angle between $\mathbf{a}_s$ and $\mathbf{b}_s$ is as close to $90°$ as possible.

$$[b_{s1}, b_{s2}, b_{s3}] = \frac{1}{n}[w_1, w_2, w_3] + f[a_{s1}, a_{s2}, a_{s3}]. \tag{F.6}$$

The product

$$(\mathbf{a}_s \times \mathbf{b}_s) \cdot d_{hkl} = V \tag{F.7}$$

should give the volume of the unit cell. For a centred lattice, the vectors $\mathbf{a}_s$ and $\mathbf{b}_s$ can be reduced such that the components match the translation vectors of the primitive cell (see Table 2.6). The volume may still be double the volume of the primitive unit cell; the surface then has a 2-D centred lattice. The translation vectors determined in that way are not necessarily the smallest vectors, since the vector $\mathbf{a}_s$ is chosen in the step direction and that depends on the choice of the decomposition of the plane (h, k, l) into two terraces.

F.1   Example 1

We want to calculate the 2-D lattice vectors for a (102) surface of a hcp lattice. The hcp lattice is primitive, of space group P6$_3$/mmc (Schoenflies symbol D$_{6h}^4$), with two atoms per unit cell. We assume lattice constants $a_1 = a_2 = 0.4$ nm; $a_3 = 0.5$ nm, $\alpha = \beta = 90°$ and $\gamma = 120°$. For the vector calculations we must use three indices.

The (102) vector in reciprocal space can be decomposed into (101) + (001) and the vector of the step edge $\mathbf{a}_s$ is (102) × (001) = [0, −1, 0], a vector in real space, according to Eq. (2.28) and Appendix C. The second translation vector is given by (cf. Eq. (2.30)):

$$[0, -1, 0] \times [b_{s1}, b_{s2}, b_{s3}] = (1, 0, 2). \tag{F.8}$$

It can be seen directly that $b_{s1} = 2$ and $b_{s3} = -1$; b_{s2} remains undefined as any vector $[2, n, -1]$ leads to the surface plane $(1, 0, 2)$. We check with $n = 0, \pm 1$, etc., to find the angle γ_s closest to 90°. The angle between $\mathbf{a}_s$ and $\mathbf{b}_s$ is given by

$$\cos \gamma_s = \frac{\sum_{i=1}^{3}\sum_{j=1}^{3} a_{si} G_{ij} b_{sj}}{|\mathbf{a}_s| \, |\mathbf{b}_s|}. \tag{F.9}$$

We find $\gamma_s = 90°$ with $n = 1$. The translation vectors of the surface unit cell are therefore:

$$\mathbf{a}_s = [0, -1, 0] \quad \text{and} \quad \mathbf{b}_s = [2, 1, -1]. \tag{F.10}$$

We could have expected a rectangular unit cell anyway, because the 3-D lattice has mirror planes normal to the translation vectors. The lengths of the translation vectors $\mathbf{a}_s$, $\mathbf{b}_s$ are calculated with the metric matrix, yielding:

$$|\mathbf{a}_s| = 0.4 \text{ nm} \quad \text{and} \quad |\mathbf{b}_s| = 0.8544 \text{ nm}. \tag{F.11}$$

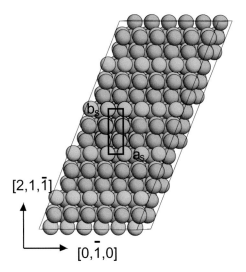

Figure F.1 The (102) surface of the hcp lattice. The surface unit cell and the lattice directions of the translation vectors $\mathbf{a}_s$ and $\mathbf{b}_s$ are shown in the figure. The brighter and darker colours refer to alternating A and B layers in the ABAB... stacking of the hcp lattice.

We can also check that the volume of the surface unit cell, $V_s = |\mathbf{a}_s|\,|\mathbf{b}_s|\,\sin\gamma_s\,d_{102}$, with a d-spacing $d_{102} = 0.22027$ nm matches the volume of the 3-D unit cell. The (102) surface and the 2-D unit cell are shown in Figure F.1.

F.2   Example 2

We want to calculate the lattice constants of the (113) surface of Si. Si has a fcc lattice and we first need the lattice vectors of the 2-D unit cell related to the centred cell with lattice constant a_0 and then we must reduce it using the shortest translation vectors of the 3-D lattice. The (113) surface can be decomposed into (111) + 2(001). The direction of the step edge is (111) × (001) = [1, −1, 0] and we can take this as translation vector $\mathbf{a}_s$. The step distance $\mathbf{w}'$ can be calculated by the vector product

$$\mathbf{w}' = [1, 1, 3] \times [1, -1, 0] = [3, 3, -2]. \tag{F.12}$$

We could use [1, 1, 3] as a vector in real space because of the cubic system. We note that in an oblique system we would have to transform it from a vector in reciprocal space to a vector in real space with the metric matrix. The vector $\mathbf{w}'$ may be a multiple of the step distance: we take $\mathbf{w} = \frac{1}{2}\mathbf{w}' = [3/2, 3/2, -1]$ as step vector and check that the vector product of the two in-plane vectors gives the correct (h, k, l):

$$[1, -1, 0] \times [3/2, 3/2, -1] = [1, 1, 3]. \tag{F.13}$$

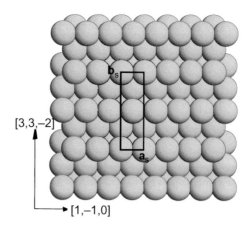

Figure F.2 The (113) surface of the fcc lattice. The 2-D unit cell of the surface is shown.

Here, $\mathbf{w}$ is not necessarily a lattice point, but in this case $\mathbf{w}$ is in fact a lattice point because of the face-centred unit cell. We now have a rectangular 2-D unit cell with translation vectors

$$\mathbf{a_s} = [1, -1, 0], \quad \mathbf{b_s} = [3/2, 3/2, -1]. \tag{F.14}$$

These vectors refer to the cubic unit cell, the layer distance being $d_{113} = a_0/\sqrt{11}$ according to Table D.1. The angle between $\mathbf{a_s}$ and $\mathbf{b_s}$ is 90° as $\mathbf{a_s} \cdot \mathbf{b_s} = 0$. The volume of the unit cell is $V_0 = |\mathbf{a_s}||\mathbf{b_s}|d_{113} = 1 \cdot a_0^3$, as it should be. We can reduce the translation $\mathbf{a_s}$ to $[1/2, -1/2, 0]$ because of the face-centred lattice; $\mathbf{b_s}$ cannot be reduced. With the reduced vectors we obtain a volume of the unit cell of ½ V_0. We know that the volume of the fcc unit cell is 4-times that of the primitive cell; our unit cell volume for the (113) surface with translation vectors

$$\mathbf{a_s} = [1/2, -1/2, 0], \quad \mathbf{b_s} = [3/2, 3/2, -1] \tag{F.15}$$

is double that of the primitive 3-D cell. That implies that the (113) surface should have a centred lattice. This is indeed the case, as can be seen in Figure F.2, where the (113) surface of the mono-atomic fcc lattice is shown for simplicity. The Si(113) surface would exhibit twice as many atoms and is not shown here. The actual Si(113) surface is in fact reconstructed.

Appendix G Symmetry Use in the Calculation of Reflection and Transmission Matrices

The use of symmetries in the multiple scattering equations for a composite layer allows the reduction of the effective number of atoms per unit cell to that in the asymmetric unit, as described in Section 5.3.1. It also reduces the size of the matrix to be inverted. For the atoms in special positions in the unit cell, symmetry adapted spherical harmonics can be used. With these functions some spherical waves vanish for symmetry reasons and some non-vanishing ones can be combined so that the number of values in the (l,m) sequence is reduced. The non-vanishing values in the spherical wave expansion are given in Table G.1 for all special positions in the 17 plane groups except for p1. For atoms in general positions, all $(l_{max}+1)^2$ values in the (l,m) sequence are required for $l = 0, \ldots, l_{max}$ and $m = -l, \ldots, l$.

The angles ϑ and φ required for the spherical harmonics are usually described in Cartesian coordinates while the atomic coordinates are conveniently given in units of translation vectors. This allows the easy identification of special positions in the unit cell and the determination of the local symmetry. It would be tedious to find the local symmetry if the coordinates were given in Cartesian coordinates (in nm or Å). We therefore need to define the position and orientation of the 2-D unit cell in Cartesian coordinates by the angle between the **a**-vector and the x-axis; the z-direction is assumed to be normal to the surface. The orientation of the unit cell can be chosen arbitrarily. We assume in the following the standard setting of the unit cell, that is, the origin at the highest rotation axis and the orientation of the mirror or glide planes as shown in Figure 2.12 or in the International Tables. The superstructure unit cell frequently has another orientation and not the standard setting, for example m $\parallel$ **b** instead of m $\parallel$ **a** in the 2-D space group pm. The orientation of the mirror planes has to be considered for the use of symmetry adapted spherical harmonics. We further assume that the unit cell may have any orientation with respect to the Cartesian coordinates. This is illustrated in Figure G.1, where the angle between the **a**-vector and the x-axis is γ_a.

The symmetry operation is applied to the set of wave vectors $\mathbf{k_g}$ generated by the layer periodicity. The conditions for non-vanishing spherical waves are given in Eq. (5.67) for a point on an n-fold axis and in Eq. (5.68) for positive m values for points on a mirror plane. The implications of symmetry operators for the spherical harmonics are repeated here in Eqs. (G.3)–(G.5) for convenience. We must consider the symmetry properties of the incoming and outgoing wave fields

Table G.1 Index selection rules and symmetry adapted spherical harmonics for special positions in all plane groups except p1. The table is divided into two parts. The first part gives the allowed values of the angular momentum components for special positions in groups without mirror planes where no symmetry adapted spherical harmonics are required; only index selection rules are listed. The second part contains all special positions in groups with mirror planes and the corresponding symmetry adapted spherical harmonics. In the table of space groups with mirror planes the angle γ_m between the **a**-vector and the mirror plane in the standard setting is given, except for pm where two possible orientations of the mirror plane are considered

Space group	Special positions	Local symmetry	Non-vanishing l	Non-vanishing m	Wyckoff positions
p2	(0, 0), (1/2, 0), (0, 1/2), (1/2, 1/2)	2	$l = 0 \bmod(2)$	$m = 0 \bmod(2)$	a, b, c, d
p3	(0, 0), (1/3, 2/3), (2/3, 1/3)	3	$l = 0 \bmod(3)$	$m = 0 \bmod(3)$	a, b, c
p4	(0, 0), (1/2, 1/2)	4	$l = 0 \bmod(4)$	$m = 0 \bmod(4)$	a, b
	(1/2, 0), (0, 1/2)	2	$l = 0 \bmod(2)$	$m = 0 \bmod(2)$	c, d
p6	(0, 0)	6	$l = 0 \bmod(6)$	$m = 0 \bmod(6)$	a
	(1/3, 2/3), (2/3, 1/3)	3	$l = 0 \bmod(3)$	$m = 0 \bmod(3)$	b
	(0, 1/2), (1/2, 0), (1/2, 1/2)	2	$l = 0 \bmod(2)$	$m = 0 \bmod(2)$	c

Space group	Special positions	Local symmetry	l	m	Symmetry adapted function	Wyckoff positions
pm	(x, 0) (x, 1/2)	$m \perp \mathbf{a}$ $\gamma_m = 0°$	all	$m \geq 0$	$\frac{1}{\sqrt{2}}\{Y_{lm} + (-1)^m Y_{l-m}\}$	a, b
		$m \perp \mathbf{b}$ $\gamma_m = 90°$			$\frac{1}{\sqrt{2}}\{Y_{lm} + Y_{l-m}\}$	
p2mm	(0, 0) (1/2, 1/2) (0, 1/2) (1/2, 0)	2mm $\gamma_m = 0°$	$l = 0$ mod(2)	$m \geq 0$ mod(2)	$\frac{1}{\sqrt{2}}\{Y_{lm} + Y_{l-m}\}$	a, b, c, d
	(x, 0) (x, 1/2)	$m \perp \mathbf{a}$ $\gamma_m = 0°$	all	$m \geq 0$	$\frac{1}{\sqrt{2}}\{Y_{lm} + (-1)^m Y_{l-m}\}$	e, f
	(0, y) (1/2, y)	$m \perp \mathbf{a}$ $\gamma_m = 90°$	all	$m \geq 0$	$\frac{1}{\sqrt{2}}\{Y_{lm} + Y_{l-m}\}$	g, h
p4mm	(0, 0) (1/2, 1/2)	4mm	$l = 0$ mod(4)	$m \geq 0$ mod(4)	$\frac{1}{\sqrt{2}}\{Y_{lm} + Y_{l-m}\}$	a, b
	(1/2, 0) (0, 1/2)	2mm $\gamma_m = 0°$	$l = 0$ mod(2)	$m \geq 0$ mod(2)	$\frac{1}{\sqrt{2}}\{Y_{lm} + Y_{l-m}\}$	c
	(x, 0) (x, 1/2)	$m \perp \mathbf{b}$ $\gamma_m = 0°$	all	$m \geq 0$	$\frac{1}{\sqrt{2}}\{Y_{lm} + (-1)^m Y_{l-m}\}$	d, e
	(x, x)	$m \perp \mathbf{a} + \mathbf{b}$ $\gamma_m = 45°$	all	$m \geq 0$	$\frac{1}{\sqrt{2}}\{Y_{lm} + (-i)^m Y_{l-m}\}$	f
p3m1	(0, 0)	3m	$l = 0$ mod(3)	$m \geq 0$ mod(3)	$\frac{1}{\sqrt{2}}\{Y_{lm} + Y_{l-m}\}$	a
	(1/3, 2/3) (2/3, 1/3)	3	$l = 0$ mod(3)	$m = 0$ mod(3)		b, c
	(x, 2x)	$m \perp \mathbf{a}$ $\gamma_m = 90°$	all	$m \geq 0$	$\frac{1}{\sqrt{2}}\{Y_{lm} + Y_{l-m}\}$	d
	(2x, x)	$m \perp \mathbf{b}$ $\gamma_m = 30°$	all	$m \geq 0$	$\frac{1}{\sqrt{2}}\{Y_{lm} + e^{im\frac{\pi}{3}}(-1)^m Y_{l-m}\}$	d

Table G.1 (*cont.*)

Space group	Special positions	Local symmetry	l	m	Symmetry adapted function	Wyckoff positions
	$(x, -x)$	$m \perp \mathbf{a} + \mathbf{b}$ $\gamma_m = -30°$	all	$m \geq 0$	$\frac{1}{\sqrt{2}}\left\{Y_{lm} + e^{-im\frac{\pi}{3}}(-1)^m Y_{l-m}\right\}$	d
p31m	$(0, 0)$	3m	$l = 0$ mod(3)	$m \geq 0$ mod(3)	$\frac{1}{\sqrt{2}}\left\{Y_{lm} + Y_{l-m}\right\}$	a
	$(1/3, 2/3)$ $(2/3, 1/3)$	3	$l = 0$ mod(3)	$m = 0$ mod(3)		b
	$(x, 0)$	$m \parallel \mathbf{a}$ $\gamma_m = 0°$	all	$m \geq 0$	$\frac{1}{\sqrt{2}}\left\{Y_{lm} + (-1)^m Y_{l-m}\right\}$	c
	$(0, x)$	$m \parallel \mathbf{b}$ $\gamma_m = 120°$			$\frac{1}{\sqrt{2}}\left\{Y_{lm} + e^{-im\frac{2\pi}{3}}(-1)^m Y_{l-m}\right\}$	
	(x, x)	$m \parallel \mathbf{a} + \mathbf{b}$ $\gamma_m = 60°$			$\frac{1}{\sqrt{2}}\left\{Y_{lm} + e^{im\frac{2\pi}{3}}(-1)^m Y_{l-m}\right\}$	
p6mm	$(0, 0)$	6mm $\gamma_m = 45°$	$l = 0$ mod(6)	$m \geq 0$ mod(6)	$\frac{1}{\sqrt{2}}\left\{Y_{lm} + Y_{l-m}\right\}$	a
	$(1/3, 2/3)$ $(2/3, 1/3)$	3m	$l = 0$ mod(3)	$m = 0$ mod(3)	$\frac{1}{\sqrt{2}}\left\{Y_{lm} + Y_{l-m}\right\}$	b
	$(0, 1/2)$ $(1/2, 0)$ $(1/2, 1/2)$	2mm	$l = 0$ mod(2)	$m \geq 0$ mod(2)	$\frac{1}{\sqrt{2}}\left\{Y_{lm} + (-1)^m Y_{l-m}\right\}$	c
	$(x, 0)$	$m \parallel \mathbf{a}$ $\gamma_m = 0°$	all	$m \geq 0$	$\frac{1}{\sqrt{2}}\left\{Y_{lm} + (-1)^m Y_{l-m}\right\}$	d
	$(0, x)$	$m \parallel \mathbf{b}$ $\gamma_m = 120°$	all	$m \geq 0$	$\frac{1}{\sqrt{2}}\left\{Y_{lm} + e^{-im\frac{2\pi}{3}}(-1)^m Y_{l-m}\right\}$	d
	(x, x)	$m \parallel \mathbf{a} + \mathbf{b}$ $\gamma_m = 60°$	all	$m \geq 0$	$\frac{1}{\sqrt{2}}\left\{Y_{lm} + e^{im\frac{2\pi}{3}}(-1)^m Y_{l-m}\right\}$	d
	$(x, 2x)$	$m \perp \mathbf{a}$ $\gamma_m = 90°$	all	$m \geq 0$	$\frac{1}{\sqrt{2}}\left\{Y_{lm} + Y_{l-m}\right\}$	e
	$(2x, x)$	$m \perp \mathbf{b}$ $\gamma_m = 30°$	all	$m \geq 0$	$\frac{1}{\sqrt{2}}\left\{Y_{lm} + e^{im\frac{\pi}{3}}(-1)^m Y_{l-m}\right\}$	e
	$(x, -x)$	$m \perp \mathbf{a} + \mathbf{b}$ $\gamma_m = -30°$	all	$m \geq 0$	$\frac{1}{\sqrt{2}}\left\{Y_{lm} + e^{-im\frac{\pi}{3}}(-1)^m Y_{l-m}\right\}$	e
pg	(x, y)	1	all	all		a
p2gg	$(0, 0)$ $(1/2, 1/2)$ $(0, 1/2)$ $(1/2, 0)$	2	$l = 0$ mod(2)	$m = 0$ mod(2)		a, b
p2mg	$(0, 0)$ $(1/2, 1/2)$ $(0, 1/2)$ $(1/2, 0)$	2	$l = 0$ mod(2)	$m = 0$ mod(2)		a, b
					$\frac{1}{\sqrt{2}}\left\{Y_{lm} + (-1)^m Y_{l-m}\right\}$	c
	$(¼, y)$ $(¾, -y)$	m $\gamma_m = 0°$	all	$m \geq 0$		

Table G.1 (*cont.*)

Space group	Special positions	Local symmetry	l	m	Symmetry adapted function	Wyckoff positions
p4gm	(0, 0)	4	$l = 0$	$m = 0$		a
	(1/2, 1/2)		mod(4)	mod(4)		
	(0, ½)	2mm	$l = 0$	$m \geq 0$	$\frac{1}{\sqrt{2}}\{Y_{lm} + Y_{l-m}\}$	b
	(½, 0)		mod(2)	mod(2)		
	$(x, x+½)$	m	$l = 0$	$m = 0$	$\frac{1}{\sqrt{2}}\{Y_{lm} + (-i)^m Y_{l-m}\}$	c
	$(x+½, x)$	m $\perp$ **a** – **b**,	mod(2)	mod(2)		
		$\gamma_m = 45°$				
	$(x, ½-x)$	$\perp$ **a** + **b**			$\frac{1}{\sqrt{2}}\{Y_{lm} + (i)^m Y_{l-m}\}$	
	$(x+½, -x)$	$\gamma_m = -45°$				
cm	(x, x)	m $\perp$ **a′** + **b′**	$l = 0$	$m = 0$	$\frac{1}{\sqrt{2}}\{Y_{lm} + (-1)^m Y_{l-m}\}$	
		$\alpha_m = 0°$	mod(2)	mod(2)		
c2mm	(0, 0)	2mm	$l = 0$	$m \geq 0$	$\frac{1}{\sqrt{2}}\{Y_{lm} + (-1)^m Y_{l-m}\}$	
	(1/2, 1/2)	$\alpha_m = 0°$	mod(2)	mod(2)		
	(1/2, 0)	2	$l = 0$	$m = 0$		
	(0, 1/2)		mod(2)	mod(2)		

Note: For the symmetry groups cm and c2mm the primitive setting of the unit cell is chosen. In LEED I(V) calculations this is more efficient than the standard centred setting. The translation vectors are $\mathbf{a'} = (\mathbf{a} + \mathbf{b})/2$ and $\mathbf{b'} = (\mathbf{a} - \mathbf{b})/2$. For the group cm the mirror line is chosen to be parallel to $\mathbf{a'} + \mathbf{b'}$ and the angle to the x-axis is $\alpha_m = 0°$. For the group c2mm the mirror lines are parallel to $\mathbf{a'} + \mathbf{b'}$ and $\mathbf{a'} - \mathbf{b'}$ and the x-axis is assumed to be parallel to $\mathbf{a'} + \mathbf{b'}$.

(repeated here from Eqs. (5.53) and (5.54)), depending on the local symmetry of the atom at position $\mathbf{r}^j$:

$$a_{lm}^j(\mathbf{k}_{g'}) = (n_{g'})^{1/2} \sum_{s=1}^{n_{g'}} e^{-i\mathbf{k}_{g_s'} \mathbf{r}^j} Y_{lm}(\mathbf{k}_{g_s'}), \tag{G.1}$$

$$b_{lm}^j(\mathbf{k}_g) = (n_g)^{1/2} t_l^j(E) \sum_{s=1}^{n_g} e^{i\mathbf{k}_{g_s} \mathbf{r}^j} (-1)^m Y_{l-m}(\mathbf{k}_{g_s}). \tag{G.2}$$

The sum over s in Eqs. (G.1) and (G.2) is the sum over all beams symmetrically equivalent to beam $\mathbf{k}_{g'}$ or $\mathbf{k}_g$, respectively, that is, the beam star, and is invariant under a symmetry operation which transforms the beams and the position vector $\mathbf{r}^j$ to $\mathbf{r}^{j'}$. For an n-fold axis with rotation angle $\Delta\varphi^{jj'} = 2\pi/n$ the rotated wave field at atom j' is given by

$$a_{lm}^{j'}(\mathbf{k}_g) = a_{lm}^j(\mathbf{k}_g) e^{im\Delta\varphi^{jj'}}, \tag{G.3}$$

which implies $m = 0 \bmod(n)$ as well as $l = 0 \bmod(n)$ due to the condition $|m| \leq l$. The azimuth of the incoming and outgoing plane waves and the orientation of the unit cell and the mirror planes are illustrated in Figure G.1. We use $\alpha_m = \gamma_m + \gamma_a$ in the following.

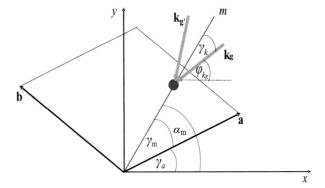

Figure G.1 Illustration of the angles used in Eqs. (G.3)–(G.5). A Cartesian coordinate system is required for the definition of the angles in the spherical harmonic expansion. A hexagonal unit cell is shown with an angle γ_a of the **a**-vector to the x-axis. The azimuthal angle of the incident beam $\mathbf{k_g}$ with the x-axis is $\varphi_{\mathbf{k_g}}$. The orientations of the mirror or glide planes are usually defined with respect to the translation vectors **a** and **b** of the surface unit cell. The angle of the mirror plane with the **a**-vector is γ_m. The atom at position $\mathbf{r}^j$ is marked in grey in the figure; the local symmetry is m (normal to the **b**-vector). The angle between the incoming beam $\mathbf{k_g}$ and the mirror plane is $\gamma_k = \gamma_m + \gamma_a - \varphi_{\mathbf{k_g}}$. The azimuthal angle of the symmetrically equivalent incident beam $\mathbf{k_{g'}}$ with the x-axis is $\varphi_{\mathbf{k_{g'}}} = 2(\gamma_m + \gamma_a) - \varphi_{\mathbf{k_g}} = 2\alpha_m - \varphi_{\mathbf{k_g}}$ and $\varphi_{\mathbf{k_{g'}}} = \alpha_m + \gamma_{\mathbf{k_{g'}}}$.

The symmetry which applies to the beam set is the local symmetry of the atom. For an atom on a mirror plane the incoming beam star consists of two equivalent beams or only one beam in the mirror plane. The azimuthal angles of the two equivalent beams are $\varphi_{\mathbf{k_g}} = \alpha_m - \gamma_{\mathbf{k_g}}$ and $\varphi_{\mathbf{k_{g'}}} = \alpha_m + \gamma_{\mathbf{k_{g'}}}$. We can multiply a_{lm}^j and $a_{lm}^{j'}$ by $e^{im\,\alpha_m}$ and obtain

$$a_{lm}^{j'}\left(\mathbf{k_g}\right)e^{im\,\alpha_m} = a_{l-m}^{j}\left(\mathbf{k_g}\right)(-1)^m e^{im\,\alpha_m}. \qquad (G.4)$$

This implies the index selection and defines the symmetry adapted function:

$$Y_{lm}^S = \frac{1}{\sqrt{2}}\{Y_{lm} + (-1)^m Y_{l-m}\}, m \geq 0. \qquad (G.5)$$

Use can be made of the symmetry properties of the spherical harmonics:

$$
\begin{aligned}
Y_{lm}(\vartheta, \varphi) &= \frac{1}{\sqrt{2\pi}}\theta_{lm}(\cos\vartheta)e^{im\varphi}, \\
\theta_{l-m}(\cos\vartheta) &= (-1)^m \theta_{lm}(\cos\vartheta), \\
Y_{lm}^*(\vartheta, \varphi) &= (-1)^m Y_{l-m}(\vartheta, \varphi), \\
Y_{lm}(\pi - \vartheta, \pi + \varphi) &= (-1)^l Y_{lm}(\vartheta, \varphi), \\
Y_{lm}(\vartheta, -\varphi) &= (-1)^m Y_{l-m}(\vartheta, \varphi).
\end{aligned}
\qquad (G.6)
$$

Here θ_{lm} are normalised Legendre polynomials.

Glide planes need only be considered in the four groups pg, p2gg, p2mg and p4gm. The other groups are symmorphic: these are groups where at least one point is invariant and the glide planes do not generate symmetry elements. For glide planes an additional phase factor occurs due to the shift vector **w** of the symmetrically equivalent position. This applies for all atom positions. For an atom on a glide plane the local symmetry is 1, the multiplicity is 2, the symmetry in reciprocal space is m and the wave field incident on the atom has symmetry m. For atoms on glide planes the use of the symmetry adapted function, Eq. (G.5), would be possible except that a phase factor -1 occurs for h or k = odd. This means that a minus sign occurs in Eq. (G.5) for the sum of spherical harmonics. For beams with h or k = even the phase factor becomes 1 and Eq. (G.6) can be applied since for these beams the phase factor becomes 1.

We do not use this possibility because the position on the glide plane is not a special position, the local symmetry is 1 and a shift away from the glide plane is allowed in a refinement procedure. For atoms in a special position this is not allowed as it would result in split positions. Therefore, for the plane group pg no index selection is included in Table G.1 and for the other three groups only special positions on rotational axes or mirror planes are used.

For incoming or outgoing beams, we may ignore whether the mirror plane or the rotation axis goes through the origin or not. Only the orientation of the mirror plane counts. For rotation axes and mirror planes the wave field must have the local symmetry of the point $\mathbf{r}^j$. As described in Eqs. (G.3)–(G.5), symmetry adapted spherical harmonics can be used for atoms in special positions in Eqs. (G.1) and (G.2) and in the vectors $\mathbf{a}^j$, $\mathbf{b}^j$ and $\mathbf{Z}^j$ in Eqs. (5.53)–(5.55). For all atoms on mirror planes, only positive values of m and $m = 0$ mod(2) are needed. For atoms in general positions with local symmetry p1 no symmetry reduction is possible.

The angles of the beams $\varphi_{\mathbf{k}_g}$ refer to the Cartesian coordinate system. The angles γ_m of the mirror lines to the **a**-axis can have the following values: $\gamma_m = 0°$, $\pm90°$, $\pm45°$, $\pm30°$, $\pm60°$, $\pm120°$, depending on the 2-D space group (see Figure 2.12).

Appendix H Symmetry Use in Tensor LEED

As described in Section 5.4.1, in Tensor LEED at first a reference structure is calculated with full multiple scattering. When symmetries are not considered, this is done by solving Eq. (5.77), which is a slightly different form of Eq. (5.50). The atomic scattering matrix $\mathbf{t}^j$ is removed from $\mathbf{b}^j$ in Eq. (5.55) and appears now as a separate factor between the vectors $\mathbf{a}^j$ and $\mathbf{Z}^j$ from Eqs. (5.74) and (5.76). This does not change the symmetry properties, so that the symmetrised vectors $\mathbf{a}^{S,j}$ and $\mathbf{Z}^{S,j}$, in which only the symmetrically independent atoms occur, can be used to calculate the reflection and transmission matrices:

$$M_{\mathbf{g}'\mathbf{g}}^{\pm\pm} = -\frac{16\pi^2 i m_e}{A k_{\mathbf{g}z}^+ \hbar^2} \sum_{j_0} n_{j_0} \mathbf{a}^{S,j_0}(\mathbf{k}_{\mathbf{g}'}) \mathbf{t}^{j_0} \mathbf{Z}^{S,j_0}(\mathbf{k}_{\mathbf{g}}) + \delta_{\mathbf{g}'\mathbf{g}}\delta_{\pm\pm}. \tag{H.1}$$

The vectors $\mathbf{a}^{S,j_0}(\mathbf{k}_{\mathbf{g}})$, $\mathbf{b}^{S,j_0}(\mathbf{k}_{\mathbf{g}})$ and $\mathbf{Z}^{S,j_0}(\mathbf{k}_{\mathbf{g}})$ are defined by

$$\mathbf{a}^{S,j_0}(\mathbf{k}_{\mathbf{g}}) = \sum_{j'}^{n_{j_0}} \mathbf{a}^{j'}(\mathbf{k}_{\mathbf{g}}) \mathbf{D}^{j'j_0}, \tag{H.2}$$

$$\mathbf{b}^{S,j_0}(\mathbf{k}_{\mathbf{g}}) = \sum_{j'}^{n_{j_0}} \mathbf{D}^{j_0 j'} \mathbf{b}^{j'}(\mathbf{k}_{\mathbf{g}}), \tag{H.3}$$

$$\mathbf{Z}^{S,j_0}(\mathbf{k}_{\mathbf{g}}) = \sum_{j'}^{n_{j_0}} \mathbf{D}^{j_0 j'} \mathbf{Z}^{j'}(\mathbf{k}_{\mathbf{g}}). \tag{H.4}$$

The sum over j_0 in Eq. (H.1) is taken over the symmetrically independent atoms in the unit cell, while n_{j_0} is the multiplicity of atom j_0. The sum over j' in Eqs. (H.2)–(H.4) is taken over the atoms which are symmetrically equivalent to atom j_0, including atom j_0. The symmetry operators $\mathbf{D}^{j_0 j'}$ are defined in Eqs. (5.59) and (5.60) and Appendix G. Equation (H.1) is valid because $\mathbf{t}^{j_0}$ is a diagonal matrix and invariant with respect to a symmetry operation. This is no longer the case when the atom j_0 is shifted by $\delta \mathbf{r}^{j_0}$; then the change in the t-matrix is defined as

$$\delta \mathbf{t}^{j_0}(-\mathbf{k}_{\mathbf{g}}', \mathbf{k}_{\mathbf{g}}) = e^{-i\mathbf{k}_{\mathbf{g}}'\delta \mathbf{r}^{j_0}} \mathbf{t}^{j_0}(\mathbf{k}_{\mathbf{g}}', \mathbf{k}_{\mathbf{g}}) e^{i\mathbf{k}_{\mathbf{g}}\delta \mathbf{r}^{j_0}} - \mathbf{t}^{j_0}(\mathbf{k}_{\mathbf{g}}', \mathbf{k}_{\mathbf{g}}), \tag{H.5}$$

and the change in the reflection matrix is

$$\delta M_{\mathbf{g'g}} = M'_{\mathbf{g'g}} - M_{\mathbf{g'g}}.$$

Here $M'_{\mathbf{g'g}}$ is calculated with the shifted atom (we henceforth omit the $\pm$ signs for convenience). Since δt^{j_0} is not invariant under a symmetry operation, we need to define a symmetry operator which transforms δt^{j_0} to the symmetrically equivalent $\delta t^{j'}$ by shifting from atom j_0 to atom j' and back. We use the symbol $\mathbf{S}$ to distinguish this new operator from the other operators $\mathbf{D}$:

$$\delta t^{j_0} = \mathbf{S}^{j_0 j'} \delta t^{j'} \mathbf{S}^{j' j_0}. \tag{H.6}$$

We thus obtain for the difference in the reflection and transmission matrices:

$$\delta M_{\mathbf{g'g}} = -\frac{16\pi^2 i m_e}{A k_{\mathrm{gz}}^+ \hbar^2} \sum_{j_0} \sum_{j'} \mathbf{a}^{j'}(\mathbf{k_{g'}}) \mathbf{D}^{j' j_0} \mathbf{S}^{j_0 j'} \delta t^{j'} \mathbf{S}^{j' j_0} \mathbf{D}^{j_0 j'} \mathbf{Z}^{j'}(\mathbf{k_g}) - M_{\mathbf{g'g}}$$

$$= -\frac{16\pi^2 i m_e}{A k_{\mathrm{gz}}^+ \hbar^2} \sum_{j_0} \sum_{j'} \mathbf{a}^{S, j_0}(\mathbf{k_{g'}}) \mathbf{S}^{j_0 j'} \delta t^{j'} \mathbf{S}^{j' j_0} \mathbf{Z}^{S, j_0}(\mathbf{k_g}) - M_{\mathbf{g'g}}$$

$$= -\frac{16\pi^2 i m_e}{A k_{\mathrm{gz}}^+ \hbar^2} \sum_{lm, l'm'} \sum_{j_0} \sum_{j'} a_{lm}^{l'}(\mathbf{k_{g'}}) D_{lm}^{j' j_0} S_{lm}^{j_0 j'} \, \delta t_{lm, l'm'}^{j'}$$

$$\cdot S_{l'm'}^{j' j_0} D_{l'm'}^{j_0 j'} Z_{l'm'}^{j'}(\mathbf{k_g}) - t^{j_0}. \tag{H.7}$$

The symmetry operators are vectors in the angular momentum expansion and we need to define $\mathbf{S}^{j_0 j'}$. The shift of the atomic t-matrix in terms of plane wave propagation is:

$$\delta t^j(\mathbf{k_{g'}}, \mathbf{k_g}) = \exp\left(-i k_{g'} \delta \mathbf{r}^j\right) \mathbf{t}^j(\mathbf{k_{g'}}, \mathbf{k_g}) e^{i k_g \delta \mathbf{r}^j} - \mathbf{t}^j(\mathbf{k_{g'}}, \mathbf{k_g}). \tag{H.8}$$

The transformation of a t-matrix from spherical to plane wave representation is given by:

$$t(\mathbf{k'}, \mathbf{k}) = 8\pi^2 \sum_{lm} \sum_{l'm'} Y_{lm}(\mathbf{k'}) t_{lm, l'm'} (-1)^{m'} Y_{l'-m'}(\mathbf{k}). \tag{H.9}$$

We need the reciprocal transformation to obtain $\delta t_{lm, l'm'}^j$:

$$\delta t_{lm, l'm'}^j(\delta \mathbf{r}^j) = \frac{1}{8\pi^2} \int \int (-1)^{l+m} Y_{l-m}(\mathbf{k_{g'}}) \delta t^j(\mathbf{k_{g'}}, \mathbf{k_g}) Y_{l'm'}(\mathbf{k_g}) d\Omega_{\mathbf{k_g}} d\Omega_{\mathbf{k_{g'}}}. \tag{H.10}$$

Inserting Eq. (H.9) into Eq. (H.10) gives

$$\delta t_{lm, l'm'}^j(\delta \mathbf{r}^j) = \frac{1}{8\pi^2} \int \int (-1)^{l+m} Y_{l-m}(\mathbf{k_{g'}})$$

$$\cdot \left\{ \exp\left(-i k_{g'}' \delta \mathbf{r}^j\right) \mathbf{t}^j\left(\mathbf{k_g'}, \mathbf{k_g}\right) \exp\left(i k_g \delta \mathbf{r}^j\right) \right\}$$

$$\cdot Y_{l'm'}(\mathbf{k_g}) d\Omega_{\mathbf{k_g}} d\Omega_{\mathbf{k_{g'}}} - t_l^j(E). \tag{H.11}$$

We also insert the spherical wave expansion of the phase factors due to shifting atom j:

$$\exp\left(i\mathbf{k}\delta\mathbf{r}^{j'}\right) = \sum_{lm} 4\pi i^l j_l(k\delta r) Y_{lm}(\Omega_{\delta r})(-1)^m Y_{l-m}(\Omega_k), \tag{H.12}$$

and obtain the t-matrix of the shifted atom in the spherical wave expansion:

$$\delta t_{lm,l'm'}\left(\delta\mathbf{r}^j\right) = \sum_{l_2=0}^{l_s} 4\pi i^{l_2} j_{l_2}\left(k\delta r^j\right) \sum_{m_2=-l_2}^{l_2} Y_{l_2 m_2}\left(\delta\mathbf{r}^j\right)(-1)^{m_2+m}$$

$$\cdot \sum_{l_3=0}^{l_{max}} \sum_{m_3=-l_3}^{l_3} \sum_{l_5=0}^{l_s} \sum_{m_5=-l_5}^{l_5} B(l_2 - m_2, l_3 m_3, l - m) t_{l_3}^j (-1)^{m_3+m_5}$$

$$\cdot B(l_3 - m_3, l_5 m_5, l'm') 4\pi i^{l_5} j_{l_5}\left(k\delta r^j\right) Y_{l_5 m_5}\left(\delta\mathbf{r}^j\right) - t_{lm,l'm'}^j \delta_{ll'}\delta_{mm'}. \tag{H.13}$$

The upper limit l_s for l_2 and l_5 is set by the value where the spherical Bessel function $j_l(k\delta r^j)$ becomes negligibly small, usually $l_s \leq 3$ or 4. We now can define a symmetry operator $\mathbf{S}^{j_0 j'}$ which transforms the shift vector $\delta\mathbf{r}^{j_0}$ to $\delta\mathbf{r}^{j'}$.

$$Y_{lm}\left(\vartheta_{r^{j'}}, \varphi_{r^{j'}}\right) = S_m^{j' j_0} Y_{lm}(\vartheta_{r^{j_0}}, \varphi_{r^{j_0}}). \tag{H.14}$$

It should be noted that the operator $S_m^{j' j_0}$ transforms the shift vector $\delta\mathbf{r}^{j_0}$ while $D_m^{j' j_0}$ applies to $a_{lm}^j(\mathbf{k_{g'}})$ or $Z_{lm}^j(\mathbf{k_g})$ and transforms the vectors $\mathbf{k_g}$ or $\mathbf{k_{g'}}$ in the opposite direction.

We finally obtain for the difference in the transmission and reflection matrices:

$$\delta M_{g'g} = -\frac{16\pi^2 i m_e}{A k_{gz}^+ \hbar^2} \sum_{j_0}^{n_{j_0}} \sum_{j'=j_0} \sum_{lm} a_{lm}^{j'}(\mathbf{k_{g'}}) D_{lm}^{j' j_0} \sum_{l_2 m_2} 4\pi i^{l_2} j_{l_2}\left(k\delta r^{j_0}\right) S_{m_2}^{j_0 j'} Y_{l_2 m_2}\left(\delta\mathbf{r}^{j'}\right)$$

$$\cdot \sum_{l'm'} \sum_{l_3 m_3} \sum_{l_5 m_5} \left\{(-1)^{l_2} B(l_2 - m_2, l_3 m_3, l - m) t_{l_3}^j (-1)^{m_3} B(l_3 - m_3, l_5 m_5, l'm')\right.$$

$$\left. \cdot 4\pi i^{l_5} j_{l_5}\left(k\delta r^{j_0}\right) S_{m_5}^{j' j_0} Y_{l_5 m_5}\left(\delta\mathbf{r}^{j_0}\right) - t_l^{j_0}\right\} D_{l'm'}^{j_0 j'} Z_{m'}^{j'}(\mathbf{k_g}) - M_{g'g}$$

$$= -\frac{16\pi^2 i m_e}{A k_{gz}^+ \hbar^2} \sum_{j_0} n_{j_0} a^{S,j_0}(\mathbf{k_{g'}}) \delta t^S\left(\delta\mathbf{r}^{j_0}\right) Z^{S,j_0}(\mathbf{k_g}) - M_{g'g}. \tag{H.15}$$

In the last line we have used the symmetrized difference from the original scattering matrix of the shifted atom:

$$\delta t^S\left(\delta\mathbf{r}^{j_0}\right) = \sum_{j'=j_0}^{n_{j_0}} \mathbf{S}^{j_0 j'} \delta t^{j'}\left(\delta\mathbf{r}^{j'}\right) \mathbf{S}^{j' j_0}, \tag{H.16}$$

where $\delta t^{j'}\left(\delta\mathbf{r}^{j'}\right)$ is defined in Eq. (H.11); $\delta t^S(\delta\mathbf{r}^{j_0})$ is invariant with respect to a symmetry operation.

The calculation of the difference in the reflection and transmission matrices requires only the symmetrically independent atoms, while the size of the matrices in the spherical wave expansion is reduced by use of symmetry adapted spherical harmonics, as described in Appendix G. The use of symmetries speeds up the calculation by a large factor and helps to make Tensor LEED a highly efficient procedure in surface structure refinement.

Appendix I Atomic Displacement Parameters

In the kinematic theory of diffraction, the influence of static or dynamic atomic displacements on the Bragg intensities is usually described by a Debye–Waller factor $T(\mathbf{q})$ in the structure factor. If c_v is the occupation factor at site v among N sites in the unit cell, then

$$F(\mathbf{q}) = \sum_{v=1}^{N} c_v f_v(\mathbf{q}) T(\mathbf{q}) e^{2\pi i \mathbf{q}(\mathbf{r}_v - \mathbf{u}_v)}. \tag{I.1}$$

$T(\mathbf{q})$ is given by the Fourier transform of the probability density function (PDF) $p(\mathbf{u})$:

$$T(\mathbf{q}) = \int p(\mathbf{u}) \exp(i\mathbf{q} \cdot \mathbf{u}) d\mathbf{q}. \tag{I.2}$$

When the Debye–Waller factor is experimentally determined, the corresponding PDF is obtained by the inverse Fourier transform of the Debye–Waller factor:

$$p(\mathbf{u}) = (2\pi)^{-3} \int T(\mathbf{q}) \exp(-i\mathbf{q} \cdot \mathbf{u}) d\mathbf{q}. \tag{I.3}$$

We can consider different cases for the PDF:

- Harmonic interatomic forces, assuming Gaussian distribution functions:
  - isotropic displacements;
  - anisotropic displacements;
- Anharmonic interatomic forces, assuming general forms of the distribution functions.

In all cases the symmetry of the distribution function is limited by the local symmetry of the atom. Different parameters are used in X-ray and neutron diffraction to describe the Debye–Waller factor. We describe here briefly the different parameters and their relation following the recommendations of the IUCr committee on Atomic Displacement Parameter Nomenclature published in K. N. Trueblood et al. [5.79].

I.1   Isotropic Displacements and Gaussian Distributions

The Debye–Waller factor $T(\mathbf{q}_\nu)$ for atom ν is:

$$T_\nu(\mathbf{q}) = \exp\left(-\frac{1}{2}\langle(\mathbf{q}\cdot\mathbf{u}_\nu)\rangle^2\right).\tag{I.4}$$

For isotropic displacements, $\mathbf{u}_\nu$ is independent of the direction and we obtain $T_\nu(\mathbf{q})$ with $|q| = 2\pi\frac{\sin\vartheta}{\lambda}$, where ϑ is the Bragg angle which is defined as the angle between the incident beam and the lattice plane:

$$\begin{aligned}T_\nu(\mathbf{q}) &= \exp\left(-8\pi^2|q|^2\langle\mathbf{u}_\nu\rangle^2\right)\\ &= \exp\left(-8\pi^2\frac{\sin^2\vartheta}{\lambda^2}\langle\mathbf{u}_\nu\rangle^2\right)\\ &= \exp\left(-B_\nu\frac{\sin^2\vartheta}{\lambda^2}\right),\end{aligned}\tag{I.5}$$

$$B_\nu = 8\pi^2\langle\mathbf{u}\rangle^2.\tag{I.6}$$

The average of the mean square displacement is

$$\langle\mathbf{u}\rangle^2 = \frac{1}{2}|u_x|^2 + \frac{1}{2}|u_y|^2 + \frac{1}{2}|u_z|^2 = \frac{3}{2}|u|^2.\tag{I.7}$$

The so-called B-factor is the most used form for isotropic displacements in X-ray diffraction, often denoted as B_{iso}, and is found in many structure analyses.

In LEED multiple scattering programs frequently the Debye temperature Θ_D is used. The relation to the mean square displacement in the high temperature limit is:

$$\langle u^2\rangle = \frac{3\hbar^2 T}{2m_a k_B \Theta_D^2},\tag{I.8}$$

where m_a is the mass of the atom. The Debye temperature Θ_D is a refinement parameter for individual atoms or layers and is completely equivalent to the B-factor in X-ray analyses. It does not correspond directly to the Debye model, but is practical for simple substrates for which tabulated values for Θ_D can be found. Most LEED programs require or have the option to read in Debye temperatures for atoms and the routines to calculate temperature dependent phase shifts mostly use Θ_D as input parameter.

I.2   Anisotropic Displacements and Gaussian Distributions

The atomic displacement vector $\mathbf{u}_\nu$ can be described in different coordinate systems and all are used in the literature. There are three main ways to describe the vector $\mathbf{u}$:

- Fractional coordinates Δx, Δy, Δz, referring to the translation vectors $\mathbf{a}$, $\mathbf{b}$, $\mathbf{c}$. The fractional coordinates are dimensionless.
- Coordinates $\Delta \xi$, $\Delta \eta$, $\Delta \zeta$, having the dimension of length and referring to the dimensionless basis vectors, having unit lengths.
- Cartesian coordinates referring to the basis vectors $\mathbf{e}_1$, $\mathbf{e}_2$, $\mathbf{e}_3$.

We briefly describe the relations between the parameters and refer the reader to the more detailed description in K. N. Trueblood et al. [5.79] or U. Schmueli [5.86]. In the following we omit the atom index ν but still assume individual displacement parameters for each inequivalent atom. Most common is to refer the diffraction vector to the reciprocal lattice $\mathbf{a}^*$, $\mathbf{b}^*$, $\mathbf{c}^*$ and the displacement vector to the basis of the lattice $\mathbf{a}$, $\mathbf{b}$, $\mathbf{c}$.

The International Union of Crystallography recommends distinguishing covariant and contravariant vectors by use of subscripts versus superscripts when indices are used. The reciprocal lattice is the contravariant coordinate system and the real lattice the covariant system. In K. N. Trueblood et al. [5.79] the reciprocal lattice vectors are therefore given by $\mathbf{a}^i$ with lengths a^i and the vectors of the real lattice are $\mathbf{a}_j$ with lengths a_j. The same convention applies to coefficients, matrices and tensors. We use this convention here to be compatible with K. N. Trueblood et al. [5.79] although we have not used it in Section 5.5, but we also add the superscript * to the reciprocal vectors to avoid confusion, that is, we use a^{i*} for the reciprocal vectors.

I.2.1 Anisotropic Displacement Parameter β^{ij}

Using

$$\mathbf{q} = 2\pi(h\mathbf{a}^* + k\mathbf{b}^* + l\mathbf{c}^*) = 2\pi \sum_{1=1}^{3} h_i \mathbf{a}^{i*} \tag{I.9}$$

and

$$\mathbf{u} = \Delta x \mathbf{a} + \Delta y \mathbf{b} + \Delta z \mathbf{c} = \sum_{i=1}^{3} \Delta x^i \mathbf{a}_i, \tag{I.10}$$

we have

$$\mathbf{q} \cdot \mathbf{u} = \sum_{i=1}^{3} \Delta x^i h_i. \tag{I.11}$$

The quantities Δx^i are dimensionless and measure the displacement in units of the translation vectors. The Debye–Waller factor becomes:

$$T(\mathbf{q}) = \exp\left(-2\pi^2 \sum_{i=1}^{3} \sum_{j=1}^{3} h_i \langle \Delta x^i \Delta x^j \rangle h_j \right) = \exp\left(-\sum_{i=1}^{3} \sum_{j=1}^{3} h_i \beta^{ij} h_j \right). \tag{I.12}$$

The tensor components β^{ij} are contravariant and cannot be used to determine the magnitude of the displacements directly from

$$\beta^{ij} = 2\pi^2 \langle \Delta x^i \Delta x^j \rangle, \tag{I.13}$$

so we need the comparison with the parameters U^{ij} where the length of the displacements is used.

I.2.2   Anisotropic Displacement Parameter U^{ij}

Using

$$\mathbf{u} = \Delta\xi a^*\mathbf{a} + \Delta\eta b^*\mathbf{b} + \Delta\xi c^*\mathbf{c} = \sum_{i=1}^{3} \Delta\xi^i a^{i*}\mathbf{a}_i, \tag{I.14}$$

we have

$$T = \exp\left(-2\pi^2 \sum_{i=1}^{3}\sum_{j=1}^{3} h_i a^{i*} \langle \Delta\xi^i \Delta\xi^j \rangle h_j a^{j*}\right) = \exp\left(-2\pi^2 \sum_{i=1}^{3}\sum_{j=1}^{3} h_i a^{i*} U^{ij} h_j a^{j*}\right). \tag{I.15}$$

Comparison with Eq. (I.12) leads to:

$$U^{ij} = \langle \Delta\xi^i \Delta\xi^j \rangle = \frac{\beta^{ij}}{2\pi^2 a^{i*} a^{j*}}. \tag{I.16}$$

I.2.3   Anisotropic Displacement Parameters in Cartesian Coordinates U_{ij}^C

The transformation to a Cartesian basis is described in Section 2.1.7. Using this we write in Cartesian coordinates both the diffraction vector

$$\mathbf{q} = h_1^C \mathbf{e}_1 + h_2^C \mathbf{e}_2 + h_3^C \mathbf{e}_3 = \sum_{i=1}^{3} h_i^C \mathbf{e}_i \tag{I.17}$$

and the displacement vector

$$\mathbf{u} = \Delta\xi_1^C \mathbf{e}_1 + \Delta\xi_2^C \mathbf{e}_2 + \Delta\xi_3^C \mathbf{e}_3 = \sum_{i=1}^{3} \Delta\xi_i^C \mathbf{e}_i. \tag{I.18}$$

We now obtain for the Debye–Waller factor:

$$T = \exp\left(-2\pi^2 \sum_{i=1}^{3}\sum_{j=1}^{3} h_i^C \langle \Delta\xi_i^C \Delta\xi_j^C \rangle h_j^C\right) = \exp\left(-2\pi^2 \sum_{i=1}^{3}\sum_{j=1}^{3} h_i^C U_{ij}^C h_j^C\right). \tag{I.19}$$

I.2.4   Relation between U^{ij} and β^{ij}

The relation between β^{ij} and U^{ij} is given in Eq. (I.16). To express β_{ij} in Cartesian coordinates we have to use the transformation matrix for the vectors $\mathbf{a}_i$ to $\mathbf{e}_i$. We use the following approach, where $\mathbf{e}_1$ is parallel to $\mathbf{a}_1$:

$$\mathbf{e}_1 = \mathbf{a}_1/a, \quad \mathbf{e}_2 = \mathbf{e}_3 \times \mathbf{e}_1, \quad \text{and} \quad \mathbf{e}_3 = \mathbf{c}^*/c. \tag{I.20}$$

With

$$\mathbf{u} = \sum_{i=1}^{3} \Delta\xi_i^C \mathbf{e}_i = \sum_{i=1}^{3} \Delta x_i \mathbf{a}_i, \tag{I.21}$$

we obtain

$$\begin{pmatrix} \Delta\xi_1^C \\ \Delta\xi_2^C \\ \Delta\xi_3^C \end{pmatrix} = \begin{pmatrix} \mathbf{e}_1 \cdot \mathbf{a} & \mathbf{e}_1 \cdot \mathbf{b} & \mathbf{e}_1 \cdot \mathbf{c} \\ \mathbf{e}_2 \cdot \mathbf{a} & \mathbf{e}_2 \cdot \mathbf{b} & \mathbf{e}_2 \cdot \mathbf{c} \\ \mathbf{e}_3 \cdot \mathbf{a} & \mathbf{e}_3 \cdot \mathbf{b} & \mathbf{e}_3 \cdot \mathbf{c} \end{pmatrix} \begin{pmatrix} \Delta x \\ \Delta y \\ \Delta z \end{pmatrix}, \tag{I.22}$$

that is,

$$\Delta\xi_i^C = \sum_{j=1}^{3} A_{ij}\Delta x_j, \quad A_{ij} = \mathbf{e}_i \cdot \mathbf{a}_j. \tag{I.23}$$

The displacements U_{ij}^C in Cartesian coordinates can now be expressed in terms of β_{ij}:

$$U_{ij}^C = \left(2\pi^2\right)^{-1} \sum_{m=1}^{3} \sum_{n=1}^{3} A_{im}A_{jn}\beta_{mn}.$$

I.2.5   Relation between U^{ij} and U_{ij}^C

In a similar way, we obtain the transformation of U^{ij} to U_{ij}^C [5.79]:

$$U_{ij}^C = \left(2\pi^2\right)^{-1} \sum_{m=1}^{3} \sum_{n=1}^{3} D_{im}D_{jn}U^{mn} \tag{I.24}$$

where

$$D_{ij} = \left(\mathbf{e}_i \cdot \mathbf{a}_j\right)a^{j*}. \tag{I.25}$$

I.2.6   Equivalent Isotropic Displacement Parameters

It is frequently necessary to use isotropic displacement parameters in a surface structure analysis as initial data for the structure refinement, for example with adsorbed molecules where the coordinates are taken from a database of non-adsorbed molecules and anisotropic parameters β^{ij} or U^{ij} are given. The six parameters of anisotropic displacement must be replaced by a single parameter. There are several ways to obtain the isotropic equivalent. In the case of Cartesian coordinates, we can simply use the average:

$$U_{eq} = \frac{1}{3}\left(U_{11}^C + U_{22}^C + U_{33}^C\right). \tag{I.26}$$

With oblique basis vectors and displacement parameters U^{ij} or β^{ij}, Eq. (I.26) cannot be applied. We must use

$$U_{eq} = \frac{1}{3}\left\langle |\mathbf{u}|^2 \right\rangle = \frac{1}{3}\langle \mathbf{u} \cdot \mathbf{u} \rangle, \tag{I.27}$$

which gives the relation to anisotropic parameters from Eqs. (I.13) and (I.15):

$$U_{eq} = \frac{1}{3}\left(2\pi^2\right)^{-1}\sum_{i=1}^{3}\sum_{j=1}^{3}\Delta x^i \Delta x^j \left(\mathbf{a}_i \cdot \mathbf{a}_j\right)$$

$$= \frac{1}{3}\left(2\pi^2\right)^{-1}\sum_{i=1}^{3}\sum_{j=1}^{3}\beta^{ij}\left(\mathbf{a}_i \cdot \mathbf{a}_j\right)$$

$$= \frac{1}{3}\sum_{i=1}^{3}\sum_{j=1}^{3}U^{ij}a^{i*}a^{j*}\left(\mathbf{a}_i \cdot \mathbf{a}_j\right). \tag{I.28}$$

This is not the only way to determine the equivalent isotropic parameter: further details can be found in K. N. Trueblood et al. [5.79] and E. Prince [5.97]. The graphical representation of the displacements by an ellipsoid is also not discussed here. This is done by most structure display programs but is not usually produced by a LEED multiple scattering program.

Appendix J Averaging over Domains

In LEED and X-ray diffraction the intensity of diffracted beams is measured in such a way that the primary beam usually illuminates an area of the surface which is large enough to contain many domains and terraces. An exception occurs with special instruments like the LEEM (see for example Section 3.6.2), where the diffracted beams from a single domain can be measured by microfocus X-ray diffractometers. We discuss here only the normal case where the primary beam covers many domains or terraces. Steps occur on every surface and the different terraces may exhibit different terminations depending on the bulk structure, symmetry group and orientation of the surface. Furthermore, antiphase domains and twin domains occur in most of the adsorbate layers and reconstructed surfaces. Averaging over symmetrically equivalent domains or structurally different domains is therefore the standard procedure in surface diffraction experiments. The question is frequently raised whether the domain averaging should be performed coherently or incoherently: we discuss this issue in this appendix.

We consider here the idealized case of a periodic substrate which exhibits steps and terraces and an adsorbate or reconstructed surface layer having different domains. This leads to broadened reflections. In real cases the substrate may not be periodic but may be polycrystalline and exhibit grains with slightly different orientations, causing additional broadening of the reflections. For a detailed discussion of the integrated intensity, see B. E. Warren [7.44].

The diffracted intensity is given by the coherent sum over all domains. F_n is the structure factor of a single unit cell; $\mathbf{R}_n$ is the vector of the origin of the n-th unit cell; N is the total number of unit cells in the illuminated area; and $\mathbf{q} = \mathbf{k}' - \mathbf{k}$ is the scattering vector. We use for simplicity the kinematic theory – the multiple scattering necessary for the calculation of LEED intensities does not change the general considerations. We omit the pre-factors for the absolute intensity and define:

$$I(\mathbf{q}) = \left| \sum_{n=1}^{N} F_n(\mathbf{q}) e^{i\mathbf{q}\mathbf{R}_n} \right|^2 . \tag{J.1}$$

The structure factors F_n include the sum over all layers so the sum over n is two-dimensional, but the vectors $\mathbf{R}_n$ include the component normal to the surface due to steps. We can describe the structure factor by an average and a deviation from the average:

$$\Delta F_n(\mathbf{q}) = F_n(\mathbf{q}) - \langle F(\mathbf{q})\rangle, \tag{J.2}$$

$$
\begin{aligned}
I(\mathbf{q}) &= \left| \sum_{n=1}^{N} (\langle F_n(\mathbf{q})\rangle + \Delta F_n(\mathbf{q}))e^{i\mathbf{q}\mathbf{R}_n} \right|^2 \\
&= \left| N\langle F(\mathbf{q})\rangle \sum_{n=1}^{N} e^{i\mathbf{q}\mathbf{R}_n} + \sum_{n=1}^{N} \Delta F_n(\mathbf{q})e^{i\mathbf{q}\mathbf{R}_n} \right|^2 \\
&= N|\langle F(\mathbf{q})\rangle|^2 \left| \sum_{n=1}^{N} e^{i\mathbf{q}\mathbf{R}_n} \right|^2 + \sum_{n=1}^{N}\sum_{n'=1}^{N} \left[\Delta F_n(\mathbf{q})\Delta F_{n'}^{*}(\mathbf{q})e^{i\mathbf{q}(\mathbf{R}_n - \mathbf{R}_{n'})} \right]. \tag{J.3}
\end{aligned}
$$

With the lattice factors $G(h)$ and $G(k)$, $\mathbf{q} = h\mathbf{a}^* + k\mathbf{b}^*$, basis vectors $\mathbf{a}^*$, $\mathbf{b}^*$ of the 2-D reciprocal lattice and Eq. (J.2), the intensity becomes:

$$I(\mathbf{q}) = N|\langle F(\mathbf{q})\rangle|^2 G(h)G(k) + N(\langle|F(\mathbf{q})|^2\rangle - |\langle F(\mathbf{q})\rangle|^2)P(\mathbf{q}), \tag{J.4}$$

$$G(h) = \left| \sum_{n=-(N_1-1)/2}^{(N_1-1)/2} e^{2\pi i nh} \right|^2 = \frac{\sin^2 N_1\pi h}{\sin^2 \pi h} \quad \text{and} \quad G(k) = \frac{\sin^2 N_2\pi k}{\sin^2 \pi k}. \tag{J.5}$$

The number of unit cells is $N = N_1 N_2$ and the structure factors represent the whole column in the direction normal to the surface. Equation (J.4) is a fundamental equation for disordered systems.

The first term in Eq. (J.4) represents a sharp peak due to the average structure factor in the periodic lattice. The second term is the diffuse part arising from the finite size effect of the domains and the irregular sequence of domains. $P(\mathbf{q})$ is the Fourier transform of the correlation function and makes the profile a function of the diffuse part, the intensity of which is proportional to the mean square deviation in the structure factor. It is usually described as a mixture of Gaussian and Lorentzian functions; for a statistical distribution of domain sizes, it is more like a Lorentzian than a Gaussian function. The calculation of profile functions for possible domain distributions is not discussed here; for our purposes, it is only relevant that the integral over the lattice function in Eq. (J.5) is unity and the integral over the profile function in reciprocal space is unity as well. Vanishing of the average implies vanishing of the probability to find the same structure factor at the same level at a large distance in the periodic lattice; then there is no central peak and the reflection is continuously broadened. For example, the average structure factor becomes zero in a sequence of antiphase domains occurring with equal probability. If the distribution function of domains becomes periodic the diffuse intensity is concentrated in satellite reflections; see also Section 4.6.

The integral over the sharp central peak and the diffuse part is measured in X-ray diffraction by rocking scans and in LEED by integration over the intensity in the vicinity of the peak; in both cases the background is subtracted. We assume that the structure factor varies slowly with $\mathbf{q}$ and is approximately constant in the integration area around the centre of the reflection. The integral over $\mathbf{q}$ is then given by

$$\int I(\mathbf{q})d\mathbf{q} = \sum_n |F_n(\mathbf{q})|^2 = N\langle |F(\mathbf{q})|^2 \rangle. \tag{J.6}$$

This corresponds to incoherent averaging. The averaging over domains therefore requires in any case incoherent averaging when the integral intensity is measured. Coherent averaging is required when only the central peak is measured.

The integral intensity is measured in a certain area around the diffraction peak. This does not represent the total intensity, as some part is distributed in the whole reciprocal space in the background. Statistically distributed point defects, thermal diffuse scattering and inelastic scattering not filtered out by the detector form the background in LEED. With X-ray diffraction, additional sources of background intensity arise from diffuse scattering in the substrate and Compton scattering. This part outside the peak area can be neglected in the integrated intensity. If the diffuse part is so broad that a reliable integration is not possible, then the surface is likely so disordered that a structure analysis assuming domains on an ordered lattice does not seem to be fruitful and further methods to analyse distribution of defects are required, for example imaging methods such as scanning microscopies.

References

[1.1] W. Friedrich, P. Knipping, and M. von Laue, "Interferenz-Erscheinungen bei Röntgenstrahlen" (transl.: "Interference Phenomena with X-rays"), *Sitzungsberichte der Königlichen Bayerischen Akademie der Wissenschaften Math. Phys. Kl.*, Band 1912,14, pp. 303–322, 1912.

[1.2] W. L. Bragg, "The Diffraction of Short Electromagnetic Waves by a Crystal", *Proc. Cambridge Philos. Soc.*, vol. 17, pp. 43–57, 1913.

[1.3] J. A. Golovchenko, J. R. Patel, D. R. Kaplan, P. L. Cowan, and M. J. Bedzyk, "Solution to the Surface Registration Problem Using X-ray Standing Waves", *Phys. Rev. Lett.*, vol. 49, pp. 560–563, 1982.

[1.4] P. Eisenberger and W. C. Marra, "X-ray Diffraction Study of the Ge(001) Reconstructed Surface", *Phys. Rev. Lett.*, vol. 46, pp. 1081–1084, 1981.

[1.5] C. Davisson and L. H. Germer, "The Scattering of Electrons by a Single Crystal of Nickel", *Nature*, vol. 119, pp. 558–560, 1927.

[1.6] C. Davisson and L. H. Germer, "Diffraction of Electrons by a Crystal of Nickel", *Phys. Rev.*, vol. 30, pp. 705–740, 1927.

[1.7] G. P. Thomson and A. Reid, "Diffraction of Cathode Rays by a Thin Film", *Nature*, vol. 119, pp. 890–890, 1927.

[1.8] J. B. Pendry, *Low Energy Electron Diffraction: The Theory and its Application to Determination of Surface Structure*, Academic Press, London, 1974, out of print.

[1.9] L. J. Clarke, *Surface Crystallography: An Introduction to Low Energy Electron Diffraction*, Wiley, Chichester, 1985, out of print.

[1.10] M. A. Van Hove, W. H. Weinberg, and C.-M. Chan, *Low-Energy Electron Diffraction: Experiment, Theory and Surface Structure Determination*, Springer, Berlin, 1986, out of print.

[1.11] R. Feidenhans'l, "Surface Structure Determination by X-ray Diffraction", *Surf. Sci. Rep.*, vol. 10, pp. 105–188, 1989.

[1.12] I. K. Robinson, "Surface Crystallography", in *Handbook on Synchrotron Radiation*, eds. D. E. Moncton and G. S. Brown, Elsevier, Amsterdam, 1991, vol. III, ch. 7, pp. 221–266.

[1.13] A. Stierle and E. Vlieg, "Surface-Sensitive X-Ray Diffraction Methods", in *Modern Diffraction Methods*, eds. E. J. Mittemeijer and U. Welzel, Wiley-VCH, Weinheim, 2013, ch. 8, pp. 221–257.

[1.14] X.-L. Zhou and S.-H. Chen, "Theoretical Foundation of X-ray and Neutron Reflectometry", *Phys. Rep.*, vol. 257, pp. 223–348, 1995.

[2.1] "Symmetry Relations between Space Groups", in *International Tables of Crystallography*, vol. A1, eds. H. Wondratschek and U. Müller, Kluwer Academic Publishers, Dordrecht, 2004.

[2.2] C. Giacovazzo, ed., *Fundamentals of Crystallography*, Oxford University Press, Oxford, 1992.

[2.3] U. Shmueli, *Theories and Techniques of Crystal Structure Determination*, Oxford University Press, Oxford, 2007.

[2.4] W. Clegg, ed., *Crystal Structure Analysis, Principles and Practice*, Oxford University Press, Oxford, 2009.

[2.5] J. P. Glusker and K. N. Trueblood, *Crystal Structure Analysis*, Oxford University Press, Oxford, 2010.

[2.6] K. Hermann, *Crystallography and Surface Structure: An Introduction for Surface Scientists and Nanoscientists*, 2nd ed., Wiley, Weinheim, 2016.

[2.7] R. W. Grosse-Kunstleve, N. K. Sauter, and P. D. Adams, "Numerically Stable Algorithms for the Computation of Reduced Unit Cells", *Acta Cryst.*, vol. A60, pp. 1–6, 2004.

[2.8] E. A. Wood, "Vocabulary of Surface Crystallography", *J. Appl. Phys.*, vol. 35, pp. 1306–1312, 1964.

[2.9] W. Moritz, J. Landskron, and M. Deschauer, "Perspectives for Surface Structure Analysis with Low Energy Electron Diffraction", *Surf. Sci.*, vol. 603, pp. 1306–1314, 2009.

[2.10] H. L. Meyerheim, T. Gloege, and H. Maltor, "Surface X-ray Diffraction on Large Organic Molecules: Thiouracil on Ag(111)", *Surf. Sci. Lett.*, vol. 442, pp. L1029–L1035, 1999.

[2.11] B. Lang, R. W. Joyner, and G. A. Somorjai, "Low Energy Electron Diffraction Studies of Chemisorbed Gases on Stepped Surfaces of Platinum", *Surf. Sci.*, vol. 30, pp. 454–474, 1972.

[2.12] M. A. Van Hove and G. A. Somorjai, "A New Microfacet Notation for High-Miller-Index Surfaces of Cubic Materials with Terrace, Step and Kink Structures", *Surf. Sci.*, vol. 92, pp. 489–518, 1980.

[2.13] D. E. Sands, *Vectors and Tensors in Crystallography*, Addison-Wesley, Boston, 1982.

[2.14] Balsac program. Available at: www.fhi-berlin.mpg.de/KHsoftware/Balsac/index.html, last accessed 2 April 2022.

[2.15] CRYSCON program. Available at: www.shapesoftware.com/00_Website_Homepage/, last accessed 2 April 2022.

[2.16] M. P. Seah and W. A. Dench, "Quantitative Electron Spectroscopy of Surfaces: A Standard Data Base for Electron Inelastic Mean Free Paths in Solids", *Surf. Interface Anal.*, vol. 1, pp. 2–11, 1979.

[3.1] H. Ibach, "Surfaces in Ultrahigh Vacuum", in *Physics of Surfaces and Interfaces*, Chapter 2.2, Springer-Verlag, Berlin, 2006.

[3.2] R. Pentcheva, W. Moritz, J. Rundgren, S. Frank, D. Schrupp, and M. Scheffler, "A Combined DFT/LEED-Approach for Complex Oxide Surface Structure Determination: $Fe_3O_4(001)$", *Surf. Sci.*, vol. 602, pp. 1299–1305, 2008.

[3.3] P. Heilmann, E. Lang, K. Heinz, and K. Müller, "Fast LEED-Intensity Measurements with a Computer Controlled Television System", *Appl. Phys.*, vol. 9, pp. 247–251, 1976.

[3.4] U. Scheithauer, G. Meyer, and M. Henzler, "A New LEED Instrument for Quantitative Spot Profile Analysis", *Surf. Sci.*, vol. 178, pp. 441–451, 1986.

[3.5] W. Telieps and E. Bauer, "An analytical reflection and emission UHV surface electron microscope", *Ultramicroscopy*, vol. 17, pp. 57–66, 1985.

[3.6] D. Yu, Ch. Math, M. Meier, M. Escher, G. Rangelov, and M. Donath, "Characterisation and application of a SPLEED–based spin polarisation analyser", *Surf. Sci.*, vol. 601, pp. 5803–5808, 2007.

[3.7] L. S. Dake, D. E. King, J. R. Pitts, and A. W. Czanderna, "Beam Effects, Surface Topography, and Depth Profiling in Surface Analysis", in *Methods in Surface Characterisation*, A. W. Czanderna, T. E. Madey and C. J. Powell, eds., Kluwer Academic Publishers, New York, 2002, vol. 5, ch. 3, pp. 97–273.

[3.8] C. Becker, "UHV Surface Preparation Methods", in *Encyclopedia of Interfacial Chemistry: Surface Science*, K. Wandelt, ed., Elsevier, Amsterdam, 2015, vol. 1.1, pp. 580–590.

[3.9] C. Crotti, E. Farnetti, T. Celestino, and S. Fontana, "Preparation of Solid-State Samples of a Transition Metal Coordination Compound for Synchrotron Radiation Photoemission Studies", *J. Electr. Spectr. Rel. Phen.*, vol. 128, pp. 141–157, 2003.

[3.10] M. Sterrer and H.-J. Freund, "Surface Science Approach to Catalyst Preparation Using Thin Oxide Films as Substrates", in *Encyclopedia of Interfacial Chemistry: Surface Science*, K. Wandelt, ed., Elsevier, Amsterdam, 2018, pp. 632–642.

[3.11] M. Lübbe and W. Moritz, "A LEED Analysis of the Clean Surfaces of α-Fe$_2$O$_3$(0001) and α-Cr$_2$O$_3$(0001) Bulk Single Crystals", *J. Phys.: Condens. Matter*, vol. 21, p. 134010, 2009.

[3.12] S. L. Cunningham and W. H. Weinberg, "Determining the Angles of Incidence in a LEED Experiment", *Rev. Sci. Instrum.*, vol. 49, pp. 752–755, 1978.

[3.13] M. A. Van Hove, W. H. Weinberg, and C.-M. Chan, *Low-Energy Electron Diffraction: Experiment, Theory and Surface Structure Determination*, Berlin, Springer, 1986, ch. 2.6.

[3.14] G. P. Price, "Techniques for Very Low Energy Electron Diffraction", *Rev. Sci. Instrum.*, vol. 51, pp. 605–609, 1980.

[3.15] A. C. Sobrero and W. H. Weinberg, "Unified Approach to Photographic Methods for Obtaining the Angles of Incidence in Low-Energy Electron Diffraction", *Rev. Sci. Instrum.*, vol. 53, pp. 1566–1572, 1982.

[3.16] F. Sojka, M. Meissner, Ch. Zwick, R. Forker, and T. Fritz, "Determination and Correction of Distortions and Systematic Errors in Low-Energy Electron Diffraction", *Rev. Sci. Instrum.*, vol. 84, p. 015111, 2013.

[3.17] F. Sojka, M. Meissner, Ch. Zwick, R. Forker, M. Vyshnepolsky, C. Klein, M. Horn-von Hoegen, and T. Fritz, "To Tilt or Not to Tilt: Correction of the Distortion Caused by Inclined Sample Surfaces in Low-Energy Electron Diffraction", *Ultramicroscopy*, vol. 133, pp. 35–40, 2013.

[3.18] C. Klein, T. Nabbefeld, H. Hattab, D. Meyer, G. Jnawali, M. Kammler, F.-J. Meyer zu Heringdorf, A. Golla-Franz, B. H. Müller, Th. Schmidt, M. Henzler, and M. Horn-von Hoegen, "Lost in Reciprocal Space? Determination of the Scattering Condition in Spot Profile Analysis Low-Energy Electron Diffraction", *Rev. Sci. Instrum.*, vol. 82, p. 035111, 2011.

[3.19] M. Henzler, "LEED-Investigation of Step Arrays on Cleaved Germanium (111) Surfaces", *Surf. Sci.*, vol. 19, pp. 159–171, 1970.

[3.20] K. M. Schindler, M. Huth, and W. Widdra, "Improved Extraction of I–V Curves from Low-Energy Electron Diffraction Images", *Chem. Phys. Lett.*, vol. 532, pp. 116–118, 2012.

[3.21] W. Moritz, J. Landskron, and M. Deschauer, "Perspectives for Surface Structure Analysis with Low Energy Electron Diffraction", *Surf. Sci.*, vol. 603, pp. 1306–1314, 2009.

[3.22] M. Deschauer, "LEED-Strukturanalyse organischer Molekülschichten: 2-Thiouracil auf Ag(111)", Dissertation, University of Munich, Dept. of Earth and Environmental Sciences, Germany, 2000.

[3.23] R. L. Park, J. E. Houston, and D. G. Schreiner, "The LEED Instrument Response Function", *Rev. Sci. Instrum.*, vol. 42, pp. 60–65, 1971.

[3.24] G.-C. Wang and M. G. Lagally, "Quantitative Island Size Determination in the Chemisorbed Layer: W(110)p(2 × 1)-O", *Surf. Sci.*, vol. 81, pp. 69–89, 1979.

[3.25] T.-M. Lu and M.G. Lagally, "The Resolving Power of a Low-Energy Electron Diffractometer and the Analysis of Surface Defects", *Surf. Sci.*, vol. 99, pp. 695–713, 1980.

[3.26] D. G. Welkie and M. G. Lagally, "Investigation of the Instrumental Response of a Vidicon based Low Energy Electron Diffraction System", *Appl. Surf. Sci.*, vol. 3, pp. 272–292, 1979.

[3.27] G. Comsa, "Coherence Length and/or Transfer Width", *Surf. Sci.*, vol. 81, pp. 57–68, 1979.

[3.28] M. Henzler, "Atomic Steps on Single Crystals: Experimental Methods and Properties", *Appl. Phys.*, vol. 9, pp. 11–17, 1976.

[3.29] M. Henzler, "LEED Studies of Defects at Surfaces and Interfaces", *Surf. Sci.*, vol. 168, pp. 744–750, 1986.

[3.30] Specs GmbH, Berlin, Germany, ErLEED 100/150, Reverse View LEED Optics, https://www.specs-group.com/, last accessed 5 April 2022.

[3.31] Scienta Omicron, Uppsala, Sweden and Taunusstein, Germany, Femto-LEED with Integral Shutter Model DLD-L800, www.scientaomicron.com/en/Components/LEED-RHEED/LEED-800-MCP, last accessed 2 April 2022.

[3.32] OCI Vacuum Microengineering, Inc., Ontario, Canada. Femto-LEED with Integral Shutter Model DLD-L800, http://ocivm.com/leed_aes_spectrometers.html#Femto-LEED-ISH, last accessed 2 April 2022.

[3.33] C. Seidel, J. Poppensieker, and H. Fuchs, "Real-Time Monitoring of Phase Transitions of Vacuum Deposited Organic Films by Molecular Beam Deposition LEED", *Surf. Sci.*, vol. 408, pp. 223–231, 1998.

[3.34] M. Horn-von Hoegen, "Growth of Semiconductor Layers Studied by Spot Profile Analysing Low Energy Electron Diffraction", *Z. Krist.*, vol. 214, pp. 591–629 and 684–721, 1999.

[3.35] Dr. Peter Kury, out-of-the-box systems GmbH, Nierenhofer Strasse 68a, 45257 Essen, Germany, www.spa-leed.com, last accessed 2 April 2022.

[3.36] P. Kirschbaum, L. Brendel, K. R. Roos, M. Horn-von Hoegen, and F.-J. Meyer zu Heringdorf, "Decay of Isolated Hills and Saddles on Si(001)", *Mater. Res. Express*, vol. 3, p. 085011, 2016.

[3.37] G. Held, S. Uremovic, C. Stellwag, and D. Menzel, "A Low-Energy Electron Diffraction Data Acquisition System for Very Low Electron Doses Based upon a Slow Scan Charge Coupled Device Camera", *Rev. Sci. Instr.*, vol. 67, pp. 378–383, 1996.

[3.38] S. Kubsky, L. Borucki, F. Gorris, H. W. Becker, C. Rolfs, W. H. Schulte, I. J. R. Baumvol, and F. C. Stedile, "Interconnected UHV Facilities for Materials Preparation and Analysis", *Nuclear Instr. Methods Phys. Res. A*, vol. 435, pp. 514–522, 1999.

[3.39] Y. Kakefuda, Y. Yamashita, K. Mukai, and J. Yoshinobu, "Compact UHV System for Fabrication and In Situ Analysis of Electron Beam Deposited Structures Using a Focused Low Energy Electron Beam", *Rev. Sci. Instrum.*, vol. 77, p. 053702, 2006.

[3.40] F. Pesty and P. Garoche, "Improving the Coherence of a Low-Energy Electron Beam by Modulation", *Appl. Phys.*, vol. 94, pp. 7910–7913, 2003.

[3.41] S. Mizuno, F. Rahman, and M. Iwanaga, "Low-Energy Electron Diffraction Patterns Using Field-Emitted Electrons from Tungsten Tips", *Jpn. J. Appl. Phys.*, vol. 45, pp. L178–L209, 2006.

[3.42] A. Janzen, B. Krenzer, P. Zhou, D. von der Linde, and M. Horn-von Hoegen, "Ultrafast Electron Diffraction at Surfaces after Laser Excitation", *Surf. Sci.*, vol. 600, pp. 4094–4098, 2006.

[3.43] S. Wall, B. Krenzer, S. Wippermann, S. Sanna, F. Klasing, A. Hanisch-Blicharski, M. Kammler, W. G. Schmidt, and M. Horn-von Hoegen, "An Atomistic Picture of Charge Density Wave Formation at Surfaces", *Phys. Rev. Lett.*, vol. 109, p. 186101, 2012.

[3.44] A. Hanisch-Blicharski, A. Janzen, B. Krenzer, S. Wall, F. Klasing, A. Kalus, T. Frigge, M. Kammler, and M. Horn-von Hoegen, "Ultra-fast Electron Diffraction at Surfaces: From Nanoscale Heat Transport to Driven Phase Transitions", *Ultramicroscopy*, vol. 127, pp. 2–8, 2013.

[3.45] B. Hafke, T. Witte, C. Brand, Th. Duden, and M. Horn-von Hoegen, "Pulsed Electron Gun for Electron Diffraction at Surfaces with Femtosecond Temporal Resolution and High Coherence Length", *Rev. Sci. Instr.*, vol. 90, p. 045119, 2019.

[3.46] M. Gulde, S. Schweda, G. Storeck, M. Maiti, H. K. Yu, A. M. Wodtke, S. Schäfer, and C. Ropers, "Ultrafast Low-Energy Electron Diffraction in Transmission Resolves Polymer/Graphene Superstructure Dynamics", *Science*, vol. 345, pp. 200–204, 2014.

[3.47] E. T. J. Nibbering, "Low-Energy Electron Diffraction at Ultrafast Speeds", *Science*, vol. 345, pp. 137–138, 2014.

[3.48] G. Storeck, S. Vogelgesang, M. Sivis, S. Schäfer, and C. Ropersa, "Nanotip-based Photoelectron Microgun for Ultrafast LEED", *Struct. Dyn.*, vol. 4, p. 044024, 2017.

[4.1] R. Mas-Ballesté, C. Gómez-Navarro, J. Gómez-Herrero, and F. Zamora, "2D Materials: To Graphene and Beyond", *Nanoscale*, vol. 3, pp. 20–30, 2011.

[4.2] U. Stahl, D. Gador, A. Soukopp, R. Fink, and E. Umbach, "Coverage-Dependent Superstructures in Chemisorbed NTCDA Monolayers: A Combined LEED and STM Study", *Sur. Sci.*, vol. 414, pp. 423–434, 1998.

[4.3] LEEDpat, graphical LEED pattern simulator, PC-based software tool to visualize and analyze LEED spot patterns of well-ordered substrates and overlayers, available for download at www.fhi-berlin.mpg.de/KHsoftware/LEEDpat/ where more information is available (executable for Windows; by K. Hermann and M. A. Van Hove), last accessed 2 April 2022.

[4.4] F. Sojka, M. Meissner, Ch. Zwick, R. Forker, and T. Fritz, "Determination and Correction of Distortions and Systematic Errors in Low-Energy Electron Diffraction", *Rev. Sci. Instr.*, vol. 84, p. 015111, 2013.

[4.5] F. Sojka, M. Meissner, Ch. Zwick, R. Forker, M. Vyshnepolsky, C. Klein, M. Horn-von Hoegen, and T. Fritz, "To Tilt or Not to Tilt: Correction of the Distortion Caused by Inclined Sample Surfaces in Low-Energy Electron Diffraction", *Ultramicroscopy*, vol. 133, pp. 35–40, 2013.

[4.6] K. Glöckler, C. Seidel, A. Soukopp, M. Sokolowski, E. Umbach, M. Böhringer, R. Berndt, and W.-D. Schneider, "Highly Ordered Structures and Submolecular Scanning Tunnelling Microscopy Contrast of PTCDA and DM-PBDCI Monolayers on Ag (111) and Ag (110)", *Surf. Sci.*, vol. 405, pp. 1–20, 1998.

[4.7] M. Horn-von Hoegen, "Growth of Semiconductor Layers Studied by Spot Profile Analysing Low Energy Electron Diffraction", *Z. Krist.*, vol. 214, pp. 591–629 and 684–721, 1999.

[4.8] Q. Chen and N. V. Richardson, "Surface Facetting Induced by Adsorbates", *Progr. Surf. Sci.*, vol. 73, pp. 59–77, 2003.

[4.9] C. Tegenkamp, "Vicinal Surfaces for Functional Nanostructures", *J. Phys.: Condens. Matter*, vol. 21, p. 013002, 2009.

[4.10] T. E. Madey, C.-H. Nien, K. Pelhos, J. J. Kolodziej, I. M. Abdelrehim, and H.-S. Tao, "Faceting Induced by Ultrathin Metal Films: Structure, Electronic Properties and Reactivity", *Surf. Sci.*, vol. 438, pp. 191–206, 1999.

[4.11] K. Hermann, *Crystallography and Surface Structure: An Introduction for Surface Scientists and Nanoscientists*, 2nd ed., Wiley, Weinheim, 2016, ch. 4.2, p. 205.

[4.12] N. Reinecke and E. Taglauer, "The Kinetics of Oxygen-Induced Faceting of Cu(115) and Cu(119) Surfaces", *Surf. Sci.*, vol. 454–456, pp. 94–100, 2000.

[4.13] P. Zeppenfeld, K. Kern, R. David, and G. Comsa, "Diffraction from Domain-Wall Systems", *Phys. Rev. B*, vol. 38, pp. 3918–3924, 1988.

[4.14] F. Timmer and J. Wollschläger, "Effects of Domain Boundaries on the Diffraction Patterns of One-Dimensional Structures", *Condens. Matter*, vol. 2, p. 7, 2017.

[4.15] S. K. Hämäläinen, M. P. Boneschanscher, P. H. Jacobse, I. Swart, K. Pussi, W. Moritz, J. Lahtinen, P. Liljeroth, and J. Sainio: "Structure and Local Variations of the Graphene Moiré on Ir(111)", *Phys. Rev. B*, vol. 88, p. 201406(R), 2013.

[4.16] X. Zhu, J. Guo, J. Zhang, and E. W. Plummer, "Misconceptions Associated with the Origin of Charge Density Waves", *Adv. Phys.: X*, vol. 2, pp. 622–640, 2017.

[4.17] R. He, J. Okamoto, Z. Ye, G. Ye, H. Anderson, X. Dai, X. Wu, J. Hu, Y. Liu, W. Lu, Y. Sun, A. N. Pasupathy, and A. W. Tsen, "Distinct Surface and Bulk Charge Density Waves in Ultrathin $1T-TaS2$", *Phys. Rev. B*, vol. 94, p. 201108(R), 2016.

[5.1] J. B. Pendry, *Low Energy Electron Diffraction*, Academic, London, 1974.

[5.2] S. Y. Tong, "Theory of Low Energy Electron Diffraction", *Progr. Surf. Sci.*, vol. 7, pp. 1–48, 1975.

[5.3] M. A. Van Hove, W. H. Weinberg, and C.-M. Chan, *Low-Energy Electron Diffraction: Experiment, Theory and Structural Determination*, Springer, Berlin, 1986.

[5.4] D. F.-T. Tuan and G. W. Loar, "Corrections of Overlapping Spheres in the Self-consistent Field X_α Scattered-Wave Method", *J. Mol. Struct.*, vol. 223, pp. 123–148, 1990.

[5.5] J. Rundgren, "Optimized Surface-Slab Excited-State Muffin-Tin Potential and Surface Core Level Shifts", *Phys. Rev. B*, vol. 68, p. 125405, 2003.

[5.6] S. Nagano and S. Y. Tong, "Multiple-Scattering Theory of Low-Energy Electron Diffraction for a Nonspherical Scattering Potential", *Phys. Rev. B*, vol. 32, pp. 6562–6570, 1985.

[5.7] J.-V. Peetz and W. Schattke, "Non-Muffin-Tin Atomic Scattering-Matrices for Semiconductor LEED-Calculations", *J. Electr. Spectr. Rel. Phen.*, vol. 68, pp. 167–173, 1994.

[5.8] O. Rubner, M. Kottcke, and K. Heinz, "Tensor LEED for the Approximation of Non-spherical Atomic Scattering", *Surf. Sci.*, vol. 340, pp. 172–178, 1995.

[5.9] Y. Joly, "Finite-Difference Method for the Calculation of Low Energy Electron Diffraction", *Phys. Rev. Lett.*, vol. 68, pp. 950–953, 1992.

[5.10] G. M. Gavaza, Z. X. Yu, L. Tsang, C. H. Chan, S. Y. Tong, and M. A. Van Hove, "Theory of Low-Energy Electron Diffraction for Detailed Structural Determination of Nanomaterials: Finite-Size and Disordered Structures", *Phys. Rev. B*, vol. 75, p. 235403, 2007.

[5.11] G. M. Gavaza, Z. X. Yu, L. Tsang, C. H. Chan, S. Y. Tong, and M. A. Van Hove, "Efficient Calculation of Electron Diffraction for the Structural Determination of Nanomaterials", *Phys. Rev. Lett.*, vol. 97, p. 055505, 2006.

[5.12] G. M. Gavaza, Z. X. Yu, L. Tsang, C. H. Chan, S. Y. Tong, and M. A. Van Hove, "Theory of Low-Energy Electron Diffraction for Detailed Structural Determination of Nanomaterials: Ordered Structures", *Phys. Rev. B*, vol. 75, p. 014114, 2007.

[5.13] G. M. Gavaza, Z. X. Yu, M. A. Van Hove, and S. Y. Tong, "Theory of Low-Energy Electron Diffraction for Nanomaterials: Subclusters, Automated Searches", *J. Phys. Condens. Matter*, vol. 20, p. 304202, 2008.

[5.14] J. C. Spence, H. C. Poon, and D. K. Saldin, "Convergent-Beam Low Energy Electron Diffraction (CBLEED) and the Measurement of Surface Dipole Layers", *Microsc. Microanal.*, vol. 10, pp. 128–133, 2004.

[5.15] E. Steinwand, J.-N. Longchamp, and H.-W. Fink, "Coherent Low-Energy Electron Diffraction on Individual Nanometer Sized Objects", *Ultramicroscopy*, vol. 111, pp. 282–284, 2011.

[5.16] W. Telieps and E. Bauer, "An Analytical Reflection and Emission UHV Surface Electron Microscope", *Ultramicroscopy*, vol. 17, pp. 57–66, 1985. See https://elmitec.de/Users_List for installed instruments and applications, last accessed 2 April 2022.

[5.17] Y. Kakefuda, Y. Yamashita, K. Mukai, and J. Yoshinobu, "Compact UHV System for Fabrication and In Situ Analysis of Electron Beam Deposited Structures Using a Focused Low Energy Electron Beam", *Rev. Sci. Instrum.*, vol. 77, p. 053702, 2006.

[5.18] S. Mizuno, F. Rahman, and M. Iwanaga, "Low-Energy Electron Diffraction Patterns Using Field-Emitted Electrons from Tungsten Tips", *Jpn. J. Appl. Phys.*, part 2, vol. 45, p. L178, 2006.

[5.19] R. Barret, M. Berry, T. F. Chan, J. Demmel, J. Donato, J. Dongarra, V. Eijkhout, R. Pozo, C. Romine, and H. Van der Vorst, *Templates for the Solution of Linear Systems: Building Blocks for Iterative Methods*, 2nd ed., SIAM, Philadelphia, PA, 1994, vol. 1, ch. 2.3.1, p. 14.

[5.20] L. Tsang and Q. Li, "Wave Scattering with UV Multilevel Partitioning Method for Volume Scattering by Discrete Scatterers", *Microwave Opt. Technol. Lett.*, vol. 41, pp. 354–360, 2004.

[5.21] D. K. Saldin and J. B. Pendry, "The Cluster Approach to LEED Calculations", *Surf. Sci.*, vol. 162, pp. 941–944, 1985.

[5.22] M. A. Van Hove, W. Moritz, H. Over, P. J. Rous, A. Wander, A. Barbieri, N. Materer, U. Starke, and G. A. Somorjai, "Automated Determination of Complex Surface Structures by LEED", *Surf. Sci. Rep.*, vol. 19, pp. 191–229, 1993.

[5.23] K. Kambe, "Theory of Low-Energy Electron Diffraction I. Application of the Cellular Method to Monatomic Layers", *Z. Naturforsch.*, vol. 22a, pp. 322–330, 1967.

[5.24] K. Kambe, "Theory of Electron Diffraction by Crystals I. Green's Function and Integral Equation", *Z. Naturforsch.*, vol. 22a, pp. 422–431, 1967.

[5.25] K. Kambe, "Theory of Low-Energy Electron Diffraction II. Cellular Method for Complex Monolayers and Multilayer", *Z. Naturforsch.*, vol. 23a, pp. 1280–1294, 1968.

[5.26] P. J. Rous and J. B. Pendry, "Diffuse LEED from Simple Stepped Surfaces", *Surf. Sci.*, vol. 173, pp. 1–19, 1986.

[5.27] D. W. Jepsen, "New Transfer-Matrix Method for Low-Energy-Electron Diffraction and Other Surface Electronic-Structure Problems", *Phys. Rev. B*, vol. 22, pp. 5701–5715, 1980.

[5.28] P. Pinkava and S. Crampin, "On the Calculation of LEED Spectra from Stepped Surfaces", *Surf. Sci.*, vol. 233, pp. 27–34, 1990.

[5.29] X.-G. Zhang and A. Gonis, "New, Real-Space, Multiple-Scattering-Theory Method for the Determination of Electronic Structure", *Phys. Rev. Lett.*, vol. 62, 1161–1164, 1989.

[5.30] J. M. MacLaren, X.-G. Zhang, and A. Gonis, "Multiple-Scattering Solutions to the Schrödinger Equation for Semi-infinite Layered Materials", *Phys. Rev. B*, vol. 40, pp. 9955–9958, 1989.

[5.31] X.-G. Zhang, M. A. Van Hove, G. A. Somorjai, P. J. Rous, D. Tobin, A. Gonis, J. M. MacLaren, K. Heinz, M. Michl, H. Lindner, K. Müller, M. Ehsasi, and J. H. Block, "Efficient Determination of Multilayer Relaxation in the Pt(210) Stepped and Densely Kinked Surface", *Phys. Rev. Lett.*, vol. 67, pp. 1298–1301, 1991.

[5.32] X.-G. Zhang, P. J. Rous, J. M. MacLaren, A. Gonis, M. A. Van Hove, and G. A. Somorjai, "A Real-Space Multiple Scattering Theory of Low-Energy Electron Diffraction: A New Approach for the Structure Determination of Stepped Surfaces", *Surf. Sci.*, vol. 239, pp. 103–118, 1990.

[5.33] W. Moritz, "Effective Calculation of LEED Iintensities Using Symmetry Adapted Functions", *J. Phys. C*, vol. 17, pp. 353–362, 1984.

[5.34] C. J. Bradley and A. P. Cracknell, *The Mathematical Theory of Symmetry in Solids*, Clarendon, Oxford, 1972.

[5.35] K. Heinz, M. Kottcke, U. Löffler, and R. Döll, "Recent Advances in LEED Surface Crystallography", *Surf. Sci.*, vol. 357–358, pp. 1–9, 1996.

[5.36] P. J. Rous and J. B. Pendry, "Tensor LEED I: A Technique for High Speed Surface Structure Determination by Low Energy Electron Diffraction. TLEED I", *Comp. Phys. Comm.*, vol. 54, pp. 137–156, 1989.

[5.37] P. J. Rous and J. B. Pendry, "Tensor LEED II: A Technique for High Speed Surface Structure Determination by Low Energy Electron Diffraction. TLEED 2", *Comp. Phys. Comm.*, vol. 54, pp. 157–166, 1989.

[5.38] Z. X. Yu and S. Y. Tong, "Surface Structure Determination by a One-Stop Search Method for the Deepest Minimum", *Phys. Rev. B*, vol. 71, p. 161404(R), 2005.

[5.39] J. B. Pendry and D. K. Saldin, "SEXAFS without X-rays", *Surf. Sci.*, vol. 145, pp. 33–47, 1984.

[5.40] M. A. Van Hove, "Solving Complex and Disordered Surface Structures with Electron Diffraction", in *Chemistry and Physics of Solid Surfaces VII*, eds. R. F. Howe and R. Vanselow, Springer-Verlag, Berlin, 1988, Springer Series in Surface Sciences, vol. 10, pp. 513–546.

[5.41] U. Starke, J. B. Pendry, and K. Heinz, "Diffuse Low-Energy Electron Diffraction", *Progr. Surf. Sci.*, vol. 52, pp. 53–124, 1996.

[5.42] C. J. Barnes, E. AlShamaileh, T. Pitkänen, and M. Lindroos, "Early Stages of Surface Alloy Formation: A Diffuse LEED *I*(*V*) Study", *Surf. Sci.*, vol. 482, pp. 1425–1430, 2001.

[5.43] H. C. Poon, M. Weinert, D. K. Saldin, D. Stacchiola, T. Zheng, and W. T. Tysoe, "Structure Determination of Disordered Organic Molecules on Surfaces from the Bragg Spots of Low-Energy Electron Diffraction and Total Energy Calculations", *Phys. Rev. B*, vol. 69, p. 035401, 2004.

[5.44] K. Heinz, U. Starke, M. A. Van Hove, and G. A. Somorjai, "The Angular Dependence of Diffuse LEED Intensities and Its Structural Information Content", *Surf. Sci.*, vol. 261, pp. 57–63, 1992.

[5.45] D. F. Ogletree, G. S. Blackman, R. Q. Hwang, U. Starke, G. A. Somorjai, and J. E. Katz, "A New Pulse Counting Low-Energy Electron Diffraction System Based on a Position Sensitive Detector", *Rev. Sci. Instrum.*, vol. 63, pp. 104–113, 1992.

[5.46] H. Ibach and S. Lehwald, "Elastic Diffuse and Inelastic Electron Scattering from Surfaces with Disordered Overlayers", *Surf. Sci.*, vol. 176, pp. 629–634, 1986.

[5.47] D. K. Saldin, J. B. Pendry, M. A. Van Hove, and G. A. Somorjai, "Interpretation of Diffuse Low-Energy Electron Diffraction Intensities", *Phys. Rev. B*, vol. 31, pp. 1216–1218, 1985.

[5.48] M. A. Van Hove, R. F. Lin, and G. A. Somorjai, "Efficient Scheme for LEED Calculations in the Presence of Large Superlattices, with Application to the Structural Analysis of Benzene Adsorbed on Rh(111)", *Phys. Rev. Lett.*, vol. 51, pp. 778–781, 1983.

[5.49] C. M. Wei and S. Y. Tong, "Direct Atomic Structure by Holographic Diffuse LEED", *Surf. Sci.*, vol. 274, pp. L577–L582, 1992.

[5.50] K. Heinz, "LEED and DLEED As Modern Tools for Quantitative Surface Structure Determination", *Rep. Prog. Phys.*, vol. 58, pp. 637–704, 1995.

[5.51] P. J. Hu and D. A. King, "A Direct Inversion Method for Surface Structure Determination from LEED Intensities", *Nature*, vol. 360, pp. 655–658, 1992.

[5.52] M. A. Mendez, C. Gluck, and K. Heinz, "Holography with Conventional LEED", *J. Phys.: Cond. Matt.*, vol. 4, pp. 999–1006, 1992.

[5.53] D. K. Saldin and X. Chen, "Three-Dimensional Reconstruction by Holographic LEED: Proper Identification of the Reference Wave", *Phys. Rev. B*, vol. 52, pp. 2941–2948, 1995.

[5.54] D. K. Saldin, K. Reuter, P. L. de Andres, H. Wedler, X. Chen, J. B. Pendry, and K. Heinz, "Direct Reconstruction of Three-Dimensional Atomic Adsorption Sites by Holographic LEED", *Phys. Rev., vol. B*, 54, pp. 8172–8176, 1996.

[5.55] S. Y. Tong, H. Huang, and X. Q. Guo, "Low-Energy Electron and Low-Energy Positron Holography", *Phys. Rev. Lett.*, vol. 69, pp. 3654–3657, 1992.

[5.56] K. Heinz, D. K. Saldin, and J. B. Pendry, "Diffuse LEED and Surface Crystallography", *Phys. Rev. Lett.*, vol. 55, pp. 2312–2315, 1985.

[5.57] P. J. Rous, J. B. Pendry, D. K. Saldin, K. Heinz, K. Müller, and N. Bickel, "Tensor LEED: A Technique for High-Speed Surface-Structure Determination", *Phys. Rev. Lett.*, vol. 57, pp. 2951–2954, 1986.

[5.58] W. Oed, U. Starke, K. Heinz, K. Müller, and J. B. Pendry, "Ordered and Disordered Oxygen and Sulfur on Ni(100)", *Surf. Sci.*, vol. 251, pp. 488–492, 1991.

[5.59] C. S. Ri and P. R. Watson, "Surface Structure of the Disordered Cl/Ti(0001) System Determined by Diffuse Low-Energy Electron Diffraction", *J. Vac. Sci. Technol.*, vol. A10, pp. 2535–2539, 1992.

[5.60] C.-S. Ri and P. R. Watson, "Identification of the Bonding Site of Chlorine on the Ti (0001) Surface", *Solid St. Comm.*, vol. 81, pp. 541–543, 1992.

[5.61] P. Hu, C. J. Barnes, and D. A. King, "Dominance of Short-Range-Order Effects in Low-Energy Electron-Diffraction Intensity Spectra", *Phys. Rev. B*, vol. 45, pp. 13595–13598, 1992.

[5.62] H. Wedler, M. A. Mendez, P. Bayer, U. Löffler, K. Heinz, V. Fritzsche, and J. B. Pendry, "Coverage-Dependent DLEED Analysis of the Adsorption Structure of K on Ni(100)", *Surf. Sci.*, vol. 293, pp. 47–56, 1993.

[5.63] L. D. Mapledoram, A. Wander, and D. A. King, "Islanding or Random Growth? The Low Coverage Growth Modes and Structure of NO on Ni{111} Studied by Diffuse ATLEED", *Surf. Sci.*, vol. 312, pp. 54–61, 1994.

[5.64] C. J. Barnes, A. Wander, and D. A. King, "A Tensor LEED Structural Study of the Coverage-Dependent Bonding of Iodine Adsorbed on Rh{111}", *Surf. Sci.*, vol. 281, pp. 33–41, 1993.

[5.65] M. Sporn, E. Platzgummer, S. Forsthuber, M. Schmid, W. Hofer, and P. Varga, "The Accuracy of Quantitative LEED in Determining Chemical Composition Profiles of Substitutionally Disordered Alloys: A Case Study", *Surf. Sci.*, vol. 416, pp. 423–429, 1998.

[5.66] I. Zasada, "On the Disordered Adsorption of CO on Pt(111) by Diffuse SPLEED", *Surf. Sci.*, vol. 498, pp. 293–306, 2002.

[5.67] I. Zasada and T. Rychtelska, "Theoretical Studies of Spin-Polarized DLEED Rotation Curves: Disordered CO Overlayer on Pt(111)", *Solid St. Comm.*, vol. 123, pp. 267–272, 2002.

[5.68] I. Zasada and T. Rychtelska, "Coverage-Dependent Diffuse Spin-Polarized Electron Diffraction Analysis of CO on Pt(111)", *Surf. Sci.*, vol. 507, pp. 362–367, 2002.

[5.69] A. Wander, G. Held, R. Q. Hwang, G. S. Blackman, M. L. Xu, P. de Andres, M. A. Van Hove, and G. A. Somorjai, "A Diffuse LEED Study of the Adsorption Structure of Disordered Benzene on Pt(111)", *Surf. Sci.*, vol. 249, pp. 21–34, 1991.

[5.70] R. Döll, C. A. Gerken, M. A. Van Hove, and G. A. Somorjai, "Structure of Disordered Ethylene Adsorbed on Pt(111) Analyzed by Diffuse LEED: Asymmetrical di-σ Bonding Favored", *Surf. Sci.*, vol. 374, pp. 151–161, 1997.

[5.71] U. Starke, K. Heinz, N. Materer, A. Wander, M. Michl, R. Döll, M. A. Van Hove, and G. A. Somorjai, "Low-Energy Electron Diffraction Study of a Disordered Monolayer of H_2O on Pt(111) and an Ordered Thin Film of Ice Grown on Pt(111)", *J. Vac. Sci. Technol.*, vol. A10, pp. 2521–2528, 1992.

[5.72] U. Starke, N. Materer, A. Barbieri, R. Döll, K. Heinz, M. A. Van Hove, and G. A. Somorjai, "A Low-Energy Electron Diffraction Study of Oxygen, Water and Ice Adsorption on Pt(111)", *Surf. Sci.*, vol. 287/288, pp. 432–437, 1993.

[5.73] M. C. Desjonquères and D. Spanjaard, *Concepts in Surface Physics*, Springer, Berlin, 1993.

[5.74] F. W. Kuhs, "Generalized Atomic Displacements in Crystallographic Structure Analysis", *Acta Cryst.*, vol. A48, pp. 80–98, 1992.

[5.75] H. Ibach and D. L. Mills, *Electron Energy Loss Spectroscopy and Surface Vibrations*, Academic Press, New York, 1982.

[5.76] J. Landskron, W. Moritz, B. Narloch, G. Held, and D. Menzel, "Analysis of Thermal Vibrations by Temperature-Dependent Low Energy Electron Diffraction: Comparison of Soft Modes of Pure and O-Coadsorbed CO on Ru(0001)", *Surf. Sci.*, vol. 441, pp. 91–106, 1999.

[5.77] H. L. Meyerheim, W. Moritz, H. Schulz, P. J. Eng, and I. K. Robinson, "Anharmonic Thermal Vibrations Observed by Surface X-ray Diffraction for Cs/Cu(001)", *Surf. Sci.*, vol. 331–333, pp. 1422–1429, 1995.

[5.78] B. T. M. Willis and A. W. Pryor, *Thermal Vibrations in Crystallography*, Cambridge University Press, London, 1975.

[5.79] K. N. Trueblood, H.-B. Bürgi, H. Burzlaff, J. D. Dunitz, C. M. Gramaccioli, H. H. Schulz, U. Shmueli, and S. C. Abrahams, "Atomic Displacement Parameter Nomenclature", *Acta Cryst.*, vol. A52, pp. 770–781, 1996.

[5.80] M. A. Krivoglaz, *The Theory of X-ray and Thermal Neutron Diffraction by Real Crystals*, Plenum Press, London, 1969.

[5.81] J. T. McKinney, E. R. Jones, and M. B. Webb, "Surface Lattice Dynamics of Silver. II. Low-Energy Electron Thermal Diffuse Scattering", *Phys. Rev.*, vol. 160, pp. 523–530, 1967.

[5.82] V. Zielasek, A. Büssenschütt, and M. Henzler, "Low-Energy Electron Thermal Diffuse Scattering from Al(111) Individually Resolved in Energy and Momentum", *Phys. Rev. B*, vol. 55, pp. 5398–5403, 1997.

[5.83] G. M. Piccini and J. Sauer, "Effect of Anharmonicity on Adsorption Thermodynamics", *J. Chem. Theory Comput.*, vol. 10, pp. 2479–2487, 2014.

[5.84] H. Brune, "Thermal Dynamics at Surfaces", *Ann. Phys.*, vol. 18, pp. 675–698, 2009.

[5.85] E. Prince, *Mathematical Techniques in Crystallography and Materials Science*, Springer-Verlag, New York, 1982.

[5.86] U. Schmueli, *Theories and Techniques of Crystal Structure Determination*, Oxford University Press, Oxford, 2007.

[5.87] M. Gierer, H. Bludau, H. Over, and G. Ertl, "The Bending Mode Vibration of CO on Ru(0001) Studied with Low-Energy Electron-Diffraction", *Surf. Sci.*, vol. 346, pp. 64–72, 1996.

[5.88] C. B. Duke and G. E. Laramore, "Effect of Lattice Vibrations in a Multiple-Scattering Description of Low-Energy Electron Diffraction. I. Formal Perturbation Theory", *Phys. Rev. B*, vol. 2, pp. 4765–4782, 1970. G. E. Laramore and C. B. Duke, "Effect of Lattice Vibrations in a Multiple-Scattering Description of Low-Energy Electron Diffraction. II. Double-Diffraction Analysis of the Elastic Scattering Cross Section", *Phys. Rev. B*, vol. 2, pp. 4783–4795, 1970.

[5.89] J. Landskron, "Analyse thermischer Schwingungen an Oberflächen mittels Beugung langsamer Elektronen", Dissertation, University of Munich, Dept. of Earth and Environmental Sciences, Germany, 1996.

[5.90] W. Moritz and J. Landskron, "Anisotropic Temperature Factors in the Calculation of Low-Energy Electron Diffraction Intensities", *Surf. Sci.*, vol. 337, pp. 278–284, 1995.

[5.91] W. Moritz and J. Landskron, "Thermal Vibrations at Surfaces Analysed with LEED", in *Solid-State Photoemission and Related Methods*, eds. W. Schattke and M. A. Van Hove, Wiley-VCH, Weinheim, 2003, pp. 433–459.

[5.92] P. L. de Andres and D. A. King, "Anisotropic and Anharmonic Effects through the t-Matrix for Low-Energy Electron Diffraction (TMAT V1.1)", *Comp.Phys. Commun.*, vol. 138, pp. 281–301, 2001.

[5.93] A. Messiah, *Quantum Mechanics*, Dover, Mineola, NY, vols. I & II, 2014.

[5.94] M. Blanco-Rey, P. L. de Andres, G. Held, and D. A. King, "A Molecular T-Matrix Approach to Calculating Low-Energy Electron Diffraction Intensities for Ordered Molecular Adsorbates", *Surf. Sci.*, vol. 579, p. 89–99, 2005.

[5.95] M. Blanco-Rey, P. de Andres, G. Held, and D. A. King, "Molecular t-Matrices for Low-Energy Electron Diffraction (TMOL v1.1)", *Comp. Phys. Commun.*, vol. 161, pp. 166–178, 2004.

[5.96] W. Riedl and D. Menzel, "ESDIAD Beam Widths", *Surf. Sci.*, vol. 207, pp. 494–451, 1989.

[6.1] P. R. Watson, M. A. Van Hove, and K. Hermann, *NIST Surface Structure Database Ver. 5.0, NIST Standard Reference Data Program*, Gaithersburg, MD, USA, 2004.

[6.2] M. A. Van Hove, P. R. Watson, and K. Hermann, "Adsorbate-Induced Changes in Surface Structure on Metals and Semiconductors", in Landolt-Börnstein, Group III: Condensed Matter, Vol. 42: *Physics of Covered Solid Surfaces*, Springer, Berlin, 2002, subvol. A, part 2, ch. 4.1, pp. 4.1-43–4.1-164.

[6.3] W. Moritz, "Elastic Scattering and Diffraction of Electrons and Positrons", in Landolt-Börnstein, Group III: Condensed Matter, Vol. 45: *Physics of Solid Surfaces*, Springer, Berlin, 2015, subvol. A, ch. 5, pp. 5.1-134–5.8-243.

[6.4] C. Giacovazzo, *Direct Phasing in Crystallography, Fundamentals and Applications*, Oxford University Press, Oxford, 1998.

[6.5] W. Clegg, A. J. Blake, R. O. Gould, and P. Main, *Crystal Structure Analysis, Principles and Practice*, Oxford University Press, Oxford, 2001.

[6.6] G. Oszlányi and A. Sütő, "Ab Initio Structure Solution by Charge Flipping", *Acta Cryst. A*, vol. 60, pp. 134–141, 2004.

[6.7] M. A. Van Hove, J. Cerdá, P. Sautet, M.-L. Bocquet, and M. Salmeron, "Surface Structure Determination by STM vs. LEED", *Progr. Surf. Sci.*, vol. 54, pp. 315–329, 1997.

[6.8] K. Heinz, L. Hammer, and S. Müller, "The Power of Joint Application of LEED and DFT in Quantitative Surface Structure Determination", *J. Phys.: Condens. Matter*, vol. 20, p. 304204, 2008.

[6.9] E. A. Soares, C. M. C. de Castilho, and V. E. de Carvalho, "Advances on Surface Structural Determination by LEED", *J. Phys.: Condens. Matter*, vol. 23, p. 303001, 2011.

[6.10] M. G. Lagally, T. C. Ngoc, and M. B. Webb, "Kinematic Low-Energy Electron-Diffraction Intensities from Averaged Data: A Method for Surface Crystallography", *Phys. Rev. Lett.*, vol. 26, pp. 1557–1559, 1971.

[6.11] D. L. Adams and U. Landman, "Further Evaluation of the Transform-Deconvolution Method for Surface-Structure Determination by Analysis of Low-Energy Electron-Diffraction Intensities", *Phys. Rev. B*, vol. 15, pp. 3775–3787, 1977.

[6.12] H. Wu and S. Y. Tong, "Surface Patterson Function by Inversion of Low-Energy Electron Diffraction I–V Spectra at Multiple Incident Angles", *Phys. Rev. Lett.*, vol. 87, p. 036101, 2001.

[6.13] S. Y. Tong and H. Wu, "Direct Inversion of Low-Energy Electron Diffraction (LEED) IV Spectra: The Surface Patterson Function", *J. Phys.: Condens. Matter*, vol. 14, pp. 1231–1236, 2002.

[6.14] J. Wang, H. S. Wu, R. So, Y. Liu, M. H. Xie, and S. Y. Tong, "Structure Determination of Indium-Induced Si(111)-In-4×1 Surface by LEED Patterson Inversion", *Phys. Rev. B*, vol. 72, p. 245324, 2005.

[6.15] T. Abukawa, T. Yamazaki, and S. Kono, "Fully Performed Constant-Momentum-Transfer-Averaging in Low-Energy Electron Diffraction Demonstrated for a Single-Domain Si(111)4×1-In Surface", *e-J. Surf. Sci. Nanotechn.*, vol. 4, pp. 661–668, 2006.

[6.16] C. Rogero, J.-A. Martin-Gago, and P. L. de Andres, "Patterson Function from Low-Energy Electron Diffraction Measured Intensities and Structural Discrimination", *Phys. Rev. B*, vol. 67, p. 073402, 2003.

[6.17] T. Kuzushita, A. Murata, A. Yamamoto, and A. Urano, "Surface Structure Analysis of Metal Adsorbed Si(111) Surfaces by Patterson Function with LEED I–V Curves", *Appl. Surf. Sci.*, vol. 254, pp. 7824–7826, 2008.

[6.18] C. Y. Chang, Z. C. Lin, Y. C. Chou, and C. M. Wei, "Direct Three-Dimensional Patterson Inversion of Low-Energy Electron Diffraction I(E) Curves", *Phys. Rev. Lett.*, vol. 83, pp. 2580–2583, 1999.

[6.19] J. Wang, R. So, Y. Liu, H. Wu, M. H. Xie, and S. Y. Tong, "Observation of a ($\sqrt{3} \times \sqrt{3}$)-R30° Reconstruction on GaN(0001) by RHEED and LEED", *Surf. Sci.*, vol. 600, pp. L169–L174, 2006.

[6.20] S. H. Xu, H. S. Wu, X. Q. Dai, W. P. Lau, L. X. Zheng, M. H. Xie, and S. Y. Tong, "Direct Observation of a Ga Adlayer on a GaN(0001) Surface by LEED Patterson Inversion", *Phys. Rev. B*, vol. 67, p. 125409, 2003.

[6.21] D. Gabor, "A New Microscopic Principle", *Nature*, vol. 161, pp. 777–778, 1948.

[6.22] S. Y. Tong, "Inversion of Low-Energy Electron Diffraction Data with No Pre-knowledge Factor beyond Optical Holography", *Adv. Phys.*, vol. 48, pp. 135–165, 1999.

[6.23] K. Heinz, U. Starke, and J. Bernhardt, "Surface Holography with LEED Electrons", *Progr. Surf. Sci.*, vol. 64, pp. 163–178, 2000.

[6.24] J. J. Barton, "Photoelectron Holography", *Phys. Rev. Lett.*, vol. 61, pp. 1356–1359, 1988.

[6.25] C. S. Fadley, M. A. Van Hove, A. Kaduwela, S. Omori, L. Zhao, and S. Marchesini, "Photoelectron and X-ray Holography by Contrast: Enhancing Image Quality and Dimensionality", *J. Phys.: Condens. Matter*, vol. 13, pp. 10517–10532, 2001.

[6.26] H. Daimon, "Overview of Three-Dimensional Atomic-Resolution Holography and Imaging Techniques: Recent Advances in Local-Structure Science", *J. Phys. Soc. Jpn.*, vol. 87, p. 061001, 2018.

[6.27] S. Omori, Y. Nihei, E. Rotenberg, J. D. Denlinger, S. D. Kevan, B. P. Tonner, M. A. Van Hove, and C. S. Fadley, "Differential Photoelectron Holography: A New Approach for Three-Dimensional Atomic Imaging", *Phys. Rev. Lett.*, vol. 88, p. 055504, 2002.

[6.28] A. Suzuki, A. Hashimoto, M. Nojima, M. Owari, and Y. Nihei, "Holographic Imaging of TiO_2 (110) Surface Structure by Differential Photoelectron Holography", *Surf. Interface Anal.*, vol. 40, pp. 1627–1630, 2008.

[6.29] D. K. Saldin and P. L. de Andres, "Holographic LEED", *Phys. Rev. Lett.*, vol. 64, pp. 1270–1273, 1990.

[6.30] D. K. Saldin, G. R. Harp, B. L. Chen, and B. P. Tonner, "Theoretical Principles of Holographic Crystallography", *Phys. Rev. B*, vol. 44, pp. 2480–2494, 1991.

[6.31] M. A. Mendez, C. Gluck, and K. Heinz, "Holography with Conventional LEED", *J. Phys.: Condens. Matter*, vol. 4, pp. 999–1006, 1992.

[6.32] P. J. Hu and D. A. King, "A Direct Inversion Method for Surface Structure Determination from LEED Intensities", *Nature*, vol. 360, pp. 655–658, 1992.

[6.33] C. M. Wei and S. Y. Tong, "Direct Atomic Structure by Holographic Diffuse LEED", *Surf. Sci.*, vol. 274, pp. L577–L582, 1992.

[6.34] D. K. Saldin and X. Chen, "Three-Dimensional Reconstruction by Holographic LEED: Proper Identification of the Reference Wave", *Phys. Rev. B*, vol. 52, pp. 2941–2948, 1995.

[6.35] D. K. Saldin, K. Reuter, P. L. de Andres, H. Wedler, X. Chen, J. B. Pendry, and K. Heinz, "Direct Reconstruction of Three-Dimensional Atomic Adsorption Sites by Holographic LEED", *Phys. Rev. B*, vol. 54, pp. 8172–8176, 1996.

[6.36] K. Reuter, J. Schardt, J. Bernhardt, K. Wedler, U. Starke, and K. Heinz, "LEED Holography Applied to a Complex Superstructure: A Direct View of the Adatom Cluster on SiC(111)-(3×3)", *Phys. Rev. B*, vol. 58, pp. 10806–10814, 1998.

[6.37] J. Schardt, J. Bernhardt, U. Starke, and K. Heinz, "Crystallography of the (3×3) Surface Reconstruction of 3C-SiC(111), 4H-SiC(0001), and 6H-SiC(0001) Surfaces Retrieved by Low-Energy Electron Diffraction", *Phys. Rev. B*, vol. 62, pp. 10335–10344, 2000.

[6.38] U. Starke, J. B. Pendry, and K. Heinz, "Diffuse Low-Energy Electron Diffraction", *Progr. Surf. Sci.*, vol. 52, pp. 53–123, 1996.

[6.39] A. Seubert and K. Heinz, "Iterative Image Recovery in Holographic LEED", *Surf. Rev. Lett.*, vol. 9, pp. 1413–1423, 2002.

[6.40] A. Seubert, K. Heinz, and D. K. Saldin, "Direct Determination by Low-Energy Electron Diffraction of the Atomic Structure of Surface Layers on a Known Substrate", *Phys. Rev. B*, vol. 67, p. 125417, 2003.

[6.41] D. Kolthoff, C. Schwennicke, and H. Pfnür, "Holographic Diffuse LEED Image Reconstruction for Simultaneous Occupation of Different Oxygen Adsorption Sites on Ni(111)", *Surf. Sci.*, vol. 529, pp. 443–454, 2003.

[6.42] H. Wu, S. Xu, S. Ma, W. P. Lau, M. H. Xie, and S. Y. Tong, "Surface Atomic Arrangement Visualization via Reference-Atom-Specific Holography", *Phys. Rev. Lett.*, vol. 89, p. 216101, 2002.

[6.43] T. Abukawa and S. Kono, "Semi-direct Method for Surface Structure Analysis Using Correlated Thermal Diffuse Scattering", *Progr. Surf. Sci.*, vol. 72, pp. 19–51, 2003.

[6.44] G. Kleinle, W. Moritz, and G. Ertl, "An Efficient Method for Surface Crystallography", *Surf. Sci.*, vol. 238, pp. 119–131, 1990.

[6.45] J. B. Pendry, "Reliability Factors for LEED Calculations", *J. Phys. C: Solid State Phys.*, vol. 13, pp. 937–944, 1980.

[6.46] Th. Sirtl, J. Jelic, J. Meyer, K. Das, W. M. Heckl, W. Moritz, J. Rundgren, M. Schmittel, K. Reuter, and M. Lackinger, "Adsorption Structure Determination of a Large Polyaromatic Trithiolate on Cu(111): Combination of LEED-I(V) and DFT-vdW", *Phys. Chem. Chem. Phys.*, vol. 15, pp. 11054–11060, 2013,

[6.47] K. Pussi, H. I. Li, H. Shin, L. N. Serkovic Loli, A. K. Shukla, J. Ledieu, V. Fournée, L. L. Wang, S. Y. Su, K. E. Marino, M. V. Snyder, and R. D. Diehl, "Elucidating the Dynamical Equilibrium of C_{60} Molecules on Ag(111)", *Phys. Rev. B*, vol. 86, p. 205406, 2012.

[6.48] E. Primorac, H. Kuhlenbeck, and H.-J. Freund, "LEED I/V Determination of the Structure of a MoO_3 Monolayer on Au(111): Testing the Performance of the CMA-ES Evolutionary Strategy Algorithm, Differential Evolution, a Genetic Algorithm and Tensor LEED Based Structural Optimization", *Surf. Sci.*, vol. 649, pp. 90–100, 2016.

[6.49] E. Zanazzi and F. Jona, "A Reliability Factor for Surface Structure Determination by Low Energy Electron Diffraction", *Surf. Sci.*, vol. 62, pp. 61–80, 1977.

[6.50] J. Philip and J. Rundgren, "Metric Distance between LEED Spectra", in *Determination of Surface Structure by LEED*, eds. P. M. Marcus and F. Jona, Plenum, New York, 1984, pp. 409–437.

[6.51] J. Rundgren, "Electron Inelastic Mean Free Path, Electron Attenuation Length, and Low-Energy Electron-Diffraction Theory", *Phys. Rev. B*, vol. 59, pp. 5106–5114, 1999.

[6.52] S. Tanuma, C. J. Powell, and D. R. Penn, "Calculations of Electron Inelastic Mean Free Paths. IX. Data for 41 Elemental Solids over the 0 eV to 30 keV Range", *Surf. Interface Anal.*, vol. 43, pp. 689–713, 2011.

[6.53] D. L. Adams, "A Simple and Effective Procedure for the Refinement of Surface Structure in LEED", *Surf. Sci.*, vol. 519, pp. 157–172, 2002.

[6.54] C. B. Duke, A. Lazarides, A. Paton, and Y. R. Wang, "Statistical Error Analysis of Surface-Structure Parameters Determined by Low-Energy Electron and Positron Diffraction: Data Errors", *Phys. Rev. B*, vol. 52, pp. 14878–14894, 1995.

[6.55] N. Mulakaluri, R. Pentcheva, M. Wieland, W. Moritz, and M. Scheffler, "Partial Dissociation of Water on $Fe_3O_4(001)$: Adsorbate Induced Charge and Orbital Order", *Phys. Rev. Lett.,* vol. 103, p. 176102, 2009.

[6.56] J. P. Hofmann, S. F. Rohrlack, F. Heß, J. C. Goritzka, P. P. T. Krause, A. P. Seitsonen, W. Moritz, and H. Over, "Adsorption of Chlorine on Ru(0001): A Combined Density Functional Theory and Quantitative Low Energy Electron Diffraction Study", *Surf. Sci.,* vol. 606, pp. 297–304, 2012.

[6.57] H. I. Li, K. Pussi, K. J. Hanna, L.-L. Wang, D. D. Johnson, H.-P. Cheng, H. Shin, S. Curtarolo, W. Moritz, J. A. Smerdon, R. McGrath, and R. D. Diehl, "Surface Geometry of C_{60} on Ag(111)", *Phys. Rev. Lett.,* vol. 103, p. 056101, 2009.

[6.58] R. Wyrwich, T. E. Jones, S. Günther, W. Moritz, M. Ehrensperger, S. Böcklein, P. Zeller, A. Lünser, A. Locatelli, T. O. Menteş, M. Á. Niño, A. Knop-Gericke, R. Schlögl, S. Piccinin, and J. Wintterlin, "LEED−I(V) Structure Analysis of the $(7\times\sqrt{3})$rect SO_4 Phase on Ag(111): Precursor to the Active Species of the Ag-Catalyzed Ethylene Epoxidation", *J. Phys. Chem. C,* vol. 122, pp. 26998–27004, 2018.

[6.59] H. Over, U. Ketterl, W. Moritz, and G. Ertl, "Optimisation Methods and Their Use in Low-Energy Electron-Diffraction Calculations", *Phys. Rev. B,* vol. 46, pp. 15438–15446, 1992.

[6.60] D. W. Marquardt, "An Algorithm for Least-Squares Estimation of Nonlinear Parameters", *J. Soc. Indust. Appl. Math.,* vol. 11, pp. 431–441, 1963.

[6.61] R. Reichelt, S. Günther, J. Wintterlin, W. Moritz, L. Aballe, and T. O. Mentes, "Low Energy Electron Diffraction and Low Energy Electron Microscopy Microspot I/V Analysis of the (4×4) O Structure on Ag(111): Surface Oxide or Reconstruction", *J. Chem. Phys.,* vol. 127, p. 134706, 2007.

[6.62] M. Lübbe and W. Moritz, "A LEED Analysis of the Clean Surfaces of α-$Fe_2O_3(0001)$ and α-$Cr_2O_3(0001)$ Bulk Single Crystals", *J. Phys.: Condens. Matter,* vol. 21, p. 134010, 2009.

[6.63] P. J. Rous, "The Tensor LEED Approximation and Surface Crystallography by Low-Energy Electron Diffraction", *Progr. Surf. Sci.,* vol. 39, pp. 3–63, 1992.

[6.64] W. H. Press, B. P. Flannery, S. A. Teukolsky, and W. T. Vetterling, *Numerical Recipes*, Cambridge University Press, Cambridge, 1986.

[6.65] R. P. Brent, *Algorithms for Minimization without Derivatives*, Prentice-Hall, Englewood Cliffs, NJ, 1973.

[6.66] H. H. Rosenbrook, "An Automatic Method for Finding the Greatest or Least Value of a Function", *Comput. J.,* vol. 3, pp. 175–184, 1960.

[6.67] G. H. Stout and L. H. Jensen, *X-ray Structure Determination*, MacMillan, New York, 1968.

[6.68] P. G. Cowell and V. E. de Carvalho, "Unconstrained Optimisation in Surface Crystallography by LEED: Preliminary Results of its Application to CdTe(110)", *Surf. Sci.,* vol. 187, pp. 175–193, 1987.

[6.69] M. J. D. Powell, "The BOBYQA Algorithm for Bound Constrained Optimization without Derivatives" (Report), Department of Applied Mathematics and Theoretical Physics, Cambridge University, DAMTP 2009/NA06, www.damtp.cam.ac.uk/user/na/NA_papers/NA2009_06.pdf, last accessed 2 April 2022.

[6.70] P. J. Rous, "A Global Approach to the Search Problem in Surface Crystallography by Low-Energy Electron Diffraction", *Surf. Sci.,* vol. 296, pp. 358–373, 1993.

[6.71] H. Szu and R. Hartley, "Fast Simulated Annealing", *Phys. Lett. A,* vol. 122, pp. 157–162, 1987.

[6.72] C. Tsallis, "Possible Generalization of Boltzmann-Gibbs Statistics", *J. Stat. Phys.*, vol. 52, pp. 479–487, 1988.

[6.73] V. B. Nascimento, V. E. de Carvalho, C. M. C. de Castilho, B. V. Costa, and E. A. Soares, "The Fast Simulated Annealing Algorithm Applied to the Search Problem in LEED", *Surf. Sci.*, vol. 487, pp. 15–27, 2001.

[6.74] E. dos R. Correia, V. B. Nascimento, C. M. C. de Castilho, E. A. Soares, and V. E. de Carvalho, "The Generalized Simulated Annealing Algorithm in the Low Energy Electron Diffraction Search Problem", *J. Phys.: Condens. Matter*, vol. 17, pp. 1–16, 2005.

[6.75] M. Kottcke and K. Heinz, "A New Approach to Automated Structure Optimization in LEED Intensity Analysis", *Surf. Sci.*, vol. 376, pp. 352–366, 1997.

[6.76] M. L. Viana, D. D. dos Reis, E. A. Soares, M. A. Van Hove, W. Moritz, and V. E. de Carvalho, "Novel Genetic Algorithm Searching Procedure for LEED Surface Structure Determination", *J. Phys.: Condens. Matter*, vol. 26, p. 225005, 2014.

[6.77] R. Döll and M. A. Van Hove, "Global Optimization in LEED Structure Determination Using Genetic Algorithms", *Surf. Sci.*, vol. 355, pp. L393–L398, 1996.

[6.78] M. L. Viana, R. Diez Muiño, E. A. Soares, M. A. Van Hove, and V. E. de Carvalho, "Global Search in Photoelectron Diffraction Structure Determination Using Genetic Algorithms", *J. Phys.: Condens. Matter*, vol. 19, p. 446002, 2007.

[6.79] M. L. Viana, W. Simões e Silva, E. A. Soares, V. E. de Carvalho, C. M. C. de Castilho, and M. A. Van Hove, "Scaling Behavior of Genetic Algorithms Applied to Surface Structural Determination by LEED", *Surf. Sci.*, vol. 602, pp. 3395–3402, 2008.

[6.80] R. Storn and K. Price, "Differential Evolution: A Simple and Efficient Heuristic for Global Optimization over Continuous Spaces", *J. Global Optimization*, vol. 11, pp. 341–359, 1997.

[6.81] N. Hansen, "The CMA Evolution Strategy: A Comparing Review", in *Towards a New Evolutionary Computation. Advances on Estimation of Distribution Algorithms*, eds. J. A. Lozano, P. Larranaga, I. Inza, and E. Bengoetxea, Springer, Berlin, 2006, pp. 75–102.

[6.82] V. B. Nascimento and E. W. Plummer, "Differential Evolution: Global Search Problem in LEED-IV Surface Structural Analysis", *Mater. Charact.*, vol. 100, pp. 143–151, 2015.

[6.83] Z. J. Zhao, J. C. Meza, and M. Van Hove, "Using Pattern Search Methods for Surface Structure Determination of Nanomaterials", *J. Phys.: Condens. Matter*, vol. 18, pp. 8693–8706, 2006.

[6.84] C. Audet and J. E. Dennis, Jr., "Pattern Search Algorithms for Mixed Variable Programming", *SIAM J. Optim.*, vol. 11, pp. 573–594, 1999.

[6.85] T. G. Kolda, R. M. Lewis, and V. Torczon, "Optimization by Direct Search: New Perspectives on Some Classical and Modern Methods", *SIAM Rev.*, vol. 45, pp. 385–482, 2003.

[6.86] M. A. Abramson, C. Audet, J. W. Chrissis, and J. G. Walston, "Mesh Adaptive Direct Search Algorithms for Mixed Variable Optimization", *Optim. Lett.*, vol. 3, pp. 35–47, 2009.

[6.87] H. Jiang, S. Mizuno, and H. Tochihara, "Adsorption Mode Change from Adlayer- to Restructuring-Type with Increasing Coverage, Evidenced by Structural Determination of c(2×2)→(4×4)→(5×5) Sequence Formed on Ni(001) by Li Deposition", *Surf. Sci.*, vol. 380, pp. L506–L512, 1997.

[6.88] K. Schwarz, "DFT Calculations of Solids with LAPW and WIEN2k", *J. Solid State Chem.*, vol. 176, pp. 319–328, 2003.

[6.89] T. L. Loucks, *Augmented Plane Wave Method*, Benjamin, New York, 1967.

[6.90] L. F. Mattheiss, "Energy Bands for Solid Argon", *Phys. Rev. A*, vol. 133, pp. A1399–A1403, 1964.

[6.91] J. B. Pendry, D. J. Titterington, S. J. Gurman, and G. Aers, *MUFPOT code*, Daresbury, 1970. No longer available.

[6.92] A. Barbieri and M. A. Van Hove, phase shift code, https://www.icts.hkbu.edu.hk/VanHove_files/leed/phshift2007.zip, last accessed 2 April 2022.

[6.93] C. G. Kinniburgh, "A LEED Study of MgO(100). II. Theory at Normal Incidence", *J. Phys. C: Solid State Phys.*, vol. 8, pp. 2382–2394, 1975.

[6.94] C. G. Kinniburgh and J. A. Walker, "LEED Calculations for the NiO(100) Surface", *Surf. Sci.*, vol. 63, pp. 274–282, 1977.

[6.95] R. Lindsay, A. Wander, A. Ernst, B. Montanari, G. Thornton, and N. M. Harrison, "Revisiting the Surface Structure of $TiO_2(110)$: A Quantitative Low-Energy Electron Diffraction Study", *Phys. Rev. Lett.*, vol. 94, p. 246102, 2005.

[6.96] S. E. Chamberlin, C. J. Hirschmugl, H. C. Poon, and D. K. Saldin, "Geometric Structure of $TiO_2(011)(2\times1)$ Surface by Low Energy Electron Diffraction (LEED)", *Surf. Sci.*, vol. 603, pp. 3367–3373, 2009.

[6.97] J. Rundgren, "Optimized Surface-Slab Excited-State Muffin-Tin Potential and Surface Core Level Shifts", *Phys. Rev. B*, vol. 68, p. 125405, 2003.

[6.98] J. Rundgren, "Elastic Electron-Atom Scattering in Amplitude-Phase Representation with Application to Electron Diffraction and Spectroscopy", *Phys. Rev. B*, vol. 76, p. 195441, 2007.

[6.99] L. Hedin and B. I. Lundqvist, "Explicit Local Exchange-Correlation Potentials", *J. Phys. C*, vol. 4, pp. 2064–2083, 1971.

[6.100] G. D. Mahan and B. E. Sernelius, "Electron–Electron Interactions and the Band Width of Metals", *Phys. Rev. Lett.*, vol. 62, pp. 2718–2721, 1989.

[6.101] V. B. Nascimento, R. G. Moore, J. Rundgren, J. Zhang, L. Cai, R. Jin, D. G. Mandrus, and E. W. Plummer, "Procedure for LEED I–V Structural Analysis of Metal Oxide Surfaces: $Ca_{1.5}Sr_{0.5}RuO_4(001)$", *Phys. Rev. B*, vol. 75, p. 035408, 2007.

[6.102] NIST Electron Inelastic-Mean-Free-Path Database 71, Version 1.1: https://www.nist.gov/srd/nist-standard-reference-database-71, last accessed 5 April 2022.

[6.103] J. P. Ruf, P. D. C. King, V. B. Nascimento, D. G. Schlom, and K. M. Shen, "Surface Atomic Structure of Epitaxial $LaNiO_3$ Thin Films Studied by in situ LEED-I(V)", *Phys. Rev. B*, vol. 95, p. 115418, 2017.

[6.104] J. Rundgren, B. E. Sernelius, and W. Moritz, "Low-Energy Electron Diffraction with Signal Electron Carrier-Wave Wavenumber Modulated by Signal Exchange–Correlation Interaction", *J. Phys. Commun.*, vol. 5, p. 105012, 2021.

[6.105] D. K. Saldin and J. C. H. Spence, "On the Mean Inner Potential in High- and Low-Energy Electron Diffraction", *Ultramicroscopy*, vol. 55, pp. 397–406, 1994.

[6.106] D. F.-T. Tuan and G. W. Loar, "Corrections of Overlapping Spheres in the Self-Consistent Field Xα Scattered-Wave Method", *J. Mol. Struct.*, vol. 223, pp. 123–148, 1990.

[6.107] D. F.-T. Tuan and G. W. Loar, "Corrections of Overlapping Spheres in the Self-Consistent Field Xα Scattered-Wave Method", *J. Mol. Struct.*, vol. 228, pp. 167–180, 1991.

[6.108] L. Vitos, H. L. Skriver, B. Johansson, and J. Kollár, "Application of the Exact Muffin-Tin Orbitals Theory: The Spherical Cell Approximation", *Comp. Mat. Sci.*, vol. 18, pp. 24–38, 2000.

[6.109] R. E. Watson, J. F. Herbst, L. Hodges, B. I. Lundqvist, and J. W. Wilkins, "Effect of Ground-State and Excitation Potentials on Energy Levels of Ni Metal", *Phys. Rev. B*, vol. 13, pp. 1463–1467, 1976.

[6.110] W. Moritz, unpublished.

[6.111] S. Walter, V. Blum, L. Hammer, S. Müller, K. Heinz, and M. Giesen, "The Role of an Energy-Dependent Inner Potential in Quantitative Low-Energy Electron Diffraction", *Surf. Sci.*, vol. 458, pp. 155–161, 2000.

[6.112] J. Vuorinen, K. Pussi, R. D. Diehl, and M. Lindroos, "Correlation of Electron Self-Energy with Geometric Structure in Low-Energy Electron Diffraction", *J. Phys.: Condens. Matter*, vol. 24, p. 015003, 2012.

[6.113] M. P. Seah and W. A. Dench, "Quantitative Electron Spectroscopy of Surfaces: A Standard Data Base for Electron Inelastic Mean Free Paths in Solids", *Surf. Interface Anal.*, vol. 1, pp. 2–11, 1979.

[6.114] A. Jablonski and C. J. Powell, "The Electron Attenuation Length Revisited", *Surf. Sci. Rep.*, vol. 47, pp. 33–91, 2002.

[6.115] C. J. Powell and A. Jablonski, "Evaluation of Calculated and Measured Electron Inelastic Mean Free Paths Near Solid Surfaces", *J. Phys. Chem. Ref. Data*, vol. 28, pp. 19–62, 1999.

[6.116] P. de Vera and R. Garcia-Molina, "Electron Inelastic Mean Free Paths in Condensed Matter Down to a Few Electronvolts", *J. Phys. Chem. C*, vol. 123, pp. 2075–2083, 2019.

[6.117] S. Tanuma, C. J. Powell, and D. R. Penn, "Calculations of Electron Inelastic Mean Free Paths. IX. Data for 41 Elemental Solids over the 50 eV to 30 keV Range", *Surf. Interface Anal.*, vol. 43, pp. 689–713, 2011.

[6.118] O. Yu. Ridzel, V. Astašauskas, and W. S. M. Werner, "Low Energy (1–100 eV) Electron Inelastic Mean Free Path (IMFP) Values Determined from Analysis of Secondary Electron Yields (SEY) in the Incident Energy Range of 0.1–10 keV", *J. Electr. Spectr. Rel. Phen.*, vol. 241, p. 146824, 2020.

[6.119] V. Zielasek, A. Büssenschütt, and M. Henzler, "Low-Energy Electron Thermal Diffuse Scattering from Al(111) Individually Resolved in Energy and Momentum", *Phys. Rev. B*, vol. 55, pp. 5398–5403, 1997.

[6.120] S. van Smaalen, "Incommensurate Crystal Structures", *Cryst. Rev.*, vol. 4, pp. 19–202, 1995.

[6.121] W. Steurer and T. Haibach, "Reciprocal-Space Images of Aperiodic Crystals", in *International Tables for Crystallography*, Kluwer Academic Publishers, Dordrecht, vol. B, ch. 4.6, pp. 486–532, 2006.

[6.122] D. Shechtman, I. Blech, D. Gratias, and J. W. Cahn, "Metallic Phase with Long-Range Orientational Order and No Translational Symmetry", *Phys. Rev. Lett.*, vol. 53, pp. 1951–1954, 1984.

[6.123] C. Janot, *Quasicrystals: A Primer*, Clarendon Press, Oxford, 1994.

[6.124] W. Steurer, "Twenty Years of Structure Research on Quasicrystals. Part I. Pentagonal, Octagonal, Decagonal and Dodecagonal Quasicrystals", *Z. Kristallogr.*, vol. 219, pp. 391–446, 2004.

[6.125] H.-R. Trebin, ed., *Quasicrystals, Structure and Physical Properties*, Wiley-VCH, Weinheim, 2003.

[6.126] B. Suck, M. Schreiber, and P. Haussler, eds., *Quasicrystals: An Introduction to Structure, Physical Properties and Applications*, Springer, Berlin, 2002.

[6.127] J. M. Dubois, *Useful Quasicrystals*, World Scientific Press, Singapore, 2005.

[6.128] W. Steurer and S. Deloudi, *Crystallography of Quasicrystals: Concepts, Methods and Structures*, Springer, Berlin, 2009.

[6.129] W. J. Boettinger, F. S. Biancaniello, G. M. Kalonji, and J. W. Cahn, "Eutectic Solidification and the Formation of Metallic Glasses", in *Proceedings of the Second International Conference on Rapid Solidification Processing: Principles and Technologies, II*, ed. R. Mehrabian, B. H. Kear and M. Cohen, Claitor's Publishing Division, Baton Rouge, 1980, pp. 50–55.

[6.130] P. Gille, P. Dreier, M. Gräber, and T. Scholpp, "Large Single-Grain AlCoNi Quasicrystals Grown by the Czochralski Method", *J. Cryst. Growth*, vol. 207, pp. 95–101, 1999.

[6.131] R. McGrath, J. Ledieu, E. J. Cox, and R. D. Diehl, "Quasicrystal Surfaces: Structure and Potential as Templates", *J. Phys.: Condens. Matter*, vol. 14, pp. R119–R144, 2002.

[6.132] R. D. Diehl, J. Ledieu, N. Ferralis, A. W. Szmodis, and R. McGrath, "Low-Energy Electron Diffraction from Quasicrystal Surfaces", *J. Phys.: Condens. Matter*, vol. 15, pp. R63–R81, 2003.

[6.133] R. Penrose, "The Role of Aesthetics in Pure and Applied Mathematical Research", *Bull. Inst. Math. & Appl.*, vol. 10, pp. 266–271, 1974.

[6.134] K. Hermann, *Crystallography and Surface Structure: An Introduction for Surface Scientists and Nanoscientists*, 2nd ed., Wiley, Weinheim, 2016.

[6.135] T. Cai, F. Shi, Z. Shen, M. Gierer, A. I. Goldman, M. J. Kramer, C. J. Jenks, T. A. Lograsso, D. W. Delaney, P. A. Thiel, and M. A. Van Hove, "Structural Aspects of the Fivefold Quasicrystalline Al-Cu-Fe Surface from STM and Dynamical LEED Studies", *Surf. Sci.*, vol. 495, pp. 19–34, 2001.

[6.136] A. L. MacKay, "Crystallography and the Penrose Pattern", *Physica*, vol. 114A, pp. 609–613, 1982.

[6.137] W. Steurer, T. Haibach, B. Zhang, S. Kek, and R. Lück, "The Structure of Decagonal $Al_{70}Ni_{15}Co_{15}$", *Acta Cryst. B*, vol. 49, pp. 661–675, 1993.

[6.138] M. Gierer, M. A. Van Hove, A. I. Goldman, Z. Shen, S.-L. Chang, C. J. Jenks, C.-M. Zhang, and P. A. Thiel, "Structural Analysis of the Fivefold Symmetric Surface of the $Al_{70}Pd_{21}Mn_9$ Quasicrystal by Low Energy Electron Diffraction", *Phys. Rev. Lett.*, vol. 78, pp. 467–470, 1997.

[6.139] M. Gierer, M. A. Van Hove, A. I. Goldman, Z. Shen, S.-L. Chang, P. J. Pinhero, C. J. Jenks, J. W. Anderegg, C.-M. Zhang, and P. A. Thiel, "Fivefold Surface of Quasicrystalline AlPdMn: Structure Determination Using Low-Energy-Electron Diffraction", *Phys. Rev. B*, vol. 57, pp. 7628–7641, 1998.

[6.140] M. Gierer, "Struktur und Fehlordnung an periodischen und aperiodischen Kristalloberflächen", Habilitation, Faculty of Geosciences, University of Munich, 2000.

[6.141] M. Gierer, A. Mikkelsen, M. Gräber, P. Gille, and W. Moritz, "Quasicrystalline Surface Order on Decagonal $Al_{72.1}Ni_{11.5}Co_{16.4}$: An Investigation with Spot Profile Analysis LEED", *Surf. Sci. Lett.*, vol. 463, pp. L654–L660, 2000.

[6.142] T. M. Schaub, D. E. Bürgler, H. J. Güntherodt, and J. B. Suck, "Quasicrystalline Structure of Icosahedral $Al_{68}Pd_{23}Mn_9$ Resolved by Scanning Tunneling Microscopy", *Phys. Rev. Lett.*, vol. 73, pp. 1255–1258, 1994.

[6.143] J. Ledieu, A. Munz, T. Parker, R. McGrath, R. D. Diehl, D. W. Delaney, and T. A. Lograsso, "Structural Study of the Five-fold Surface of the $Al_{70}Pd_{21}Mn_9$ Quasicrystal", *Surf. Sci.*, vol. 433–435, pp. 666–671, 1999.

[6.144] J. Ledieu, R. McGrath, R. D. Diehl, T. A. Lograsso, D. W. Delaney, Z. Papadopolos, and G. Kasner, "Tiling of the Surface of $Al_{70}Pd_{21}Mn_9$", *Surf. Sci.*, vol. 492, pp. L729–L734, 2001.

[6.145] X. Wu, S. W. Kycia, C. G. Olson, P. J. Benning, A. I. Goldman, and D. W. Lynch, "Electronic Band Dispersion and Pseudogap in Quasicrystals: Angular-Resolved Photoemission Studies on Icosahedral $Al_{70}Pd_{21.5}Mn_{8.5}$", *Phys. Rev. Lett.*, vol. 75, pp. 4540–4543, 1995.

[6.146] Z. Shen, W. Raberg, M. Heinzig, C. J. Jenks, V. Fournée, M. A. Van Hove, T. A. Lograsso, D. Delaney, T. Cai, P. C. Canfield, I. R. Fisher, A. I. Goldman, M. J. Kramer, and P. A. Thiel, "A LEED Comparison of Structural Stabilities of the Three High-Symmetry Surfaces of Al-Pd-Mn Bulk Quasicrystals", *Surf. Sci.*, vol. 450, pp. 1–11, 2000.

[6.147] A. R. Kortan, R. S. Becker, F. A. Thiel, and H. S. Chen, "Real-Space Atomic Structure of a Two-Dimensional Decagonal Quasicrystal", *Phys. Rev. Lett.*, vol. 64. pp. 200–203, 1990.

[6.148] M. Erbudak, H.-U. Nissen, E. Wetli, M. Hochstrasser, and S. Ritsch, "Real-Space Imaging of Pentagonal Symmetry Elements by 2-keV Electrons in the Icosahedral Quasicrystal $Al_{70}Mn_9Pd_{21}$", *Phys. Rev. Lett.*, vol. 72, pp. 3037–3040, 1994.

[6.149] P. A. Thiel, "Quasicrystal Surfaces", *Annu. Rev. Phys. Chem.*, vol. 59, pp. 129–152, 2008.

[6.150] N. Ferralis, K. Pussi, E. J. Cox, M. Gierer, J. Ledieu, I. R. Fisher, C. J. Jenks, M. Lindroos, R. McGrath, and R. D. Diehl, "Structure of the Tenfold d-Al-Ni-Co Quasicrystal Surface", *Phys. Rev. B*, vol. 69, p. 153404, 2004.

[6.151] K. Pussi, N. Ferralis, M. Mihalkovic, M. Widom, S. Curtarolo, M. Gierer, C. J. Jenks, P. Canfield, I. R. Fisher, and R. D. Diehl, "Use of Periodic Approximants in a Dynamical LEED Study of the Quasicrystalline Tenfold Surface of Decagonal Al-Ni-Co", *Phys. Rev. B*, vol. 73, p. 184203, 2006.

[6.152] S.-L. Chang, W. B. Chin, C.-M. Zhang, C. J. Jenks, and P. A. Thiel, "Oxygen Adsorption on a Single-Grain, Quasicrystal Surface", *Surf. Sci.*, vol. 337, pp. 135–146, 1995.

[6.153] M. Gierer and H. Over, "Complex Surface Structures Studied by Low-Energy Electron Diffraction", *Z. Kristallogr.*, vol. 214, pp. 14–56, 1999.

[6.154] W. Steurer, "Reflections on Symmetry and Formation of Axial Quasicrystals", *Z. Kristallogr.*, vol. 221, pp. 402–411, 2006.

[6.155] W. Steurer, "Quasicrystals: What Do We Know? What Do We Want to Know? What Can We Know?" *Acta Cryst. A*, vol. 74, pp. 1–11, 2018.

[6.156] W. Moritz, S. Günther, and K. Pussi, "Quantitative LEED Studies on Graphene", in *Encyclopedia of Interfacial Chemistry: Surface Science and Electrochemistry*, ed. K. Wandelt, Elsevier, Amsterdam, 2018, vol. 4, pp. 370–377.

[6.157] S. Hagstrom, H. B. Lyon, and G. A. Somorjai, "Surface Structures of the Clean Platinum (100) Surface", *Phys. Rev. Lett.*, vol. 15, pp. 491–493, 1965.

[6.158] D. G. Fedak and N. A. Gjostein, "Structure and Stability of the (100) Surface of Gold", *Phys. Rev. Lett.*, vol. 16, pp. 171–174, 1966.

[6.159] P. Heilmann, K. Heinz, and K. Müller, "The Superstructures of the Clean Pt(100) and Ir (100) Surfaces", *Surf. Sci.*, vol. 83, pp. 487–497, 1979.

[6.160] K. Heinz, E. Lang, K. Strauss, and K. Müller, "Metastable 1×5 Structure of Pt(100)", *Surf. Sci.*, vol. 120, pp. L401–L404, 1982.

[6.161] D. Gibbs, G. Grübel, D. M. Zehner, D. L. Abernathy, and S. G. J. Mochrie, "Orientational Epitaxy of the Hexagonally Reconstructed Pt(001) Surface", *Phys. Rev. Lett.*, vol. 67, pp. 3117–3120, 1991.

[6.162] K. Kuhnke, K. Kern, G. Comsa, and W. Moritz, "Top-Layer Superstructures of the Reconstructed Pt(100) Surface", *Phys. Rev. B*, vol. 45, pp. 14388–14392, 1992.

[6.163] G. Ritz, M. Schmid, P. Varga, A. Borg, and M. Rønning, "Pt(100) Quasihexagonal Reconstruction: A Comparison between Scanning Tunneling Microscopy Data and Effective Medium Theory Simulation Calculations", *Phys. Rev. B*, vol. 56, pp. 10518–10526, 1997.

[6.164] R. Hammer, K. Meinel, O. Krahn, and W. Widdra, "Surface Reconstruction of Pt(001) Quantitatively Revisited", *Phys. Rev. B*, vol. 94, p. 195406, 2016.

[6.165] M. A. Van Hove, R. J. Koestner, P. C. Stair, J. P. Bibérian, L. L. Kesmodel, I. Bartoš, and G. A. Somorjai, "The Surface Reconstructions of the (100) Crystal Faces of Iridium, Platinum and Gold: I. Experimental Observations and Possible Structural Models", *Surf. Sci.*, vol. 103, pp. 189–217, 1981.

[6.166] D. Gibbs, B. M. Ocko, D. M. Zehner, and S. G. J. Mochrie, "Structure and Phases of the Au(001) Surface: In-Plane Structure", *Phys. Rev. B*, vol. 42, pp. 7330–7344, 1990.

[6.167] B. M. Ocko, D. Gibbs, K. G. Huang, D. M. Zehner, and S. G. J. Mochrie, "Structure and Phases of the Au(001) Surface: Absolute X-ray Reflectivity", *Phys. Rev. B*, vol. 44, pp. 6429, 1991.

[6.168] D. L. Abernathy, S. G. J. Mochrie, D. M. Zehner, G. Grübel, and D. Gibbs, "Orientational Epitaxy and Lateral Structure of the Hexagonally Reconstructed Pt(001) and Au(001) Surfaces", *Phys. Rev. B*, vol. 45, pp. 9272–9291, 1992.

[6.169] R. Hammer, A. Sander, S. Förster, M. Kiel, K. Meinel, and W. Widdra, "Surface Reconstruction of Au(001): High-Resolution Real-Space and Reciprocal-Space Inspection", *Phys. Rev. B*, vol. 90, p. 035446, 2014.

[6.170] J. Perdereau, J. P. Bibérian, and G. E. Rhead, "Adsorption and Surface Alloying of Lead Monolayers on (111) and (110) Faces of Gold", *J. Phys. F: Met. Phys.*, vol. 4, pp. 798–806, 1974.

[6.171] A. R. Sandy, S. G. J. Mochrie, D. M. Zehner, K. G. Huang, and D. Gibbs, "Structure and Phases of the Au(111) Surface: X-ray-Scattering Measurements", *Phys. Rev. B*, vol. 43, pp. 4667–4687, 1991.

[6.172] F. Besenbacher, J. V. Lauritsen, T. R. Linderoth, E. Lægsgaard, R. T. Vang, and S. Wendt, "Atomic-Scale Surface Science Phenomena Studied by Scanning Tunneling Microscopy", *Surf. Sci.*, vol. 603, pp. 1315–1327, 2009.

[6.173] H. Zajonz, A. P. Baddorf, D. Gibbs, and D. M. Zehner, "Structure of Pseudomorphic and Reconstructed Thin Cu Films on Ru(0001)", *Phys. Rev. B*, vol. 62, pp. 10436–10444, 2000.

[6.174] H. Zajonz, D. Gibbs, A. P. Baddorf, and D. M. Zehner, "Nanoscale Strain Distribution at the Ag/Ru(0001) Interface", *Phys. Rev. B*, vol. 67, p. 155417, 2003.

[6.175] B. Narloch and D. Menzel, "The Geometry of Xenon and Krypton on Ru(001): A LEED IV Investigation", *Surf. Sci.*, vol. 412/413, pp. 562–579, 1998.

[6.176] R. D. Diehl and R. McGrath, "Structural Studies of Alkali Metal Adsorption and Coadsorption on Metal Surfaces", *Surf. Sci. Rep.*, vol. 23, pp. 43–171, 1996.

[6.177] H. I. Li, K. J. Franke, J. I. Pascual, L. W. Bruch, and R. D. Diehl, "Origin of Moiré Structures in C_{60} on Pb(111) and Their Effect on Molecular Energy Levels", *Phys. Rev. B*, vol. 80, p. 085415, 2009.

[6.178] S. Marchini, S. Günther, and J. Wintterlin, "Scanning Tunneling Microscopy of Graphene on Ru(0001)", *Phys. Rev. B*, vol. 76, p. 075429, 2007.

[6.179] D. Martoccia, P. R. Willmott, T. Brugger, M. Björck, S. Günther, C. M. Schlepütz, A. Cervellino, S. A. Pauli, B. D. Patterson, S. Marchini, J. Wintterlin, W. Moritz, and T. Greber, "Graphene on Ru(0001): A 25×25 Supercell", *Phys. Rev. Lett.*, vol. 101, p. 126102, 2008.

[6.180] W. Moritz, B. Wang, M.-L. Bocquet, T. Brugger, T. Greber, J. Wintterlin, and S. Günther, "Structure Determination of the Coincidence Phase of Graphene on Ru (0001)", *Phys. Rev. Lett.*, vol. 104, p. 136102, 2010.

[6.181] S. K. Hämäläinen, M. P. Boneschanscher, P. H. Jacobse, I. Swart, K. Pussi, W. Moritz, J. Lahtinen, P. Liljeroth, and J. Sainio: "Structure and Local Variations of the Graphene Moiré on Ir(111)", *Phys. Rev. B*, vol. 88, p. 201406(R), 2013.

[6.182] J. Fabien, T. Zhou, N. Blanc, R. Felici, J. Coraux, and G. Renaud, "Topography of the Graphene/Ir(111) Moiré Studied by Surface X-ray Diffraction", *Phys. Rev. B*, vol. 91, p. 245424, 2015.

[6.183] S. van Smaalen, *Incommensurate Crystallography*, Oxford University Press, New York, 2007.

[6.184] P. Zeller and S. Günther, "What Are the Possible Moiré Patterns of Graphene on Hexagonally Packed Surfaces? Universal Solution for Hexagonal Coincidence Lattices, Derived by a Geometric Construction", *New J. Phys.*, vol. 16, p. 083028, 2014.

[6.185] P. Zeller, X. Ma, and S. Günther, "Indexing Moiré Patterns of Metal-Supported Graphene and Related Systems: Strategies and Pitfalls", *New. J. Phys.*, vol. 19, p. 013015, 2017.

[6.186] S. Günther and P. Zeller, "Moiré Patterns of Graphene on Metals", in *Encyclopedia of Interfacial Chemistry: Surface Science and Electrochemistry*, ed. K. Wandelt, Elsevier, Amsterdam, vol. 4, pp. 295–307, 2018.

[6.187] K. Hermann, "Periodic Overlayers and Moiré Patterns: Theoretical Studies of Geometric Properties", *J. Phys.: Condens. Matter*, vol. 24, p. 314210, 2012.

[6.188] *International Tables of Crystallography*, Kluwer Academic Publishers, Dordrecht, 2004, vol. C, 3rd ed., ch. 9.8: Incommensurate and commensurate modulated structures.

[6.189] W. Steurer, "3 + D-Dimensional Patterson and Fourier Methods for the Determination of One-Dimensionally Modulated Structures", *Acta Cryst. A*, vol. 43, pp. 36–42, 1987.

[7.1] I. K. Robinson, "Surface Crystallography", in *Handbook on Synchrotron Radiation*, eds. G. S. Brown and D. E. Moncton. Amsterdam, Elsevier, 1991, vol. 3, ch. 7, pp. 221–266.

[7.2] I. K. Robinson, P. Eng, and R. Schuster, "Origin of the Surface Sensitivity in Surface X-ray Diffraction", *Acta Physica Polonica A*, vol. 86, pp. 513–520, 1994.

[7.3] R. Feidenhans'l, "Surface Structure Determination by X-ray Diffraction", *Surf. Sci. Rep.*, vol. 10, pp. 105–188, 1989.

[7.4] A. Stierle and E. Vlieg, "Surface-Sensitive X-ray Diffraction Methods", in *Modern Diffraction Methods*, eds. E. J. Mittemeijer and U. Welzel, Wiley-VCH, Weinheim, 2013, ch. 8, pp. 221–257.

[7.5] M. Sauvage-Simkin, "X-ray Diffraction of Surface Structures", in *Landolt-Börnstein, Group III: Condensed Matter*, Landolt-Börnstein, Heidelberg, 2015, vol. 45, Physics of Solid Surfaces, subvol. A, ch. 4, pp. 100–132.

[7.6] H. Dosch, *Critical Phenomena at Surfaces and Interfaces: Evanescent X-ray and Neutron Scattering*, Springer Verlag, Heidelberg, 1992.

[7.7] A. S. Disa, F. J. Walker, and Ch. H. Ahn, "High-Resolution Crystal Truncation Rod Scattering: Application to Ultrathin Layers and Buried Interfaces", *Adv. Mater. Interfaces*, vol. 7, p. 1901772, 2020.

[7.8] J. Zegenhagen and A. Kazimirov, eds., Review articles on the X-ray standing wave technique, principles and applications, in *Series on Synchrotron Radiation Techniques and Applications*, World Scientific, Singapore, 2013, vol. 7, parts I and II.

[7.9] G. Renaud, R. Lazzari, and F. Leroy, "Probing Surface and Interface Morphology with Grazing Incidence Small Angle X-ray Scattering", *Surf. Sci. Rep.*, vol. 64, pp. 255–380, 2009.

[7.10] T. Li, A. J. Senesi, and B. Lee, "Small Angle X−ray Scattering for Nanoparticle Research", *Chem. Rev.*, vol. 116, pp. 11128–11180, 2016.

[7.11] E. Chason and T. M. Mayer, "Thin Film and Surface Characterization by Specular X-ray Reflectivity", *Crit. Rev. Solid State Mater. Sci.*, vol. 22, pp. 1–67, 1997.

[7.12] M. Tolan and W. Press, "X-ray and Neutron Reflectivity", *Z. Kristallogr.*, vol. 213, pp. 319–336, 1998.

[7.13] J. Als-Nielsen, D. Jacquemain, K. Kjaer, E. Leveiller, M. Lahav, and L. Leiserowitz, "Principles and Applications of Grazing-Incidence X-ray and Neutron-Scattering from Ordered Monolayers at the Air-Water-Interface", *Phys. Rep.*, vol. 246, pp. 252–313, 1994.

[7.14] J. Stöhr, "Geometry and Bond Lengths of Chemisorbed Atoms and Molecules: NEXAFS and SEXAFS", *Z. Physik B - Condensed Matter*, vol. 61, pp. 439–445, 1985.

[7.15] P. Eisenberger, P. Citrin, R. Hewitt, and B. Kincaid, "Sexafs: New Horizons in Surface Structure Determinations", *Crit. Rev. Solid State Mater. Sci.*, vol. 10, pp. 191–207, 1981.

[7.16] D. C. Koningsberger, *X-ray Absorption: Principles, Applications, Techniques of EXAFS, SEXAFS and XANES*, Wiley InterScience, Weinheim, 1988.

[7.17] J. Haase, "Surface Structure Determinations Using X-ray Absorption Spectroscopy", *J. Chem. Soc., Faraday Trans.*, vol. 92, pp. 1653–1667, 1996.

[7.18] G. Hähner, "Near Edge X-ray Absorption Fine Structure Spectroscopy As a Tool to Probe Electronic and Structural Properties of Thin Organic Films and Liquids", *Chem. Soc. Rev.*, vol. 35, pp. 1244–1255, 2006.

[7.19] H. Ade and A. P. Hitchcock, "NEXAFS Microscopy and Resonant Scattering: Composition and Orientation Probed in Real and Reciprocal Space", *Polymer*, vol. 49, pp. 643–675, 2008.

[7.20] C. S. Fadley, Y. Chen, R. E. Couch, H. Daimon, R. Denecke, J. D. Denlinger, H. Galloway, Z. Hussain, A. P. Kaduwela, Y. J. Kim, P. M. Len, J. Liesegang, J. Menchero, J. Morais, J. Palomares, S. D. Ruebush, E. Rotenberg, M. B. Salmeron, R. Scalettar, W. Schattke, R. Singh, S. Thevuthasan, E. D. Tober, M. A. Van Hove, Z. Wang, and R. X. Ynzunza, "Diffraction and Holography with Photoelectrons and Fluorescent X-rays", *Progr. Surf. Sci.*, vol. 54, pp. 341–387, 1997.

[7.21] J. Bansmann, S. H. Baker, C. Binns, J.A. Blackman, J.-P. Bucher, J. Dorantes-Dávila, V. Dupuis, L. Favre, D. Kechrakos, A. Kleibert, K.-H. Meiwes-Broer, G. M. Pastor, A. Perez, O. Toulemonde, K. N. Trohidou, J. Tuaillon, and Y. Xie, "Magnetic and Structural Properties of Isolated and Assembled Clusters", *Surf. Sci. Rep.*, vol. 56, pp. 189–275, 2005.

[7.22] D. Sander, "The Magnetic Anisotropy and Spin Reorientation of Nanostructures and Nanoscale Films", *J. Phys.: Condens. Matter*, vol. 16, pp. R603–R636, 2004.

[7.23] W. G. Stirling and M. J. Cooper, "X-ray Magnetic Scattering", *J. Magn. Magn. Mat.*, vol. 200, pp. 755–773, 1999.

[7.24] R. Comin and A. Damascelli, "Resonant X-ray Scattering Studies of Charge Order in Cuprates", *Annu. Rev. Condens. Matter Phys.*, vol. 7, pp. 369–405, 2016.

[7.25] J. Fink, E. Schierle, E. Weschke, and J. Geck, "Resonant Elastic Soft X-ray Scattering", *Rep. Prog. Phys.*, vol. 76, p. 056502, 2013.

[7.26] X-ray Data Booklet. Compiled and edited by A. C. Thompson, Lawrence Berkeley National Laboratory, University of California, Berkeley, CA, http://xdb.lbl.gov, last accessed 2 April 2022.

[7.27] NIST, tables of X-ray scattering factors, www.nist.gov/pml/xcom-photon-cross-sections-database, last accessed 2 April 2022.

[7.28] X.-L. Zhou and S.-H. Chen, "Theoretical Foundation of X-ray and Neutron Reflectometry", *Phys. Rep.*, vol. 257, pp. 223–348, 1995.

[7.29] International Tables of Crystallography, "Dispersion Corrections for Forward Scattering", in *Mathematical, Physical and Chemical Tables*, Kluwer Academic Publishers, Dordrecht, 2004, vol. C, 3rd ed., ch. 4.2, pp. 191–258, table 4.2.6.8.

[7.30] M. Born and E. Wolf, *Principles of Optics*, 5th ed., Pergamon, New York, 1986.

[7.31] G. H. Vineyard, "Grazing-Incidence Diffraction and the Distorted-Wave Born Approximation for the Study of Surfaces", *Phys. Rev. B*, vol. 26, pp. 4146–4159, 1982.

[7.32] O. Robach, Y. Garreau, K. Aïd, and M. B. Véron-Jolliot, "Corrections for Surface X-ray Diffraction Measurements Using the Z-axis Geometry: Finite Size Effects in Direct and Reciprocal Space", *J. Appl. Cryst.*, vol. 33, pp. 1006–1018, 2000.

[7.33] I. K. Robinson, "Crystal Truncation Rods and Surface Roughness", *Phys. Rev. B*, vol. 33, pp. 3830–3836, 1986.

[7.34] S. Brennan and P. Eisenberger, "A Novel X-ray Scattering Diffractometer for Studying Surface Structures under UHV Conditions", *Nuclear Instr. Methods Phys. Res.*, vol. 222, pp. 164–167, 1984.

[7.35] P. H. Fuoss and I. K. Robinson, "Apparatus for X-ray Diffraction in Ultrahigh Vacuum", *Nuclear Instr. Methods Phys. Res.*, vol. 222, pp. 171–176, 1984.

[7.36] T. T. Fister, P. H. Fuoss, D. D. Fong, J. A. Eastman, C. M. Folkman, S. O. Hruszkewycz, M. J. Highland, H. Zhou, and P. Fenter, "Surface Diffraction on a ψ-Circle Diffractometer Using the χ-Axis Geometry", *J. Appl. Cryst.*, vol. 46, pp. 639–643, 2013.

[7.37] E. Vlieg, "Integrated Intensities Using a Six-Circle Surface X-ray Diffractometer", *J. Appl. Cryst.*, vol. 30, pp. 532–543, 1997.

[7.38] K. W. Evans-Lutterodt and M.-T. Tang, "Angle Calculations for a '2+2' Surface X-ray Diffractometer", *J. Appl. Cryst.*, vol. 28, pp. 318–326, 1995.

[7.39] E. Vlieg, "A (2 + 3)-Type Surface Diffractometer: Mergence of the z-Axis and (2 + 2)-Type Geometries", *J. Appl. Cryst.*, vol. 31, pp. 198–203, 1998.

[7.40] M. Takahasi and J. Mizuki, "An Additional Axis for the Surface X-ray Diffractometer", *J. Synchrotron Rad.*, vol. 5, pp. 893–895, 1998.

[7.41] O. Bunk and M. M. Nielsen, "Angle Calculations for a z-Axis/(2S+2D) Hybrid Diffractometer", *J. Appl. Cryst.*, vol. 37, pp. 216–222, 2004.

[7.42] C. M. Schlepütz, S. O. Mariager, S. A. Pauli, R. Feidenhans'l, and P. R. Willmott, "Angle Calculations for a (2+3)-Type Diffractometer: Focus on Area Detectors", *J. Appl. Cryst.*, vol. 44, pp. 73–83, 2011.

[7.43] P. R. Willmott, D. Meister, S. J. Leake, M. Lange, A. Bergamaschi, M. Böge, M. Calvi, C. Cancellieri, N. Casati, A. Cervellino, Q. Chen, C. David, U. Flechsig, F. Gozzo, B. Henrich, S. Jäggi-Spielmann, B. Jakob, I. Kalichava, P. Karvinen, J. Krempasky, A. Lüdeke, R. Lüscher, S. Maag, C. Quitmann, M. L. Reinle-Schmitt, T. Schmidt, B. Schmitt, A. Streun, I. Vartiainen, M. Vitins, X. Wang, and R. Wullschleger, "The Materials Science Beamline Upgrade at the Swiss Light Source", *J. Synchrotron Rad.*, vol. 20, pp. 667–682, 2013.

[7.44] B. E. Warren, *X-ray Diffraction*, Dover, New York, 1990.

[7.45] W. Clegg, ed., Various articles, in *Crystal Structure Analysis, Principles and Practice*, 2nd ed., Oxford University Press, Oxford, 2009.

[7.46] R. Shayduk, "Use of Linear Scans to Obtain $|F_{hkl}|^2$ and the Integrated Intensity", *J. Appl. Cryst.*, vol. 43, pp. 1121–1123, 2010.

[7.47] C. Schamper, H. L. Meyerheim, and W. Moritz, "Resolution Correction for Surface X-ray Diffraction at High Beam Exit Angles", *J. Appl. Cryst.*, vol. 26, pp. 687–696, 1993.

[7.48] N. Jedrecy, "Coupling between Spatial and Angular Variables in Surface X-ray Diffraction: Effects on the Line Shapes and Integrated Intensities", *J. Appl. Cryst.*, vol. 33, pp. 1365–1375, 2000.

[7.49] X. Torrelles and J. Rius, "Faster Acquisition of Structure-Factor Amplitudes in Surface X-ray Diffraction Experiments Instruments", *J. Appl. Cryst.*, vol. 37, pp. 395–398, 2004.

[7.50] C. M. Schlepütz, R. Herger, P. R. Willmott, B. D. Patterson, O. Bunk, Ch. Brönnimann, B. Henrich, G. Hülsen, and E. F. Eikenberry, "Improved Data Acquisition in Grazing-Incidence X-ray Scattering Experiments Using a Pixel Detector", *Acta Cryst. A*, vol. 61, pp. 418–425, 2005.

[7.51] S. O. Mariager, S. L. Lauridsen, A. Dohn, N. Bovet, C. B. Sørensen, C. M. Schlepütz, P. R. Willmott, and R. Feidenhans'l, "High-Resolution Three-Dimensional Reciprocal-Space Mapping of InAs Nanowires", *J. Appl. Cryst.*, vol. 42, pp. 369–375, 2009.

[7.52] J. Drnec, T. Zhou, S. Pintea, W. Onderwaater, E. Vlieg, G. Renaud, and R. Felici, "Integration Techniques for Surface X-ray Diffraction Data Obtained with a Two-Dimensional Detector", *J. Appl. Cryst.*, vol. 47, pp. 365–377, 2014.

[7.53] S. J. Leake, M. L. Reinle-Schmitt, I. Kalichava, S. A. Pauli, and P. R. Willmott, "Cluster Method for Analysing Surface X-ray Diffraction Data Sets Using Area Detectors", *J. Appl. Cryst.*, vol. 47, pp. 207–214, 2014.

[7.54] S. Roobol, W. Onderwaater, J. Drnec, R. Felici, and J. Frenken, "BINoculars: Data Reduction and Analysis Software for Two-Dimensional Detectors in Surface X-ray Diffraction", *J. Appl. Cryst.*, vol. 48, pp. 1324–1329, 2015.

[7.55] C. Nicklin, "Capturing Surface Processes", *Science*, vol. 343, pp. 739–740, 2014.

[7.56] J. Gustafson, M. Shipilin, C. Zhang, A. Stierle, U. Hejral, U. Ruett, O. Gutowski, P.-A. Carlsson, M. Skoglundh, and E. Lundgren, "High-Energy Surface X-ray Diffraction for Fast Surface Structure Determination", *Science*, vol. 343, pp. 758–761, 2014.

[7.57] J. H. Lee, I. C. Tung, S.-H. Chang, A. Bhattacharya, D. D. Fong, J. W. Freeland, and H. Hong, "In Situ Surface/Interface X-ray Diffractometer for Oxide Molecular Beam Epitaxy", *Rev. Sci. Instr.*, vol. 87, p. 013901, 2016.

[7.58] H. Hong and T.-C. Chiang, "A Six-Circle Diffractometer System for Synchrotron X-ray Studies of Surfaces and Thin Film Growth by Molecular Beam Epitaxy", *Nuclear Instr. Methods Phys. Res. A*, vol. 572, pp. 942–947, 2007.

[7.59] J. Rubio-Zuazo, P. Ferrer, A. López, A. Gutiérrez-León, I. da Silva, and G. R. Castro, "The Multipurpose X-ray Diffraction End-Station of the BM25B-SpLine Synchrotron Beamline at the ESRF", *Nuclear Instr. Methods Phys. Res. A*, vol. 716, pp. 23–28, 2013.

[7.60] J. Rubio-Zuazo and G. R. Castro, "Beyond Hard X-ray Photoelectron Spectroscopy: Simultaneous Combination with X-ray Diffraction", *J. Vac. Sci. Technol. A*, vol. 31, p. 031103, 2013.

[7.61] M.-C. Saint-Lager, A. Bailly, P. Dolle, R. Baudoing-Savois, P. Taunier, S. Garaudée, S. Cuccaro, S. Douillet, O. Geaymond, G. Perroux, O. Tissot, J.-S. Micha, O. Ulrich, and F. Rieutord, "New Reactor Dedicated to In Operando Studies of Model Catalysts by Means of Surface X-ray Diffraction and Grazing Incidence Small Angle X-ray Scattering", *Rev. Sci. Instrum.*, vol. 78, p. 083902, 2007.

[7.62] O. H. Seeck, C. Deiter K. Pflaum, F. Bertam, A. Beerlink, H. Franz, J. Horbach, H. Schulte-Schrepping, B. M. Murphy, M. Greve, and O. Magnussen, "The High-Resolution Diffraction Beamline P08 at PETRA III", *J. Synchrotron Rad.*, vol. 19, pp. 30–38, 2012.

[7.63] B. M. Murphy, M. Greve, B. Runge, C. T. Koops, A. Elsen, J. Stettner, O. H. Seeck, and O. M. Magnussen, "A Novel X-ray Diffractometer for Studies of Liquid–Liquid Interfaces", *J. Synchrotron Rad.*, vol. 21, pp. 45–56, 2014.

[7.64] C. Nicklin, T. Arnold, J. Rawle, and A. Warne, "Diamond Beamline I07: A Beamline for Surface and Interface Diffraction", *J. Synchrotron Rad.*, vol. 23, pp. 1245–1253, 2016.

[7.65] Website of SPring-8, beamline BL13XU, www.spring8.or.jp/wkg/BL13XU/instrument/lang-en/INS-0000000396/instrument_summary_view, last accessed 2 April 2022.

[7.66] P. Eisenberger and W. C. Marra, "X-ray Diffraction Study of the Ge(001) Reconstructed Surface", *Phys. Rev. Lett.*, vol. 46, pp. 1081–1085, 1981.

[7.67] Y. Fujii, K. Yoshida, T. Nakamura, and K. Yoshida, "A Compact Ultrahigh Vacuum X-ray Diffractometer for Surface Glancing Scattering Using a Rotating-Anode Source", *Rev. Sci. Instrum.*, vol. 68, pp. 1975–1979, 1997.

[7.68] M. Albrecht, H. Antesberger, W. Moritz, H. Plöckl, M. Sieber, and D. Wolf, "Six-Circle Diffractometer for Surface Diffraction Using an In-vacuum X-ray Detector", *Rev. Sci. Instrum.*, vol. 70, pp. 3239–3243, 1999.

[7.69] The diffractometer was developed by H. L. Meyerheim at the Max-Planck-Institut für Mikrostrukturphysik, www.mpi-halle.mpg.de, last accessed 2 April 2022.

[7.70] Ga-Jet Source, www.bruker.com/products/x-ray-diffraction-and-elemental-analysis/x-ray-diffraction/components/xrd-components/sources.html, last accessed 2 April 2022.

[7.71] List of Synchrotron Radiation Facilities, https://en.wikipedia.org/wiki/List_of_synchrotron_radiation_facilities, last accessed 2 April 2022.

[7.72] H. Tajiri, "Progress in Surface X-ray Crystallography and the Phase Problem", *Jpn. J. Appl. Phys.*, vol. 59, p. 020503, 2020.

[7.73] E. Vlieg, "ROD: A Program for Surface X-ray Crystallography", *J. Appl. Cryst.*, vol. 33, pp. 401–405, 2000. The program is available at www.esrf.fr, last accessed 2 April 2022.

[7.74] U. Shmueli, *Theories and Techniques of Crystal Structure Determination*, Oxford University Press, Oxford, 2007.

[7.75] C. Giacovazzo, ed., *Fundamentals of Crystallography*, Oxford University Press, Oxford, 1992.

[7.76] C. Giacovazzo, *Direct Phasing in Crystallography, Fundamentals and Applications*, Oxford University Press, Oxford, 1998.

[7.77] J. S. Wu and J. C. H. Spence, "Phase Extension in Crystallography Using the Iterative Fienup-Gerchberg-Saxton Algorithm and Hilbert Transforms", *Acta Cryst. A*, vol. 59, pp. 577–583, 2003.

[7.78] A. van der Lee, "Charge Flipping for Routine Structure Solution", *J. Appl. Cryst.*, vol. 46, pp. 1306–1315, 2013.

[7.79] J. Rius, C. Miravitlles, and R. Allmann, "A Tangent Formula Derived from Patterson-Function Arguments. IV. The Solution of Difference Structures Directly from Superstructure Reflections", *Acta Cryst. A*, vol. 52, pp. 634–639, 1996.

[7.80] Y. Takeuchi, "The Investigation of Superstructures by Means of Partial Patterson Functions", *Z. Kristallogr.*, vol. 135, pp. 120–136, 1972.

[7.81] M. J. Buerger, *Vector Space and its Application in Crystal-Structure Investigation*, Wiley, New York, 1959, pp. 310–329.

[7.82] T. Takahashi, K. Sumitani, and S. Kusano, "Holographic Imaging of Surface Atoms Using X-ray Diffraction", *Surf. Sci.*, vol. 493, pp. 36–41, 2001.

[7.83] L. D. Marks, N. Erdman, and A. Subramanian, "Crystallographic Direct Methods for Surfaces", *J. Phys.: Condens. Matter*, vol. 13, p. 10677, 2001.

[7.84] L. D. Marks, W. Sinkler, and E. Landree, "A Feasible Set Approach to the Crystallographic Phase Problem", *Acta Cryst. A*, vol. 55, pp. 601–612, 1999.

[7.85] C. Kumpf, L. D. Marks, D. Ellis, D. Smilgies, E. Landemark, M. Nielsen, R. Feidenhans'l, J. Zegenhagen, O. Bunk, J. H. Zeysing, Y. Su, and R. L. Johnson, "Subsurface Dimerization in III-V Semiconductor (001) Surfaces", *Phys. Rev. Lett.*, vol. 86, pp. 3586–3589, 2001.

[7.86] Y. Yacoby, R. Pindak, R. MacHarrie, L. Pfeiffer, L. Berman, and R. Clarke, Direct Structure Determination of Systems with Two-Dimensional Periodicity, *J. Phys.: Condens. Matter*, vol. 12, pp. 3929–3938, 2000.

[7.87] J. Rius, C. Miravitlles, and R. Allmann, "A Tangent Formula Derived from Patterson–Function Arguments. IV.The Solution of Difference Structures Directly from Superstructure Reflections", *Acta Cryst. A*, vol. 52, pp. 634–639, 1996.

[7.88] J. Rius, "Derivation of a New Tangent Formula from Patterson–Function Arguments", *Acta Cryst. A*, vol. 49, pp. 406–409, 1993.

[7.89] H. Vogler, "Röntgenstrukturanalysen der Ge(113)-c(3×1)- und InSb(111)-(2×2)-Oberflächen und Anwendung der Maximum Entropie Methode zur Oberflächenstrukturanalyse", Dissertation, University of Munich, Department of Earth and Environmental Sciences, Germany, 1997.

[7.90] D. K. Saldin, R. Harder, H. Vogler, W. Moritz, and I. K. Robinson, "Solving the Structure Completion Problem in Surface Crystallography", *Comp. Phys. Commun.*, vol. 137, pp. 12–24, 2001.

[7.91] M. Björck, C. M. Schlepütz, S. A. Pauli, D. Martoccia, R. Herger, and P. R. Willmott, "Atomic Imaging of Thin Films with Surface X-ray Diffraction: Introducing DCAF", *J. Phys.: Condens. Matter*, vol. 20, p. 445006, 2008.

[7.92] I. K. Robinson, M. Tabuchi, S. Hisadome, R. Oga, and Y. Takeda, "Perturbation Method of Analysis of Crystal Truncation Rod Data", *J. Appl. Cryst.*, vol. 38, pp. 299–305, 2005.

[7.93] S. A. Pauli, S. J. Leake, M. Björck, and P. R. Willmott, "Atomic Imaging and Direct Phase Retrieval Using Anomalous Surface X-ray Diffraction", *J. Phys.: Condens. Matter*, vol. 24, p. 305002, 2012.

[7.94] V. M. Kaganer, M. Albrecht, A. Hirnet, M. Gierer, W. Moritz, B. Jenichen, and K. H. Ploog, "Solving the Phase Problem for Surface Crystallography: Indirect Excitation via Bulk Reflection", *Phys. Rev. B*, vol. 61, pp. R16355–R16358, 2000.

[7.95] Y. Yacoby, C. Brooks, D. Schlom, J. O. Cross, D. A. Walko, C. N. Cionca, N. S. Husseini, A. Riposan, and R. Clarke, "Structural Changes Induced by Metal Electrode Layers on Ultrathin BaTiO3 Films", *Phys. Rev. B*, vol. 77, p. 195426, 2008.

[7.96] Y. Yacoby, H. Zhou, R. Pindak, and I. Božović, "Atomic-Layer Synthesis and Imaging Uncover Broken Inversion Symmetry in $La_{2-x}Sr_xCuO_4$ Films", *Phys. Rev. B*, vol. 87, p. 014108, 2013.

[7.97] M. Sowwan, Y. Yacoby, J. Pitney, R. MacHarrie, M. Hong, J. Cross, D. A. Walko, R. Clarke, R. Pindak, and E. A. Stern, "Direct Atomic Structure Determination of Epitaxially Grown Films: Gd_2O_3 on GaAs(100)", *Phys. Rev. B*, vol. 66, p. 205311, 2002.

[7.98] H. Baltes, Y. Yacoby, R. Pindak, R. Clarke, L. Pfeiffer, and L. Berman, "Measurement of the X-ray Diffraction Phase in a 2D Crystal", *Phys. Rev. Lett.*, vol. 79, pp. 1285–1288, 1997.

[7.99] X. Torrelles, J. Rius, A. Hirnet, W. Moritz, M. Pedio, R. Felici, P. Rudolf, M. Capozi, F. Boscherini, S. Heun, B. H. Mueller, and S. Ferrer, "Real Examples of Surface Reconstructions Determined by Direct Methods", *J. Phys.: Condens. Matter*, vol. 14, pp. 4075–4086, 2002.

[7.100] X. Torrelles, J. Rius, F. Boscherini, S. Heun, B. H. Mueller, S. Ferrer, J. Alvarez, and C. Miravitlles, "Application of X-ray Direct Methods to Surface Reconstructions: The Solution of Projected Superstructures", *Phys. Rev. B*, vol. 57, pp. R4281–R4284, 1998.

[7.101] A. Hirnet, K. Schroeder, S. Blügel, X. Torrelles, M. Albrecht, B. Jenichen, M. Gierer, and W. Moritz, "Novel Sb Induced Reconstruction of the (113) Surface of Ge", *Phys. Rev. Lett.*, vol. 88, p. 226102, 2002.

[7.102] M. Pedio, R. Felici, X. Torrelles, P. Rudolf, M. Capozi, J. Rius, and S. Ferrer, "Study of C_{60}/Au(110)-p(6×5) Reconstruction from In-Plane X-ray Diffraction Data", *Phys. Rev. Lett.*, vol. 85, pp. 1040–1043, 2000.

[7.103] D. K. Saldin, R. J. Harder, V. L. Shneerson, and W. Moritz, "Phase Retrieval Methods for Surface X-ray Diffraction", *J. Phys.: Condens. Matter*, vol. 13, pp. 10689–10707, 2001.

[7.104] D. K. Saldin, R. J. Harder, V. L. Shneerson, and W. Moritz, "Surface X-ray Crystallography with Alternating Constraints in Real and Reciprocal Space: The Case of Mixed Domains", *J. Phys.: Condens. Matter*, vol. 14, pp. 4087–4100, 2002.

[7.105] D. K. Saldin and V. L. Shneerson, "Direct Methods for Surface Crystallography", *J. Phys.: Condens. Matter*, vol. 20, pp. 304208, 2008.

[7.106] P. F. Lyman, V. L. Shneerson, R. Fung, R. J. Harder, E. D. Lu, S. S. Parihar, and D. K. Saldin, "Atomic-Scale Visualization of Surfaces with X-rays", *Phys. Rev. B*, vol. 71, p. 081402(R), 2005.

[7.107] J. R. Fienup, "Phase Retrieval Algorithms: A Comparison", *Appl. Opt.*, vol. 21, pp. 2758–2769, 1982.

[7.108] P. F. Lyman, V. L. Shneerson, R. Fung, and S. S. Parihar, "Structure and Stability of Sb/Au (1 1 0)-c (2×2) Surface Phase", *Surf. Sci.*, vol. 600, pp. 424–435, 2006.

[7.109] R. Fung, V. L. Shneerson, P. F. Lyman, S. S. Parihar, H. T. Johnson-Steigelman, and D. K. Saldin, "Phase and Amplitude Recovery and Diffraction Image Generation Method: Structure of Sb/Au(110)–($\sqrt{3}\times\sqrt{3}$)R54.7° from Surface X-ray Diffraction", *Acta Cryst. A*, vol. 63, pp. 239–250, 2007.

[7.110] X. Torrelles, J. Zegenhagen, J. Rius, T. Gloege, L. X. Cao, and W. Moritz, "Atomic Structure of a Long-Range Ordered Vicinal Surface of SrTiO$_3$", *Surf. Sci.*, vol. 589, pp. 184–191, 2005.

[7.111] V. Elser, "Solution of the Crystallographic Phase Problem by Iterated Projections", *Acta Cryst. A*, vol. 59, pp. 201–209, 2003.

[7.112] C. M. Schlepütz, M. Björck, E. Koller, S. A. Pauli, D. Martoccia, Ø. Fischer, and P. R. Willmott, "Structure of Ultrathin Heteroepitaxial Superconducting YBa$_2$Cu$_3$O$_{7-x}$ Films", *Phys. Rev. B*, vol. 81, p. 174520, 2010.

Index

Printed in the United States
by Baker & Taylor Publisher Services